Environmental Hazards Methodologies for Risk Assessment and Management

Environmental Hazards Methodologies for Risk Assessment and Management

Edited by

Nicolas R. Dalezios

IWA PUBLISHING

Published by **IWA Publishing**
Alliance House
12 Caxton Street
London SW1H 0QS, UK
Telephone: +44 (0)20 7654 5500
Fax: +44 (0)20 7654 5555
Email: publications@iwap.co.uk
Web: www.iwapublishing.com

First published 2017

British Library Cataloguing in Publication Data
A CIP catalogue record for this book is available from the British Library

ISBN 9781780407128 (Paperback)
ISBN 9781780407135 (eBook)

"to an ever improving and sustainable world"

Contents

Chapter 2

About the editor

Nicolas R. Dalezios: Professor of Agrometeorology and Remote Sensing, University of Thessaly, Volos Grecce and Agricultural University of Athens, Greece (2011-present). President of the Council of the Agricultural University of Athens, Greece. Professor and founding Director of the Laboratory of Agrometeorology, University of Thessaly, Volos Hellas (1991–2011). Postgraduate studies in Meteorology (Athens, 1972), Hydrological Engineering (Univ. of Delft, 1974), Ph.D. in Civil Eng. (Univ. of Waterloo, Canada, 1982). He has a long standing research record in agrometeorology, agrohydrology, remote sensing, modelling, environmental hazards, risk assessment, climate variability/change. Author or co-author in over 280 refereed scientific and technical publications. Member of Editorial Boards and reviewer in about 30 Web of Science (ISI) International Scientific journals. Author and Editor of 2 books. Editor or co-editor in over 15 edited publications. Co-author in 25 book chapters published by Taylor and Francis group, IWA.

List of Contributors

Dr. Alan Ager
USDA Forest Service, Rocky Mountain Research Station, 72510 Coyote Road, Pendleton, OR 97801 USA.
aager@fs.fed.us

Dr. Martin Bednarik
Faculty of Natural Sciences, Comenius University, Ilkovičova 6, 842 15 Bratislava 4, Slovakia.
mbednarik@fns.uniba.sk

Mgr. Marek Biskupič
Avalanche Prevention Center, Dr. J. Gašperíka 598, 033 01 Liptovský Hrádok, Slovakia and Institute for Environmental Studies, Charles University, Ovocnýtrh 3-5, 116 36 Praha 1, Czech Republic.
biskupic@hzs.sk

Prof. Marco Borga
Department of Land, Environment, Agriculture and Forestry, University of Padova, Legnaro, Italy.
marco.borga@unipd.it

Dr. Jean-Dominique Creutin
Université de Grenoble/CNRS, Laboratoire d'étude des Transferts en Hydrologie et Environnement, LTHE, UMR 5564, Grenoble F-38041, France.
jean-dominique.creutin@ujf-grenoble.fr

Prof. Nicolas R. Dalezios
University of Thessaly, Volos, Greece and Agricultural University of Athens, Athens Greece.
dalezios.n.r@gmail.com

Mrs. Antonia Daleziou
MPH, National School of Public Health, Athens, Greece.
antonia_dal@hotmail.com

Mrs. Elisa Destro
Department of Land, Environment, Agriculture and Forestry, University of Padova, Legnaro, Italy.
elisa.destro@studenti.unipd.it

Prof. Saeid Eslamian
Isfahan University of Technology, Isfahan, Iran.
prof.s.eslamian@gmail.com

Dr. Mark Finney
Forest Service, USDA, Missoula, USA.
mfinney@fs.fed.us

Prof. Stefan Greiving
Technical University of Dortmund, Germany.
stefan.greiving@tu-dortmund.de

Dr. Orestis Kairis
Dept. of Natural Resources and Agricultural Engineering, Agricultural University of Athens, Athens, Greece.
kairis@aua.gr

Prof. K. Kalabokidis
Dept. of Geography, Aegean University, 81100 Mytilene, Greece.
kalabokidis@aegean.gr

Prof. Giannis Kassaras
Dept. of Geology, National and Kappodistrian Univ. of Athens, Athens, Greece.
kassaras@geol.uoa.gr

Mrs. Danai Kazantzidou-Firtinidou
Dept. of Geology, National and Kappodistrian Univ. of Athens, Athens, Greece.
dkazantzidou@geol.uoa.gr

Prof. Costas Kosmas
Dept. of Natural Resources and Agricultural Engineering, Agricultural University of Athens, Athens, Greece.
ckosm@aua.gr

Dr. Terry W. Krauss
Accredited Consulting Meteorologist, AMEC Met-Ocean Services, 79 Irving Crescent, Red Deer, Alberta, Canada T4R 3S3.
twkrauss@gmail.com

Prof. Costas Lalenis
University of Thessaly,Volos, Greece.
klalenis@uth.gr

Dr. Pavel Liščák
State Geological Institute of Dionýy Štúr, Mlynská dolina 1, 81704 Bratislava, Slovakia.
pavel.liscak@geology.sk

Prof. John M. Logan
(public safety), Univ. of Oregon, USA.
jmllogan@aol.com

Dr. Peter F. Moore
Consultant, Forest Fire Management and Disaster Risk Reduction, Forest Policy and Resources Division, FAO-Forestry Department, Viale delle Terme di Caracalla, I-00153 Rome, Italy.
Peter.Moore@fao.org

Prof. Panagiotis Nastos
Department of Geology, National and Kappodistrian University of Athens, Athens, Greece.
nastos@geol.uoa.gr

Dr. Efthymios Nikolopoulos
Department of Civil and Environmental Engineering, University of Connecticut, Storrs, CT, USA.
efthymios.nikolopoulos@uconn.edu

Dr. Augusto Neri
Istituto Nazionale di Geofisica e Vulcanologia, Sezione di Pisa, via della Faggiola 32, 56126 Pisa, and via di Vigna Murata 605, 00143 Roma.
augusto.neri@ingv.it

Dr. Pavol Nejedlik
Earth Science Institute of Slovak Academy of Science, Dubravska cesta 9, 84505 Bratislava, Slovakia.
nejedlik@yahoo.com

Prof. Spiros Rapsomanikis
Dept. of Environmental Engineering, Democritus University of Thrace, Xanthi, Greece.
rapso@env.duth.gr

Dr. Josef Richnayvsky
Avalanche Prevention Center, Dr. J. Gašperíka 598, 033 01 Liptovský Hrádok, Slovakia.
slp@hzs.sk

Dr. Nicos V. Spyropoulos
Dept. of Natural Resources and Agricultural Engineering, Agricultural University of Athens, Athens, Greece.
nicospir@gmail.com

Prof. Ana M. Tarquis
CEIGRAM, Dept. of Applied Mathematics, Technological Univ. of Madrid, Spain.
tarquis61@gmail.com
anamaria.tarquis@upm.es

Dr. Aikaterini Trepekli
Dept. of Civil Engineering, Democritus University of Thrace, Xanthi, Greece.
atrepekl@civil.duth.gr

Prof. Cees van Westen
ITC, Univ. of Twente, Twente, The Netherlands.
westen@itc,nl
c.j.vanwesten@utwente.nl

Dr. George E. Vougioukalakis
Institute of Geology and Mineral Exploration, S, Lui street 1, Olympic Village, Aharne 13677, Athens, Greece.
gvoug@igme.gr

Preface

Increasing climate variability and climate change, lead to increases in climate extremes. Under a changing climate, the role of several sectors of the economy, such as agriculture, as provider of environmental and ecosystem services, will further gain importance. On the other hand, natural disasters play a major role in several sectors of the economy, including agriculture, energy, health, transportation, tourism, and the economic cost associated with all natural disasters has increased significantly. Current scientific projections point, among others, to changes in climate extremes, mainly floods and droughts, in many areas of the world. Environmental degradation is one of the major factors contributing to vulnerability, because it directly magnifies the risk of natural disasters. Vulnerability of the environment can be reduced through adaptation measures and tools to increasing climate variability. In order to ensure environmental sustainability, a better understanding of the natural disasters and their impacts, is essential.

From the beginning of 21st century, there is an awareness of risk in the environment along with a growing concern for the continuing potential damage caused by hazards. Moreover, besides physical protection, a synthesis of anti-hazard measures starts being considered, including land use management, better planning for response and recovery and emergency warnings. Further, a global program to reduce the losses from natural hazards was adopted in December 1989 by the U.N. General Assembly declaring the 1990s as the International Decade for Natural Disaster Reduction (IDNDR). The challenge to researchers within the hazards community was to use their skill and adopt a wider perspective involving global change for a safer and sustainable environment. In addition, the World Conference on Disaster Reduction has adopted several strategic goals resulting in a number of priorities for action during the following 10 years (2005–2015). Recently, the Sendai Framework for Disaster Risk Reduction 2015–2030 has been adopted, which requires a better understanding of risk in all its dimensions of vulnerability, exposure and hazards. The Sendai Framework recognizes the importance of science and technology for disaster risk reduction. The goal is to prevent new and reduce existing disaster risk through the implementation of integrated and inclusive economic, environmental, technological, educational, structural, legal, social, health, cultural, political and institutional measures that prevent and reduce hazard exposure and vulnerability to disaster, increase preparedness for response and recovery, and thus strengthen resilience.

The current scientific trend focuses on the relationship between climate change and extreme weather and climate events, the impacts of such events, and the strategies to manage the associated risks. Thus, it has been recognized that a holistic and integrated approach to environmental hazards needs to be attempted using common methodologies, such as risk analysis, which involves risk management and risk assessment. Indeed, risk management means reducing the threats posed by known hazards, whereas at the same time accepting unmanageable risks and maximizing any related benefits. Risk assessment constitutes the first part within the risk management framework and involves evaluating the importance of a risk, either quantitatively, or qualitatively. Nevertheless, the risk management framework also includes a fourth step, risk governance, i.e. the need for a feedback of all the risk assessment undertakings. However, there is a lack of such feedback, which constitutes a serious deficiency in the reduction of environmental hazards at the present time.

The purpose of the book can be summarized within the following points, which also constitute advantages of the book:

(1) To present current quantitative methodologies of environmental hazards, i.e. forecasting-nowcasting (before), monitoring (during) and assessment (after).
(2) To incorporate these methodologies within the corresponding components of the risk management framework.
(3) To develop a holistic and integrated methodological procedure for risk management of environmental hazards.

This book attempts primarily to provide a text and also to serve as a cook book on environmental hazards for senior undergraduate students, graduate students, researchers and professionals of environmental science, environmental economics and management, physical and natural sciences, applied sciences, engineering, geography, geology, agriculture, ecology and similar fields. As already mentioned, the emphasis is placed on methodological approaches and procedures for the three main temporal stages in the study of environmental hazards, namely before, during and after, based on geoinformatic technologies and data, as well as simulation. This approach is considered a contemporary and innovative procedure and constitutes a current research trend in the field of environmental hazards.

The subject of the book, besides comprehensive and conceptual descriptions, consists of analysis and presentation of quantitative methodologies of environmental hazards within the risk management framework and, in particular, within the three components of risk assessment, namely risk identification, risk estimation and risk evaluation. Specifically, the book covers hydrometeorological hazards (floods, droughts, storms, hail, desertification), biophysical hazards (frost, heatwaves, epidemics, forest fires), geophysical hazards including geological hazards (landslides, snow and ice avalanches), tectonic hazards (earthquakes, volcanoes), and technological hazards. Nevertheless, the need for such a book comes from my own experience in teaching such courses and conducting research on the subject for several decades.

Nicolas R. Dalezios
Volos, Greece
August 2016

Acknowledgements

This book owes its completion to generous assistance by many sources. As editor, I wish to acknowledge the comprehensive and thoughtful work conducted by all the chapter contributors, who are listed in a separate file. Moreover, being also contributor to several chapters, I wish to state that this book constitutes the outcome of diachronic upgrading of University lectures in several similar courses. Indeed, there have been incorporated valuable help and comments given over the years by present and former colleagues and associates in the teaching and learning of environmental hazards. In addition, a number of individuals have provided useful unpublished information.

It is also mentioned that the successful completion of the book depends heavily on the editorial advice by IWA Publishing, London office. In particular, Maggie Smith and Mark Hammond have been very precious in offering continuous encouragement and practical advice. Finally, my family has provided a relaxed home environment appropriate for writing.

Needless to say, there has been every effort to identify and acknowledge the original sources, however, if there have been any omissions or errors by chance, the editor and the publisher apologize to those related to.

Nicolas R. Dalezios
Volos, Greece
August, 2016

Part 1

Prolegomena

Chapter 1

Environmental hazards concepts

Nicolas R. Dalezios and Costas Lalenis

1.1 CONCEPTS AND SCOPE OF ENVIRONMENTAL HAZARDS

1.1.1 Concepts of environmental hazards and disasters

Disasters are defined by the United Nations International Strategy for Disaster Risk Reduction (UNISDR, 2005) as "a serious disruption of the functioning of a community or a society causing widespread human, material, economic or environmental losses which exceed the ability of the affected community or society to cope using its own resources". Similarly, a hazard is defined as "a potentially damaging physical event, phenomenon or human activity that may cause the loss of life or injury, property damage, social and economic disruption or environmental degradation". This event has a probability of occurrence within a specified period of time and within a given area, and has a given intensity (UNISDR, 2005).

Environmental degradation is one of the major factors contributing to the vulnerability of several sectors, such as environment, society, economy and agriculture, because it directly magnifies the risk of natural disasters. In order to ensure sustainability in the above sectors, a better understanding of the natural disasters and their impacts is essential. A comprehensive assessment of impacts of natural disasters on different sectors requires a multidisciplinary, multi-sectoral and integral approach involving several components and factors. Priority should be given to supporting applied research, since research is necessary to understand the physical and biological factors contributing to disasters. Community-wide awareness and capacity building programs on natural disasters, mainly for farmers and stakeholders should also be included in any research effort. Programs for improving prediction and early warning methods, as well as dissemination of warnings should be expanded and intensified. Moreover, efforts are required to determine the impact of disasters on natural resources. It has been recognized that, although hazards may impact at the local scale, there are increasing international consequences due to global factors, such as climate change, poverty or the rise of mega-cities. Indeed, the losses from natural disasters justify a concern about the sustainability of continuous population growth and wealth development. Nevertheless, the challenge in hazard research is to improve the skills and adopt a broader perspective including global change with the objective to create a safer future environment.

1.1.2 Scope of hazards and disasters

Global warming is likely to bring significant changes in the world's climate over the coming decades. It is probable that the most significant physical consequences and effects will be experienced in countries highly dependent on natural resource use, which influence activities, such as agricultural development, forestry, wetland reclamation and river management. The overall result is likely to widen the gap between developed and developing regions and countries, since the impact is expected to be most severe on ecosystems already under stress and for regions, which have few spare resources for mitigating or adapting to climate change. The world land-use data of FAO, show that 70% of the global land use is for agriculture, rangeland and forestry. Indeed, agriculture is essential source of income in most developing countries. Moreover, agricultural production, along with other sectors of economy, is highly depended on weather and climate, and is adversely affected by weather and climate-related disasters.

Recent research findings suggest that variability of climate, if encompassing more intense and frequent extremes, such as major large-scale hazards like droughts, heatwaves or floods, results in the occurrence of natural disasters that are beyond our socio-economic planning levels. This is expected to stretch regional response capabilities beyond their capacity and will require new adaptation and preparedness strategies. Disaster prevention and preparedness should become a priority and rapid response capacities to climate change need to be accompanied by a strategy for disaster prevention. Nevertheless, each type of extreme events has its own particular climate, cultural and environmental setting, and mitigation activities must use these settings as a foundation of proactive management (IPCC, 2012). There is an urgent need to assess the forecasting skills for natural disasters affecting agriculture in order to determine those where greater research is necessary. It is well known that lack of good forecast skill is a constraint to improve adaptation, management and mitigation.

In this book the environmental hazards affecting several sectors of the economy are considered under increasing climate variability. The emphasis is on methodologies. It should be stated that current natural disaster management is crisis driven. It is thus realised that there is an urgent research need for a more risk-based management approach to natural disaster planning, which would include a timely and user-oriented early warning systems. Disaster risk zoning is also an essential component of natural disaster mitigation and preparedness strategies. GIS and remote sensing and, in general, geoinformatics are increasingly employed due to the complex nature of databases to facilitate strategic and tactical applications at the farm and policy levels. Therefore, additional research is required to incorporate GIS, remote sensing, simulation models and other computational techniques into an integrated multi-hazard risk management framework for sustainable environment, which includes early warnings of natural disasters (Sivakumar *et al.* 2005). There should also be more research attention to the impacts of potentially increasing frequency and severity of extreme events associated with global change and appropriate mitigation strategies. Moreover, a revision on existing conventional approaches is required in order to include the major recent research advancements in understanding these complex interrelationships. This revision should emphasize on prediction, monitoring and early warning methods, as well as vulnerability and impact assessment techniques and preparedness and mitigation strategies. In addition, there is a current need for more research into the physical behaviour of the climate system to develop better mitigation strategies.

1.2 A TYPOLOGY AND CLASSIFICATION OF HAZARDS

Hazards can be single, sequential or combined in their origin and effects. Each hazard is characterized by its location, area affected (size or magnitude), intensity, speed of onset, duration and frequency. Hazards can be classified in several ways. A possible classification is between natural, human-induced and human-made hazards. Natural hazards are natural processes or phenomena in the Earth's system (lithosphere, hydrosphere, biosphere or atmosphere) that may constitute a damaging event (e.g. earthquakes, volcanic eruptions, hurricanes). A classification relates natural hazards to the main controlling factors of the hazards leading to a disaster. They may be hydro-meteorological hazards, which include floods and wave surges, storms, droughts and related disasters, such as desertification. Also geophysical hazards result from anomalies in the Earth's surface or subsurface, such as earthquakes, tsunamis and volcanic eruptions, and also include geological hazards, namely landslides and snow avalanches, as well as technological hazards. Moreover, biophysical hazards, such as extreme temperatures, namely frosts and heatwaves, forest/scrub fires, as well as biological hazards related to epidemics and insect infestations. In addition, human-induced hazards are those resulting from modifications of natural processes in the Earth's system caused by human activities, which accelerate or aggravate the damage potential, such as land degradation, landslides, or forest fires. Human-made hazards originate from industrial or technological accidents, dangerous procedures, infrastructure failures or certain human activities, which may cause the loss of life or injury, property damage, social and economic disruption or environmental degradation, such as industrial pollution, nuclear activities and radioactivity, toxic wastes and dam failures, as well as transport, industrial or technological accidents such as explosions, fires and oil spills. Figure 1.1 shows the interactions for multi-hazards risk assessment (van Westen, 2013; CAPRA, 2009).

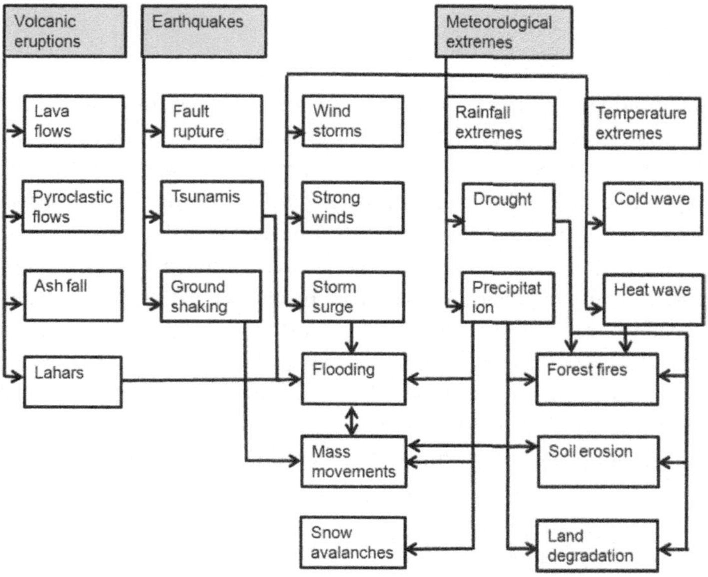

Figure 1.1 Interactions for multi-hazard risk assessment (from Van Westen, 2013).

The term environmental hazards can include a broad range of hazard types from natural or geophysical events, through technological or man-made events to social or human behavior events. Moreover, the scale of hazards can range based on whether the impacts are intense and local or diffuse and wide-spread within the society. Similarly, it is important to distinguish between voluntary or involuntary hazards. In addition, based on the World Meteorological Organization (WMO, 2006), some natural hazards are weather events, such as tropical and extra-tropical cyclones, tornadoes, thunderstorms, lightning, hailstorms, high winds, snow storms, freezing rain, dense fog, thermal extremes and drought. Others are related to weather, climate and water, namely floods and flash floods, storm surges, high waves at sea, sand- or dust storms, forest or bush fires, smoke and haze, landslides and mudslides, avalanches and desert locust swarms. Each hazard is in some way unique. Tornadoes and flash floods are short-lived, violent events, affecting a relatively small area. Others, such as droughts, develop slowly, but can affect most of a continent and entire populations for months or even years. An extreme weather event can involve multiple hazards at the same time or in quick succession. In addition to high winds and heavy rain, a tropical storm can result in flooding and mudslides. In temperate latitudes, severe summer weather, in the form of thunder and lightning storms or tornadoes, can be accompanied by heavy hail and flash floods. Winter storms with high winds and heavy snow or freezing rain can also contribute to avalanches on some mountain slopes and to high runoff or flooding later on in the melt season.

1.2.1 Hydrometeorological hazards

1.2.1.1 Storms

Heavy precipitation typically occurs with moist deep convection. The excess water vapor in rising air parcels condenses to form a cloud. The heat released through this condensation can help to sustain the convection by warming the air further and making it rise still higher, which causes more water vapor to condense, so the process feeds on itself. Doswell *et al.* (1996) have stated that in order to produce moist deep convection three ingredients are needed: 1) the environmental lapse rate must be conditionally unstable, 2) there must be enough lifting so that a parcel will reach its level of free convection, 3) there must be enough moisture present that a rising parcel's associated moist adiabat has a level of free convection. In mid-latitudes, convective precipitation is associated with cold fronts (often behind the front), squall lines, and warm fronts in very moist air. Graupel and hail indicate convection. Besides, warm rain, precipitation produced solely through condensation and accretion of liquid, is known to be important in the tropics (Houze, 1977). However, the warm rain process may also play a critical role in heavy convective precipitation events in middle latitudes as well, resulting in many flash floods and landslides. A number of researchers have noted the importance of convection and especially, mesoscale convective systems in producing warm season precipitation. A recent research campaign concerns the COnvective Precipitation Experiment (COPE), which was a joint UK-US field campaign held during the summer of 2013 in the southwest peninsula of England, designed to study convective clouds that produce heavy rain leading to flash floods. The clouds form along convergence lines that develop regularly due to the topography. The overarching goal of COPE is to improve quantitative convective precipitation forecasting by understanding the interactions of the cloud microphysics and dynamics and thereby to improve NWP model skill for forecasts of flash floods. Besides, WRF simulations were carried out to examine the sensitivity of the rainfall distribution in and around the urban area to different urban land surface model representations and urban land-use scenarios. Simulation results suggest that urbanization plays an important role in precipitation distribution, even in settings characterized by strong large-scale forcing.

1.2.1.2 Droughts

Drought is a natural, casual and temporary state of continuous decline in precipitation and water availability in relation to normal values, spanning a considerable period and covers a wide area. It is discriminated into meteorological, hydrological and agricultural drought. It is a local phenomenon identified by the severity or intensity, duration, frequency and extent. Drought impacts concern a variety of sectors of economy, environment and society of the affected area (Dalezios *et al.* 2014). The identification of dry areas has been considered two millennia ago. The classical Greek thought acknowledged that the latitude affects the arid, temperate and cold zones of the earth. There was a perception that the arid climates in small latitudes were dry. The evaluation of drought is accomplished by the drought indices, the most important of which and widely used are the Aridity Index (AI), which is based on the ratio of annual precipitation and potential evapotranspiration rates, the Standardized Precipitation Index (SPI), which is based on the probability of precipitation for any time scale (McKee *et al.* 1993), Palmer Drought Severity Index (PDSI), which is a soil moisture algorithm calibrated for relatively homogeneous regions (Palmer, 1965) and Reclamation Drought Index (RDI), which is based on a calculation of drought at the river basin level, incorporating temperature as well as precipitation, snowpack, stream flow and reservoir levels as input. Reconnaissance Drought Index (RDI) (Tsakiris *et al.* 2007) is one of the most recent developments in the field of meteorological drought indices. Essentially, it relates precipitation to the potential evapotranspiration at a location, and can be considered as an extension of the SPI. The development of Earth observation satellites from the 1980s onwards promoted the drought monitoring and detection. Certainly the Normalized Difference Vegetation Index (NDVI), a widely used index, has also been applied to drought monitoring (Dalezios *et al.* 2014).

1.2.1.3 Floods

Floods can be devastating disasters that anyone can be affected at almost any time. Indeed, flooding has been one of the most costly disasters in terms of both human casualties and property throughout the last century. Specifically, major floods in China caused the death of almost 2 million people in 1887, about 4 million in 1931 and almost 1 million in 1938. Similarly, the 1993 flood on the upper Mississippi river and the Midwest caused the death of only 47 people, but the economic loss was estimated between 15 to 20 billion dollars. Hazards associated with flooding can be divided into primary hazards that occur due to contact with water, secondary effects that occur because of the flooding, such as disruption of services, health impacts, such as famine and disease, and tertiary effects, such as changes in the position of river channels.

In order to reduce the risk due to flooding, three main approaches are considered to flood prediction. At first, determination is undertaken of the probability and frequency of high discharges of streams that cause flooding. Secondly, floods can be modeled and maps can be produced to determine the extent of possible flooding that may occur in the future. Thirdly, since the main causes of flooding are abnormal amounts of rainfall and sudden melting of snow or ice, storms and snow levels can be monitored to provide short-term flood prediction. The timing and magnitude of floods are needed for design purposes. In most of the cases, it is also necessary to categorize the flood flows according to the flood producing mechanisms, such as in flood frequency studies. Indeed, the categorization of flood flows in various physical types should provide a better and reliable estimate of the magnitude of design floods, which, in turn, is necessary for the design of hydrotechnical projects.

Flood frequency analysis is used to predict design floods for sites along a river. The frequency of occurrence of floods of different magnitude can be estimated by a variety of methods depending on the availability of hydrometric data (Loukas *et al.* 2002). Under normal conditions, observed annual peak flow discharge data are used to calculate statistical information, which then constitute the basis to construct frequency distributions, which delineate the likelihood of various discharges as a function of recurrence interval or exceedance probability. The actual choice of a design flood magnitude with its assessed return period depends both on the expected life of the scheme and on the degree of protection required. However, in ungauged watersheds the flood flow is estimated by various methods, which require the estimation of rainfall of particular critical duration and return period. This leads to the design storm concept, which is still the dominant design method in hydrological engineering.

1.2.1.4 Desertification

The United Nations Convention to Combat Desertification (UNCCD) has defined desertification as "the land degradation in arid, semi-arid and dry sub-humid areas, resulting from various factors, including climatic variations and human activities" considering climate change as an additional factor of desertification. Desertification has been and still is a controversial issue. In the previous decades, this was largely due to the lack of a common understanding of "what to measure" and "how to measure it". During the 1980's the need for a general and flexible approach to combat desertification became more keenly felt. Although desertification processes are frequently grouped into physical, chemical, biologic and anthropogenic, the mechanism of desertification is characterized by the reduction of available soil water to the growing plants resulting to critical low plant productivity. Desertification is the consequence of a series of important degradation processes in semi-arid and arid regions, where water is the main limiting factor of land use performance on ecosystems. Undoubtedly desertification is associated with degradation processes with the distinction that desertification is defined as land degradation in arid areas while land degradation not only refers to the arid regions (UNCCD, 2009).

A holistic view of land degradation issue defines it as the deterioration in physical and chemical properties of the soils that occur due to changes in environmental state. Soil degradation can be divided into natural which refers mainly to soil formation factors (pan formation, calcification, leaching, etc.) and anthropogenic comprising concepts related to cultivation practices (compaction, runoff, crusting, etc.), land management (decline in soil organic matter, nutrient depletion, nutrient toxicity, etc.) and non-rational management of natural resources (pollution, salinization, etc.). Irrespective of the scientific classifications and approaches the main causes of land degradation and of subsequent desertification in human biological time are human-induced (Salvati *et al.* 2015). This perspective has been accepted at EU level since 2001 when a significant turn in research priorities was noted indicating the important role of local and national stakeholders in conducting research programs that explore the incentives of land use decisions leading many times to unsustainable practices, which damage environment.

1.2.2 Biophysical hazards

1.2.2.1 Heatwaves

Heatwave is commonly considered as a period of abnormally and uncomfortably hot weather with high air humidity. Typically, a heatwave lasts at least two days (Koppe *et al.* 2004). Nevertheless, a clear definition of heatwaves has not yet been addressed by the World Meteorological Organization. Despite the fact that the heatwave concerns a meteorological phenomenon, it could not be assessed without reference to the related impacts on humans. So, it would be better to take into account the human sensation of heat against determining specific thresholds of meteorological parameters. Robinson (2001) considers a heatwave as an extended period of uncommonly high atmosphere-related heat stress, which causes temporary modifications in lifestyle habits and adverse health related problems affecting humans. It is very likely that heatwaves will occur with a higher frequency and duration by late 21th century, due to global warming (Nastos & Kapsomenakis, 2015). A recent research has given evidence that 'Mega-heatwaves' such as the 2003 and 2010 events broke the 500-yr long seasonal temperature records over approximately 50% of Europe (Barriopedro *et al.* 2011). In summer 2003, in a large area of central Europe, temperatures exceeded the 1961–1990 mean by about 3°C, corresponding to an excess of up to 5 standard deviations (Schär *et al.* 2004). In major cities of Europe, the daily maximum temperature exceeded 35°C for more than a week, causing about 70,000 excess deaths in parts of southern, western and central Europe (Robine *et al.* 2006).

1.2.2.2 Wildfires

The frequency of large wildfires and the total area burned have been steadily increasing, with global warming being a major contributing factor. Drier conditions will increase the probability of fire occurrence. Longer fire seasons will result as spring runoff occurs earlier, summer heat builds up more quickly, and warm conditions extend further into fall. More fuel for forest fires will become available because warmer and drier conditions are conducive to widespread beetle and other insect infestations, resulting in broad ranges of dead and highly combustible trees. Increased frequency of lightning is expected as thunderstorms become more severe. Heat waves, droughts, and cyclical climate changes such as El Niño can also have a dramatic effect on the risk of wildfires. Although, more than four out of every five wildfires are caused by people. There is a variety of fire danger rating systems used worldwide, including the Canadian Forest Fire Weather Index System (CFFWIS) used in Canada (van Wagner, 1987), the National Fire Danger Rating System (NFDRS) used in the USA (Deeming *et al.* 1977) and the McArthur Forest Fire Danger Index (FFDI) used in Australian forests (Mc Arthur, 1967). In Europe, some well-known indices include the Finnish Fire Index (FFI), developed by the Finnish Meteorological Institute (Venäläinen & Heikinheimo, 2003); the Portuguese index (ICONA, 1988); and the Italian index (IREPI) proposed by Bovio *et al.* (1984).

1.2.2.3 Frost

Frost is a meteorological phenomenon that occurs when the air temperature near the earth's surface drops to 0°C or below. Frost is a natural environmental risk with impacts on various human activities. Technically, the word "frost" refers to the formation of ice crystals on surfaces, either by freezing of dew or a phase change from vapor to ice. The word frost is used by the public to describe the condition when plants experience freezing injury. There are several definitions of frost, such as the following: (1) when the surface temperature drops below 0°C; and the existence of low air temperature that causes damage or death to the plants. (2) The occurrence of a temperature less than or equal to 0°C measured in a "Stevenson-screen" shelter at a height between 1.25 and 2.0 m. (3) Frost is defined as the condition which exists when the air temperature near the earth's surface drops below 0°C (Kalma *et al.* 1992).

Frost as a hazard means that protoplasm in plants functions at varying temperatures for each plant within restricted ranges. Indeed, frost implies near 0°C temperatures, but some plants cease protoplasm functioning at temperatures higher or lower than 0°C. Thus, a plant may suffer "chilling" injury, but not as a result of frost or freezing. Whatever the process of formation or product, frost is a symptom of a climatic situation in which temperatures have been reduced through radiation or advection to the freezing stage. Thus frost, particularly hoar frost, is a symptom of or forerunner to plant or crop loss. Because frost can freeze plant tissue, which marks the end of growth for the plant, it is significant that this can occur at either end of the growing season. Since destruction is involved, it is evident that frost is a natural hazard along with other climatic hazards, such as excessive precipitation, tornadoes, or hail.

1.2.2.4 Biological hazards

Biological hazards are organisms or products of organisms that present a health hazard to humans. Biological hazards can be encountered anywhere in the environment, including home, school or work. Exposure to biological hazards in the workplace may result in a significant amount of occupationally associated disease, although the biological hazards are not always recognized in the work setting. Biological hazards can be put into different categories. The most common biological hazards include: Bacteria – microscopic organisms that live in soil, water, organic matter or the bodies of plants and animals and are characterized by lack of a distinct nucleus and the inability to photosynthesize. Viruses – a group of pathogens that consist mostly of nucleic acids and that lack cellular structure. Viruses are totally dependent on their hosts for replication. Fungi – any of a major group of lower plants that lack chlorophyll and live on dead or other living organisms. Examples of different types of biological hazards: bacteria – Escherichia coli (E. coli), Mycobacterium tuberculosis (TB), tetanus; viruses – common cold, influenza, measles, SARS, Hantavirus, rabies; fungi – athlete's foot, mould, rusts, mildew, smut, yeast, mushrooms. Other concerns from the environment may include: insect stings/bites; allergic reactions; e.g. peanuts, pollen, bee stings; poisonous plants/animals; e.g. poison ivy, cobras, belladonna.

1.2.3 Geophysical hazards

1.2.3.1 Geological hazards

The assessment of landslide hazard and risk is predominantly important in urbanised zones included into land use plans which assume certain socioeconomic and technological progress of a region. Especially in the last decade a number of scientific works have been devoted to statistical approaches to landslide hazard using map algebra tools implemented within the geographic information systems. This methodology includes the assessment of landslide hazard, identification of the elements at risk, assessment of vulnerability and the assessment of landslide risk. The application of Geographic Information Systems (GIS) represents an innovative, modern and prospective approach in the assessment of hazards and risk for larger territorial units. The emphasis is on the theoretical description of the most common quantitative methods used for landslide hazard or susceptibility assessment. Four basic axioms are needed before starting to assess landslides hazard: (1) landslide will occur under identical geological, geomorphological, hydrogeological and climatic conditions as in the past; (2) the triggering factors of landslides may be defined and analysed; (3) the degree of landslide hazard must be quantifiable; (4) slope deformations must be classifiable.

Snow avalanches are typical mass downslope movements and are a part of bigger group called natural hazards. Mass of snow moving downslope driven by gravitational force is usually referred to as snow avalanche. Snow avalanches can range from only few meters long almost harmless sluffs to several kilometres long, disastrous avalanches capable of destroying a whole village. Topography of the slope and meteorological situation are the most important factors and thus determine when and where the naturally triggered avalanches will be released. On the other hand, human trigged avalanches constitute an additional factor. There are two basic types of snow avalanche classification based on the following: (1) type of triggering mechanism and motion (loose snow, slab and glide avalanche); (2) type of snow in avalanche (wet, powder and mixed avalanches). Despite the fact that snow avalanches do not have the impact of large natural hazards as earthquakes, floods and volcanic eruptions, they still can be a threat to humans and infrastructure in mountainous areas. Worldwide annual statistics count approximately 250 fatalities caused by snow avalanches.

1.2.3.2 Earthquakes

Earthquake hazards comprise any natural phenomenon associated with earthquakes. Earthquakes are known as the shaking of the Earth's surface, producing significant impacts on both physical and urban environments with severe socioeconomic aspects. Most of the earthquakes are generated at the boundaries of the lithospheric plates, which float over the mantle's asthenosphere, converging or diverging. Friction caused by the plates interaction builds up stresses that, when released produce ground faulting that radiates through the lithosphere producing complex seismic waves, which in turn, can affect the near built environment.

Although earthquakes mainly control the morphology of the Earth's surface, they would unlikely be considered of major significance in the absence of their effects on the anthropogenic environment, primary and secondary. Primary earthquake hazards at a site regard the ground shaking due to the passage of the seismic waves (dynamic deformation) and/or the ground displacements (static deformation) in the vicinity of the causative fault. Secondary earthquake hazards are the after-earthquake effects caused by the primary ones and may often be more catastrophic. Such hazards are ground failures, fire, landslides, rock and snow avalanches, liquefaction, flooding, tsunamis and seiche that have been frequently reported to follow the occurrence of strong earthquakes. The measure of earthquake hazards at a site mainly depends on the size and type of the seismic rupture, its distance from the site and the geological structure between the source and the site's surface that may impact the seismic energy.

Given that earthquake prediction is still infeasible, the major task in seismological research is the understanding of earthquake phenomena and their consequences on the natural and anthropogenic environment, with the purpose of mitigating them by providing valid and timely information, to be used for effective earthquake planning and decision-making processes. Primary and secondary effects are related with vulnerability which is defined as a set of prevailing or consequential physical and sociopolitical conditions that affect a community's ability to mitigate, prepare for, or respond to an earthquake hazard (ADPC, 2003). Earthquake hazards, structural vulnerabilities and exposed values, when combined, yield a region's exposure to seismic risk.

1.2.3.3 Volcanoes

Volcanoes are classified as active, dormant or extinct. It is stated that all volcanoes, which erupted within the last 25,000 years, should be considered as at least potentially active. Although only 5 per cent of eruptions cause human fatalities, the relative infrequency of hazardous volcanic events constitutes one of their most dangerous features. Moreover, the distribution and behavior of volcanoes is controlled by the global geometry of plate tectonics, whereas active volcanoes exist in every continent, except Australia. All volcanoes are formed from magma, which is the molten material within the Earth's crust. There is no agreed international scale against which to measure the size of individual volcanic eruptions, however, a semi-quantitative volcanic explosivity index (VEI) is used. VEI combines the total volume of ejected products, the height of the erupted cloud, the duration of the main eruptive phase and several other factors into a simple 0–8 scale of increasing hazard.

For example, a VEI = 5 eruption occurs every 10 years on the average, whereas with VEI = 7 every 100 years.

Primary volcanic hazards are associated with the products ejected by the volcanic eruption. A dense cloud of lava fragments is finally ejected to form a turbulent mixture of hot gases and pyroclastic material. Moreover, there are secondary volcanic hazards, which are ground deformations and occur widely as volcanoes grow from within by magma intrusion and as new layers of lava and pyroclastic material accumulate on the surrounding slopes. At the present time, GPS technology contributes to real-time measurements of this process. It should be stated that there is no known method of preventing volcanic eruptions. Indeed, lava flows moving at relatively slow speeds are the volcanic hazard over which most physical control can be exerted.

1.2.3.4 Technological hazards

Technological hazards include potential hazards in buildings with exterior stone panels on vertical faces, especially tall ones in urban centers, as well as transportation and industrial hazards. The focus on buildings is on marble, granite and sandstone panels and their susceptibility to loose strength when subjected to environmental conditions such as such heat gain and wind loads. This may lead to bowing and fracturing in marble panels, and a further loss of strength, and fracturing in granite and sandstones. Recognition of these features and their association with strength measurements on panels leads to criteria for potential panel replacement. This may involve the entire building or selected areas at periodic intervals. In either case emergency plans if whole panels or even pieces fall need to be in place. In all cases continuing monitoring is critical.

The recent increase of industrial and natural disasters frequency and severity has stimulated a great awareness on behalf of both the general public and local governments, in environmental information management systems that will be able to collect, process, visualize and interpret geospatial data and value-added application workflows in order to support decision making in case of emergency. Since 1900 there have been 35,000 natural catastrophic events on the planet resulting more than 8 million life loses. In order to make efficiently handling, forecasting, mitigation and prevention of such disasters new open, scalable and distributed service platforms need to be created. Instead of leaving this task to specialized and stand-alone and fragmented systems, integrating such services in global web mapping platforms has a huge potential to create a totally new area or market of applications for such platforms that will be addressed not only to the general public but also to governments and local authorities. Geospatial data, real time and historical measurements, wireless sensors, traffic information, population distribution data, and value added services can all be integrated into current web mapping services by creating a new set of APIs and enhancing already available and paid for, geospatial data that these services currently incorporate.

1.3 CAUSES – FACTORS – FEATURES – DRIVERS OF HAZARDS

Hazards impacts refer to a multitude of drivers that may turn physical causes of hazards, such as reduced average precipitation for drought, deficient soil moisture and low water levels, into disaster events for vulnerable populations and economies. Some apparent disaster trends are attributed to socio-economic factors rather than the frequency and magnitude of geophysical processes. There are certainly several causes and drivers, which justify the increasing trend in disaster impact, despite the fact that the frequency of geophysical events remains almost the same and there are many positive efforts to reduce disasters. These types of risk causes and drivers may include land pressure, climate change, economic growth, technological innovation, population growth, urbanization, inequalities, political change, social expectations, risk governance capacities and global interdependence, among others. A very brief description of the main risk drivers is outlined below, which constitutes a current research subject due to its global significance.

1.3.1 Features and characteristics of hazards

For assessing and monitoring hazards, several features are usually detected. Indeed, conventional and/or remote sensing data and methods can be used to delineate the spatial and temporal variability of several features in quantitative terms (Dalezios *et al.* 2014). A description of some key features follows.

Severity: severity or intensity of hazard is defined as escalation of the phenomenon into classes from mild, moderate, severe and extreme. The severity is usually determined through indicators and indices, which include the above mentioned classes. In some climate hazards, such as droughts, the phenomenon evolves gradually in regions severely affected, and there is a seasonal and annual shift of the so-called epicenter, which is the area of maximum severity.

Periodicity: is considered the recurrence interval of hazard.

Duration: duration of a hazard episode is defined as the time interval from the start and end time.

Onset: the beginning of a hazard is determined by the occurrence of a hazard episode. The beginning can be assessed through modelling or indicators or indices reaching certain threshold value.

End time: end time of a hazard episode signifies the termination based again on threshold values of indicators or indices. It is often difficult to determine the onset and the ending of a hazard, such as drought, and on what criteria these determinations should be made.

Areal extent: areal extent of hazard is considered the spatial coverage of the phenomenon as is quantified in classes by indicators or indices. Areal extent varies in time and remote sensing has contributed significantly in the delineation of this parameter by counting the number of pixels in each class. From a planning perspective, the spatial characteristics of hazard may have serious implications for several sectors of the economy, such as agriculture, energy, transportation, health, recreating and tourism.

1.3.2 Factors and drivers of hazards

Disaster risk arises when hazards interact with physical, social, economic and environmental vulnerabilities. The impact of disaster can be transferred from one region to another. This, compounded by increasing vulnerability related to several factors, such as population growth, land pressure, urbanization, social inequality, climate change, political change, economic growth, technological innovation, social expectations, global interdependence, environmental degradation, competition for scarce resources and the impact of epidemics, points to a future, where disasters could increasingly threaten, among others, the sustainable development of several regions (Smith, 2013; Dalezios *et al.* 2016). Sustainable development, socio-economic improvement, good governance, and disaster risk reduction are mutually supportive objectives. A brief presentation of several factors and drivers of hazards follows.

Population growth. The world population is steadily increasing, thus, the number of people exposed to hazards is equally increasing, since about 90 percent of the population growth takes place in the less developed countries, which face high human vulnerability.

Urbanization. Urban and economic development is seldom planned to take adequate measures for water management and conservation. This results into increasing water demand for sectors, such as intensive agriculture, urban development and tourism, among others, which constitutes both agricultural and hydrological drought risk driver.

Land pressure. A portion of the world's population (about 20 percent) lives in areas suffering severe environmental degradation, with the majority depending on agriculture. In such cases, rural land pressure in terms of inappropriate water and soil management constitutes a drought risk driver. Typical examples are: water-intensive crops, soil erosion, shifting production patterns, deforestation, as well as overgrazing and rangeland management.

Climate variability and change. The main effect of increasing climate variability and/or change is temperature increase and precipitation reductions in several climate zones around the world [9]. Climate scenarios indicate an increase in the reoccurrence interval of droughts leading, for example, to significant crop losses or even failure, as well as impact on small-scaling farming (Sivakumar *et al.* 2005).

Weak risk governance. Capacities is a drought risk driver. There is a need to develop national drought risk management policies, to put high priority to drought, to avoid fragmented responsibilities for drought risk management at local or regional level, to establish drought insurance compensations and to avoid conflict and excess water use. Technological innovations may be helpful in drought forecasting and early warning, as well as in seasonal climatic forecasting.

Technological innovation. The technological advancements of the developed countries are expected to help preventing disasters through better early warning systems and probably safer infrastructure. Nevertheless, the dependence of the society to advanced technology may increase the risk of technology failure and potential for disaster. There are numerous examples of trends that enhance the vulnerability to hazard, such as new high-rise buildings, nuclear reactors, mobile homes, among others.

Global interdependence. The world economy works against the developing countries by increasing hazard vulnerability. Indeed, most of their export belong to primary commodities, where market prices either remain very unstable or have fallen over several decades. The developing countries have little chance to process and market their products, whereas at the same time are dependent on importation of highly priced manufactured goods from developed countries. This cycle becomes acute when natural disasters destroy local products. For example, the Sahelian drought disrupts not only local economies, but may also affects neighboring regions creating international refugees with a global impact.

Political change. The developed countries gradually reduce the internal welfare and their international commitment. Moreover, in eastern Europe and the former Soviet Union, health care, education and social provision has replaced state influence by free-market ideas, which strongly affect the poor members of the society. On the other hand, the developed countries have reduced the development aid, leading to higher vulnerability to disasters.

Economic growth. Economic growth has increased the risk to catastrophic property damage in the developed countries. Capital development is expected to be faced with an increasing amount of property for each hazard, unless measures are taken for risk reduction within urban or industrial areas. It is observed that growth has occurred in areas vulnerable to natural hazards. Indeed, the increased leisure time has resulted in the construction of many resort houses in environmentally susceptible areas.

Social expectations. It is anticipated that rising social expectations may cause an increase in hazard vulnerability, especially in developed countries. It is stated that the same level of security of services is expected by the public from most weather dependent systems, such as water supply or energy supply. However, several industrial or technological improvements allow for less scope for an effective response to environmental hazard.

Inequality. Disaster vulnerability is closely associated with the economic gap between rich and poor. Indeed, poverty and rural vulnerability constitute hazard risk drivers, which may result, for example, into lack of irrigation and water storage, as well as expansion of intensive cash crop production.

1.4 DIACHRONIC EVOLUTION AND TRENDS OF HAZARDS

It is clear that broad-based research into natural hazards has begun at the middle of 20th century. Until then, hazards had been viewed as isolated geophysical events, which were to be considered through engineering works, such as dams. Gradually, by the early 1970s, research on natural hazards was basically fragmented. Indeed, engineers and geologists continued with direct efforts for improved prediction of extreme natural events and the construction of physical work designed for resistance. At the same time, hazard-based approach was developed on the concept of human ecology and the mitigation of losses by providing several human adjustments. Moreover, a more disaster-based approach was adopted with emphasis on understanding the role of human behavior during community crisis and the need for preparedness improvement. Similarly, during the 1970s, extreme natural events, such as the prolonged Sahelian drought, have exposed the vulnerability of several, even developed, countries to climate variability. Moreover, during the 1980s, emphasis was given to the relationships between under-development and hazard impact especially in the third world. As a result, existing multidisciplinary research widened even further reflecting the complexity and diversity of the field and affecting also an earliest distinction between "natural" and "man-made" hazards.

From the beginning of 21st century, there is an awareness of risk in the environment along with a growing concern for the continuing potential damage caused by hazards.

Gradually, a more integrated approach to environmental hazards has been attempted using common methodologies, such as risk analysis. Besides physical protection, a synthesis of anti-hazard measures starts being considered, including land use management, better planning for response and recovery and emergency warnings. Moreover, a global program to reduce the losses from natural hazards was adopted in December 1989 by the U.N. General Assembly declaring the 1990s as the International Decade for Natural Disaster Reduction (IDNDR). The challenge to researchers within the hazards community was to use their skill and adopt a wider perspective involving global change for a safer and sustainable environment for all, now and in the future (Smith, 2013). Recently, the World Conference on Disaster Reduction (UNISDR, 2005) has adopted several strategic goals resulting in a number of priorities for action during the next 10 years (2005–2015). Finally, the 2012 special report on Managing the Risks of Extreme Events and Disasters to Advance Climate Change Adaptation (SREX) from the IPCC focuses on the relationship between climate change and extreme weather and climate events, the impacts of such events, and the strategies to manage the associated risks (IPCC, 2012). Understanding of extreme events and disasters is a pre-requisite for the development of adaptation strategies in the context of climate change and risk reduction within the disaster risk management framework (EM-DAT, 2009). Extreme events will have greater impacts on sectors with closer links to climate, such as agriculture and food security. Figure 1.2 shows the number of reported disasters, along with the number of people killed and the number of people affected over the period 1900–2009 (van Westen, 2013).

The Sendai Framework for Disaster Risk Reduction 2015–2030 has been recently adopted (UNISDR, 2015), which requires a better understanding of risk in all its dimensions of vulnerability, exposure and hazards. It aims to ensure that the multi-hazard management of disaster risk is factored into development at all levels, as well as within and across all sectors. The Sendai Framework is wide in scope. It applies to the risk of small-scale and large-scale, frequent and infrequent, sudden and slow-onset disasters, caused by natural or man-made hazards as well as related environmental, technological and biological hazards and risks. The Sendai Framework recognizes the importance of science and technology for disaster risk reduction. The goal is to prevent new and reduce existing disaster risk through the implementation of integrated and inclusive economic, environmental, technological, educational, structural, legal, social, health, cultural, political and institutional measures that prevent and reduce hazard exposure and vulnerability to disaster, increase preparedness for response and recovery, and thus strengthen resilience.

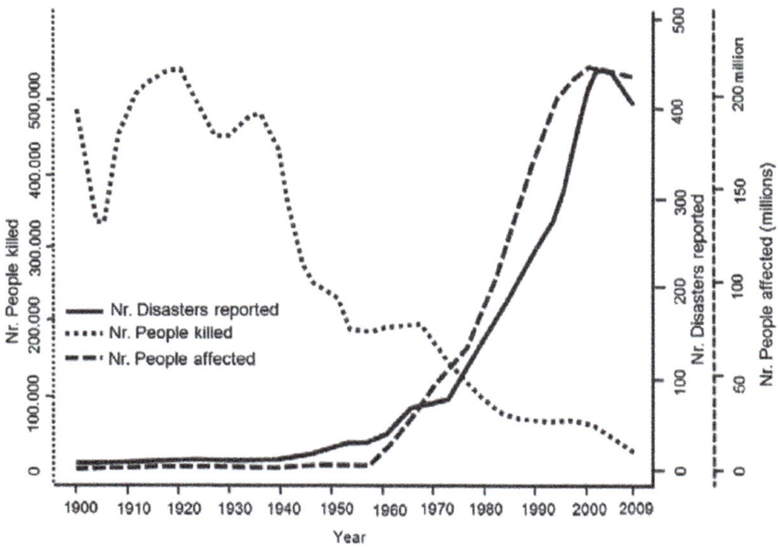

Figure 1.2 Number of reported disasters, number of people killed and number of people affected over the period 1900–2009 (from Van Westen, 2013).

1.5 HAZARD AND RISK ANALYSIS

1.5.1 Hazard and risk concepts

Hazard is an inescapable part of life. Hazard is defined as "*a potentially damaging physical event, phenomenon or human activity that may cause the loss of life or injury, property damage, social and economic disruption or environmental degradation*". Hazards can include latent conditions that may represent future threats and can have different origins: natural (geological, hydrometeorological and biological) or induced by human processes (environmental degradation and technological hazards) (UNISDR, 2005). **Risk** is sometimes taken as synonymous with hazard (UNISDR, 2005), but risk has the additional implication of the chance of a particular hazard actually occurring. Thus, risk is the actual exposure of something of human value to a hazard and is often regarded as the product of probability and loss. Therefore, hazard (or cause) may be defined as "*a potential threat to humans and their welfare*" and risk (or consequence) as "*the probability of a hazard occurring and creating loss*" (Smith, 2013). Unlike hazard and risk, a **disaster** is an actual happening, rather than a potential threat, thus, a disaster may be defined as "*the realization of hazard*". A more detailed disaster definition is "*an event, concentrated in time and space, in which a community experiences severe danger and disruption of its essential functions, accompanied by widespread human, material or environmental losses, which often exceed the ability of the community to cope without external assistance*" (Smith, 2013).

The term **environmental hazard** has the advantage of including a wide variety of hazard types ranging from "*natural*" (geophysical) events, through "*technological*" (man-made) events to "*social*" (human behavior) events. Specifically, it is possible to use the following working definition of environmental hazards: "*Extreme geophysical events, biological processes and major technological accidents, characterized by concentrated releases of energy or materials which pose a large unexpected threat to human life and can cause significant damage to goods and environment*" (Smith, 2013).

Vulnerability is defined as "*The conditions determined by physical, social, economic and environmental factors or processes, which increase the susceptibility of a community to the impact of hazards*" (UNISDR, 2005). The concept of vulnerability, like risk and hazard, indicates a possible future state. Most approaches to reduce system-scale vulnerability can be viewed as expressions of either resilience or reliability. **Resilience** is defined as "*The capacity of a system, community or society potentially exposed to hazards to adapt, by resisting or changing in order to reach and maintain an acceptable level of functioning and structure*". This is determined by the degree to which the social system is capable of organizing itself to increase this capacity for learning from past disasters for better future protection and to improve "risk reduction measures" (UNISDR, 2005). **Reliability**, on the other hand, reflects the frequency with which protective devices against hazard fail.

The **climate change** research community has identified sensitivity, adaptation and vulnerability as dimensions for consideration of climate change impacts, including changes in variability and extremes. **Sensitivity** means the degree to which a system is affected, either positively or negatively by climate related stimuli. **Adaptation** identifies the adjustment capacity of a system under climate change to moderate damage, take advantage of opportunities or otherwise cope with the consequences. **Vulnerability** identifies the degree to which a system is susceptible or unable to cope with adverse effects of climate. A distinction should be made between **climate change**, which is a long term change to a different climate regime, through the accretion of small changes of trend and **climate variability**, which is the swing from one extreme to another along the progression of weather to a new stable set of climate conditions.

1.5.2 Risk management framework

A holistic and integrated approach to environmental hazards has been gradually considered using common methodologies, such as risk analysis. Indeed, through risk analysis there is an attempt to investigate and better understand hazards with the objective to develop pro-active measures and procedures before a crisis. Risk analysis includes the components, which constitute the risk management framework (Dalezios & Eslamian, 2016). Risk management covers the risk assessment component, along with risk governance, which involves a feedback by all the affected parties. Indeed, risk assessment consists of three steps, namely risk identification, risk estimation and risk evaluation (Figure 1.3). Risk management means reducing the threats posed by known hazards, whereas at the same time accepting unmanageable risks and maximizing any related benefits.

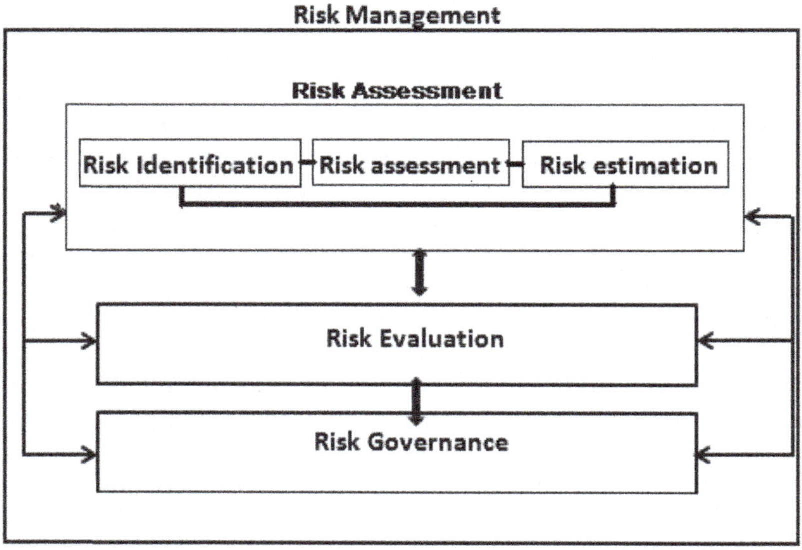

Figure 1.3 Components of risk management (from Dalezios *et al.* 2014).

1.5.2.1 Risk identification and quantification

Risk identification involves hazard quantification, event monitoring including early warning systems, statistical inference and the development of a database, which includes recorded historical information on hazards and their effect.

Database Development. A database is developed, which constitutes the input data to risk analysis and is based on recorded historical environmental data of the study areas. At first, digital information is collected on **environmental factors,** such as geology, geomorphology, soil, topography, hydrology, meteorology, agronomy, land use, land cover, GIS and similar aspects, which are used in susceptibility assessment. Information is also collected on **triggering factors** leading to hazards**,** which are used in hazard assessment. Also a **hazard inventory** is developed based on recorded historical disaster events. This hazard inventory is used in susceptibility and hazard assessments. Finally, the exposed **elements at risk** are identified and recorded, which are used in exposure analysis and vulnerability assessment.

Risk Quantification. Hazards are quantified either numerically or by modelling or even by using indicators and/or indices, depending on the type of hazard. Furthermore, the potential damage caused by a disaster requires forecasting and monitoring in the affected region.

Susceptibility Assessment. This assessment involves initiation and spreading analyses, which are used in risk estimation and hazard assessment. Initiation analysis includes hazard inventory and heuristic, statistical and physical hazard modelling based on environmental and triggering factors. Spreading analysis includes empirical, analytical and numerical hazard modelling.

1.5.2.2 Risk estimation and vulnerability assessment

The assessment of probability distributions along with quantification and analysis of the exposed elements at risk constitutes the basis of this component. Also estimation of hazard magnitude-duration-frequency and areal extent relationships are considered (e.g. Dalezios *et al.* 2000). Vulnerability assessment also includes testing, selecting and mapping indicators.

Risk Estimation for Hazard Assessment. Risk estimation involves the risk of hazard events, i.e. the probability of such event. As already mentioned, risk estimation also includes hazard magnitude-duration-frequency and areal extent relationships with the associated costs. The changes in environmental conditions, reflected in the changes of land use patterns, also form the input in the models that are used for hazard assessment, resulting in a number of possible hazard scenarios, with associated probabilities and indications of magnitude, frequency and spatial extent. This part contributes to exposure analysis.

Exposure Analysis. This part is essentially a GIS analysis and involves spatial overlap of hazard footprints for the elements at risk. The elements at risk are considered in terms of type, temporal variation and quantification, i.e. quantity and economic value. This part contributes to risk assessment and, in particular, to the quantification of the amount of elements at risk.

Vulnerability Assessment. An inventory is considered of exposed elements at risk, and their characterization in terms of aspects that can be used for the assessment of vulnerability. Uncertainty analysis of vulnerability is also considered. The uncertainty in vulnerability approaches are usually evaluated based on historical damage catalogues, modelling and expert assessment. Indices for vulnerability assessment that include uncertainty levels are also considered.

1.5.2.3 Risk assessment

The aim is to integrate the techniques for probabilistic hazard assessment, which incorporate the uncertainty due to future environmental changes, with the results of the exposure and vulnerability analyses into Quantitative Risk Assessment (QRA).

Risk Assessment. This part involves specific risk scenarios, which consist of a combination of probabilistic hazard scenarios with scenarios of exposed elements at risk and their vulnerabilities. A probabilistic scenario seems the most feasible, since large uncertainties are involved in predicting changes in risk. Hazard assessment consists of temporal probability in terms of duration and time of onset, hazard intensity and spatial extent through exposure analysis. Vulnerability refers to the degree of loss to each type of elements at risk as related to hazard intensity, where exposure means the spatial overlay of hazard and each element at risk. The term elements at risk refers to the type, the temporal variation, quantification, as well as the location of the elements at risk through exposure as described above. This part contributes to quantitative risk assessment.

Quantitative Risk Assessment (QRA). The combination of hazard scenarios along with vulnerability scenarios and quantification of elements at risk develops the QRA. In particular, the combination of all the specific risks leads to the total risk, as already mentioned, which constitutes the Quantitative Risk Assessment (QRA). If risks cannot be quantified, the qualitative risk assessment should be used involving indices. This part contributes to risk evaluation.

1.5.2.4 Risk evaluation

The aim of this component is to analyze all the risk management options and optimal tools based on the previous risk scenarios results in order to achieve risk reduction.

Risk evaluation. Risk evaluation consists of the loss associated to each event and involves Environmental Impact Assessment (EIA) and Strategic Environmental Assessment (SEA), land use planning, cost-benefit analysis of adaptation options for the development of mitigation measures and emergency preparedness plans.

Development of Decision Support System (DSS). All the risk management options are used to evaluate the required changes in risk management approaches and are integrated into a Web-based platform for risk management in terms of a multi-scale and interactive DSS. For the development of the DSS, there are certain phases that are followed starting from the intelligence phase involving the problem analysis, then the design phase for generation of alternatives and then the decision phase. Specifically, the decision phase involves several methods including economic techniques, such as cost benefit analysis (CBA), physical planning approaches, social impact assessment, environmental impact assessment (EIA) and Multi Criteria Evaluation (MCE), which consists of certain steps, namely formulation of objectives and procedure development for vulnerability assessment using indicators for social and physical vulnerability as well as capacity.

1.5.2.5 Risk governance

Risk governance is an integration of all the rules, processes and mechanisms implemented and communicated within the risk management framework through the developed DSS for risk reduction. In fact, sound risk evaluation requires both good science and good judgment. However, at the present time the lack of a feedback constitutes a serious deficiency in the reduction of environmental hazards (Smith, 2013). Thus, there is a need for feedback of all the risk assessment exercises. This could justify the level of public awareness and response by the Authorities.

Feedback of risk reduction. The effectiveness of risk reduction measures is based on successful risk governance. It is necessary to consider both the QRA and the relevant aspects of risk perception. The target is to achieve an agreement with local stakeholders and end-users on risk reduction measures.

Dissemination of results and public awareness. The effectiveness of risk communication strategies is analyzed and suitable information is developed for different stakeholders. Dissemination tools and activities are employed for public awareness and information spreading.

In summary, figure 1.4 presents a description of the main steps of natural hazards and risks management for the development of expert (knowledge-based) systems (from Tacnet & Curt, 2013). Moreover, figure 1.5 shows expert systems and information systems produced by experts for risk management (from Tacnet & Curt, 2013).

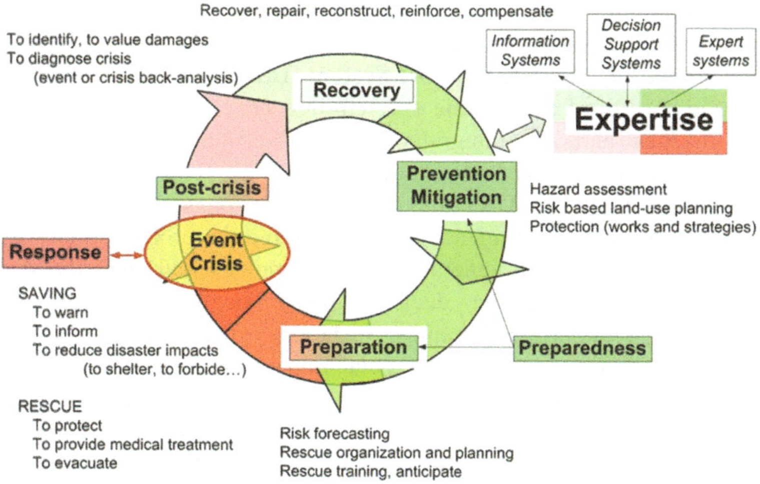

Figure 1.4 Description of main steps of natural hazards and risks management for the development of expert (knowledge-based) systems (from Tacnet & Curt, 2013).

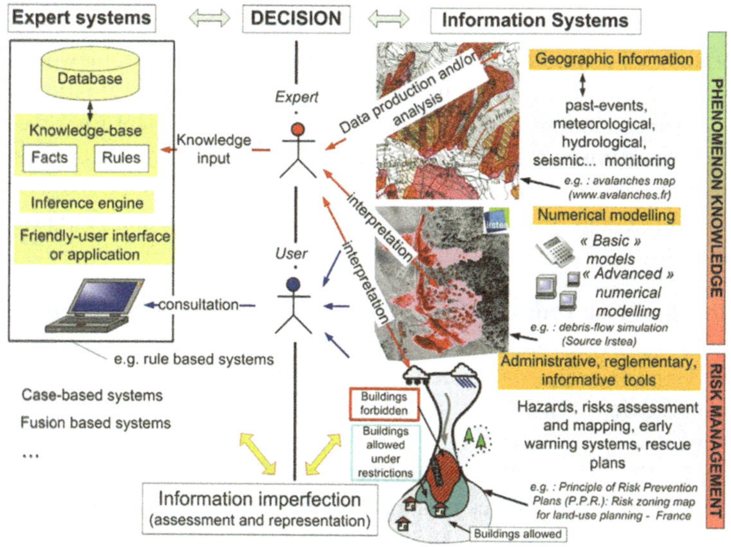

Figure 1.5 Expert systems and information produced by experts for risk management (from Tacnet & Curt, 2013).

1.6 LEGAL AND INSTITUTIONAL ASPECTS OF HAZARDS

The legal and institutional framework concerning Disaster Risk Deduction (DRR) and risk management, in general, is gradually expanding on the international,

European, and national/regional/local level, focusing on prevention, mitigation and preparedness as equal priorities to response (EC DG ECHO, 2013). On the international level, the International Strategy for Disaster Reduction (ISDR), launched in 2000 by the United Nations General Assembly (resolution A/54/219) stressed the need for coordination of Academia and NGOs to achieve substantive reduction in disaster losses and to build resilient nations and communities as an essential condition for sustainable development. Consequently, the World Conference on Disaster Reduction (WCDR) in Kobe, Japan, at 2005, adopted the Hyogo Framework for Action (HFA) 2005–2015: *Building the resilience of nations and communities to disasters* –a ten year strategy for national governments. The coordination of actions aimed at promoting the DRR policy agenda lies with the secretariat of ISDR, the United Nations International Strategy for Disaster Reduction (UNISDR) (Innocenti *et al.* 2010). The program is based on three strategic objectives:

(1) The integration of DRR into sustainable development policies and planning.
(2) The development and strengthening of institutions, mechanisms and capacities to build resilience to hazards.
(3) The systematic incorporation of risk reduction approaches into the implementation of emergency preparedness, response and recovery programs.

The role of both, developed and developing countries to develop strategies for adaptation to climate change was emphasized in the International Panel on Climate Change (IPCC) fourth assessment synthesis report (IPCC, 2007). In this report, the fact that climate change poses a threat of disasters was also underlined. These findings were used in the UNFCCC Bali Action Plan (INFCCC, 2007) in which DRR was recognized as a vital dimension in adaptation to climate change. UNISDR has been collaborating with equivalent European bodies, and in particular with EC Directorate General for Research (DG Research) in organizing workshops (October 2009 and July 2010), exchange of information and increasing public awareness (Innocenti *et al.* 2010; UNISDR, 2007a; UNISDR, 2007b; UNISDR, 2008). The European Commission policies related to DRR are described in DG ECHO Thematic Policy Document no 5 "Disaster Risk Reduction: increasing resilience by reducing disaster risk in humanitarian action" (DG ECHO, 2013). According to it, DG ECHO's policy focuses on the following objectives:

(1) Maximize the effectiveness, efficiency and relevance of DG ECHO-funded DRR actions;
(2) Strengthen DG ECHO's preparedness to respond rapidly to unfolding humanitarian crises;
(3) Increase the coherence of DG ECHO's decision-making;
(4) Enhance coherence with other DG ECHO policies;
(5) Inform partners and other relevant stakeholders about DG ECHO's policy on DRR.

The above were encapsulated in DIPECHO (Disaster Preparedness ECHO) program in which a comprehensive people-centered approach to DRR was developed, including engaging with institutions at all levels (ECHO, 2013), while the Council Regulation No 1257/96 provided the basis for DG ECHO's mandate in Disaster Risk Reduction.

Two key policy documents can further exhibit the EU's commitment to Disaster Risk Reduction. These are: the European Consensus on Development (European Council, 2006), and the Consensus on Humanitarian Aid. The integrated approach on DRR adopted by the European Commission has been also expressed in the White Paper "Adapting to climate change: Towards a European framework for action" (2009), which includes DRR as an integral part of the adaptation strategy. This was complimented by the following initiatives: on 23 February 2009, the Commission adopted the Communication "EU Strategy for Supporting Disaster Risk Reduction in Developing Countries", alongside the Communication on a "Community Approach on the Prevention of Natural and Man-Made Disasters" addressing disaster risk within the EU. This was later complemented by the Communication on Resilience of October 2012. Generally, the EU Strategy towards DRR:

(1) Defines responsibilities amongst Commission Services and Member States;
(2) Positions the EU strategy towards the Hyogo Framework for Action (2005–2015) "Building the Resilience of Nations and Communities to Disasters" 6 and 7;
(3) Advocates for more effective cooperation between the humanitarian and development actors within the EU.

Since 2010, DG ECHO has combined in its DRR policies, Civil Protection and Humanitarian Aid (ECHO….). Furthermore, the EU fully supports the commitments made at the Busan High Level Forum on Aid Effectiveness (2011), which recognizes the importance of partnering to strengthen resilience and reduce vulnerability among people and societies at risk of disasters. Partnerships are considered as of crucial importance to DRR, and they include all organizations eligible under the regulations of the Framework Partnership Agreement (FPA) and the Financial and Administrative Framework Agreement (FAFA). Although not eligible for DG ECHO funds, state actors and local civil society organizations are considered essential stakeholders for DRR actions. Regarding Risk Assessment Commitment, currently, the EU policy context is set out in the following (Marra, 2014):

(1) The Union Civil Protection Mechanism (Decision No. EU/1313/2013), in article 5 1.(b) – COM, to support/promote MS risk assessment activities, in article 5 1.(c) – COM in cross-sectorial overview and map, and in article 6(a) – MS deadline December 2015
(2) *Ex-ante conditionality for Cohesion policy (deadline 2016), Thematic Objective 5: Promoting climate change adaptation, risk prevention and management.*

Currently, the Sendai Framework for Disaster Risk Reduction 2015–2030 has been adopted by Member States on 18th March 2015, at the World Conference on Disaster Risk Reduction (DRR) held in Japan, and was endorsed by the UN General Assembly in June 2015. The Sendai Framework promotes shifting focus from managing disasters to managing risks (UNISDR, 2015).

In terms of harmonization of EU policies and institutional framework with the national/regional/local legal frameworks of member countries, authorities for disaster management, such as those involved with civil protection and civil defense, are usually urged to move forward the risk reduction agenda, encouraging and initiating the collaboration and coordination with relevant ministries and actors responsible for DRR (Sapountzaki *et al.* 2011; Sapountzaki & Dandoulaki, 2006). Still, in many countries the two often operate in isolation, which limits the contribution that one sector can have for the other. Many European governments have improved their internal policy coordination on adaption and DRR. This was possible mostly due to the establishment of their National Platforms for DRR (as provided in the HFA implementation). National Platforms provide national mechanisms where actors dealing with different aspects of risk reduction (e.g. line ministries, NGOs, Academia, private sector, etc.) share information and take coordinated actions to influence national strategic investments and research in implementing risk reduction activities as a means to adapt to climate change. Currently, the European Commission has relevant information from 20 participating States (EC DG ECHO, 2013), and the accruing challenges, which have to be faced, and accordingly transferred in the institutional and legal frameworks on both, the European and the national levels, concern data comparability, clear terminology, clear methodology, and use of the risk matrix developed.

1.7 SUMMARY

In this introductory chapter, a comprehensive and conceptual description of hazards has been presented. At first, the basic concepts and definitions of hazards has been described, along with the scope and objectives of disaster reduction. This has been followed by a classification and conceptual description of environmental hazards, which are covered in this book. Then, the major features and drivers of environmental hazards have been presented. This has been followed by a brief review of the diachronic trends of environmental hazards, along with research and applications. The next section has covered a brief presentation of a major component of the book, namely the risk management framework. Finally, the main legal and institutional aspects of environmental hazards are considered.

REFERENCES

Asian Disaster Preparedness Center (ADPC) (2003). Third Regional Training Course on Earthquake Vulnerability Reduction for Cities. Dhaka, Bangladesh.

Barriopedro D., Fischer E. M., Luterbacher J., Trigo R. M. and García-Herrera R. (2011). The hot summer of 2010: redrawing the temperature record map of Europe. *Science*, **332**, 220–224. doi: 10.1126/science.1201224.

Bovio G., Quaglino A. and Nosenzo A. (1984). Individuazione di un indice di previsione per il Pericolo di Incendi Boschivi. *Monti e Boschi Anno*, **35**(4), 39–44.

Businger S. and Reed R. (1989). Cyclogenesis in cold air masses. *Weather Forecast*, **20**, 133–156.

CAPRA (2009). Central American Probabilistic Risk Assessment (CAPRA). World Bank. www.ecapra.org.

Dalezios N. R. and Eslamian S. (2016). Regional design storm for Greece within the flood risk management framework. *International Journal of Hydrology Science and Technology (I.J.H.S.T)*, **6**(1), 82–102.

Dalezios N. R., Loukas A., Vasiliades L. and Liakopoulos H. (2000). Severity-duration-frequency analysis of droughts and wet periods in Greece. *Hydrological Sciences Journal*, **45**(5), 751–769.

Dalezios N. R., Blanta A., Spyropoulos N. V. and Tarquis A. M. (2014). Risk identification of agricultural drought for sustainable Agroecosystems. *Natural Hazards Earth System Sciences*, **14**, 2435–2448.

Dalezios N. R., Tarquis A. M. and Eslamian S. (2016). Drought assessment and risk analysis. In: Book Chapter in Vol. 1 of 3-Volume Handbook of Drought and Water Scarcity (HDWS), S. Eslamian (ed.), Taylor and Francis (in press).

Deeming J. E., Burgan R. E. and Cohen J. D. (1977). The National Fire-Danger Rating System–1978, USDA Forest Service General technical Report INT-39. Intermountain Forest and Range Experiment Station, Ogden, UT.

Doswell C. A., Brooks H. E. and Maddox R. A. (1996). Flash flood forecasting: an ingredients based methodology. *Weather Forecast*, **11**, 560–581.

Emanuel K. (2005). Genesis and maintenance of "Mediterranean hurricanes". *Advances Geosciences*, **2**, 217–220.

EM-DAT (2009). The OFDA/CRED International Disaster Database. Université Catholique de Louvain, Brussels, Belgium. http://www.emdat.be.

European Commission, DG ECHO (2013). Disaster Risk Reduction. Thematic Policy Document no 5, Brussels, pp. 5–10.

Fujita T. T. (1973). Tornadoes around the world. *Weatherwise*, **26**, 56–83.

Golden J. H. (2003). Waterspouts. In: Encyclopedia of Atmospheric Sciences, J. R. Holton (ed.), Academic Press, Oxford, pp. 2510–2525. ISBN 9780122270901, http://dx.doi.org/10.1016/B0-12-227090-8/00451-6.

Houze R. A. (1977). Structure and dynamics of a tropical squall-line system. *Monthly Weather Review*, **105**(12), 1540–1567.

ICONA (1988). Experimentacion de un nuevo sistema para determinacion del peligro de incendios forestales derivado de los combustibles: instrucciones de calculo. Instituto Nacional para la Conservacion de la Naturaleza, Madrid, Spain.

Innocenti D. and Albrito P. (2010). Reducing the risks posed by natural hazards and climate change: the need for a participatory dialogue between the scientific community and policy makers. *Environmental Science & Policy*, **14** (2011), 730–733. Elsevier.

IPCC (2007). Fourth Assessment Synthesis Report. http://195.70.10.65/pdf/assessment-report/ar4/syr/ar4_syr.pdf.

IPCC (2012). Managing the Risks of Extreme Events and Disasters to Advance Climate Change Adaptation. Special Report of IPCC, 594p.

Kalma J. D., Laughlin G. P., Caprio J. M. and Hamer P. J. C. (1992). Advances in Bioclimatology. Springer, Berlin. The Bio Climatology of Frost, Vol. 2.

Koppe C., Jendritzky G., Kovats S. and Menne B. (2004). Heat-waves: risks and responses. Regional Office for Europe, Health and Global Environmental Change, Series No. 2. Kopenhagen, Denmark.

Loukas A., Vasiliades L. and Dalezios N. R. (2002). Potential climate change impacts on flood producing mechanisms in Southern British Columbia, Canada using the CGCMa1 simulation results. *Journal of Hydrology*, **259**(1–4), 163–188.

Marra M. (2014). EU disaster risk assessment policy and its application in Cohesion Policy 2014–2020. Ex ante conditionalities. In: JASPERS Networking Platform workshop "Promoting climate change adaptation, risk prevention and management in the Water Sector", European Commission, Brussels.

Matsangouras I. T., Nastos P. T., Bluestein H. B. and Sioutas M. V. (2014). A climatology of tornadic activity over Greece based on historical records. *Internal Journal of Climatology*, **34**, 2538–2555.

Mc Arthur A. G. (1967). Fire Behaviour in Eucalypt Forests. Department of National Development, Forestry and Timber Bureau Leaflet No. 107, Canberra, Australia.

McKee T. B., Doesken N. J. and Kleist J. (1993). The relationship of drought frequency and duration to timescales. Preprints, *Eighth Conf. on Applied Climatology*, Anaheim, CA, Amer. Meteor. Soc., pp. 179–184.

Miglietta M. M., Mastrangelo D. and Conte D. (2015). Influence of physics parameterization schemes on the simulation of a tropical-like cyclone in the Mediterranean Sea. *Atmospheric Research*, **153**, 360–375.

Nastos P. T. and Dalezios N. R. (2016). Preface: Advances in Meteorological Hazards and Extreme Events. Special Issue of NHESS. Nastos, Dalezios and Ulbrich (eds), **16**, 1–10.

Nastos P. T. and Kapsomenakis J. (2015). Regional climate model simulations of extreme air temperature in Greece. Abnormal or common records in the future climate? *Atmospheric Research*, **152**, 43–60.

Nastos P. T., Karavana-Papadimou K. and Matzangouras I. T. (2015). Tropical-like cyclones in the Mediterranean: Impacts and composite daily means and anomalies of synoptic conditions. In: Proceedings of the 14th International Conference on Environmental Science and Technology, Rhodes, Greece, 3–5 September 2015, CEST2015_00407.

Palmer W. C. (1965). Meteorological drought. Weather Bureau Research Paper No. 45, US Department of Commerce, Washington, DC. 58 pp.

Robine J. M., Cheung S. L. K., Le Roy S., van Oyen H., Griffiths C., Michel J.-P. and Herrmann F. R. (2006). Death toll exceeded 70,000 in Europe during the summer of 2003. *C. R. Biol.*, **331**, 171–178.

Robinson P. J. (2001). On the definition of heat waves. *Journal of Applied Meteorology*, **40**, 762–775.

Salvati L., Mavrakis A., Colantoni A., Mancino G. and Ferrara A. (2015). Complex Adaptive Systems, soil degradation and land sensitivity to desertification: a multivariate assessment of Italian agro-forest landscape. *Science of The Total Environment*, **521–522**, 15 July 2015, 235–245.

Sapountzaki K. and Dandoulaki M. (2006). Coping with Seismic Risk in Greece: The Traditional Merits of the System and the Challenges of the Future. ARMONIA Book Publication, Athens. Also in http://www.researchgate.net/publication/ /235752068.

Sapountzaki K., Wanczura S., Casertano G., Greiving S., Xanthopoulos G., Ferrara F. (2011). Disconnected policies and actors and the missing role of spatial planning throughout the risk management cycle. Nat Hazards, Springer publications, DOI 10.1007/s11069-011-9843-3.

Schär C., Vidale P. L., Lüthi D., Frei C., Häberli C., Liniger M. A. and Appenzeller C. (2004). The role of increasing temperature variability in European summer heatwaves. *Nature*, **427**, 332–336.

Simpson J. S., Morton B. R., McCumber M. C. and Penc R. S. (1986). Observations and mechanisms of GATE waterspouts. *Journal of the Atmospheric Sciences*, **43**, 753–782.

Sivakumar M. V. K., Motha R. P. and Das H. P. (eds). (2005). Natural Disaster and Extreme Events in Agriculture. Springer, Berlin, Heidelberg, New York, ISBN-10 3-540-22490-4, 367p.

Smith K. (2013). Environmental Hazards: Assessing Risk and Reducing Disaster. Springer, Berlin, Heidelberg, New York, 6th edn, 478p.

Tacnet J.-M. and Curt C. (2013). Expert (knowledge-based) systems for disaster management. In: Encyclopedia of Natural Hazards, P. T. Bobrowsky (ed.), Springer, UK, pp. 300–305.

Tsakiris G., Pangalou D. and Vangelis H. (2007). Regional drought assessment based on the Reconnaissance Drought Index (RDI). *Water Resources Management*, **21**(5), 821–833.

UNCCD – United Nations Convention to Combat Desertification (2009). Summary for Decision Makers. Revitalizing the UNCCD, p. 3.

United Nations Framework on Convention on Climate Change (UNFCCC) (2007). Bali Action Plan (1/Cp.13), report of the Conference of the Parties (COP 13). http://unfccc.int/resource/docs/2007/cop13/eng/06a01.pdf.

UNISDR (2005). Hyogo framework for Action 2005–2015. Building the Resilience of Nations and Communities to Disasters. United Nations, International Strategy for Disaster Reduction, Geneva, Switzerland. http://www.unisdr.org/eng/hfa/hfa.htm.

UNISDR (2007a). Words In to Action: A Guide for Implementing the Hyogo Framework. United Nations, Geneva, Switzerland. Available at: http://www.preventionweb.net/english/professional/publications/v.php?id=594.

UNISDR (2007b). Hyogo Framework for Action 2005–2015: Building the Resilience of Nations and Communities to Disasters. United Nations, Geneva, Switzerland. Available at: http://www.preventionweb.net/english/professional/publications/v.php?id=1037&pid:22.

UNISDR (2008). Briefing Note 01—Climate Change and Disaster Risk Reduction. United Nations, Geneva, Switzerland. Available at: http://www.preventionweb.net/english/professional/publications/v.php?id=4146.

UNISDR (2015). Reading the Sendai Framework for Disaster Risk Reduction 2015–2030. UNISDR, Geneva, Switzerland, 34p.

van Wagner C. E. (1987). Development and structure of a Canadian forest fire weather index system. Forestry Tech. Rep. 35, Canadian Forestry Service, Ottawa.

Van Westen C. J. (2013). Remote sensing and GIS for natural hazards assessment and disaster risk management. In: Treatise on Geomorphology, J. F. Shroder (ed.), Volume 3. Academic Press, San Diego, pp. 259–298.

Venäläinen A. and Heikinheimo M. (2003). The Finnish forest fire index calculation system. In: Early warning systems for natural disaster reduction, J. Zschau and A. Kuppers (ed.), Springer, pp. 645–648.

World Meteorological Organization (2006). Preventing and mitigating natural disasters. WMO-No. 993, ISBN 92-63-10993-1.

Zhang X., Hegerl G., Zwiers F. W. and Kenyon J. (2005). Avoiding inhomogeneity in percentile-based indices of temperature extremes. *Journal of Climate*, **18**, 1641–1651.

Chapter 2

Multi-hazard risk assessment and decision making

Cees J. van Westen and Stefan Greiving

The earth is shaped by endogenic processes, caused by forces from within the earth, resulting in hazardous events like earthquakes or volcanic eruptions, and exogenic processes, caused by forces related to the earth's atmosphere, hydrosphere, geosphere, biosphere and cryosphere and their interactions. Anthropogenic activities have had a very important influence on a number of these processes, especially in the last two hundred years, for instance through the increase of greenhouse gasses, leading to global warming, but also through dramatic changes in the land cover and land use, and overexploitation of scarce resources. The above mentioned processes from endogenic, exogenic and anthropogenic nature may lead to potentially catastrophic events, even in locations that may be far away. For instance earthquakes might trigger landslides which may lead to landslide-dammed lakes that may break out and cause flooding downstream. Or the dams of large reservoirs in mountains, constructed for hydropower, irrigation or drinking water, may fail under an earthquake or extreme rainfall event and cause a similar flood wave.

These potentially harmful events are called hazards. They pose a level of threat to life, health, property, or environment. They may be classified in different ways, for instance according to the main origin of the hazard in geophysical, meteorological, hydrological, climatological, biological, extra-terrestrial and technological (See Table 2.1, from Guha-Sapir *et al.* 2016). Such classifications are always a bit arbitrary, and several hazard types could be grouped in different categories, e.g. landslides could be caused by earthquakes, extreme precipitation and human interventions.

Hazards have a number of characteristics that should be understood in order to be able to assess and subsequently reduce their potential damage. Hazards with certain magnitudes may occur with certain frequencies, as small events may occur often, and large events seldom. In order to be able to establish a magnitude-frequency

Table 2.1 Classification of hazard types as used by the International Disaster Database EM-DAT (Guha-Sapir *et al.* 2016), which is based on and adapted from the he IRDR Peril Classification and hazard Glossary (IRDR, 2014).

Main Group	Main Sub-group	Main Type	Sub-Type
Natural	Geophysical: A hazard originating from solid earth. This term is used interchangeably with the term geological hazard.	Earthquake	Ground shaking, tsunami
		Mass movement	
		Volcanic	Ash fall, lahar, pyroclastic flow, lava flow
	Meteorological: A hazard caused by short-lived, micro- to meso-scale extreme weather and atmospheric conditions that last from minutes to days.	Storm	Extra-tropical storm, tropical storm, convective storm
		Extreme temperature	Cold wave, heat wave, severe winter conditions
		Fog	
	Hydrological: A hazard caused by the occurrence, movement, and distribution of surface and subsurface freshwater and saltwater.	Flood	Coastal flood, riverine flood, flash flood, ice jam flood.
		Landslide	Avalanche (snow, debris), mudflow, rockfall
		Wave action	Rogue wave, seiche
	Climatological: A hazard caused by long-lived, meso- to macro-scale atmospheric processes ranging from intra-seasonal to multi-decadal climate variability.	Drought	
		Glacial Lake outburst	
		Wildfire	Forest Fire, land fire (bush, pasture)

	Biological: A hazard caused by the exposure to living organisms and their toxic substances or vector-borne diseases that they may carry. Examples are venomous wildlife and insects, poisonous plants, and mosquitoes carrying disease-causing agents such as parasites, bacteria, or viruses (e.g. malaria).	Epidemic	Viral , bacterial, parasitic, fungal, prion disease
		Insect infestation	Grasshopper, locust
		Animal accident	
	Extraterrestrial: A hazard caused by asteroids, meteoroids, and comets as they pass near-earth, enter the Earth's atmosphere, and/or strike the Earth, and by changes in interplanetary conditions that effect the Earth's magnetosphere, ionosphere, and thermosphere.	Impact	
		Space weather	Energetic particles, geomagnetic storm
Technological	Industrial accident		Chemical spills, collapse, explosion, fire, gas-leak, poisoning, radiation, other
	Transport accident		Air, road, rail, water
	Miscellaneous accident		Collapse, explosion, fire, other.

relationship for hazard events, it is generally necessary to collect historical data (e.g. from seismographs, meteo-stations, stream gauges, historical archives, remote sensing, field investigations etc.) and carry out statistical analysis (e.g. using extreme event analysis such as Gumbel analysis) (Van Westen *et al.* 2008). The magnitude of the hazard gives an indication of the size of the event, or the energy released, whereas the intensity of a hazard refers to the spatially varying effects. For example earthquake magnitude refers to the energy released by the ruptured fault (e.g. measured on the Richter scale) whereas the intensity refers to the amount of ground shaking which varies with the distance to the epicentre (e.g. measured on Modified Mercalli scale). The magnitude of floods may be measured as the discharge in the main channel at the outlet of a watershed before leaving the mountainous area, whereas the intensity may be measured as the water height or velocity which is spatially distributed, and depends on the local terrain. For some types of hazards there is no unique intensity scale defined, e.g. for landslides (Corominas *et al.* 2015).

These events may be potentially harmful to people, property, infrastructure, economy and activities, but also to the environment, which are all grouped together under the term Elements-at-risk or assets. Also the term exposure is used to indicate those elements-at-risk that are subject to potential losses. Important elements-at-risk that should be considered in analysing potential damage of hazards are population, building stock, essential facilities and critical infrastructure. Critical infrastructure consists of the primary physical structures, technical facilities and systems, which are socially, economically or operationally essential to the functioning of a society or community, both in routine circumstances and in the extreme circumstances of an emergency (UN-ISDR, 2009). Elements-at-risk have a certain level of vulnerability, which can be defined in a number of different ways. The general definition is that vulnerability describes the characteristics and circumstances of a community, system or asset that make it susceptible to the damaging effects of a hazard (UN-ISDR, 2009). There are many aspects of vulnerability, related to physical, social, economic, and environmental conditions (see for example Birkmann, 2006). When considering physical vulnerability only, it can be defined as the degree of damage to an object (e.g. building) exposed to a given level of hazard intensity (e.g. water height, ground shaking, and impact pressure).

2.1 RISK

Risk is defined as the probability of harmful consequences, or expected losses (deaths, injuries, property, livelihoods, economic activity disrupted or environment damaged) resulting from interactions between natural or human-induced hazards and vulnerable conditions (UN-ISDR, 2009; EC, 2011). Risk can presented conceptually with the following basic equation indicated in Figure 2.1. Risk can presented conceptually with the following basic equation:

$$\text{Risk} = \text{Hazard} \times \text{Vulnerability} \times \text{Amount of elements-at-risk} \tag{2.1}$$

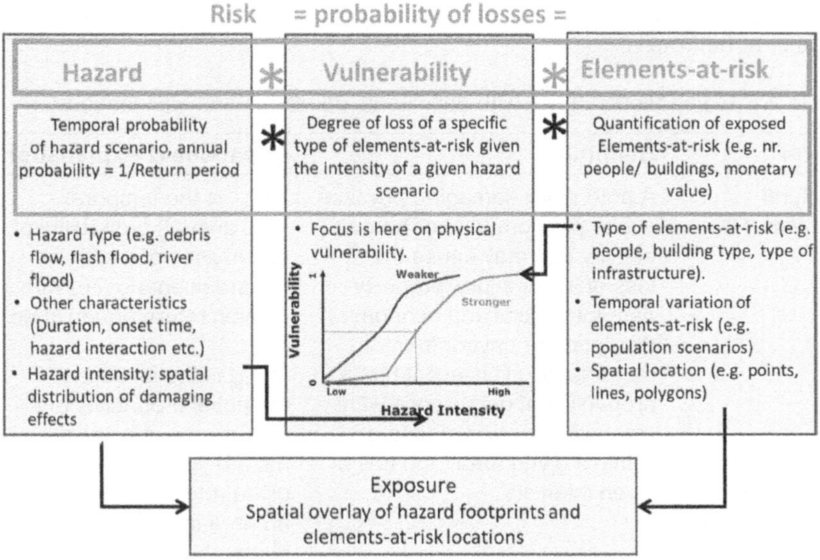

Figure 2.1 Schematic representation of risk as the multiplication of hazard, vulnerability and quantification of the exposed elements-at-risk. The various aspects of hazards, vulnerability and elements-at-risk and their interactions are also indicated. This framework focuses on the analysis of physical losses, using physical vulnerability data.

The equation given above is not only a conceptual one, but can also be actually calculated with spatial data in a GIS to quantify risk from geomorphological hazards. The way in which the amount of elements-at-risk are characterized (e.g. as number of buildings, number of people, economic value) also defines the way in which the risk is presented. Table 2.2 gives a more in-depth explanation of the various components involved. In order to calculate the specific risk equation 2.1 can be modified in the following way:

$$R_S = P_{(T:\mathrm{Hs})} \times P_{(L:\mathrm{Hs})} \times V_{(\mathrm{Es|Hs})} \times A_{\mathrm{ES}} \tag{2.2}$$

in which:

$P_{(T:\mathrm{Hs})}$ is the temporal (e.g. annual) probability of occurrence of a specific hazard scenario (H_s) with a given return period in an area;

$P_{(L:\mathrm{Hs})}$ is the locational or spatial probability of occurrence of a specific hazard scenario with a given return period in an area impacting the elements-at-risk;

$V_{(\mathrm{Es|Hs})}$ is the physical vulnerability, specified as the degree of damage to a specific element-at-risk E_s given the local intensity caused due to the occurrence of hazard scenario H_S;

A_{Es} is the quantification of the specific type of element at risk evaluated (e.g. number of buildings).

Table 2.2 Components of risk with definitions, equations and explanations.

Term	Definition	Equations & Explanation
Natural hazard (*H*)	A potentially damaging physical event, phenomenon or human activity that may cause the loss of life or injury, property damage, social and economic disruption or environmental degradation. This event has a probability of occurrence within a specified period of time and within a given area, and has a given intensity.	$P_{(T:HS)}$ is the temporal (e.g. annual) probability of occurrence of a specific hazard scenario (H_s) with a given return period in an area; $P_{(L:HS)}$ is the locational or spatial probability of occurrence of a specific hazard scenario with a given return period in an area impacting the elements-at-risk
Elements-at-risk (*E*)	Population, properties, economic activities, including public services, or any other defined values exposed to hazards in a given area". Also referred to as "assets".	E_s is a specific type of elements-at-risk (e.g. masonry buildings of 2 floors)
Vulnerability (*V*)	The conditions determined by physical, social, economic and environmental factors or processes, which increase the susceptibility of a community to the impact of hazards. Can be subdivided in physical, social, economical, and environmental vulnerability.	$V_{(Es\|Hs)}$ is the physical vulnerability, specified as the degrees of damage to E_S given the local intensity caused due to the occurrence of hazard scenario H_S It is expressed on a scale from 0 (no damage) to 1 (total loss)
Amount of elements-at-risk (A_E)	Quantification of the elements-at-risk either in numbers (of buildings, people etc), in monetary value (replacement costs etc), area or perception (importance of elements-at-risk).	A_{ES} is the quantification of the specific type of element at risk evaluated (e.g. number of buildings)
Consequence (*C*)	The expected losses (of which the quantification type is determined by A_E) in a given area as a result of a given hazard scenario.	C_S is the "specific consequence", or expected losses of the specific hazard scenario which is the multiplication of $V_S \times A_{ES}$

Table 2.2 Components of risk with definitions, equations and explanations (*Continued*).

Term	Definition	Equations & Explanation
Specific risk (R_S)	The expected losses in a given area and period of time (e.g. annual) for a specific set of elements-at-risk as a consequence of a specific hazard scenario with a specific return period.	$R_S = H_S \times V_S \times A_{ES}$ $R_S = H_S \times C_S$ $R_S = P_{(T:Hs)} \times P_{(L:Hs)} \times V_{(Es\|Hs)} \times A_{ES}$
Total risk (R_T)	The probability of harmful consequences, or expected losses (deaths, injuries, property, livelihoods, economic activity disrupted or environment damaged) resulting from interactions between natural or human-induced hazards and vulnerable conditions in a given area and time period. It is calculated by first analyzing all specific risks. It is the integration of all specific consequences over all probabilities.	$R_T \approx \Sigma(R_T) = \Sigma(H_S \times V_S \times A_{ES})$ Or better: $R_T = \int(V_S \times A_{ES})$ – For all hazard types – For all return periods – For all types of elements-at-risk. It is normally obtained by plotting consequences against probabilities, and constructing a risk curve. The area below the curve is the total risk.

The term risk mapping is often used as being synonymous with risk analysis in the overall framework of risk management. Risk assessments (and associated risk mapping) include: a review of the technical characteristics of hazards such as their location, intensity, frequency and probability; the analysis of exposure and vulnerability including the physical social, health, economic and environmental dimensions; and the evaluation of the effectiveness of prevailing and alternative coping capacities in respect to likely risk scenarios (UN-ISDR, 2009; EC, 2011; ISO 31000). In the framework of natural hazards risk assessment, the term risk mapping also indicates the importance of the spatial aspects of risk assessment. All components of the risk equation (Figure 2.1) are spatially varying and the risk assessment is carried out in order to express the risk within certain areas. To be able to evaluate these components there is a need to have spatially distributed information. Computerized systems for the collection, management, analysis and dissemination of spatial information, so-called Geographic Information Systems (GIS) are used to generate the data on the various risk components, and to analyse the risk (OAS, 1991; Coppock,

1995; Cova, 1999; van Westen, 2013). Hazard data is generally the most difficult to generate. For each hazard type (e.g. flooding, debris flow, rock fall) so-called **hazard scenarios** should be defined, which are hazard events with a certain magnitude/intensity/frequency relationship (e.g. flood depth maps for 10, 50 and 100 year return periods). Different types of modelling approaches are required for the hazard scenario analysis, depending on the hazard type, scale of analysis, availability of input data, and availability of models. Generally speaking a separate analysis is required to determine the probability of occurrence for a given magnitude of events, followed by an analysis of the initiation of the hazard (e.g. hydrological modelling or landslide initiation modelling), and of the run-out or spreading of the hazard (e.g. hydrodynamic modelling or landslide run-out modelling). Overviews of hazard and risk assessment methods for landslides for example can be found in Corominas *et al.* (2014), and for floods in Prinos (2008). Elements-at-risk data are very often based on building footprint maps, which represent the location of buildings, with attributes related to their use, size, type and number of people during different periods of the year (e.g. daytime, night-time). Remote Sensing is often used to extract these building maps if existing cadastral maps are not available. For other elements-at-risk like transportation infrastructure and land cover maps also remote sensing data are used as important inputs. Vulnerability data is often collected in the form of vulnerability curves, fragility curves or vulnerability matrices, which indicate the relationship between the levels of damage to a particular type of element-at-risk (e.g. single storey masonry building) given intensity levels of a particular hazard type (e.g. debris flow impact pressure). Generation of vulnerability curves is a complicated issue, as they can be generated empirically from past damage event for which intensity and damage is available for many elements-at-risk, or through numerical modelling (Roberts *et al.* 2009). Table 2.2 gives an overview of the various components of risk.

Risk can be presented in a number of different ways, depending on the objectives of the risk assessment (Birkmann, 2007). Risk can be expressed in absolute or relative terms. Absolute population risk can be expressed as individual risk (the annual probability of a single exposed person to be killed) or as societal risk (the relation between the annual probability and the number of people that could be killed). Absolute economic risk can be expressed in terms of Average Annual Loss, Maximum Probable Loss, or other indices that are calculated from a series of loss scenarios, each with a relation between frequency and expected monetary losses (Jonkman *et al.* 2002). It is also possible to differentiate between direct risk (which is the risk directly resulting from the impact of the hazard) and indirect risk (which may occur later as a consequence of the direct impact). Some examples of direct risk are the destruction of physical objects (e.g. buildings, transportation infrastructure), and examples of indirect losses are loss of revenues and economic production, disruption of transportation networks leading to longer travel time etc. A significant component of the losses are intangible (difficult or impossible to quantify), for example the societal or psychological impact of disaster events.

2.2 MULTI-HAZARD RISK

One of the difficult issues in natural hazards risk assessment is how to analyse the risk for more than one hazard in the same area, and the way they interact. Figure 2.2 shows an illustration of how different sets- of triggering factors can cause a number of different hazards. There are many factors that contribute to the occurrence of hazardous phenomena, which are either related to the environmental setting (topography, geomorphology, geology, soils etc.) or to anthropogenic activities (e.g. deforestation, road construction, tourism). Although these factors contribute to the occurrence of the hazardous phenomena and therefore should be taken into account in the hazard and risk assessment, they are not directly triggering the events. For these, there is a need for triggering phenomena, which can be of meteorological or geophysical origin (earthquakes, or volcanic eruptions). A generally accepted definition of multi-hazard still does not exist. In practice, this term is often used to indicate all relevant hazards that are present in a specific area, while in the scientific context it frequently refers to "more than one hazard". Likewise, the terminology that is used to indicate the relations between hazards is unclear. Many authors speak of interactions (Tarvainen *et al.* 2006; de Pippo *et al.* 2008; Marzocchi *et al.* 2009; Zuccaro & Leone, 2011; European Comission, 2011), while others call them chains (Shi, 2002), cascades (Delmonaco *et al.* 2006a; Carpignano *et al.* 2009; Zuccaro & Leone, 2011; European Comission, 2011), domino effects (Luino, 2005; Delmonaco *et al.* 2006a; Perles & Cantarero, 2010; van Westen, 2010; European Comission, 2011), compound hazards (Alexander, 2001) or coupled events (Marzocchi *et al.* 2009).

Compared to single processes, standard approaches and methodological frameworks for multi-hazard risk assessment are less common in the literature (Kappes *et al.* 2012), which is related to the complex nature of the interaction between the hazards, and the difficulty to quantify these.

2.2.1 Independent events

The simplest approach is to consider that the hazards are independent and caused by different triggers. This means that the expected losses from one hazard type are independent from the losses expected from the other hazard type. If that is the case, the risk can be calculated by adding the average annual losses for the different types of hazard. This would be the case for example for earthquake hazard and flood hazard. They have different triggering mechanisms, which do not directly interact. Therefore, earthquake hazard is independent of flood hazard and may be analysed separately. Also the risk may be analysed separately and the resulting losses could be added. Other examples of independent hazard are for instance technological hazards and flood hazards. Many of the existing software tools for multi-hazard risk assessment (See section 2.9) deal with these hazard independently, and sum up the losses. However, when these apparent independences are examined in detail, the relation may be more complicated. For instance, an earthquake may trigger landslides that may block a river leading to flooding, which makes that the earthquake and flood risk cannot be considered entirely independent. Even flooding and technological hazards cannot be considered completely independent: during flood events there may be a higher risk of technological accidents (Glade & Alexander, 2013).

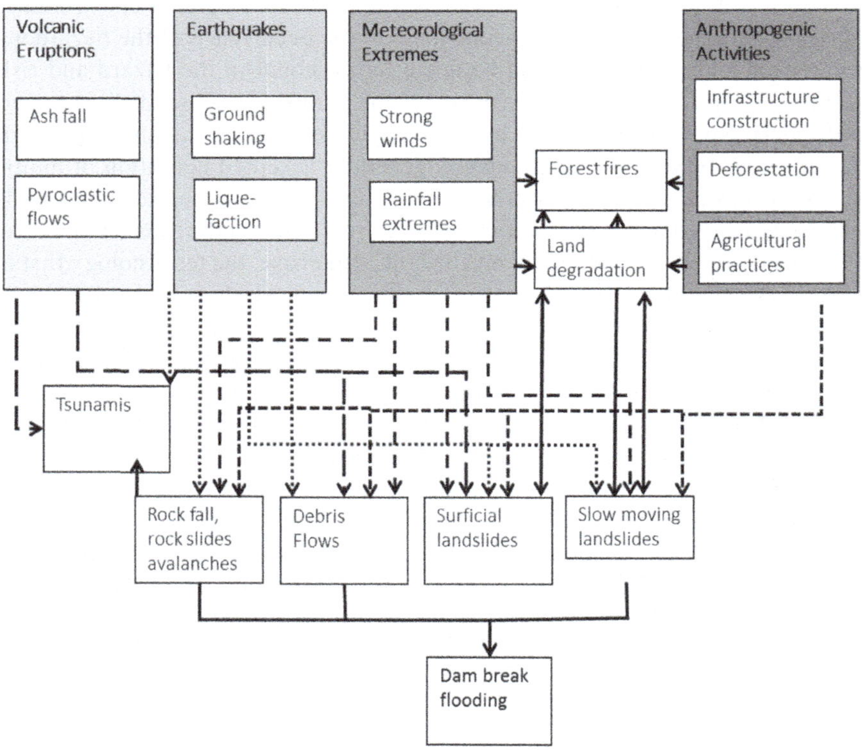

Figure 2.2 Schematic representation of multi-hazards interactions between the main triggering events (volcanic eruptions, Earthquakes, Meteorological extremes, and anthropogenic activities) and secondary hazards.

2.2.2 Coupled events

The second multi-hazard relationship is between different hazard types that are triggered by the same triggering event. These are what we would call **coupled**

events (Marzocchi *et al.* 2009). The temporal probability of occurrence of such coupled events is the same as it is linked to the probability of occurrence of the triggering mechanism. For analysing the spatial extent of the hazard, one should take into account that when such coupled events occur in the same area and the hazard footprints overlap, the processes will interact, and therefore the hazard modelling for these events should be done simultaneously, which is still very complicated. In order to assess the risk for these multi-hazards, the consequence modelling should therefore be done using the combined hazard footprint areas, but differentiating between the intensities of the various types of hazards and using different vulnerability-intensity relationships. When the hazard analyses are carried out separately, the consequences of the modelled scenarios cannot be simply added up, as the intensity of combined hazards may be higher than the sum of both or the same areas might be affected by both hazard types, leading to overrepresentation of the losses, and double counting. Examples of such types of coupled events is the effect of an earthquake on a snow-covered building (Lee & Rosowsky, 2006) and the triggering of landslides by earthquakes occurring simultaneously with ground shaking and liquefaction (Delmonaco *et al.* 2006b; Marzocchi *et al.* 2009). Within multi-hazard risk assessment the best way to treat coupled risk is to take the maximum of the risks that are coupled. For example, during the same tropical storm a village may be hit by flash floods or debris flows. Once it is hit by one type there is damage, and buildings cannot be destroyed twice during the same event.

2.2.3 One hazard changes conditions for the next

A third type of interrelations is the influence one hazard exerts on the **disposition** of a second hazard, though without triggering it (Kappes *et al.* 2010). An example is the "fire-flood cycle" (Cannon & De Graff, 2009): forest fires alter the susceptibility to debris flows and flash floods due to their effect on the vegetation and soil properties. This problem highlights the fact that the conditions that make certain areas more susceptible to hazards may change constantly. For instance, land cover and land use have a large effect on hydro-meteorological hazards, such as flooding and landslides. When these change as a result of other hazards (like forest fires), also the susceptibility to landslides, debris flows or floods increases. Many of the hazard relations are of this type. For instance, volcanic eruptions may lead to the deposition of volcanic ash, which will increase the susceptibility to landslides and flooding. Earthquakes may trigger landslides, and the landslide scars that are unvegetated may lead to increased erosion and debris flows. It is very difficult to take this type of relationship into account before one particular hazard has changed the conditions that make the terrain more susceptible to the second hazard. The practice is to update a multi-hazard risk assessment each time after the occurrence of a major hazard event (like a volcanic eruption, major earthquake or hurricane).

2.2.4 Domino or cascading hazards

The fourth type of hazard relationships consists of those that occur in **chains**: one hazard causes the next. These are also called **domino** effects, **concatenated**, or **cascading** hazards. These are the most problematic types to analyse in a multi-hazard risk assessment. Hazard may occur in sequence, where one hazard may trigger the next. These hazard chains or domino effects are extremely difficult to quantify over certain areas, although good results have been obtained at a local level (e.g. Peila & Guardini, 2008). The best approach for analysing such hazard chains is to use so-called event-trees (See section 2.3.2). However, it is often very difficult to apply such event-trees in a spatial manner, where in fact different parts of an area may require different event-trees. This is true for instance for the chain: earthquake-landslide-damming-dam break flood. Each part of the terrain has a different susceptibility to landslides. But also each earthquake, which a given depth and magnitude, may trigger different landslide patterns. If a landslide may be generated, the next step is to evaluate whether the size is large enough to dam a river. This also depends on the location of the landslide with respect to the river, the width of the river and the river discharge. Once the river is dammed it depends on the type of material in the dam and the strength of the river, whether the landslide dam is broken fast or whether there is a possibility for a lake to develop, which may cause more severe flooding when the dam breaks later. This sequence is described by Fan *et al.* (2012).

2.2.5 Example of multi-hazard chain: Layou Valley landslides in Dominica, Caribbean

The Layou-Carholm landslide, located on the Caribbean island of Dominica, represents an example of a multi-hazard situation that achieved climactic proportions in 1997 and 2011. The Layou River, with a length of 17 km is one of the largest watersheds in Dominica (70 km^2) and drains about 9% of the land (ACOE, 2004). The Layou Tuff forms vertical walls along the lower Matthieu and Layou Rivers through these reaches. This welded tuff resulted from ignimbrite eruptions in the Late Pleistocene (Roobol & Smith, 2004). Landslides were common in the area, with specific reports occurring between 1987 and 1997. There is an eyewitness account of a slide following Hurricane Hugo in 1989 and also following Hurricanes Iris, Luis and Marilyn in 1995. There was a major change to the pattern of small landslides. Dramatic slumping occurred between November 18 and 25, 1997. Two major slides blocked the river and created a natural dam. The dam was breached on November 21 with mudflows reaching the sea accompanied by extensive flooding of the lower river valley. The larger of the Layou flood events which happened on November 28, 1997, measured 1,325,000 m^3. A wall of material estimated at 50 feet high was washed downstream. The riverbed rose dramatically in its lower reaches. This elevation was estimated at 10 meters at the location of the bridge. The river had dried up between 18 and 20 November 1997 and then flooded on November 21. Further landslides occurred on November 25, 1997 and October 8 and 11, 1998 with subsequent dam breaks being significant events. End-of-the-year measurements show that the lake depth increased from 22 m in 1998 to nearly 40 m in 2008 (DeGraff *et al.* 2010). The maximum volume estimate is 3,611,985 m^3, assuming failure by overtopping and complete draining of the lake (Breheny, 2007). A major dam break event occurred on 28/06/2011. The road along the Layou River to Pont Casse was closed, due to flood hazard. R Also in later years the Layou valley was heavily affected by floods and landslides. In August 2015, during tropical storm Erika, severe flooding damaged the road in a number of places (Figure 2.3).

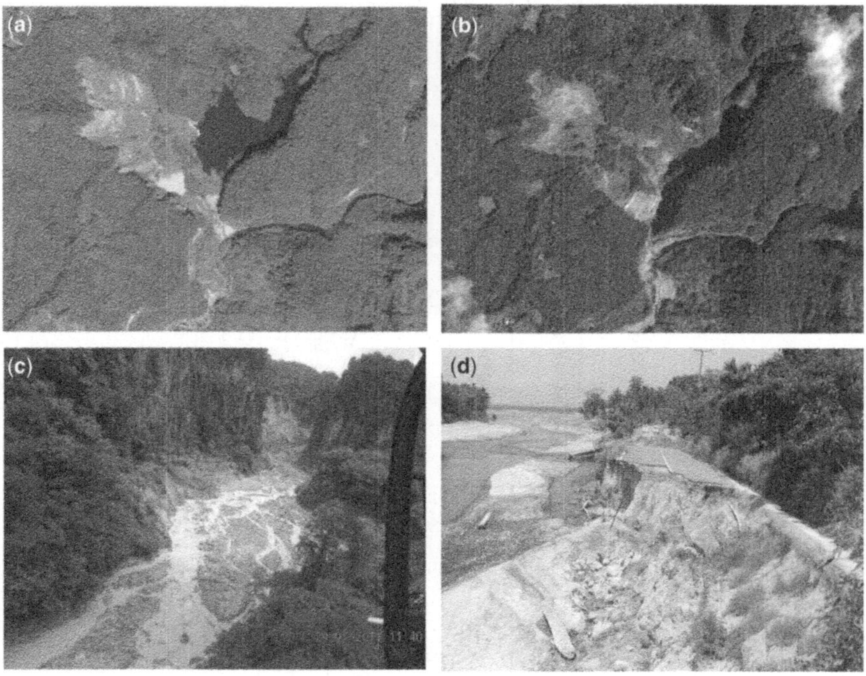

Figure 2.3 Mathieu landslide dam development. **(a)** Carholm landslide blocking the Mathieu River and forming a lake in 1997 (Satellite image from 3-8-2005), **(b)** Google Earth image from 21-12-2012 after the breaching of the landslide dam in 2011 **(c)** View of the river valley just below the breaching point. **(d)** Downstream part of the valley where the road was washed out by heavy flooding on 27 August 2015 during Tropical Storm Erika.

Table 2.3 shows the main multi-hazard relationships for a number of hazards occurring in the Caribbean countries.

Table 2.3 Main hazard types and their interactions. The relationship should be read starting from the left and reading horizontally (Source: www.charim.net).

	Earthquake	**Volcanic Eruption**	**Tsunami**	**Storm Surge**	**River Flooding**	**Landslides**	**Forest Fires**
Earthquake	–	Independent	Chain	Independent	Independent	Chain	Independent
Volcanic eruption	Independent	–	Chain	Independent	Disposition	Disposition	Chain
Tsunami	Caused by	Caused by	–	Independent	Independent	Chain along coast	Independent
Storm surge	Independent	Independent	Independent	–	Chain	Chain	Independent
River flooding	Independent	Independent	Independent	Coupled	–	Coupled	Independent
Landslides	Caused by	Independent	Independent	Coupled	Coupled	–	
Forest Fires	Independent	Coupled	Independent	Independent	Disposition	Disposition	–

2.3 RISK ANALYSIS APPROACHES

Risk assessment is a process to determine the probability of losses by analysing potential hazards and evaluating existing conditions of vulnerability that could pose a threat or harm to property, people, livelihoods and the environment on which they depend (UN-ISDR, 2009). ISO 31000 (2009) defines risk assessment as a process made up of three processes: risk identification, risk analysis, and risk evaluation. Risk identification is the process that is used to find, recognize, and describe the risks that could affect the achievement of objectives. Risk analysis is the process that is used to understand the nature, sources, and causes of the risks that have been identified and to estimate the level of risk. It is also used to study impacts and consequences and to examine the controls that currently exist. Risk evaluation is the process that is used to compare risk analysis results with risk criteria in order to determine whether or not a specified level of risk is acceptable or tolerable.

Risk mapping for natural hazard risk can be carried out at a number of scales and for different purposes. Table 2.4 and Figure 2.4 give a summary. In the following sections four methods of risk mapping will be discussed: Quantitative risk assessment (QRA), Event-Tree Analysis (ETA), Risk matrix approach (RMA) and Indicator-based approach (IBA).

Table 2.4 Indication of scales of analysis with associated objectives and data characteristics.

Scale of Analysis	Scale	Possible Objectives	Possible Approaches
International, Global	<1:1 million	Prioritization of countries/ regions; Early warning	Simplified RMA & IBA
Small: provincial to national scale	<1:100,000	Prioritization of regions; Analysis of triggering events; Implementation of national programs; Strategic environmental assessment; Insurance	Simplified EVA, RMA & IBA
Medium: municipality to provincial level	1:100000 to 1:25000	Analyzing the effect of changes; Analysis of triggering events; Regional development plans	RMA/IBA
Local: community to municipality	1:25000 to 1:5000	Land use zoning; Analyzing the effect of changes; Environmental Impact Assessments; Design of risk reduction measures	QRA/EVA/ RMA IBA
Site-specific	1:5000 or larger	Design of risk reduction measures; Early warning systems; detailed land use zoning	QRA/EVA/ RMA

Approaches: QRA = Quantitative risk assessment, EVA = Event-Tree Analysis, RMA = Risk matrix approach, IBA = Indicator-based approach.

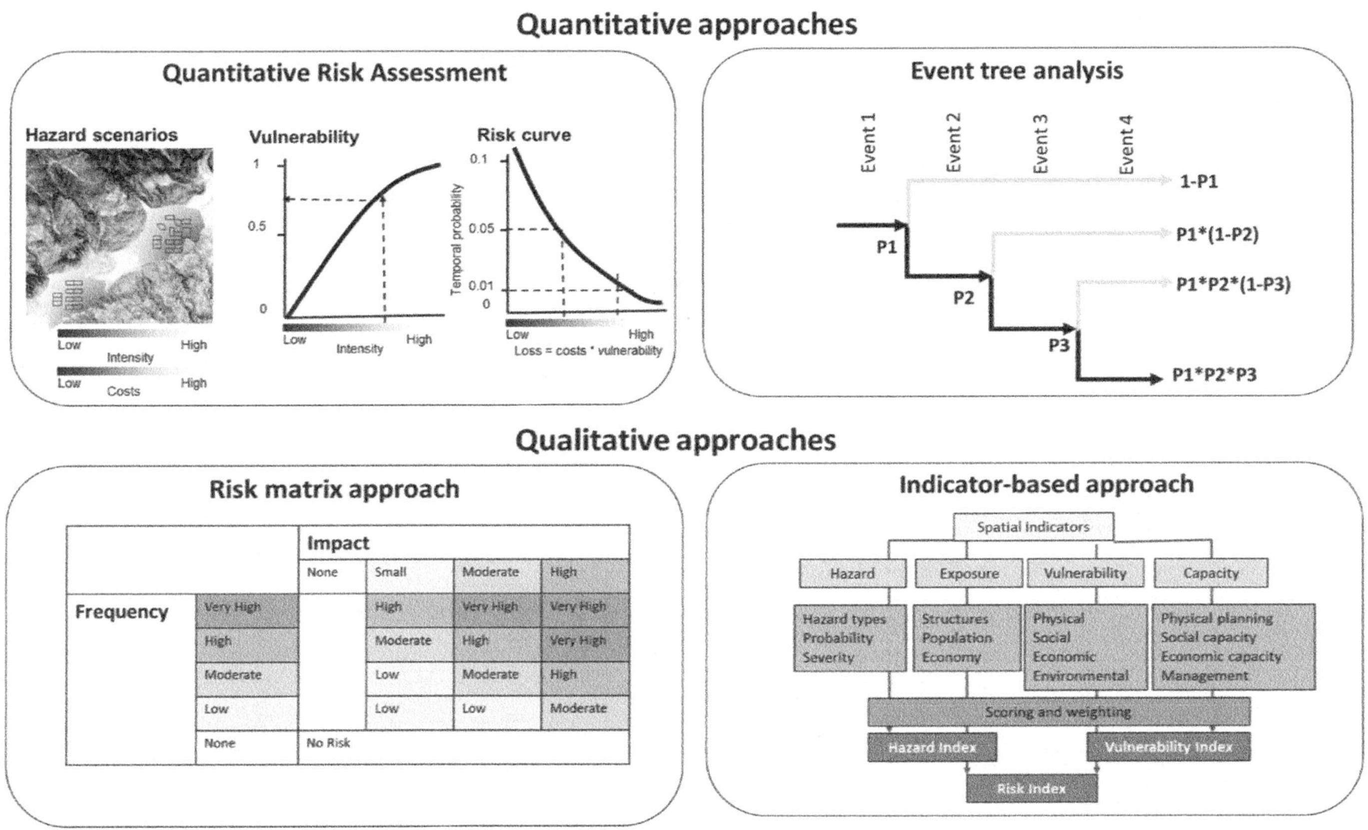

Figure 2.4 Components relevant for risk assessment, and the four major types of risk mapping that are presented in this section.

2.3.1 Quantitative risk assessment

If the various components of the risk equation can be spatially quantified for a given set of hazard scenarios and elements-at-risk, the risk can be analysed using the following equation:

$$\text{Risk} = \sum_{\text{All hazards}} \left(\int_{P_T=0}^{P_T=1} P_{(T|\text{HS})} \times \left(\sum_{\text{All EaR}} \left(P_{(S|\text{HS})} \times \left(A_{(\text{ER}|\text{HS})} \times V_{(\text{ER}|\text{HS})} \right) \right) \right) \right) \tag{2.3}$$

In which:

$P_{(T|\text{HS})}$ = the temporal probability of a certain hazard scenario (HS). A hazard scenario is a hazard event of a certain type (e.g. flooding) with a certain magnitude and frequency;

$P_{(S|\text{HS})}$ = the spatial probability that a particular location is affected given a certain hazard scenario;

$A_{(\text{ER}|\text{HS})}$ = the quantification of the amount of exposed elements-at-risk, given a certain hazard scenario (e.g. number of people, number of buildings, monetary values, hectares of land) and

$V_{(\text{ER}|\text{HS})}$ = the vulnerability of elements at risk given the hazard intensity under the specific hazard scenario (as a value between 0 and 1).

The method is schematically indicated in Figure 2.5. GIS operations are used to analyse the exposure as the intersection between the elements-at-risk and the hazard footprint area for each hazard scenario. For each element-at-risk also the level of intensity is recorded through a GIS-overlay operation. These intensity values are used in combination with the element-at-risk type to find the corresponding vulnerability curve, which is then used as a lookup table to find the vulnerability value. The way in which the amount of elements-at-risk are characterized (e.g. as number of buildings, number of people, economic value) also defines the way in which the risk is calculated. The multiplication of exposed amounts and vulnerability should be done for all elements-at-risk for the same hazard scenario. The results are multiplied with the spatial probability that the hazard footprint actually intersects with the element-at-risk for the given hazard scenario $P_{(S|\text{HS})}$ to account for uncertainties in the hazard modelling. The resulting value represents the losses, which are plotted against the temporal probability of occurrence for the same hazard scenario in a so-called *risk curve*. This is repeated for all available hazard scenarios. At least three individual scenarios should be used, although it is preferred to use at least 6 events with different return periods (FEMA, 2004) to better represent the risk curve. The area under the curve is then calculated by integrating all losses with their respective annual probabilities. It is possible to create risk curves for the entire study area, or for different spatial units, such as administrative units, census tracks, road or railway sections etc.

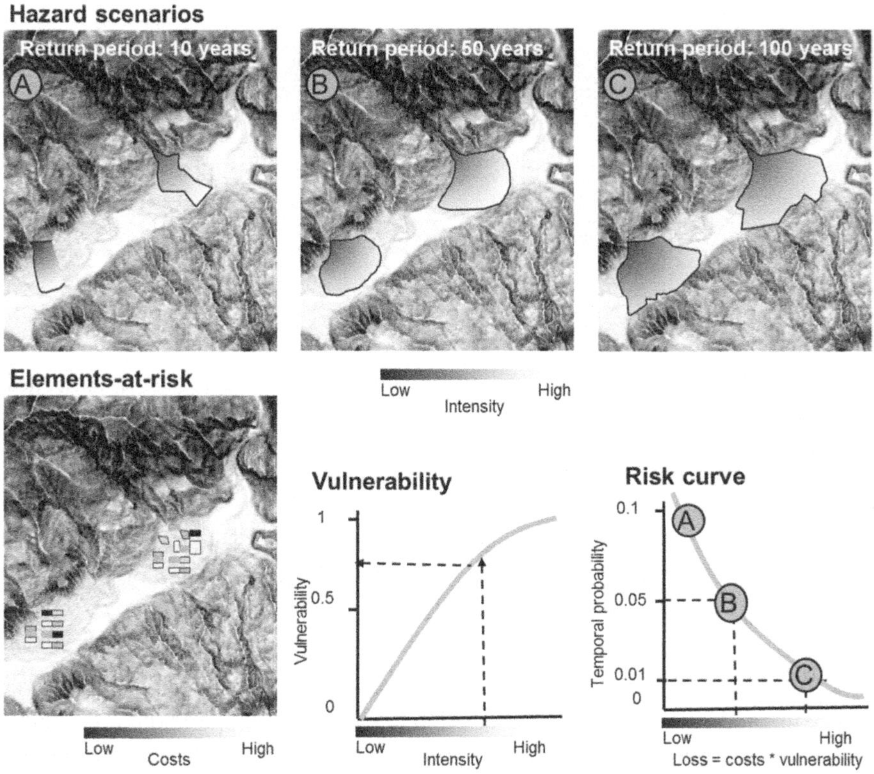

Figure 2.5 Schematic representation of Quantitative Risk Assessment.

The components that are involved in risk assessment have a high degree of uncertainty. Aleatory uncertainty is associated with the variation of the input data used in the risk assessment. For example the variations in soil characteristics used to model landslide probability, surface characteristics, building characteristics etc. These are normally incorporated in probabilistic risk analysis (Bedford & Cook, 2001), which calculates thousands of hazard and risk scenarios taking the variations of the input factors and calculating exceedance probabilities using techniques, such as Monte Carlo simulation. Epistemic uncertainty refers to uncertainty associated with incomplete or imperfect knowledge about the processes involved, and lack of sufficient data. This is often a serious problem as there may not be enough data available to determine individual hazard scenarios, or there are no vulnerability curves for the types of elements-at-risk within the study area. Probabilistic risk assessment takes into account all possible hazard scenarios and the uncertainty of the input factors, by running thousands of loss scenarios, and calculate eventually the loss exceedance curve. For a number of hazards, such as landslides or flooding, it is very complicated to develop a large number of hazard scenarios due to the large epistemic uncertainty caused by lack of data. In such cases uncertainty can be taken into account using the method illustrated in Figure 2.6. In this method data are used showing the range of possible values for the temporal probability, spatial probability, intensity of the hazard, value of the elements-at-risk and vulnerability. The uncertainty range in the temporal probability of the hazard scenario is reflected by a range of possible values on the Y-axis of the risk curve. The uncertainty in the hazard intensity (e.g. water height for flooding, impact pressure for landslides) combined with the uncertainty in the vulnerability curve will results in larger uncertainty ranges in vulnerability, which are then multiplied with the uncertainty range of the quantification of elements-at-risk (e.g. building costs). This then gives a range of values for the expected losses. Thus, instead of a single point in the risk curve, each hazard scenario will result in a rectangle, defined by the variation in probability and losses. The upper right corners of the rectangle are connected to provide the most pessimistic risk curve, and the lower right corners are connected to provide the most optimistic risk curve. When calculating the area under the curves it is then possible to show the range in annual expected losses.

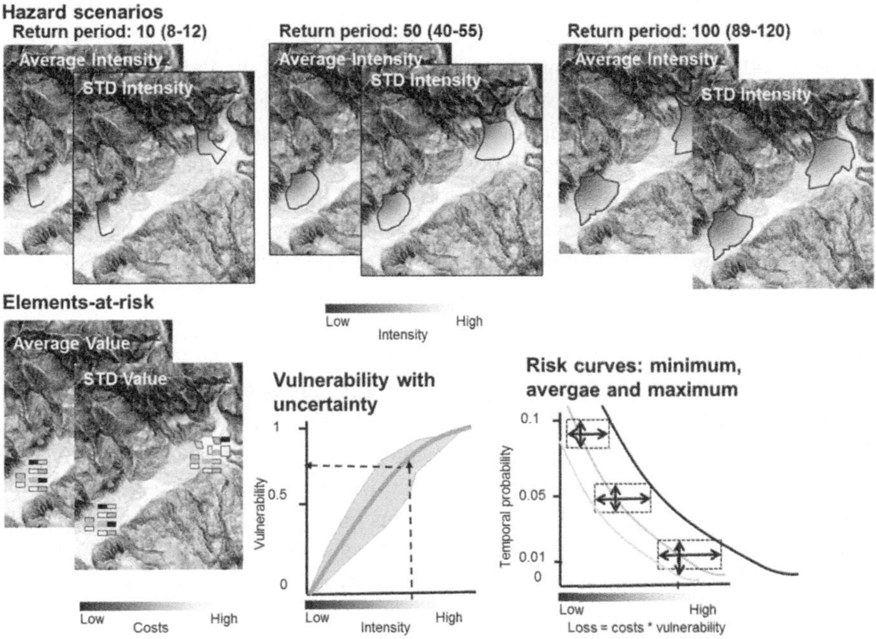

Figure 2.6 Method for including uncertainty in Quantitative Risk Analysis in cases where it is not possible to define many hazard scenarios.

Figure 2.7 gives an example of a quantitative risk assessment. In this simple example we are taking a flood situation. The figure shows a cross section through a flood plain. There are three hazard scenarios, which have been modelled using a flood model. They have different return periods (10 years, 20 years and 50 years). In this simple example there are 3 elements at risk only (buildings) that are of two types. Building A and building B are wooden and relatively weak buildings. They have also lower replacement values. They are located in different elevations. Building C is a concrete building, which is stronger. It is also located at a higher elevation than building B. It is also larger and more expensive. In the exposure analysis, there is overlaying the flood heights with the building heights and the water height is calculated for each hazard scenario and for each building. For the 10 years return period, building A is not flooded, and building C only 0.1 meter. For the 20 year return period, all buildings are flooded, but with different degrees. For the 50 year return period, all buildings are flooded, building B and C very much. For the vulnerability analysis, there is a need for vulnerability curves, which are related for each type of building the degree of loss to a building with a given water height. These curves are generated from past event damage assessment, by correlating the water height with damage. For example: building B has an exposed intensity of 5.6 for the scenario of 20 years return period. And it is a wooden building so we take the value of 5.6 on the X-axis of the vulnerability curve, representing the flood depth. Because it is a wooden building, the curve for the wooden buildings is looked up, and then the damage value on the y-axis is read. This is done for all buildings and for all return periods. The replacement values (amount) are filled in, and the replacement values (amount) are multiplied with the vulnerability to calculate losses. The losses for the buildings are summed up for the same hazard scenario (return period). The annual probability is calculated: 1 divided by the return period. The probability is plotted for each scenario against the losses, and fit a curve through the points, which links all probabilities with all losses. The area below the curve represents the Average Annual Losses. It is the integration of all losses over all probabilities.

2.3.2 Event-tree approaches

As mentioned in section 2.2.4 a number of hazards may occur in chains: these are also called domino effects, or concatenated hazards. These are the most problematic types to analyse in a multi-hazard risk assessment. The best approach for analysing such hazard chains is to use so-called event-trees. An event tree is a system, which is applied to analyse all the combinations (and the associated probability of occurrence) of the parameters that affect the system under analysis. All the analysed events are linked to each other by means of nodes (See Figure 2.4) all possible states of the system are considered at each node and each state (branch of the event tree) is characterized by a defined value of probability of occurrence. Figure 2.8 gives an example of an event tree for a situation where a rockfall in a lake may trigger a flood wave that would impact a village (from Lacasse *et al.* 2008).

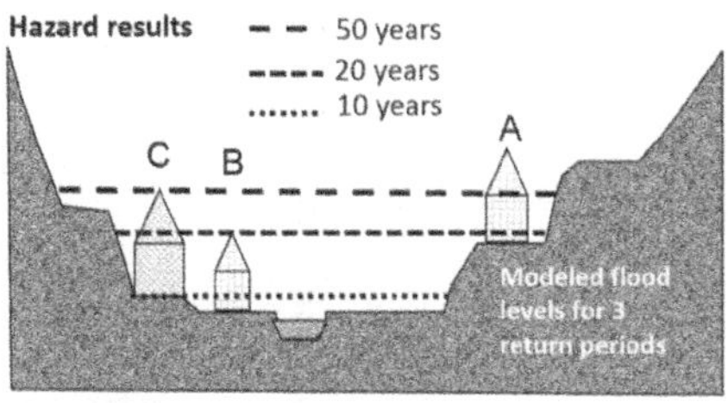

Exposure results

House	10 years	20 years	50 years
A (wood)	0	0.8	2.0
B (wood)	1.2	5.6	6.8
C (concrete)	0.1	4.5	5.7

Vulnerability results

	Return Period	Vulnerability
A	10	0
B	10	0.2
C	10	0.05
A	20	0.1
B	20	1
C	20	0.5
A	50	0.4
B	50	1
C	50	0.8

Vulnerability curves

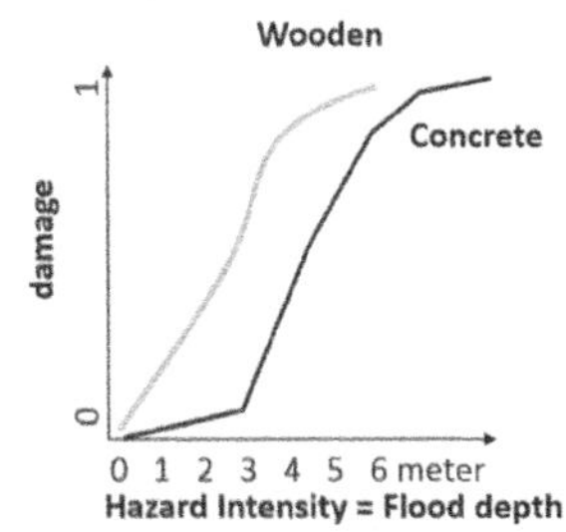

Loss calculation results

	Return Period	Vulnerability	Amount	V*A (loss)	Aggregate loss	Probability
A	10	0	100000	0	20000	0.1
B	10	0.2	50000	10000		
C	10	0.05	200000	10000		
A	20	0.1	100000	10000	160000	0.05
B	20	1	50000	50000		
C	20	0.5	200000	100000		
A	50	0.4	100000	40000	250000	0.02
B	50	1	50000	50000		
C	50	0.8	200000	160000		

Risk curve

	Aggregate loss	Probability
A	20000	0.1
B		
C		
A	160000	0.05
B		
C		
A	250000	0.02
B		
C		

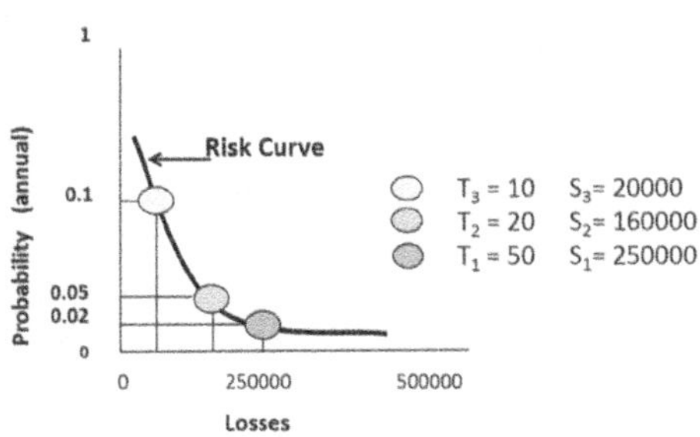

Average Annual Risk: Area under curve

$$\text{Average Annual Risk} = \frac{1}{T_1}\cdot S_1 + \left(\frac{1}{T_2} - \frac{1}{T_1}\right)\cdot\frac{S_1 + S_2}{2} + \left(\frac{1}{T_3} - \frac{1}{T_2}\right)\cdot\frac{S_2 + S_3}{2}.$$

Figure 2.7 Schematic presentation of the steps involved in quantitative risk analysis. See text for explanation.

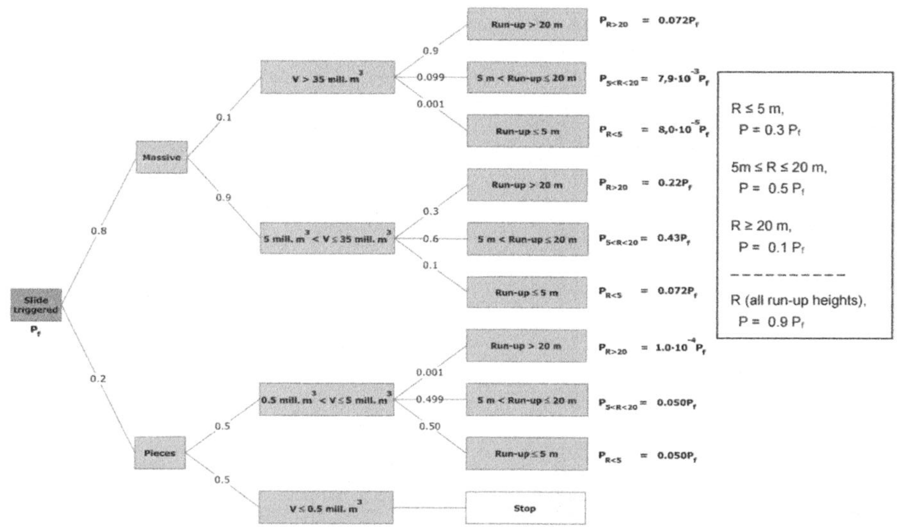

Figure 2.8 Bayesian Event tree for tsunami propagation, given that rock slide in Åknes has occurred (V = rockslide volume, R = run-up height). From Lacasse *et al.* (2008).

2.3.3 Risk matrix approach

Risk assessments are often complex and do not allow to develop a full numerical approach, since many aspects are not fully quantifiable or have a very large degree of uncertainty. This may be related to the difficulty to define hazard scenarios, map and characterize the elements-at-risk, or define the vulnerability using vulnerability curves. In order to overcome these problems, the risk is often assessed using the so-called risk matrices or consequences-frequency matrices (CFM), which are diagrams with consequence and frequency classes on the axes (See Figure 2.4). They permit to classify risks based on expert knowledge with limited quantitative data (Haimes, 2008; Jaboyedoff *et al.* 2014). The risk matrix is made of classes of frequency of the hazardous events on one axis, and the consequences (or expected losses) on the other axis. Instead of using fixed values, the use of classes allows for more flexibility and incorporation of expert opinion. Such methods have been applied extensively in natural hazard risk assessment, e.g. in Switzerland (Figure 2.9 from Jaboyedoff *et al.* 2014). This approach also permits to visualize the effects and consequences of risk reduction measures and to give a framework to understand risk assessment. The system depends on the quality of the group of experts that are formed to identify the hazard scenarios, and that carry out the hazard filtering and ranking in several sub-stages characterized by frequency (probability) and impact classes and their corresponding limits (Haimes, 2008).

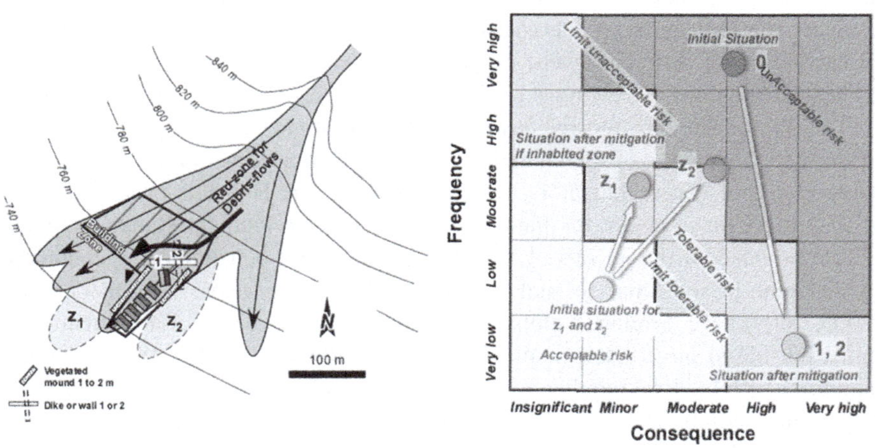

Figure 2.9 Example of potential building area in a high hazard area and illustration of the proposed solutions. The risk matrix is used to represent the degree of risk. The scope of tolerable risk (yellow) is between the limits of tolerance and of acceptability. The initial situation 0 is a combination of very high frequency of debris flows with a high impact. After construction of a deflection dike or wall the frequency doesn't change but the impact decreases considerably. The areas Z1 and Z2 on the other hand will get a higher frequency of occurrence and higher consequence as a result of the mitigation works (Jaboyedoff *et al.* 2014).

2.3.4 Indicator-based approach

There are many situations where the (semi)-quantitative methods for risk mapping are not appropriate. This could be because the data are lacking to be able to quantify the components, such as hazard frequency, intensity, and physical vulnerability. For instance, when the risk assessment is carried out over large areas, or in areas with limited data. Another reason is that one would like to take into account a number of different components of vulnerability that are not incorporated in (semi-) quantitative methods, such as social vulnerability, environmental vulnerability and capacity. In those cases, it is common to follow an indicator-based approach to measure risk and vulnerability through selected comparative indicators in a quantitative way in order to be able to compare different areas or communities. The process of disaster risk assessment is divided into a number of components, such as hazard, exposure, vulnerability and capacity (See Figure 2.4), through the so-called criteria tree, which list the subdivision into objectives, sub-objectives and indicators (Figure 2.11). Data for each of these indicators are collected at a particular spatial level, for instance by administrative units. These indicators are then standardized (e.g. by reclassifying them between 0 and 1), weighted internally within a sub-objective and then the various sub-objectives are also weighted amongst themselves. Although the individual indicators normally consist of quantitative data (e.g. population statistics), the resulting vulnerability, hazard and risk results are scaled between 0 and 1. These relative data allows to compare the indicators for the various administrative units. These methods can be carried out at different levels, ranging from local communities (e.g. Bollin & Hidajat, 2006) cities (Greiving *et al.* 2006) to countries (Van Westen *et al.* 2012).

The approaches are mostly based on the development of the so-called risk indices, and on the use of spatial multi-criteria evaluation. One of the first attempts to develop global risk indicators was done through the Hotspots project (Dilley *et al.* 2005). In a report for the Inter-American Development Bank, Cardona (2005) proposed different sets of complex indicators for benchmarking countries in different periods and to make cross-national comparisons. Peduzzi *et al.* (2005, 2009) have developed global indicators, not on the basis of administrative units, but based on gridded maps. The Disaster Risk Index (UN-ISDR, 2005b) combines both the total number and the percentage of killed people per country in large- and medium-scale disasters associated with droughts, floods, cyclones and earthquakes. In the DRI, countries are indexed for each hazard type according to their degree of physical exposure, their degree of relative vulnerability, and their degree of risk. Also at local scale, risk indices are used, often in combination with spatial multi criteria evaluation (SMCE). Castellanos and Van Westen (2007) present an example of the use of SMCE for the generation of a landslide risk index for the country of Cuba, generated by combining a hazard index and a vulnerability index. Van Westen *et al.* (2012) developed a similar approach for national scale vulnerability and risk assessment for the country of Georgia (Figure 2.10 and

Figure 2.11). The hazard index is made using indicator maps related to triggering factors (earthquakes and rainfall) and environmental factors. The vulnerability index was made using five key indicators: housing condition and transportation (physical vulnerability indicators), population (social vulnerability indicator), production (economic vulnerability indicator) and protected areas (environmental vulnerability indicator). The indicators were based on polygons related to political-administrative areas, which are mostly at municipal level. Each indicator was processed, analysed and standardized according to its contribution to hazard and vulnerability. The indicators were weighted using direct, pair wise comparison and rank ordering weighting methods and weights were combined to obtain the final landslide risk index map. The results were analysed per physiographic region and administrative units at provincial and municipal levels. Another example at the local level is presented by Villagrán de León (2006), which incorporates 3 dimensions of vulnerability, the scale or geographical level (from human being to national level), the various sectors of society, and 6 components of vulnerability. The method uses matrices to calculate a vulnerability index, which was grouped in qualitative classes (high, medium and low).

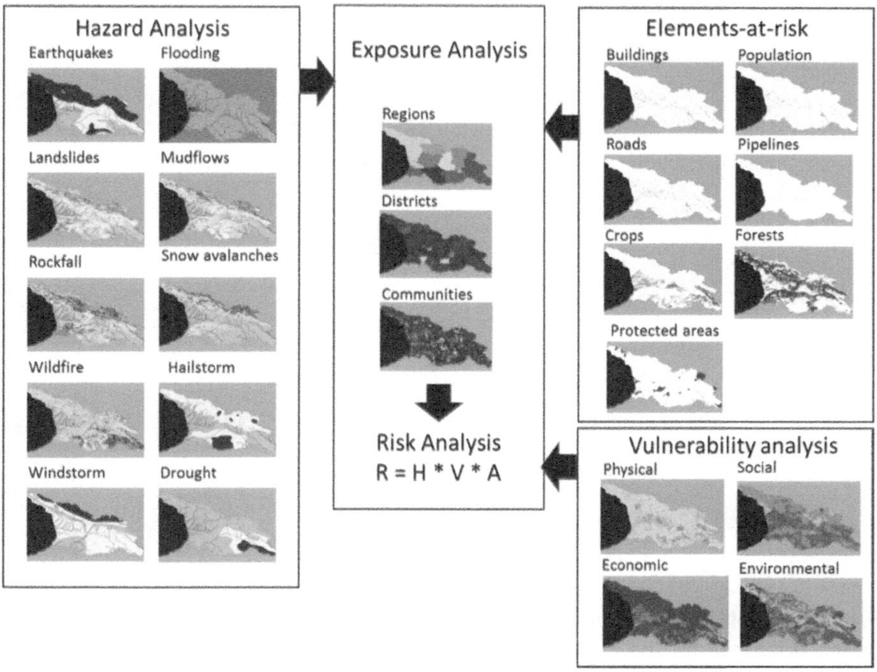

Figure 2.10 National scale multi-hazard risk assessment using an indicator-based approach for the country of Georgia (Caucasus). The method is developed in a web-based platform (http://drm.cenn.org/index.php/en/) and a risk atlas (https://issuu.com/grammallc/docs/atlas_of_risk?pageNumber=1&e=5243266/2932778).

Figure 2.11 National scale multi-hazard risk assessment using an indicator-based approach for the country of Georgia (Caucasus). A multi-hazard map was generated using a hazard criteria tree, which was combined with a vulnerability criteria tree. The vulnerability criteria tree is composed of a number of main groups: physical vulnerability, social vulnerability, environmental vulnerability and economic vulnerability. These are subdivided into subgroups and eventually in a number of spatial indicators, which are standardized and weighted. (Van Westen *et al.* 2012).

2.4 RISK ANALYSIS AND DECISION MAKING: A CASE STUDY

After the introduction parts, the second half of this chapter deals with the application of risk analysis in decision making. This is done through illustrating the procedures with a hypothetical case study.

The overall aim of the second part of this chapter is to illustrate how to analyse possible changes in risk to multi-hazards. These changes may be related to possible risk reduction measures, but also to possible future scenarios related to land use change, population change, and climate change, and the effect of possible intervention alternatives on top of these possible future scenarios. This is illustrated in Figure 2.12, which has four possible workflows:

(A) **Analysing the current level of risk.** In this workflow the stakeholders (e.g. local authorities) are interested to know the current level of risk in their municipality. They request expert organizations to provide them with hazard maps, asset maps, and vulnerability information, and use this information in risk modelling. They use the results in order to carry out a risk evaluation.

(B) **Analysing the best alternatives for risk reduction.** In this workflow the stakeholders want to analyse the best risk reduction alternative, or combination of alternatives. They define the alternatives, and request the expert organizations to provide them with updated hazard maps, assets information and vulnerability information reflecting the consequences of these alternatives. Once these hazard and asset maps are available for the alternatives, the new risk level is analysed, and compared with the existing risk level to estimate the level of risk reduction. This is then evaluated against the costs (both in terms of finances as well as in terms of other constraints) and the best risk reduction scenario is selected.

(C) **The evaluation of the consequences of possible future scenarios.** Possible future scenarios can be formulated that project possible changes related to climate, land use change or population change due to global and regional changes, and which are only partially under the control of the local planning organizations. Stakeholders would like to evaluate the effect of these changes on the hazard and assets (again here the updated maps should be provided by expert organizations) and how these would translate into different risk levels.

(D) **The evaluation how different risk reduction alternatives will perform under different future scenarios.** (trends of climate change, land use change and population change). This is the most complicated workflow, as it requires to calculate the present risk level, the effect of different risk reduction alternatives, and the overprinting of these on the scenarios. For each of these combinations of alternatives & scenarios new hazard, assets and risk maps need to be made.

These four different workflows will be presented more in detail in the coming sections. First, the study area and case study data set will be presented.

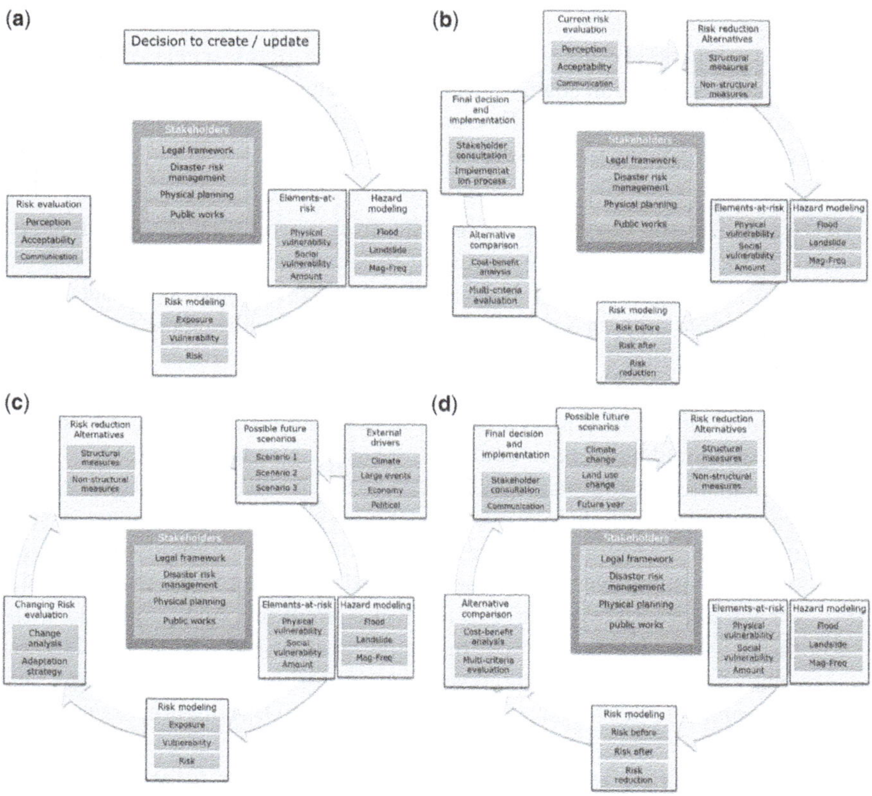

Figure 2.12 Four situations where risk assessment is used in decision making. (**a**) Analysing the current level of risk. (**b**) Analysing the best risk reduction alternative; (**c**) Analysing changing risk due to possible future scenarios. (**d**) Analysing the best "change-proof" alternative: the alternative that behaves best under possible future scenarios. See text for more explanation.

2.4.1 The case study data set

The case study data set was based on an original dataset from Nocera Inferiore, located between Naples and Salerno in Italy. This data set was prepared for the EU FP7 project SAFELAND (2011a, 2011b). The procedure and results for the hazard and risk assessment of the existing situation were reported by Ferlisi *et al.* (2016). The various risk reduction alternatives and the involvement of various stakeholders in the selection of the optimal measures was presented by Linnerooth-Bayer *et al.* (2016). Narasimhan *et al.* (2016) presented results on Cost-Benefit Analysis for some of the risk reduction measures. The Nocera dataset is used as a start, but it has modified the data and methods so that it allows to show the various procedures for multi-hazard risk assessment dealing with a hypothetical case study of a mountainous slope along the coast of a Caribbean island.

Therefore, the original hazard maps are modified to reflect the situation for the various alternatives, and scenarios, and additional tsunami hazard maps are added. Building attributes were modified by us, and all land parcel maps have been generated by ourselves based on available high resolution images. It was decided to choose a hypothetical case because of the difficulty in getting the right data for achieving all objectives in the same study area. This analysis requires local scale hazard intensity maps, detailed element-at-risk maps, vulnerability curves, risk reduction alternatives and future scenarios. Many of these data are still not available for a single area. Nevertheless, it is hoped that by following the examples in this chapter, readers get a better idea of the procedure and can eventually also apply it in their own situation, once data is available. There is also a series of GIS exercises, with descriptions, open source GIS software and a spatial dataset, available that follow the procedures step-by-step and downloaded from http://www.charim.net/use/41. The available maps are illustrated in Figure 2.13 and 2.14, respectively.

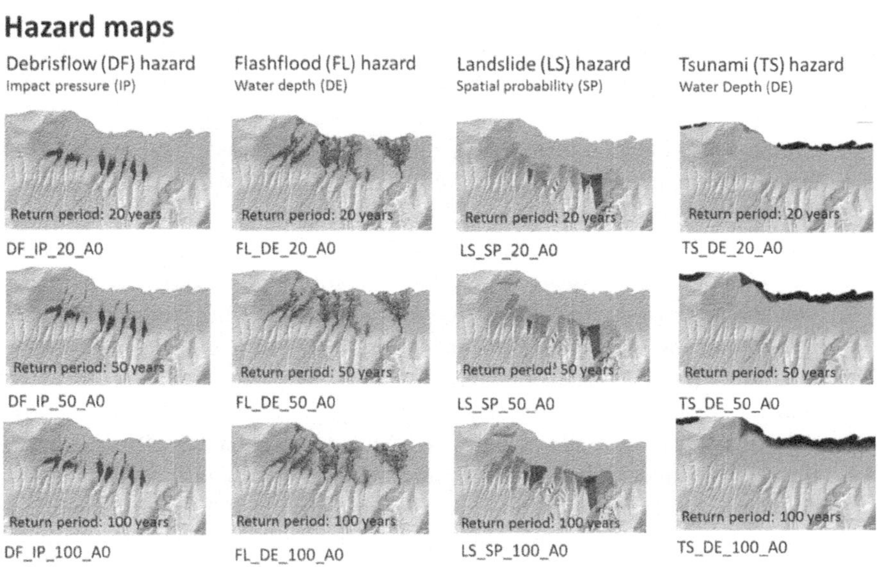

Figure 2.13 Hazard intensity maps for 4 hazard maps (debrisflows, floods, landslides and tsunami) and 3 return periods.

2.4.2 Hazard input data

The case study area is a mountainous area along the coast, with steep forested slopes in the south part, and a fault-related mountain front with triangular facets, from which landslides may be triggered. The steep slopes have a number of steep gulleys, and landslides may form in the upper parts that could lead to debris flows. During heavy rainfall also flash floods may occur in these gulleys, which may affect the flatter areas in the north. The flatter northern part of the area has agricultural, and residential areas. The area is affected by landslides, debris flows, flash floods and possibly also by tsunami. Frequency estimation of hazardous events was carried out based on the analysis of historical hazard data and rainfall data, and hazard assessment for landslides, floods and debris flows were carried out for different return periods using various models, such as OpenLISEM, SEEP/W, Slope/W, TRIGRS and FLO-2d (Ferlisi *et al.* 2016). The debris flow, flood and tsunami maps have intensity data (impact pressure for debris flows, and depth for flooding and tsunami). The landslide hazard maps do not have intensity maps, but only spatial probability maps indicating the chance that a particular area will be affected by a landslide.

The hazard maps have a specific code, which consist of the following components:

- **Hazard type:** the name of the hazard map start with two characters referring to the hazard type: LS = landslides, MF = mudflow, FL = Flood, DF = Debris flows, TS = Tsunami
- **Return Period:** this is the average frequency with which the events are expected to occur. This is based on the analysis of the magnitude and frequency of the triggering rainfall, or of the events themselves (e.g. flood discharge, or the number of landslides occurring in a particular period). The following return periods were used: 20, 50 and 100 years. For some specific situation also longer return periods (200 years) were used.
- **Intensity:** the intensity indicates the spatially distributed effect of the hazard event. This can be water depth (DE) for flooding or tsunami, or impact pressure (IP) for debris flows. These have been modelled using specific hazard modelling software. These models require quite a lot of input data and assumptions. In this chapter we will not deal with the methods how the hazard maps were created. For some types of hazards (e.g. landslides) it may also not be possible to generate intensity maps, as data or models are lacking.
- **Spatial probability:** the spatial probability indicates the chance that a particular location would actually be affected by the hazard. This could be the result of uncertainty in the flood modelling or runout modelling. Or it could also represent (in the absence of an intensity map) the likelihood that a particular area will be affected by landslides based on the area of the units, divided by the area of landslides that have occurred in the past. In this way, it can be used to reclassify the so-called landslide susceptibility maps into spatial probability maps.

2.4.3 Input data: elements-at-risk

It is possible to use four types of elements-at-risk maps: building footprints, land parcels (land use related units), line elements (e.g. transportation lines) and point elements (e.g. individual objects). In the case study, the work involves only building footprints and land parcels as elements-at-risk types (Figure 2.14). Each of them have information on:

- **Land use:** land use is one of the most important characteristics and it is used to model a number of other attributes (e.g. population).
- **Element-at-risk types:** for land parcels this would be the same as the land use type (e.g. forest, residential, commercial etc.) and for building footprints this would be the construction types. Different types of elements-at-risk can be affected differently by hazard events. For the risk analysis this is important as this provides the link to vulnerability curves, which will be explained later.
- **Value:** this is the replacement value of the elements-at-risk in monetary units (Euros, US dollars etc.). These are mostly estimated by multiplying unit costs (e.g. per m^2) with the area of land parcels or floor space for building footprints.
- **People:** the number of people that might be present in the element-at-risk. Here it is relevant to decide whether to take the maximum number of people or the people present at a given time, or to use specific population scenarios (in case of dealing with rapid events, the time of day/year is also important for the population loss estimation).

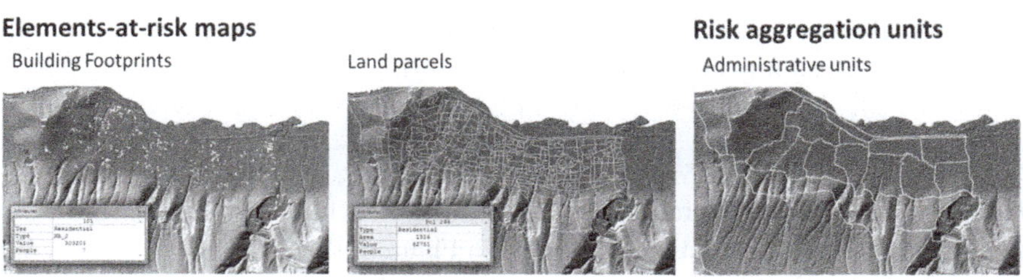

Figure 2.14 Elements-at-risk maps: building footprints and land parcels with related attributes, and administrative units.

The building footprint map and the land parcel maps have the same number of people for the parcels in which buildings are located. For the other parcels, population values per m^2 are used and multiplied these with the area of the land parcel, so that an estimate can be obtained of the total number of people per parcel. For the economic assessment, the values of the buildings are taken from the building footprint map, and used these for the value of the land parcels. For the parcels without buildings, an estimation is conducted based on the value per m^2 and multiplied this with the area.

2.4.4 Input data: vulnerability curves

Vulnerability curves are also very important components in the risk analysis. A vulnerability curve expresses the relation between the hazard intensity (e.g. water depth) and the degree of damage which is expressed between 0 and 1 for a specific type of element-at-risk. Vulnerability curves are derived from past disaster events by correlating observed intensities with observed damage and deriving average regression lines from these. Vulnerability curves may also be derived through computer modelling (e.g. finite element models where a particular building is exposed to a particular intensity and the effect is calculated) or through expert opinion. For this case study, a number of vulnerability curves have been developed for all the combinations of the hazard intensity types and the elements-at-risk types. Existing curves have been used for the literature, but there is a need for a lot of changes, as the curves were not for all of the units. For the analysis these curves should be implemented in GIS, as tables (See Figure 2.15). Curves have been developed for buildings, and land parcels, and separate curves for the physical losses (required for the economic risk analysis) and for the population losses (people killed). The codes in the table indicate the various aspects of the vulnerability curves. For example: VUL_FL_DE_LP_PH: Vulnerability curves for flooding, expressed in water depth, for land parcels, and showing the physical vulnerability.

2.4.5 Input data: administrative units

For the calculation of risk, there is also need for an administrative unit map, as there is going to be an aggregation of the losses for particular units, and the decision making is based on the risk within these units. The administrative unit map contains 19 administrative units (Figure 2.14).

2.5 ANALYSING THE CURRENT LEVEL OF RISK

This section will further present and discuss the workflow for multi-hazard risk analysis of the current situation, illustrated in Figure 2.12A.

2.5.1 Stakeholders

Central in the whole process are the stakeholders. The envisaged users of the system are organizations involved in spatial planning, planning of risk reduction measures, or emergency preparedness and response. They work in a country with a specific legislation and planning process organizations that have different mandates. These could be subdivided into:

- *Government departments responsible for the construction, monitoring, maintenance and protection of critical infrastructure (e.g. the Ministry of Public Works). Their mandate is to:*

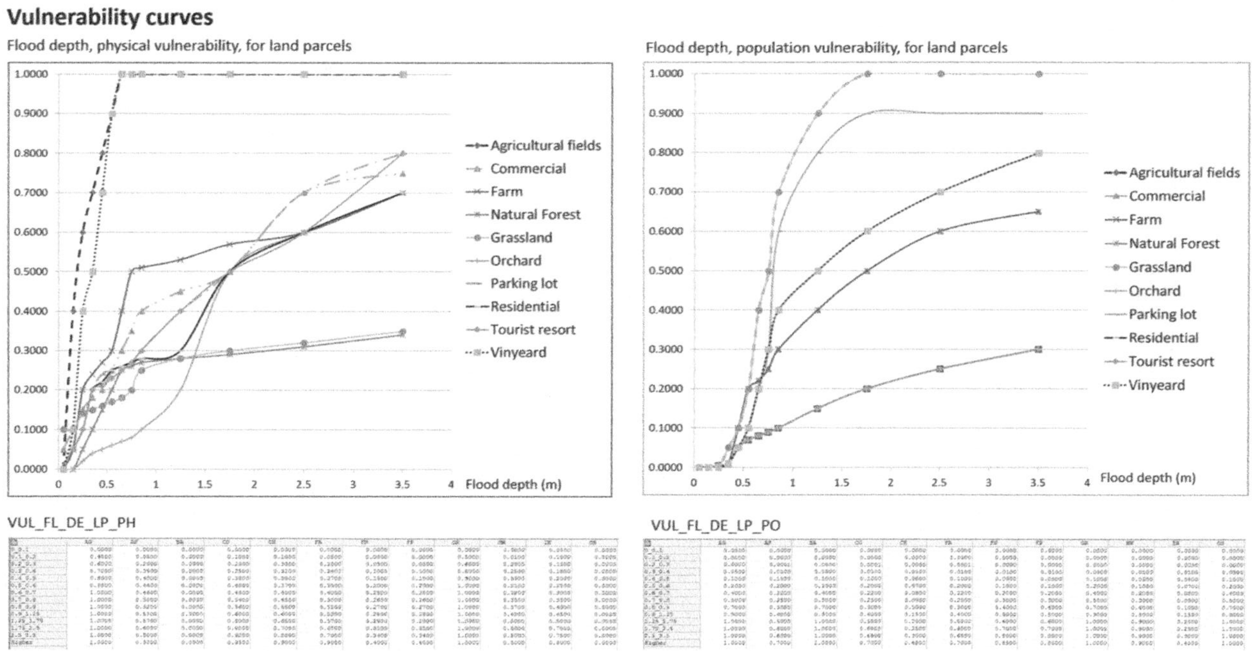

Figure 2.15 Vulnerability curves and related vulnerability tables. Here only some examples are shown.

- Plan the (re)location of critical infrastructure (roads, buildings and other critical infrastructure);
- Design guidelines for construction of roads and buildings in potentially dangerous areas;
- Design of structural and non-structural mitigation measures against flooding and landslides and other hazards;
- Design of non-structural mitigation measures (e.g. watershed management);

- *Physical planning departments responsible with the mandate to make land development plans at different scales. Their mandate is to*:
    - Develop national or regional physical development plans;
    - Develop local development plans;
    - Develop guidelines for building construction in hazard areas;
    - Evaluate relocation options for settlements in endangered areas;
    - Develop zoning maps with relevant hazard information for building control;
    - Provide relevant hazard and risk information for land subdivision;
- *National Emergency Management Organizations with the mandate to*:
    - Design disaster response plans;
    - Organize volunteers for disaster response planning;
    - Develop and management Early Warning systems;
    - Shelter planning and management.

2.5.2 Hazard modelling and elements-at-risk/ vulnerability assessment

Most government organizations normally have a few persons capable of visualizing spatial data using GIS, but are not sufficiently capable of carrying out the actual spatial hazard and risk analysis required as the basis for their work. Therefore, they will work with external consultants that will carry out this type of analysis for them, and they have to specify the exact Terms of Reference of the work of the consultants. These consultants may work on hazard modelling and the generation of elements-at-risk maps. This should be done for a specific scale of analysis:

- *National scale (for the entire country) with output at a scale of 1*:25.000–1:1.000.000 depending on the size of a country;
- *Local scale (for specific areas) with output at a scale of 1*:5.000 to 1:10.000 for specific planning areas (such as settlement areas). Our case study is at local scale;
- *Site-investigation scale (1*:1000) for specific problem sites.

Part of this work should be done by the government organizations themselves as they have the mandate to collect spatial information on the elements-at-risk. For instance the Public works Department should develop a spatial database of the roads including all relevant characteristics related to road type, culverts, bridges, drainage, road cuts and embankment fills, and slope stabilization works. The Department of Physical Planning, together with other relevant Departments is responsible for collecting and maintaining a national building database with the relevant characteristics of buildings. In the establishment of these databases external consultants may play an important role, however on the long run for the respective government organizations would have the mandate to maintain and update these databases, e.g. using web-based GIS solutions (e.g. http://www.charim-geonode.net/).

2.5.3 Risk analysis

The crucial stage in the analysis of risk, which use the available information to estimate the risk to people, property, or the environment, from hazards. Also this type of work generally is not carried out by the government organizations themselves, but rather by consultants, that have the right expertise to carry out this type of analysis for one or more types of hazards, in combination with one or more types of elements-at risk. This work is done at the appropriate scale related to the objectives of the stakeholders. The risk assessment can be subdivided into the following components:

- **Exposure analysis.** In this analysis hazard scenarios worked out by the hazard assessment consultants for different return periods (e.g. once in 20, 50 and 100 years) are combined spatially with the elements-at-risk and the number of elements-at-risk exposed to a certain hazard intensity is calculated. Also for each element-at-risk the (maximum) intensity is calculated given a certain return period.
- **Vulnerability analysis.** The results from the exposure analysis in terms of the maximum intensity per return period are then used in combination with vulnerability curves or matrices for the respective elements-at-risk types. Through the vulnerability curves a translation is made from the intensities of the hazard to the expected degree of loss for the elements at risk.
- **Loss analysis.** The results from the vulnerability assessment are then used in combination with the quantification of the elements-at-risk to calculate the expected losses. In the case of economic losses, the replacements costs for the elements-at-risk are used, resulting in specific losses per return period. In the case of population losses the number of people are used in combination with the population vulnerability resulting from the vulnerability assessment.
- **Risk assessment.** The resulting losses for different return periods are summed up for given administrative units if the hazards are independent and integrated for the different return periods to provide the average annual losses. These are used as the basis for the risk evaluation and for the formulation of possible risk mitigation measures.

Figure 2.16 gives a detailed overview of the steps involved in loss analysis and risk analysis. The **Loss analysis** has to be done for each combination of a hazard map (for a given return period) and an elements-at-risk map, and for population and/or economic losses. Each loss estimation requires a number of steps, and doing this repetitive analysis manually is very time consuming. Therefore, an automated script is developed in GIS, which combines a number of calculations and operations, and uses parameters. Figure 2.16 shows the input screen of the script. Some of these (alternative, scenario and future year) will be discussed later. This makes that the same script can be used in many different situations. It follows a number of steps:

- Overlay the element-at-risk map with the hazard intensity map;
- The resulting cross table (joint frequency table) contains all combinations of the land parcel/building code and the intensity values (e.g. water depths). The results are classified, according to the classification of the hazard intensity, so that the result is in the form of classes, which can be used to join with the vulnerability tables;
- As land parcels are sometimes large and only part of them might be actually exposed to hazard intensity the script calculates the losses first for the parts of the land parcels with the same intensity;
- In order to know which fraction of each land parcel has a certain intensity, the script reads in in the area of the whole land parcel from the attribute table of the land parcel map, and then calculates the fraction of the land parcel (Area of the unit in the joint frequency table divided by the area of the entire land parcel);
- The joint frequency table is joined with the attribute table of the land parcels/buildings and the *amount* column (either value or people) is entered, depending on the input provided by the user. Also the land use is joined;
- The joint frequency table is then joined with the vulnerability table (of the hazard type indicated) and the vulnerability values for all lands use types are read;
- The vulnerability for each record is calculated by taking the vulnerability value of the column that has the same land use code as in the record;
- A column is calculated that has an indication whether is dealing with a spatial probability map. This is done by creating a column SPCheck and then checking if the entered value is SP (Spatial Probability) or not. If this is the case, the spatial probability is used, otherwise a value of 1;
- The script then calculates the loss by multiplying the amount × vulnerability × spatial probability;
- In order to bring back the information at the level of the land parcels, the script aggregates the loss for the land parcels;
- The script also aggregates the loss for the administrative units , and for the entire study area;

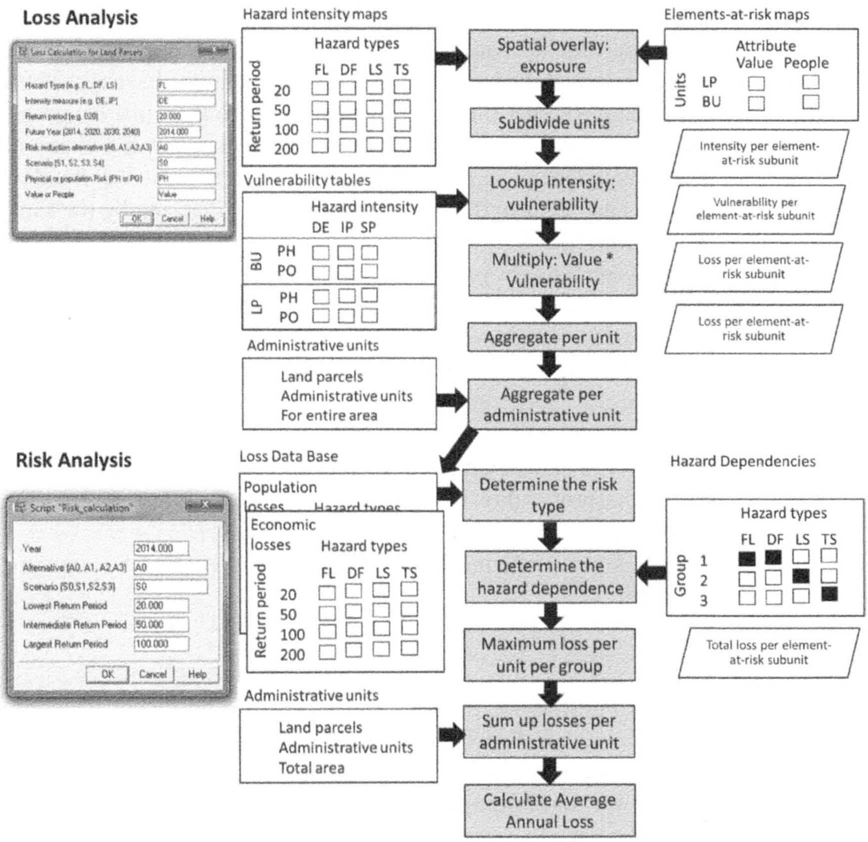

Figure 2.16 Procedure for loss analysis of all combinations of hazard scenarios for specific return periods with elements-at-risk, and the combination in the subsequent risk analysis. The following abbreviations are used: FL = Flooding, DF = Debris flow, LS = Landslides, TS = Tsunami, DE = Water Depth, IP = Impact Pressure, SP = Spatial probability, BU = Building footprints, LP = Land parcels. The two input screen are those used in the loss analysis and risk analysis.

The loss analysis results in a loss database, with the calculated losses for each combination of hazard type, return period, and elements-at-risk type. The user can then select which combinations of losses are used in the subsequent risk analysis. The risk analysis can be carried out if the loss estimation was done for at least three return periods of the hazard type(s) that were selected.

The risk analysis can also be done using a script, that uses a number of parameters, and which has the following steps:

- *The risk type is determined*: either economic risk or population risk is calculated;

- The types of hazard and their dependencies are selected. Here the user can interactively select which hazard types are taken into account. This allows to carry out single risk or multi-hazard risk. In the case of multi-hazard risk, the user has to evaluate whether the hazard types belong to the same triggering event group (See section on Multi-hazard risk in this chapter). For example, tsunami hazard is independent of the other three hazards, whereas flooding and debris flows can occur in the same area, and are triggering by the same event (extreme rainfall).
- *Based on the dependencies that are defined, the script will either sum up the losses for different independent hazards, or take the maximum for dependent hazards in the same group, to avoid double counting of losses. This is done at the lowest aggregation level*: the individual land parcels or building footprints;
- After that the multi-hazard losses are aggregated for the administrative units;
- Annual probability is calculated from the return periods, and the losses are plotted against the probability in loss curves (for economic risk) and FN curves (for population).
- Average annual loss is calculated for economic risk using the equation presented in Figure 2.7.

2.5.4 Risk evaluation

The results of the risk analysis for the case study area are presented in Figure 2.17, both for land parcels, as well as for building footprints. As can be seen from this figure the results differ when we use only building footprints, or land parcels. Economic losses are lower for buildings, because the risk analysis for land parcels also takes into account many other assets (e.g. roads, agriculture, and forest). Also the population risk for land parcels is larger, as it takes into account the persons not present in buildings. After estimating the risk, it is important to determine whether the risk is too high, and where the risk is too high. This is called the risk evaluation stage, and is the stage at which values and judgements enter the decision process, explicitly or implicitly, by including consideration of the importance of the estimated risks and the associated social, environmental, and economic consequences, in order to identify a range of alternatives for managing the risks.

Important considerations in this respect are:

- **Risk perception among stakeholders.** Risk perception is about how individuals, communities, or authorities perceive/judge/evaluate/rank the level of risk, in relation to many factors;
- **Risk communication.** An important component in determining the risk perception is the communication between the stakeholders of the levels of risk. Do government organizations actively involve other stakeholders in the consultation process of the actual level of risk and the possible risk reduction measures?

- **Risk acceptability.** An acceptable risk is a risk, which the society or impacted individuals are prepared to accept. Actions to further reduce such risk are usually not required unless reasonably practicable measures are available at low cost in terms of money, time and effort. The definition of acceptability levels is a responsibility of the national or local government in a country. Risk acceptability depends on many factors, and differs from country to country. Therefore, it is also not possible to simply export them to other countries. Risk acceptability levels are generally done on this basis of individual risk levels or societal risk levels (using so-called F-N curves), some of which are shown in Figure 2.14, based on Ho (2009).

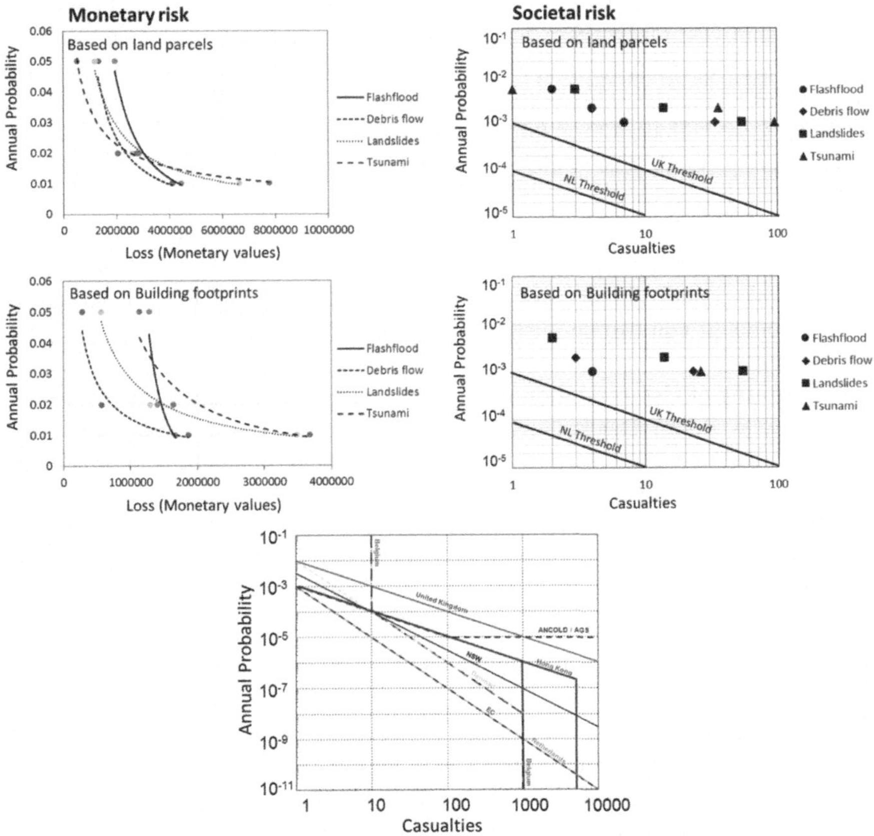

Figure 2.17 Results of the multi-hazard risk analysis for the existing situation in the case study area. Left: economic risk. Right: societal risk with risk acceptability threshold of the United Kingdom (UK) and the Netherlands (NL). Results are shown for two different elements-at-risk data sets: land parcels and building footprints. Below the risk acceptability thresholds for a number of different countries are presented (Source: Ho, 2009).

2.6 ANALYSING THE BEST PLANNING ALTERNATIVE

Once the multi-hazard risk for the current situation is considered unacceptable, which is the case in the case study, a new workflow is introduced for the evaluation of optimal risk reduction measures (Figure 2.12B).

2.6.1 Defining possible planning alternatives

In this workflow the stakeholders want to analyse the best planning alternative, or combination of alternatives. They define the alternatives, and request the expert organizations to provide them with updated hazard maps, elements-at-risk information and vulnerability information reflecting the consequences of these alternatives. Once these hazard and asset maps are available, the new risk level is analysed, and compared with the existing risk level to estimate the level of risk reduction. This is then evaluated against the costs (both in terms of finances, as well as in terms of other constraints) and the best risk reduction scenario is selected. The planning of risk reduction measures (alternatives) involves:

- **Disaster response planning:** focusing on analysing the effect of certain hazard scenarios in terms of number of people, buildings and infrastructure affected. It can also be used as a basis for the design of early warning systems;
- **Planning of risk reduction measures:** which can be engineering measures (such as dikes, check-dams, sediment catchment basins), but also non-structural measures, such as relocation planning, strengthening/protection of existing buildings etc.;
- **Spatial planning:** focusing on where and what types of activities are planned and preventing that future development areas are exposed to natural hazards.

The methods for risk reduction planning can be subdivided into:

- **Structural measures** refer to any physical construction to reduce or avoid possible impacts of hazards, which include engineering measures and construction of hazard-resistant and protective structures and infrastructure. The strategy is to modify or reduce the hazard;
- **Non-structural measures** refer to policies, awareness, knowledge development, public commitment, and methods and operating practices, including participatory mechanisms and the provision of information, which can reduce risk and related impacts. With the aim of modifying the susceptibility of hazard damage and disruption and/or modifying the impact of hazards on individuals and the community.

The planning alternatives that are evaluated may be designed without considering the possible impact of hazard and risk, and in these situation the analysis is carried out to evaluate the impact of the different alternatives on the hazard and risk (will it increase or decrease).

There are mostly different planning alternatives that can be formulated, and each of them may have advantages and disadvantages. The aim of this analysis to quantify their advantages and disadvantages in terms of hazard and risk reduction, and to evaluate these against the costs for implementation through a cost-benefit analysis. Also other criteria that cannot be quantified can be used in deciding the best alternative, using a multi-criteria evaluation.

For example, for the case study are the stakeholders defined three possible sets of risk reduction measures, which are presented in Table 2.5 and Figure 2.18.

Table 2.5 Characterization of risk reduction alternatives in the case study area. The alternatives are also shown in Figure 2.18.

	Alternative 1	**Alternative 2**	**Alternative 3**
Name	**Engineering Alternative**	**Eco-DRR Alternative**	**Relocation Alternative**
Characteristics	• Constructing six retention basins, designed to retain maximum 100 year events; • Soil removal work in selected areas for slope stabilization; • Expropriation of land and existing buildings where construction will take place; • Early Warning System;	• Planting of protection forest at the foot of the steep slope and on fans of the side streams; • Construction of drainage channels and 10 water tanks; • Expropriation of land and existing buildings where construction will take place • Soil removal work in selected areas for slope stabilization;	• Relocation of buildings with people in the most dangerous sectors; • Compensation of owners of buildings or construction of new buildings elsewhere; • Expropriation of existing buildings • Lawsuits of residents that do not want to leave the area; • Early Warning System for other areas;
Time of construction	3 years	2 years	4–10 years, depending on the resistance

(*Continued*)

Table 2.5 Characterization of risk reduction alternatives in the case study area. The alternatives are also shown in Figure 2.18 (*Continued*).

	Alternative 1	**Alternative 2**	**Alternative 3**
Name	**Engineering Alternative**	**Eco-DRR Alternative**	**Relocation Alternative**
Year when benefit starts	4th year	From third year increasing to maximum after 15 years	From fourth year increasing until the tenth year
Annual maintenance	Cleaning retention basins, Slope monitoring. Starts at fourth year	Forest management, cleaning water tanks and channels, slope monitoring. Starts in third year.	Monitoring of areas, prevention of illegal resettlements. Starts at fourth year.
Does the hazard change?	Yes, new hazard maps are needed for landslides, debris flows and floods. Events with return periods of 100 years or less are retained. Higher return periods may overtop the structures. No effect on tsunami	Yes, new hazard maps are needed for landslides, debris flows and floods. Smaller events are completely retained by protection forest, and larger events are retained, but there is still a certain hazard. No effect on tsunami	No.
Do the elements-at-risk change?	Only those units where the retention basins are constructed	Only those land parcels where the protection forest will be constructed.	Yes, the relocation areas will change. A new land parcel map is needed.
Items to estimate costs	Construction costs retention basin: unit costs × number; Soil removal: unit costs × area; Expropriation costs: unit cost × area;	Tree planting: unit cost × area; Soil removal: unit costs × area; Water tanks: unit cost × number; Expropriation costs: unit cost × area;	Number of buildings to relocate; Compensation per building; Number of lawsuits and value of legal costs; Early warning system;

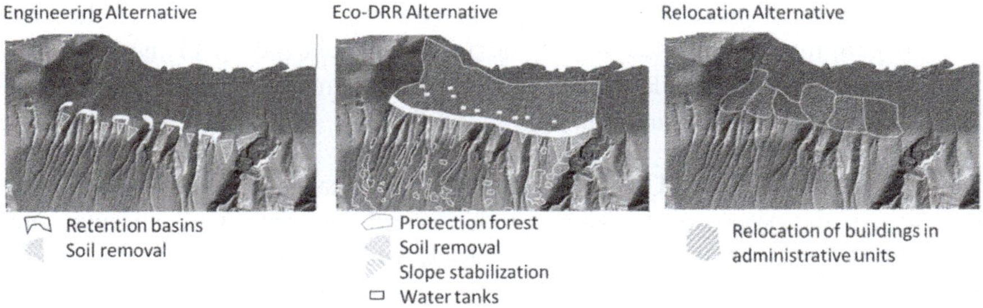

Figure 2.18 Three sets of risk reduction measures presented as three alternatives. The characteristics of these alternatives are given in Table 2.5.

2.6.2 Re-analysing hazards and elements-at-risk

The implementation of certain structural or non-structural risk mitigation measures might lead to a modification of the hazard, exposure and vulnerability. Risk is a function of Hazard × vulnerability of exposed elements-at-risk × the quantification of the elements-at-risk (Figure 2.1). Therefore, there are several possibilities that risk mitigation measures will influence:

- The **hazard.** In terms of the probability (or return period) of specific hazard events, the spatial distribution of the hazard and the intensity of the hazards. For instance, the construction of a flood wall along a river may reduce the area that will be flooded. For certain lower return periods the flood wall may retain all flood water, and therefore the intensity (flood height) outside of the flood wall will become zero for these return periods. For more extreme events the flood intensity may become lower as a result of the flood wall. Therefore, it is required to re-analyse the hazard given the implementation of the risk reduction measure.
- The **exposure of elements-at-risk.** The number of elements-at-risk might change as a result of the risk mitigation measure, or planning alternative. For instance, if one of the alternatives involves relocation, the number of exposed elements-at-risk will decrease, whereas the hazard might stay the same. In other planning alternatives the effect of future development on the number of exposed elements-at-risk might also be evaluated.
- The **vulnerability of the elements-at-risk.** The type of elements-at-risk might change as a result of implementing the planning alternative. For instance, when retrofitting is considered, the number of elements-at-risk might be the same, as well as the hazard, but the vulnerability of these elements-at-risk might decrease, leading to a lower risk level. The same can be said for the implementation of an Early Warning System. It will decrease the number of exposed people, but also their vulnerability, if they would move to shelters where they are better protected.

- The **quantification of the elements-at-risk.** might change. This might refer to planning alternatives where the value of the exposed elements-at-risk changes, e.g. they could increase when more expensive housing is considered in a certain planning alternative.

Therefore, experts should evaluate together with the stakeholders what would be the effect of the proposed alternative on the hazard, elements-at-risk location and characteristics and the vulnerability. If needed new hazard modelling should be carried out, or new elements-at-risk maps should be made representing the new situation.

2.6.3 Analyse risk reduction

After re-analysing the hazard, elements-at-risk and vulnerability for the situation after the implementation of the planning alternative, the next step is to analyse the resulting level of risk, and compare this with the current risk level. The difference between the average annual losses before and after the implementation of the planning alternative, provides information on the risk reduction. This should be done for all the possible planning alternatives. The risk reduction should be done preferable both in terms of economic risk reduction (reduction in the average annual losses in monetary values) as well as in population risk reduction (reduction in the expected casualties or exposed people).

The analysis of risk requires a repetitive procedure which has to be carried out for each hazard scenarios (different hazard types and return periods) in combination with elements-at-risk types, and then also for each possible alternative. This requires the use of the same automated procedures using scripts in GIS described in section 2.5. Figure 2.19 shows the results of the analysis of the three risk reduction alternative for the case study. It is clear from the figure that alternative 1 (engineering solutions) and 2 (Eco-DRR) result in more risk reduction than alternative 3 (relocation).

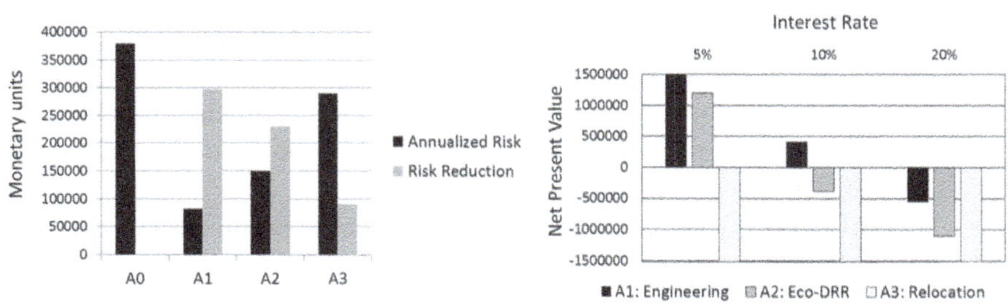

Figure 2.19 Left: Results of the multi-hazard risk analysis for the current situation (A0) and the three risk reduction alternatives: A1 (Engineering alternative), A2 (Eco-DRR alternative) and A3 (Relocation alternative). Right: Results of the Cost-Benefit Analysis showing the Net Present Value (NPV) for the three alternatives for different interest rates.

2.6.4 Compare alternatives

Once the effects of the various planning alternatives are analysed, in terms of their risk reduction, the next step is to compare them and decide which of the alternatives would be the best to implement. This could be done through different methods:

- **Cost-benefit analysis.** Here both the benefits and the costs can be quantified. The benefit of a risk reduction alternative is represented by its annual risk reduction in monetary values, which was calculated in the previous step (risk after implementation minus current risk) (Figure 2.19). The costs for the planning alternative can be quantified, as well in terms of their investment costs, maintenance costs, project life time etc. (See Table 2.5). Cost-benefit analysis can be carried out by calculating relevant indicators, such as the Net Present Value, Internal Rate of Return or Cost-Benefit ratio.
- **Cost-effectiveness analysis.** This is carried out when the costs can be quantified and compared, but the benefits in terms of risk reduction cannot be quantified in monetary values. This is the case for instance when population risk is calculated, as it is generally considered not ethical to represent human lives in monetary values.
- **Multi-Criteria Evaluation.** When both the costs and benefits cannot be quantified in monetary values, or when additional to cost-benefit or cost-effectiveness also other non-quantifiable indicators are used, a (spatial) multi-criteria evaluation is generally considered the best option. In this analysis also social, ecological, cultural and other criteria can be incorporated in the decision making process.

The comparison of alternatives is generally carried in a process, where the other stakeholders are also involved before a decision is taken on the optimal alternative that will be implemented.

2.6.5 Final decision and implementation

The last step of this workflow related to the selection of the optimal planning alternative in relation to the reduction of risk to hydro-meteorological hazards is the consultation with the various stakeholders involved (Figure 2.12B). This includes public hearings with the population, private sector, non-governmental organizations, and various social network groups (e.g. communities, churches). The stakeholders have the opportunity to request adjustment to the proposed plan of action, and if these adjustments are considered valid, and substantial, a new round of evaluation might be needed if the change of expected hazard and risk impact is substantial. Once the plan is approved, the procedures will start for the implementation of the plan.

2.7 ANALYSING POSSIBLE FUTURE SCENARIOS

Risk is changing continuous, as both hazard and elements-at-risk change. Over a longer period of time this may result in considerable changes in multi-hazard risk.

Figure 2.12C shows the workflow to analyse changing risk due to possible future scenarios.

2.7.1 Identification of possible future scenarios

The scenarios are related to possible changes related to climate, land use change or population change due to global and regional changes, and which are only partially under the control of the local planning organizations. The stakeholders might like to evaluate how these trends have an effect on the hazard and elements-at-risk and how these would translate into different risk levels. The possible future could be of the following types:

- **Climate change scenarios.** In that case, the stakeholders require the involvement of experts that would indicate which climate change scenarios would be evaluated and what would be the expected effects in terms of changes in frequency and magnitude of hydro-meteorological triggers would be expected, such as changes in sea-level and extreme precipitation;
- **Land use change scenarios.** In this case, the stakeholders require the involvement of experts that would indicate possible land use changes based on macro-economic and political developments which would be translated into local changes. For instance, scenarios could be envisaged where an increase in tourism would be translated into possible future expansion of tourist facilities would be evaluated. The future land use scenarios would also involve possible changes in population which should also be taken into account.
- **Future planning scenarios.** In the national physical development plans also possible future developments will be outlined and priorities for development indicated which have implications for the spatial distribution of land use and population.

Also combinations of these drivers might be considered. The possible future changes should be expressed for certain years in the future, for instance for 2020 and 2030 and are considered as a basis for long term planning. Table 2.6 shows four possible future scenarios for the case study area, which are combining land use changes and climate changes.

2.7.2 Re-analysing hazards and elements-at-risk for possible future scenarios

The possible future scenarios might lead to a modification of the hazard, exposure and vulnerability in certain future years from now. Therefore it is required to re-analyse:

- The hazard. Possible future scenarios of climate change might lead to a change in the frequency and magnitude of triggering events for floods and landslides. Therefore, a new magnitude-frequency analysis might be required, that take into account changing trends in frequencies of extreme events. The same hazard event

Table 2.6 Definition of four possible future scenarios for the case study.

Scenario		**1**	**2**	**3**	**4**
Name		**Business as Usual**	**Risk-Informed Planning**	**Worst Case**	**Most Realistic**
Land use change		Rapid growth without taking into account the risk information	Rapid growth that takes into account the risk information and extends the alternatives in the planning	Rapid growth without taking into account the risk information	Rapid growth that takes into account the risk information and extends the alternatives in the planning
Climate change		No major change in climate expected	No major change in climate expected	Climate change expected, leading to more frequent extreme events	Climate change expected, leading to more frequent extreme events
2020	Return period: 20	*20 (±5)*	*20 (±5)*	*17 (±6)*	*17 (±6)*
	Return period: 50	*50 (±10)*	*50 (±10)*	*45 (±12)*	*45 (±12)*
	Return period: 100	*100 (±20)*	*100 (±20)*	*90 (±23)*	*90 (±23)*
	Return period: 200	*200 (±40)*	*200 (±40)*	*180 (±44)*	*180 (±44)*
2030	Return period: 20	*20 (±5)*	*20 (±5)*	*14 (±7)*	*14 (±7)*
	Return period: 50	*50 (±10)*	*50 (±10)*	*35 (±14)*	*35 (±14)*
	Return period: 100	*100 (±20)*	*100 (±20)*	*75 (±26)*	*75 (±26)*
	Return period: 200	*200 (±40)*	*200 (±40)*	*150 (±49)*	*150 (±49)*
2040	Return period: 20	*20 (±5)*	*20 (±5)*	*11 (±8)*	*11 (±8)*
	Return period: 50	*50 (±10)*	*50 (± 10)*	*25 (±16)*	*25 (±16)*
	Return period: 100	*100 (±20)*	*100 (±20)*	*55 (±30)*	*55 (±30)*
	Return period: 200	*200 (±40)*	*200 (±40)*	*110 (±53)*	*110 (±53)*

that now has an average return period of 50 years, might have an average return period which is much smaller in a number of years from now. Also the intensities of the hazard might change for instance due to changes in land use, which might affect the hazardous processes (e.g. deforestation scenarios). In the case study (Table 2.6), this is represented by a decrease in return periods in future years, and the increasing standard deviations illustrate the increasing level of uncertainty.

- The exposure of elements-at-risk. The possible land use scenarios might lead to substantial changes in land use/land cover, which also has an important effect on the number of elements-at-risk within the various land use classes. The analysis of future changes in land use/land cover is generally carried out based on land parcel maps, rather than on the basis of building footprints maps, as it is generally very difficult to translate the land use changes directly into possible locations of buildings. Figure 2.20 shows the possible land use maps for 2020, 2030 and 2040 for the case study area, following the possible scenarios described in Table 2.6.

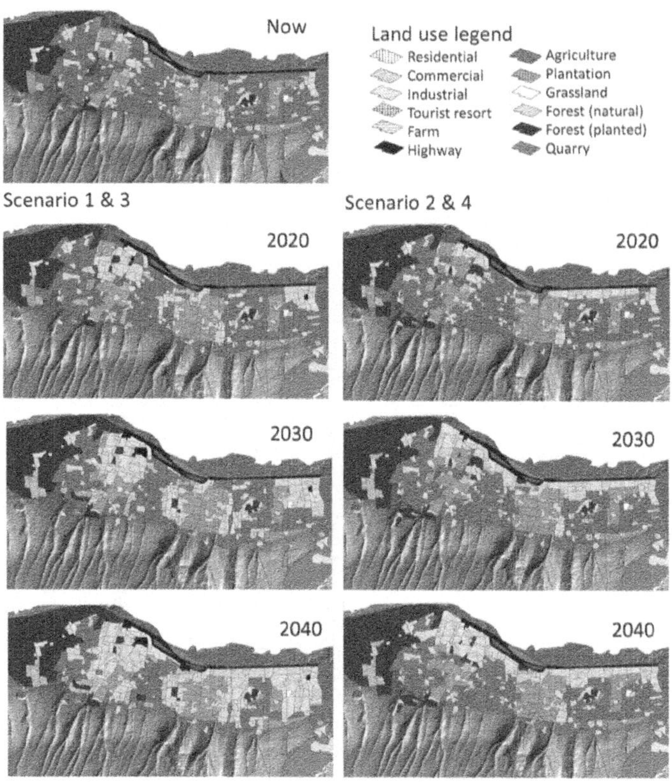

Figure 2.20 Possible land use changes for 2020, 2030 and 2040 for the possible future scenarios described in Table 2.6.

Therefore, experts should evaluate together with the stakeholders what would be the effect of the possible future scenarios on the hazard, elements-at-risk location and characteristics and the vulnerability. If needed new hazard modelling should be carried out, or new elements-at-risk maps should be made representing the new situation.

2.7.3 Analyse possible changes risk for possible future scenarios

After analysing the hazard, elements-at-risk and vulnerability for (a) future year(s) given certain possible future scenarios, the next step is to analyse the resulting change in risk, and compare this with the current risk level. The difference between the current average annual losses and those in a future year under a given change scenario provides information for decision makers on the possible negative consequences of climate change and land use change scenarios. They can be used as a basis for designing appropriate strategies for adaptation. The risk reduction should be done preferable both in terms of changes in economic risk (average annual losses in monetary values), as well as in population risk reduction (expected casualties or exposed people). It is also important to incorporate the uncertainty levels in this type of analysis, thus providing a range of change rather than concrete values. Figure 2.21 shows the differences in risk for the four scenarios for the study area.

2.7.4 Changing risk evaluation

After assessing the possible changes in risk that might result from a number of possible future scenarios related to climate change and land use change, stakeholders should analyse these changes carefully in terms of:

- **Spatial location of changes in risk.** Some areas might be much more impacted by these possible future changes than others. Based on the outcomes of the analysis stakeholders could then prioritize certain areas for critical interventions.
- **Critical sectors.** Changes in risk could be analysed for different sectors of society, such as economy, agriculture, tourism, education, transportation etc.
- **Development of adaptation strategies.** The analysis of the expected level of changes in risk and areas where an increase in risk is expected according to the possible scenarios, should lead to the formulation of adaptation strategies that aim to reduce these possible impacts through planning alternatives that could be implemented now.

When the change of risk is compared for the four scenarios, the risk can be observed including the climate change effects (by reducing the return periods as indicated in Table 2.6) and has a very significant increasing trend. When the effect of land use changes (as shown in Figure 2.17) are examined, it can be seen that the worst case scenario indeed has the largest increase in risk, and the risk informed planning scenario has a lower risk. However, this last scenario focuses on the avoidance of the most hazardous areas for flooding, debris flows and landslides, and the development takes place mostly along the coast (See Figure 2.17). Therefore, tsunami hazard is also included in the analysis, as shown in Figure 2.18, Scenario 2 and 4 have the highest risk levels. This stresses the importance of including all hazard types in multi-hazard risk assessment.

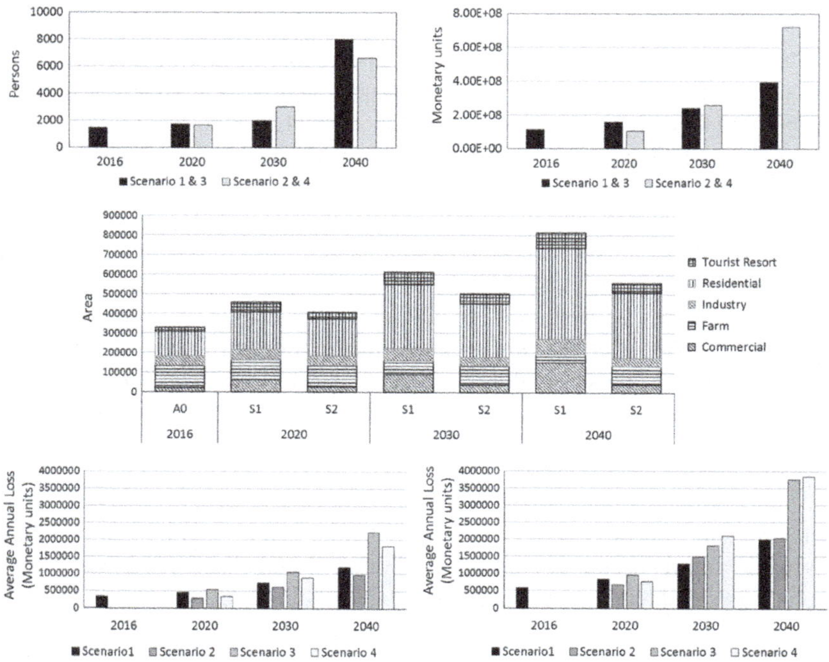

Figure 2.21 Top: changes in population and economic values from 2016 to 2040 according to two land use change scenarios. Middle: changes in some of the land use types from 2016 to 2040 according to two land use change scenarios. Below left: Changes in multi-hazard risk from 2016 to 2040 for the four possible future scenarios, without including tsunami risk. Lower right: Changes in multi-hazard risk from 2016 to 2040 for the four possible future scenarios, including tsunami risk.

2.8 ANALYSING PLANNING ALTERNATIVES UNDER POSSIBLE FUTURE SCENARIOS

The analysis as shown in Figure 2.21 shows that it is relevant to analyse how multi-hazard risk might change in future according to different possible trends. However, it also demonstrates the need to implement risk reduction measures now. In the last workflow, illustrated in Figure 2.12D, it is analysed which of the risk reduction alternatives performs best under different possible future scenarios.

2.8.1 Selection of alternatives, scenarios and future years

The evaluation how different risk reduction alternatives will lead to risk reduction under different future scenarios (trends of climate change, land use change and population change) is the most complicated workflow, as it requires to calculate the present risk level, the effect of different risk reduction alternatives, and the overprinting of these on the scenarios. For each of these combinations of alternatives and scenarios new hazard, assets and risk maps need to be made. This type of analysis allows stakeholders to make the most optimal "change proof" selection of planning alternatives. This type of analysis is entirely based on experts and consultants, which should evaluate both the effects of the planning alternatives, as well as the associated effects of possible future scenarios on hazard, vulnerability and risk. Such type of analysis could be applied to specific critical areas, such as the capitals or important critical infrastructure. Table 2.7 combines all combinations in a matrix. This table indicates the combination of the four scenarios (S1, S2, S3, S4 as described in Table 2.6) and the three risk reduction alternatives (A1, A2, A3 described in Table 2.5) in three future years (2020, 2030, 2040). Due to the large number of input maps it is important to use the coding of the files in a similar way, so that it is possible to use the calculation script to analyse the loss and multi-hazard risk all combinations. Therefore, for example: LP_2020_A1_S2 refers to the land parcels for future year 2020 under alternative A1 (Engineering solutions) and for scenario S2 (risk informed planning).

2.8.2 Re-analysing hazards and elements-at-risk for alternatives/scenarios

The combination of the implementation of certain planning alternatives (structural or non-structural risk mitigation measures) in combination with certain possible future scenarios will certainly lead to a modification of the hazard, exposure and vulnerability. This is why both the hazard maps and the elements-at-risk maps should be updated for each combination.

- The **hazard.** In terms of the probability (or return period) of specific hazard events, the spatial distribution of the hazard and the intensity of the hazards. For instance, the construction of the retention basins reduce the area that will be impacted by flooding and or debris flows. The retention basins are designed for a specific maximum discharge (e.g. 100 years). With climate change scenarios the same discharge may occur more frequently, and events with higher return periods may overtop the basins. Therefore, it is required to re-analyse the hazard given the implementation of the risk reduction measure and the possible future climate change/land use change scenario. In the case study area, different hazard intensity maps are considered for the various risk reduction measures, and change the return periods for future years following Table 2.6.

Table 2.7 Combination of four scenarios, for future years 2020, 2030 and 2040, with three risk reduction alternatives. Each combination is a situation where multi-hazard risk is analysed.

Scenario: Possible Future Trends	Alternative: Risk Reduction Options	Now 2014	Future Years		
			2020	2030	2040
S0 (Without including any future trends)	A0 (no risk reduction)	2014_A0_S0	No future trends are taking into account, and all hazards, elements at risk and vulnerabilities are considered constant in future.		
	A1 Engineering	2014_A0_S1			
	A2 Ecological	2014_A0_S2			
	A3 Relocation	2014_A0_S3			
S1 Business as usual	A0 (no risk reduction)	Does not exist: use existing situation	2020_A0_S1	2030_A0_S1	2040_A0_S1
	A1 Engineering		2020_A1_S1	2030_A1_S1	2040_A1_S1
	A2 Ecological		2020_A2_S1	2030_A2_S1	2040_A2_S1
	A3 Relocation		2020_A3_S1	2030_A3_S1	2040_A3_S1
S2 Risk informed planning	A0 (no risk reduction)	Does not exist: use existing situation	2020_A0_S2	2030_A0_S2	2040_A0_S2
	A1 Engineering		2020_A1_S2	2030_A1_S2	2040_A1_S2
	A2 Ecological		2020_A2_S2	2030_A2_S2	2040_A2_S2
	A3 Relocation		2020_A3_S2	2030_A3_S2	2040_A3_S2
S3 Worst case (Rapid growth + climate change)	A0 (no risk reduction)	Does not exist: use existing situation	2020_A0_S3	2030_A0_S2	2040_A0_S3
	A1 Engineering		2020_A1_S3	2030_A1_S3	2040_A1_S3
	A2 Ecological		2020_A2_S3	2030_A2_S3	2040_A2_S3
	A3 Relocation		2020_A3_S3	2030_A3_S3	2040_A3_S3
S4 Climate resilience (informed planning under climate change)	A0 (no risk reduction)	Does not exist: use existing situation	2020_A0_S4	2030_A0_S3	2040_A0_S4
	A1 Engineering		2020_A1_S4	2030_A1_S3	2040_A1_S4
	A2 Ecological		2020_A2_S4	2030_A2_S3	2040_A2_S4
	A3 Relocation		2020_A3_S4	2030_A3_S3	2040_A3_S4

- The **exposure of elements-at-risk.** The number of elements-at-risk might change as a result of the risk mitigation measure, or planning alternative, and also as a result of the possible future scenario. For instance, if one of the alternatives involves relocation, the number of exposed elements-at-risk will decrease, whereas the hazard might stay the same. However, under certain land use scenarios the pressure on the land might be so high that previously related areas might become occupied again. In other planning alternatives, the effect of future development on the number of exposed elements-at-risk might also be evaluated. Figure 2.22 shows the land use maps for one scenario (business as usual). Each combination of possible future scenario, future year and risk reduction alternative requires a separate land use map. From this figure it can be observed that under this scenario, the relocation alternative will not be very effective, as the areas that are relocated in 2016 will become occupied again in later years. The different land use situations is combined with increases in economic values and population numbers in the various scenarios (As was shown in Figure 2.22). These will also have an important influence on the estimated multi-hazard risk.

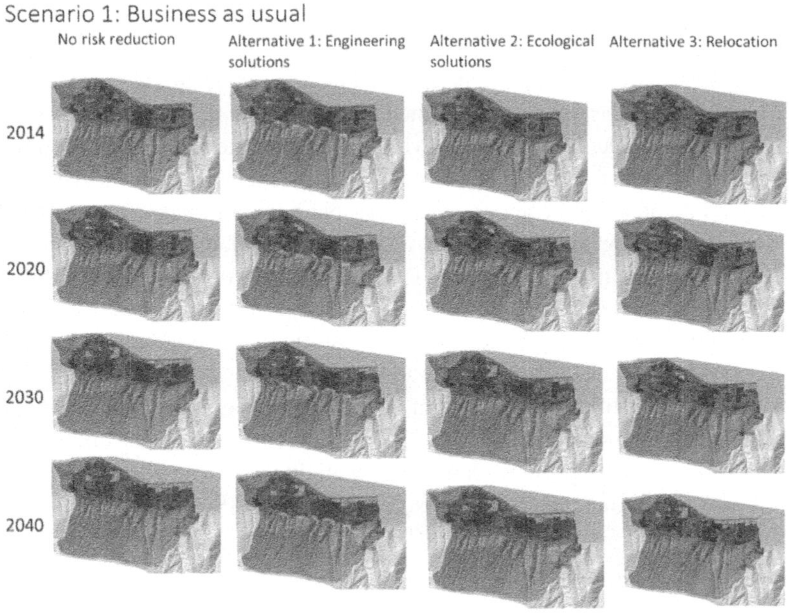

Figure 2.22 Land parcel maps for different years in a possible future scenario (Business as usual) without risk reduction measures (left) and three risk reduction alternatives.

Therefore, experts should evaluate together with the stakeholders what would be the effect of the proposed alternatives and scenarios on the hazard, elements-at-risk location and characteristics and the vulnerability for a given future year. If needed new hazard modelling should be carried out, or new elements-at-risk maps should be made representing the new situation.

2.8.3 Analyse risk reduction for alternatives/scenarios

After re-analysing the hazard, elements-at-risk and vulnerability for the specific combinations of planning alternative, possible future scenario and future year, the next step is to analyse the resulting level of risk, and compare this with the current risk level. The difference between the average annual losses before and after the implementation of the planning alternative, provides information on the risk reduction. This should be done for all the possible planning alternatives/scenario combinations. The risk reduction should be done preferable both in terms of economic risk reduction (reduction in the average annual losses in monetary values), as well as in population risk reduction (reduction in the expected casualties or exposed people).

2.8.4 Compare alternatives under different scenarios

Once the effect of the various planning alternatives has been analysed, under different future years and future scenarios, in terms of their risk reduction, the next step is to compare them and decide which of the alternatives would be the best to implement. In the cost-benefit analysis both the benefits and the costs can be quantified. The benefit of a risk reduction alternative is represented by its annual risk reduction in monetary values, which was calculated in the previous step (risk after implementation minus current risk). However, whereas the benefit would remain constant in the analysis which was presented earlier under "Analysing planning alternative", when the risk reduction of planning alternatives are analysed for different future years under possible change scenarios, the risk reduction might also change considerably over time. The costs for the planning alternative can be quantified, as well in terms of their investment costs, maintenance costs, project life time etc. Cost-benefit analysis can be carried out by calculating relevant indicators, such as the Net Present Value, Internal Rate of Return or Cost-Benefit ratio. When possible future changes are taken into account, the cost-benefit ratios of the various alternatives might be quite different than if no future changes are considered, which might lead to the selection of another planning alternative that may be the most "change proof" (See Figure 2.23).

2.8.5 Final decision and implementation

The last step of this workflow related to the selection of the optimal planning alternative in relation to the reduction of risk to hydro-meteorological hazards is the consultation with the various stakeholders involved. This includes public hearings with the population, private sector, non-governmental organizations, and various social network groups (e.g. communities, churches). The stakeholders have the opportunity to request adjustment to the proposed plan of action, and if these adjustments are considered valid, and substantial, a new round of evaluation might be needed if the change of expected hazard and risk impact is substantial. Once the plan is approval the procedures will start for the implementation of the plan.

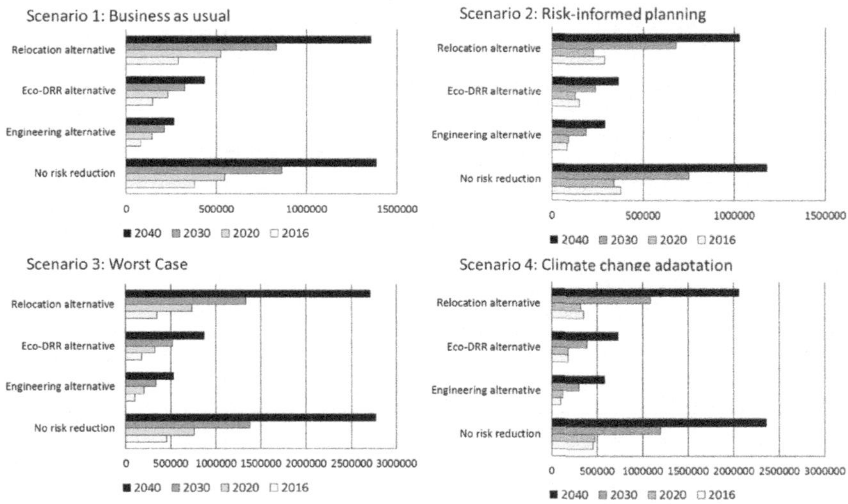

Figure 2.23 Changes in Average Annual Loss for four scenarios and three risk reduction alternatives for future years.

Here, the environmental impact assessment comes into play (Greiving, 2004). The recently amended Environmental Impact Assessment Directive (2014/52/EU) stats with Art 3 § 2: "Precautionary actions need to be taken for certain projects which, because of their vulnerability to major accidents, and/or natural disasters (such as flooding, sea level rise, or earthquakes) are likely to have significant adverse effects on the environment. For such projects, it is important to consider their vulnerability (exposure and resilience) to major accidents and/or disasters, the risk of those accidents and/or disasters occurring and the implications for the likelihood of significant adverse effects on the environment."

In this context, a broad involvement of the public is requested (see Art 6 § 2 Environmental Impact Assessment Directive and Art. 6 §§ 1–6 Strategic Environmental Assessment Directive) that needs to be adjusted to the different steps of analysis of a risk assessment as shown by Figure 2.24.

However, in many European countries, a legal basis for hazard (and partly risk) zoning exists which means that both, the methods chosen for delineating the zones, but also the legal consequences for subsequent planning activities are laid down by law. Hence, there is no discussion possible about suitable risk assessment alternatives. Table 2.9 gives an overview about existing zoning models and discusses their advantages and disadvantages.

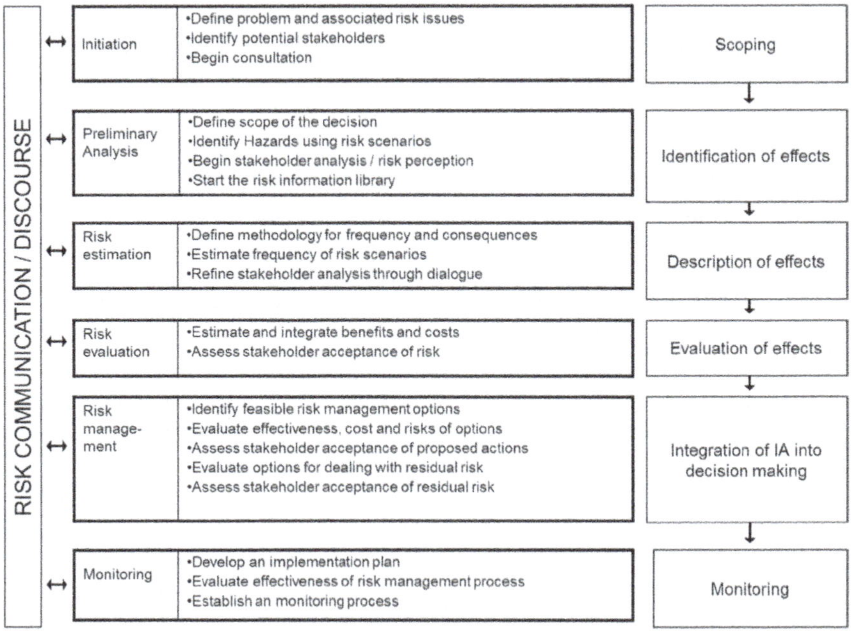

Figure 2.24 Integration of risk assessment into environmental assessments.

2.9 SUMMARY AND CONCLUSIONS

2.9.1 Which method to choose?

The four methods for risk assessment that were treated in the previous sections all have certain advantages and disadvantages, which are summarized in Table 2.8. The Quantitative Risk Assessment method is the best for evaluating several alternatives for risk reduction, through a comparative analysis of the risk before and after the implementation followed by a cost-benefit analysis. The event-tree analysis is the best approach for analysing complex chains of events and the associated probabilities. Qualitative methods for risk assessment are useful as an initial screening process to identify hazards and risks. They are also used when the assumed level of risk does not justify the time and effort of collecting the vast amount of data needed for a quantitative risk assessment, and where the possibility of obtaining numerical data is limited. The risk matrix approach is often the most practical approach as basis for spatial planning, where the effect of risk reduction methods can be seen as changes in the classes within the risk matrix. The indicator-based approach, finally, is the best when there is not enough data to carry out a quantitative analysis, but also as a follow-up of a quantitative analysis as it allows to take into account other aspects than just physical damage.

The decision depends among other factors in particular from the spatial scale of the project, plan or program, the risk assessment was done for (see Table 2.8). In this

Table 2.8 Advantages and disadvantages of the four risk assessment methods discussed.

Method	Advantages	Disadvantages	Suitability for Specific Spatial Scales
Quantitative risk assessment (QRA)	Provides quantitative risk information that can be used in Cost-benefit analysis of risk reduction measures.	Very data demanding. Difficult to quantify temporal probability, hazard intensity and vulnerability.	Normally used as basis for investments in structural mitigation measures on project level
Event-tree analysis	Allow modelling of a sequence of events, and works well for domino effects	The probabilities for the different nodes are difficult to assess, and spatial implementation is very difficult due to lack of data.	Normally used as basis for plan approval procedures of dangerous facilities (e.g. nuclear power plants, chemical establishments) on project level
Risk matrix approach	Allows to express risk using classes instead of exact values, and is a good basis for discussing risk reduction measures.	The method doesn't give quantitative values that can be used in cost-benefit analysis of risk reduction measures. The assessment of impacts and frequencies is difficult, and one area might have different combinations of impacts and frequencies.	Basis for hazard zoning in many countries like Austria France, Italy and Switzerland. Good fit for regional and local spatial planning as basis for keeping hazard prone areas free of further development
Indicator-based approach	Only method that allows to carry out a holistic risk assessment, including social, economic and environmental vulnerability and capacity.	The resulting risk is relative and doesn't provide information on actual expected losses.	Suitable for comparing the level of risk on national level (see e.g. World Risk Index)

Table 2.9 Hazard zoning models in Europe.

Model	**Coordinated Zoning in General Land-use Plan (Applied e.g. in Finland, Poland, Germany Except Floods)**	**Specific Hazard Map in General Land-use Plan with Binding Effects (Applied e.g. in Austria, France, Italy, Switzerland, Germany – Floods Only)**	**Independent Map without Binding Effects (Applied e.g. in Greece, Spain, and U.K.)**
Description	Consideration of the hazard prone areas during the compiling or review of the local land-use plan (informed i.e by Strategic Environmental Assessment)	The hazard zones are displayed as a separate map which has a direct effect on land ownership rights	Definition of hazard zones within the scope of expert planning – objections may be raised to decisions that are made on this basis
Advantages	At the local level, no additional instruments are needed; hazards are weighted- up against other concerns and interests	The hazard can be considered in a uniform manner for the whole municipality. Definition of hazard zones can be applied directly in building approval procedures	A simple alteration of a hazard zone plan is possible. Suitable for a cooperative strategy that aims at influencing existing building structures by individual building protection
Disadvantages	Land-use plans only contain information about hazard prone areas when a specific reference is made. An alternation of the danger situation means the plan must be adapted accordingly	An alteration of the risk means that the complete zoning plan has to be adapted accordingly. For legally binding effects, a very exact evidence basis is needed	Not effective if private stakeholders do not want to follow the advise

context, the subsidiarity principle plays a considerable role. Art. 5 § 2 of the Strategic Environmental Assessment Directive (2001/42/EC) lays down: "The environmental report [...] shall include [...] the level of detail in the plan or programme, its stage in the decision-making process and the extent to which certain matters are more appropriately assessed at different levels in that process in order to avoid duplication of the assessment." This means that a quantitative risk assessment would not fit to the scope of a strategic environmental assessment at the level of preparatory plans or programs, but should be used as evidence basis for an environmental impact assessment at the project level, as long as the project may be threatened by any kind of natural or technological hazard.

2.9.2 Tools for multi-hazard assessment

The analysis of risk requires a repetitive procedure which has to be carried out for each hazard scenarios (different hazard types and return periods) in combination with elements-at-risk types, and then also for each possible alternative. This requires the use of automated procedures using Geographic Information Systems.

Risk assessment is computationally intensive. It can be carried out using conventional GIS systems, although it is advisable to use specific software tools. Loss estimation has been carried out in the insurance sector since the late 1980's using geographic information systems. Since the end of the 1980's risk modelling has been developed by private companies resulting in a range of proprietary software models for catastrophe modelling for different types of hazards. Unfortunately, these are not publicly available, which is a major obstacle to the development of risk assessment for many parts of the world by government organizations. The best initiative for publicly available loss estimation thus far has been HAZUS developed by the Federal Emergency Management Agency (FEMA) together with the National Institute of Building Sciences (Schneider & Schauer, 2006). The first version of HAZUS was released in 1997 with a seismic loss estimation focus, and was extended to multi-hazard losses in 2004, incorporating also losses from floods and windstorms. HAZUS was developed as a software tool under ArcGIS. Several other countries have adapted the HAZUS methodology to their own situation. The HAZUS methodology has also been the basis for the development of several other software tools for loss estimation. One of these is called SELENA. Also an interesting example is RiskScape developed in New Zealand (Schmidt *et al.* 2011).

Another interesting development has been going on in the development of standalone software modules for multi-hazard risk assessment, which are not running as a component of an existing GIS. A good example of this is the CAPRA Probabilistic Risk Assessment Program supported by the World Bank (CAPRA, 2013). The methodology focuses on the development of probabilistic hazard assessment modules, for earthquakes, hurricanes, extreme rainfall, and volcanic hazards, and the hazards triggered by them, such as flooding, windstorms, landslides and tsunamis.

Another recent development is towards Open Source web-based modules for multi-hazard risk assessment.

A tool which is currently under development as part of the Global Earthquake Initiative (GEM), called OpenQuake (https://www.globalquakemodel.org/openquake/support/documentation/), is most probably going to be the standard for earthquake loss estimation, as there are also plans to expand it into a multi-hazard risk assessment tool.

2.9.3 Development of a spatial decision support system

Within the framework of the EU FP7 Marie Curie Project CHANGES (www.changes-itn.eu) and the EU FP7 Copernicus project INCREO (http://www.increo-fp7.eu) a spatial decision support system was developed with the aim to analyse the effect of risk reduction planning alternatives on reducing the risk now and in the future, and support decision makers in selecting the best alternatives. The Spatial Decision Support System is composed of a number of integrated components. The Risk Assessment component allows to carry out spatial risk analysis, with different degrees of complexity, ranging from simple exposure (overlay of hazard and assets maps) to quantitative analysis (using different hazard types, temporal scenarios and vulnerability curves) resulting into risk curves. The platform does not include a component to calculate hazard maps, and existing hazard maps are used as input data for the risk component. The second component of the SDSS is a risk reduction planning component, which forms the core of the platform. This component includes the definition of risk reduction alternatives (related to disaster response planning, risk reduction measures and spatial planning) and links back to the risk assessment module to calculate the new level of risk if the measure is implemented, and a cost-benefit (or cost-effectiveness/Spatial Multi Criteria Evaluation) component to compare the alternatives and make decision on the optimal one. The third component of the SDSS is a temporal scenario component, which allows to define future scenarios in terms of climate change, land use change and population change, and the time periods for which these scenarios will be made. The component doesn't generate these scenarios but uses input maps for the effect of the scenarios on the hazard and assets maps. The last component is a communication and visualization component, which can compare scenarios and alternatives, not only in the form of maps, but also in other forms (risk curves, tables, graphs). The envisaged users of the platform are organizations involved in planning of risk reduction measures, and that have staff capable of visualizing and analysing spatial data at a municipal scale. The Decision Supper System RiskChanges is accessible at: http://www.charim.net/use_case/46.

ACKNOWLEDGEMENTS

The following persons have developed the original hazard maps: Leonardo Cascini, Settimio Ferlisi, Sabatino Cuomo, and Giovanna De Chiara. We also would like to thank Anna Scolobig from IIASA for her work on the risk reduction alternatives (which we have taken as they were) the stakeholder involvement and the stakeholder roleplay exercise. Hari Narasimhan (ETH) and Emile Dopheide are thanked for their input in the cost-benefit analysis. Also we would like to thank Andrea Tripodi for his work in the development of the case study. Luc Boerboom and Ziga Malek are thanked for their input in the thinking about possible future scenarios. Kaixi Zhang is thanked for her feedback on the risk calculation method. Vera Andrejchenko, Julian Berlin, Irina Cristal, Kaixi Zhang and Roya Olyazadeh are thanked for developed the RiskChanges Spatial Decision Support System. Funding was provided through the EU FP7 Marie Curie Project CHANGES (www.changes-itn.eu) the EU FP7 Copernicus project INCREO (http://www.increo-fp7.eu), and the World Bank GFDRR project CHARIM (http://www.charim.net)

REFERENCES

ACOE (ARMY CORPS OF ENGINEERS) (2004). Water Resources Assessment of Dominica, Antigua, Barbuda, St. Kitts and Nevis: Mobile District and Topographic Engineering Center Report, U.S. Army Corps of Engineers, Mobile, AL, p. 94.

Alexander D. (2001). Encyclopedia of Environmental Science, Chapter Natural Hazards. Kluwer Academic Publishers.

Bedford T. and Cooke R. M. (2001). Probabilistic Risk Analysis: Foundations and Methods. Cambridge University Press, Cambridge, UK.

Birkmann J. Ed. (2006). Measuring Vulnerability to Natural Hazards: Towards Disaster Resilient Societies. UNU Press, Tokyo.

Birkmann J. (2007). Risk and vulnerability indicators at different scales: applicability, usefulness and policy implications. *Environmental Hazards*, **7**, 20–31.

Bollin C. and Hidajat R. (2006). Community-based disaster risk index: pilot implementation in Indonesia. In: Measuring Vulnerability to Natural Hazards—Towards Disaster Resilient Societies, J. Birkmann (ed.), UNU-Press, Tokyo.

Breheny P. (2007). Hydrogeologic/Hydrological Investigation of Landslide Dam and Impounded Lake in the Matthieu River Valley, Commonwealth of Dominica, West Indies. Unpublished M.S. Dissertation, University of Leeds, United Kingdom.

Cannon S. and DeGraff J. (2009). The increasing wildfire and post-fire debris-flow threat in Western USA, and implications for consequences of climate change. In: Landslides – Disaster Risk Reduction, K. Sassa and P. Canuti (eds), Springer Verlag, Berlin, pp. 177–190.

CAPRA (2013). Probabilistic Risk Assessment Program. http://www.ecapra.org/.

Carpignano A., Golia E., Di Mauro C., Bouchon S. and Nordvik J.-P. (2009). A methodological approach for the definition of multi-risk maps at regional level: first application. *Journal of Risk Research*, **12**, 513–534.

Coppock J. T. (1995). GIS and natural hazards: an overview from a GIS perspective. In: Geographical Information Systems in Assessing Natural Hazards, A. Carrara and F. Guzzetti (eds), Volume 5 of the series Advances in Natural and Technological Hazards Research, Kluwer, Dordrecht, pp. 21–34.

Corominas J., van Westen C. J., Frattini P., Cascini L., Malet J.-P., Fotopoulou S., Catani F., van den Eeckhaut M., Mavrouli O., Agliardi F., Pitilakis K., Winter M. G., Pastor M., Ferlisi S., Tofani V., Hervas J. and Smith J. T. (2014). Recommendations for the quantitative analysis of landslide risk: open access. *Bulletin of Engineering Geology and the Environment IAEG*, **73**(2), 209–263.

Cova T. J. (1999). GIS in emergency management. In: Geographical Information Systems: Principles, Techniques, Applications, P. A. Longley (ed.), 2nd edn, Management Issues and Applications.

DeGraff J. V. (2012). Solving the dilemma of transforming landslide hazard maps into effective policy and regulations. *Natural Hazards Earth System Sciences*, **12**, 53–60.

DeGraff J. V. and Rogers C. T. (2003). An unusual landslide dam event in Dominica, West Indies. *Landslide News*, (14/15), 8–11.

DeGraff J. V., James A. and Breheney P. (2010). The formation and persistence of the Matthieu Landslide-Dam Lake, Dominica, W.I. *Environmental & Engineering Geoscience*, **16**(2), 73–89.

Delmonaco G., Margottini C. and Spizzichino D. (2006a). ARMONIA Methodology for Multi-Risk Assessment and the Harmonisation of Different Natural Risk Maps. Deliverable 3.1.1, ARMONIA.

Delmonaco G., Margottini C. and Spizzichino D. (2006b). Report on New Methodology for Multi-Risk Assessment and the Harmonisation of Different Natural Risk Maps. Deliverable 3.1, ARMONIA.

De Pippo T., Donadio C., Pennetta M., Petrosino C., Terlizzi F. and Valente A. (2008). Coastal hazard assessment and mapping in Northern Campania, Italy. *Geomorphology*, **97**, 451–466.

EC (2011). Risk Assessment and Mapping Guidelines for Disaster Management. European Commission Commission staff working paper, European Union.

Fan X., van Westen C. J., Xu Q., Gorum T. and Dai F. (2012). Analysis of landslide dams induced by the 2008 Wenchuan earthquake. *Journal of Asian Earth Sciences*, **57**, 25–37.

FEMA (2004). HAZUS-MH. FEMA's Methodology for Estimating Potential Losses from Disasters. US Federal Emergency Management Agency. http://www.fema.gov/plan/prevent/hazus/index.shtm.

Ferlisi S., De Chiara G. and Cascini L. (2016). Quantitative risk analysis for hyperconcentrated flows in Nocera Inferiore (Southern Italy). Submitted to Natural Hazards—special issue: landslide risk Nocera Inferiore. *Nat Hazards*, **81**, S89–S115.

Garcia-Aristizabal A. and Marzocchi W. (2013). Software for Multi-Hazard Assessment. Deliverable 3.5 of the EU Matrix Project: New Methodologies for Multi-Hazard and Multi-Risk Assessment Methods for Europe. http://matrix.gpi.kit.edu/downloads/MATRIX-D3.05.pdf.

Glade, T. and Alexander, D.E. (2013). Classification of natural disasters. In: Encyclopedia of Natural Hazards, Bobrowsky, P.T. (ed.), Dordrecht, Netherlands, Springer.

Granger K., Jones T. G., Leiba M. and Scott G. (1999). Community Risk in Cairns: A Multi-Hazards Risk Assessment. Technical Report, Australian Geological Survey Organisation (AGSO). http://www.ga.gov.au/corporate_data/33548/33548.pdf.

Greiving S. (2004). Risk assessment and management as an important tool for the EU Strategic Environmental Assessment. *DISP*, **157**, 11–17.

Greiving S., Fleischhauer M. and Lückenkötter J. (2006). A Methodology for an integrated risk assessment of spatially relevant hazards. *Journal of Environmental Planning and Management*, **49**(1), 1–19.

Guha-Sapir D., Below R. and Hoyois P. H. (2016). EM-DAT: The CRED/OFDA International Disaster Database. www.emdat.be. Université Catholique de Louvain, Brussels, Belgium.

Haimes Y. Y. (2009) Risk Modeling, Assessment, and Management. 3rd edn. John Wiley & Sons, Hoboken, New Jersey, p. 1009.

Ho K. (2009). Risk Acceptability Criteria. Lecture Held at the International LARAM School. University of Salerno, Salerno, Italy.

IRDR (2014). Integrated Research on Disaster Risk. Peril Classification and Hazard Glossary (IRDR DATA Publication No. 1). Integrated Research on Disaster Risk, Beijing. http://www.irdrinternational.org/wp-content/uploads/2014/04/IRDR_DATA-Project-Report-No.-1.pdf.

ISO 31000 (2015). Risk Management – Principles and Guidelines. http://www.iso.org/iso/home/standards/iso31000.htm.

Jaboyedoff M., Aye Z. C., Derron M.-H., Nicolet P. and Olyazadeh R. (2014). Using the consequence – frequency matrix to reduce the risk: examples and teaching. In International Conference Analysis and Management of Changing Risks for Natural Hazards, 18–19 November 2014, Padua, Italy.

Jonkman S. N., Van Gelder P. and Vrijling H. (2002). An overview of quantitative risk measures and their application for calculation of flood risk. In: λμ13 – ESREL 2002 European Conference.

Kappes M. S., Keiler M., Von Elverfeldt K. and Glade T. (2012). Challenges of analyzing multi-hazard eisk: a review. *Natural Hazards*, **64**(2), 1925–1958.

Lacasse S., Eidsvik U., Nadim F., Hoeg K. and Blikra L. H. (2008). Event tree analysis of Aknes rock slide hazard. In: IV Geohazards Quebec, 4th Canadian Conf. on Geohazards, pp. 551–557.

Lee K. and Rosowsky D. (2006). Fragility analysis of woodframe buildings considering combined snow and earthquake loading. *Structural Safety*, **28**, 289–303.

Linnerooth-Bayer J. and Pratt A. (2016). Introduction to the special issue on rethinking participatory processes: the case of landslide risk in Nocera Inferiore. *Nature Hazards*, **81**, 1–6.

Linnerooth-Bayer J., Scolobig A., Ferlisi S., Cascini L. and Thompson M. (2015). Expert engagement in participatory processes: translating stakeholder discourses into policy options. Submitted to Natural Hazards—special issue: landslide risk Nocera Inferiore. *Nature Hazards*, **81**, 69–88.

Luino F. (2005). Sequence of instability processes triggered by heavy rainfall in the Northern Italy. *Geomorphology*, **66**, 13–39.

Marzocchi W., Mastellone M. and Di Ruocco A. (2009). Principles of Multi-Risk Assessment: Interactions Amongst Natural and Man-Induced Risks. European Commission. http://cordis.europa.eu/documents/documentlibrary/106097581EN6.pdf.

Narasimhan H., Ferlisi S., Cascini L., De Chiara G. and Faber M. H. (2016). A cost–benefit analysis of mitigation options for optimal management of risks posed by flow-like phenomena. *Nat Hazards*, **81**, S117–S144.

OAS (1991) Geographic Information Systems in Natural Hazard Management. Organization of American States. Primer on Natural Hazard Management in Integrated Regional Development Planning. https://www.oas.org/dsd/publications/Unit/oea66e/ch05.htm.

Peila D. and Guardini C. (2008). Use of the event tree to assess the risk reduction obtained from rockfall protection devices. *Natural Hazards Earth System Sciences*, **8**, 1441–1450.

Perles Roselló M. and Cantarero Prados F. (2010). Problems and challenges in analyzing multiple territorial risks: methodological proposals for multi-hazard mapping. *Boletín de la Asociación de Geógrafos Espanoles*, **52**, 399–404.

Prinos P. (2008). Review of Flood Hazard Mapping. European Community Sixth Framework Programme for European Research and Technological Development. FLOODsite.

Roberts N. J., Nadim F. and Kalsnes B. (2009). Quantification of vulnerability to natural hazards. *Georisk: Assessment and Management of Risk for Engineered Systems and Geohazards*, **3**(3), 164–173.

Roobol M. J. and Smith A. L. (2004). Geologic Map of Dominica, West Indies: Geology Department, University of Puerto Rico at Mayaguez. Electronic document, Available at http://www.caribbeanvolcanoes.com/dominica/content/dominicamap.pdf.

SafeLand (2011a) Deliverable 2.11 QRA Case Studies at Selected "hotspots". Synthesis of critical issues. Edited for the EU-funded SafeLand Project by Corominas J and Mavrouli O. https://www.ngi.no/eng/Projects/SafeLand.

SafeLand (2011b) Deliverable 5.7 Design and Testing: A Risk Communication Strategy and a Deliberative Process for Choosing a Set of Mitigation and Prevention Measures. Edited for the EU-funded SafeLand Project by Scolobig A and Bayer J. https://www.ngi.no/eng/Projects/SafeLand.

Schmidt J., Matcham I., Reese S., King A., Bell R., Henderson R., Smart G., Cousins J., Smith W. and Heron D. (2011). Quantitative multi-risk analysis for natural hazards: a framework for multi-risk modelling. *Natural Hazards*, **58**(3), 1169–1192.

Schneider P. J. and Schauer B. A. (2006). HAZUS, its development and its future. *Natural Hazards Review*, **7**(2), 40–44.

Shi P., Shuai J., Chen W. and Lu L. (2010). Study on the risk assessment and risk transfer mode of large scale disasters. In: The 3rd International Disaster and Risk Conference IDRC, Davos, Switzerland.

Talbot, J. (2016). What's Right with Risk Matrices? http://www.jakeman.com.au/media/whats-right-with-risk-matrices.

Tarvainen T., Jarva J. and Greiving S. (2006). Spatial pattern of hazards and hazard interactions in Europe. In: Natural and Technological Hazards and Risks Affecting the Spatial Development of European Regions, P. Schmidt-Thomé (ed.), vol. **42**, pp. 83–91.

UN-ISDR (2009). Terminology on Disaster Risk Reduction. https://www.unisdr.org/we/inform/terminology.

Van Westen C. J. (2013). Remote sensing and GIS for natural hazards assessment and disaster risk management. In: Treatise on Geomorphology, J. Shroder and M. P. Bishop (eds), vol. 3, Academic Press, San Diego, CA, Remote Sensing and GIScience in Geomorphology, pp. 259–298.

Van Westen C. J., Castellanos Abella E. A. and Sekhar L. K. (2008). Spatial data for landslide susceptibility, hazards and vulnerability assessment: an overview. *Engineering Geology*, **102**(3–4), 112–131.

Van Westen C. J., Quan Luna B. and Vargas Franco R. D. (2010). Development of training materials on the use of geo – information for multi – hazard risk assessment in a mountainous environment. In: J.-P. Malet, T. Glade and N. Casagli (eds), Mountain Risks: Bringing Science to Society : Proceedings of the Mountain Risks International Conference, 24–26 November 2010, Firenze, Italy.CERG, Strasbourg. ISBN 2-95183317-1-5. pp. 469–475.

Van Westen C. J., Straatsma M. W., Turdukulov U. D., Feringa W. F., Sijmons K., Bakhtadze K., Janelidze T. and Kheladze N. (2012). Atlas of Natural Hazards and Risks of Georgia: E-book. Caucasus Environmental NGO Network (CENN), University of Twente Faculty of Geo-Information and Earth Observation (ITC), Tbilisi. ISBN: 978-9941-0-4310-9.

Zuccaro G. and Leone M. (2011). Volcanic crisis management and mitigation strategies: a multi-risk framework case study. *Earthzine*, https://earthzine.org/2011/03/21/volcanic-crisis-management-and-mitigation-strategies-a-multi-risk-framework-case-study/.

Part 2

Hydrometeorological Hazards Methodologies

Chapter 3

Storms

Terry W. Krauss and Nicolas R. Dalezios

3.1 STORM CONCEPTS

A towering cumulus cloud with strong updrafts and downdrafts, accompanied by lightning and thunder, is called a thunderstorm. It is driven by convection, in which moist warm air rises aloft within a cooler environment. Thunderstorms may be accompanied by various conditions including high winds, heavy rain, hail, and at times tornadoes. A thunderstorm is generally called "severe" if it meets one of the following criteria: hailstones greater than approximately 2.5 cm diameter, winds gusting in excess of 90 km per hour (50 knots), or formation of a tornado or damaging winds. Each convective storm cell goes through a life cycle that could be divided into three stages: (1) The cell is dominated by an updraft and called the "cumulus" stage. (2) As the storm strengthens, a downdraft is developed adjacent to the updraft, and this is referred to as the "mature" stage. (3) After about 20–30 min, the mature stage is transitioned into a final "dissipating" stage containing only a downdraft.

The thunderstorm is variously referred to as a "convective storm", "convective cell", or "convection", since it is created by a violent upsurge of warm, buoyant, convective air currents from the lower atmosphere to altitudes as great as around 15 kilometers above mean sea level (MSL). The intense, buoyantly forced rising motions in the thunderstorm, generated by the temperature difference between the warm cloud and the relatively cold surrounding environment, namely convective instability, produce thick cloud and heavy precipitation. Adopting the definition of "mesoscale" proposed by Orlanski (1975), individual thunderstorms are typically meso-γ-scale in horizontal dimension (2–20 km), whereas an organized mesoscale convective system (MCS) is typically meso-β-scale (20–200 km) or larger in width.

The term "mesoscale" refers to a size or scale, which is equivalent to the distance spanned by a concentrated group of storms. Small-scale storms have horizontal scales of the order of tens of kilometers and exhibit lifetimes of an hour or two. For these smaller storms the effects of the Earth's rotation are not very important. Any rotation that develops in the individual thunderstorms is primarily a result of tilting of horizontally-oriented vorticity in the form of vertical shear of the horizontal wind into the vertical direction. MCSs, whose lifetimes are of the order of many hours or days, however, are strongly influenced by the rotation of the Earth, because air drawn into the storms comes from a large area in the surrounding region.

3.2 CLASSIFICATION OF STORMS

Most environmental hazards are atmospheric in origin. Indeed, the extremes of most individual atmospheric elements or events can constitute a hazard. There are many valid ways to classify storms. For instance, according to a meteorological approach, atmospheric phenomena are often differentiated based on their scale, i.e. microscale, mesoscale and synoptic scale. Similarly, a natural hazards approach, also called a geophysical process, typically focuses either on particular phenomena, such as hurricanes, floods, or blizzards, or on decision processes of various aspects, such as warning and response, risk assessment, or mitigation. Ideally, a combination of these approaches can be adopted. However, the most severe hazards originate from compound or synergistic effects, such as blizzards, which are produced when wind combines with snow. Table 3.1 presents an indicative list of single and combined atmospheric elements or events. It is also true that atmospheric-related hazards originate from a variety of causes and occur on a large range of scales. Typical examples are tornadoes, which are microscale phenomena with life cycles of the order of hours, whereas tropical cyclones are mesoscale events, which are considered part of the global circulation of the atmosphere lasting usually for days. Similarly, large-scale hazards, such as excessive seasonal rainfall, can last for several months and significantly affect the circulation of the atmosphere.

Table 3.1 Indicative list of single and combined atmospheric events or elements.

Atmospheric Events Single Events	Combined Events
Excess rainfall	Hurricanes
Freezing rain ("glaze")	"Glaze" storms
Hail	Thunderstorms
Heavy snowfalls	Blizzards
High windspeeds	Tornadoes
Extreme temperatures	Heat/cold stress

Convective storms can be classified as isolated storms or MCSs. Indeed, large-scale storms are differentiated by their physical characteristics into tropical cyclones, extratropical cyclones and MCSs. The thunderstorm is the basic element or "building block" of the MCS. MCSs include squall lines and mesoscale convective complexes (MCC). It is also known that many types of rainstorms and hailstorms exist. Specific types of isolated storms include single-cell storms, multicell storms, and supercell storms.

3.2.1 Single-cell, multi-cell, supercell and squall lines

Single-cell storms represent small rain showers and thunderstorms typically covering an area of a few kilometres or less and having a lifetime of less than an hour. Storms that form in environments with low wind shear have patterns of propagation that are often quite irregular and seemingly random. These storms involve a single major updraft, which grows and dies. Precipitation grows in the updraft until it becomes sufficient to overcome the upward motion, then falls out, creating a downdraft and an outflow of cold air, which spreads out and cuts off the warm moist environmental air from the updraft, which then decays. These storms do not process large amounts of air and normally do not individually represent a significant flooding hazard.

Multi-cell storms consist of several convective cells or single-cell storms at different stages in their lifetime, within the same complex and possibly including supercells (Figure 3.1a). Thus, the multi-cell system may have a lifetime much longer than that of any individual cell. The areal extent of the system may be much larger than that of a single cell and thus may cover a larger portion of the drainage area for a small watershed. Multi-cell systems may interact with their environment through the low-level inflow of moist air and the wind shear, i.e. the change in the horizontal wind speed and direction with height, so that development of new cells occurs systematically in a direction that allows free access to fresh, warm, moist unstable inflow air. Cell movement is largely determined by the winds in the mid-cloud layers causing the precipitation to fall out some distance away from new development. Multi-cell storms, thus, can be much larger than single-cell storms. They may also become more vigorous. They may organize in such a way as to protect mature cells from the effects of wind shear, which tends to tilt the updraft over and to disconnect it from its lower-level supply of moisture. Multi-cells are also able to take advantage of the exhaust-like effects of strong winds aloft to strengthen the updraft. These storms may process much more air mass than single-cell storms.

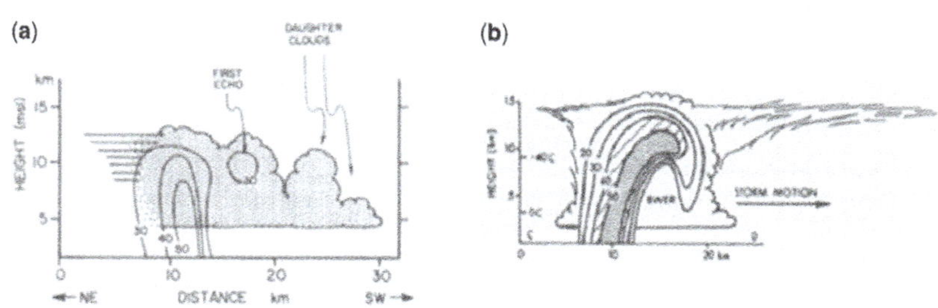

Figure 3.1 Schematic vertical cross-section of (**a**) multi-cell storm (from Dennis *et al.* 1970) and (**b**) supercell (from Chisholm & Renick, 1972).

Supercell storms occur under conditions of strong environmental wind shear in both direction and speed, and strong instability. In this case, a very large single rotating cell may develop with a single main updraft 5 km or more in diameter with updraft speeds in excess of 50 m s^{-1}. The change in wind direction and speed with height helps to separate the cloud updraft from the downdraft, which leads to longer cell lifetimes of the order of several hours. Supercells are large and intense and process vast amounts of air and moisture. Supercells often occur in conditions with strong winds aloft and so may move rapidly. Frontal boundaries, gust fronts or topography may also alter storm movement. The great strength of the updraft in supercells may cause most of the moisture to be transported to high altitudes and out into the storm anvil. However, if the low-level inflow air is rich in moisture, very large volumes of rainfall may be produced. Known as quasi-steady-state storms, they are the most severe type of thunderstorms. Supercell storms, unlike multi-cell storms, generally feature a single, large, rotating updraft zone (see Figure 3.1b). They have extremely high damage potential, as they can cause not only strong wind gusts and heavy precipitation but also large hailstones and tornados.

Ordinary squall lines. MCSs organize into a variety of configurations, the easiest to recognize being the squall line, which consists of thunderstorm cells arranged along a line that could extend up to several hundred kilometers in length. Squall-line thunderstorm systems occur throughout the tropics and mid-latitudes. At the surface there is normally a sharp roll-like line of clouds followed by a sudden wind squall or gust of 12 to 25 m/s, where behind the surface squall a heavy downpour starts, which may produce as much as 30 mm of rain in 30 minutes in the tropics. An ordinary squall line is composed of two scales of motion: (1) the cumulus scale, having a horizontal dimension on the order of 2 to 10 km, and (2) the mesoscale, characterized by air motions on a scale of 20 to 200 km. The precipitation from squall lines reflects the presence of these two scales of motion, as about 60 per cent of the precipitation is in the form of intense showers in the convective region and 40 per cent is in the form of steady, stratiform rainfall from the trailing stratiform anvil region. Tropical squall lines are somewhat bow-shaped rather than being perfectly straight lines. As a result, the low-level surface squall travels faster than the upper-level cloud debris. Figure 3.2a illustrates a low-level radar depiction of an ordinary tropical squall line. A vertical cross-section through the squall line is illustrated in Figure 3.2b. This shows the strong convective updrafts with very high radar reflectivities along the leading edge with pronounced flow from the front of the storm at low levels toward the rear of the storm at middle and upper levels.

Figure 3.3a illustrates a tropical squall line with a stratiform anvil that extends well in advance of the convective line. It appears that such a leading stratiform region develops when the vertical shear of the horizontal wind is large at upper levels, which is a more common feature of middle-latitude squall lines. Indeed, a feature that distinguishes middle-latitude, continental squall lines from their tropical, oceanic counterparts is the strength of the convective-scale updrafts and downdrafts. The middle-latitude squall line shown in Figure 3.3b exhibits a leading line of thunderstorms with a pronounced trailing anvil region. As a result, downdrafts are stronger and extend over greater depths. Also, due to the stronger surface heating over the interior continental regions and stronger wind shears in middle latitudes, convective updrafts in squall-line storms can be very intense, being as high as 25 to 40 m/s. This means that the mass transports and, resultant flow structures of middle-latitude, continental squall lines are more strongly influenced by thunderstorm-scale drafts and flow structures than the corresponding tropical squall lines.

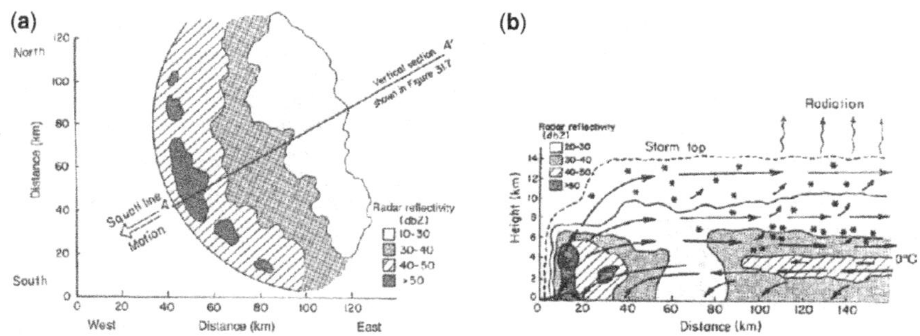

Figure 3.2 Radar reflectivities of tropical squall line (**a**) horizontal (PPI), (**b**) vertical (RHI).

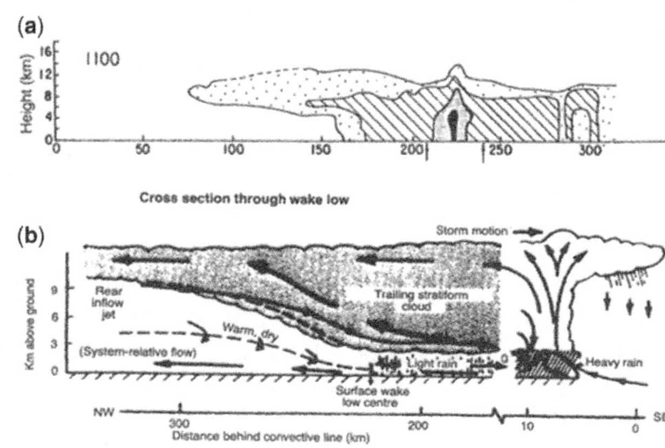

Figure 3.3 Vertical structure of squall system radar echoes (**a**) tropical, (**b**) middle-latitude.

Pre-frontal squall lines. The largest and most violent squall lines are the pre-frontal lines that form in middle latitudes. Typically, they form along, or ahead of, a cold front associated with a vigorous, mid-latitude cyclone. Often, the line forms in the warm sector of the cyclone, where the low-level jet brings warm, moist air into the region ahead of the cold front. A schematic cross-section perpendicular to a severe pre-frontal squall line is shown in Figure 3.4. The flow structure is similar to the ordinary squall line shown in Figure 3.3b, except that the storm is deeper and the updraft and downdraft strengths are much greater. Due to the strong shear in the pre-frontal air mass, the pre-frontal squall line exhibits a more pronounced leading stratiform-anvil cloud than the ordinary squall line. At low levels, a layer of warm, moist air feeds the thunderstorm updrafts. Intense updrafts extend through much of the troposphere and even penetrate to great heights in the lower stratosphere. Some of the updraft air encounters large quantities of precipitation.

Figure 3.4 Vertical structure of pre-frontal squall line.

Derecho. A "derecho" is an extremely violent type of squall line. A ***derecho*** (pronounced "deh-REY-cho" in English) is a widespread, long-lived wind storm. Derechos are associated with bands of rapidly moving showers or thunderstorms generally known as squall lines, or quasi-linear convective systems. By definition, if the swath of wind damage extends for more than 240 miles (about 400 kilometers), includes wind gusts of at least 58 mph (93 km/h) along most of its length, and several, well-separated 75 mph (121 km/h) or greater gusts, then the event may be classified as a derecho. Although a derecho can produce destruction similar to that of a tornado, the damage typically occurs in one direction along a relatively straight path. As a result, the term "straight-line wind damage" sometimes is used to describe derecho damage.

3.2.2 Mesoscale convective systems (MCS)

An MCS is defined as a complex of thunderstorms that becomes organized on a scale larger than the individual thunderstorms, but smaller than extratropical cyclones and normally persists for several hours or more. The overall cloud and precipitation pattern of an MCS may be round or linear in shape, and includes weather systems, such as tropical cyclones, squall lines, lake-effect snow events, polar lows and MCCs, and generally form near weather fronts (Maddox, 1980). MCSs are usually characterized by persistency and widespread convective activity. MCSs appear in many forms, ranging from a relatively disorganized mass of convective cells, to highly organized convective lines. It is well known that MCSs often produce severe weather that over half of the annual warm season precipitation is associated with MCSs, and that MCSs are often prolific producers of cloud-to-ground lightning. If the net direction of movement of the system is very slow, then continuous heavy rains over a period of a few hours may result and create a serious flash flooding potential.

Figure 3.5 shows a conceptual model of the structure of an MCS as viewed by radar in horizontal (PPI: Plan Position Indicator, left column) and vertical (RHI: Range Height Indicator, right column) sections, respectively. Indeed, Figure 3.5 presents the life cycle of an MCS, namely (a) initial, (b) organizing, (c) mature, and (d) dissipating stages, respectively. This is a tropical, oceanic MCS, which shares several key morphological features of the middle-latitude, continental MCSs. Specifically, the hatched and solid filled areas of panels at left identify regions of light, moderate and heavy precipitation, respectively. Moreover, the outside contour of panels at right corresponds to the weakest detectable radar reflectivity echo, whereas the inner contours correspond to successively higher radar reflectivity or precipitation content values. Heavy arrows in panels at left indicate the direction of low-level winds, whereas arrows in panels at right indicate the low- and upper-level wind directions.

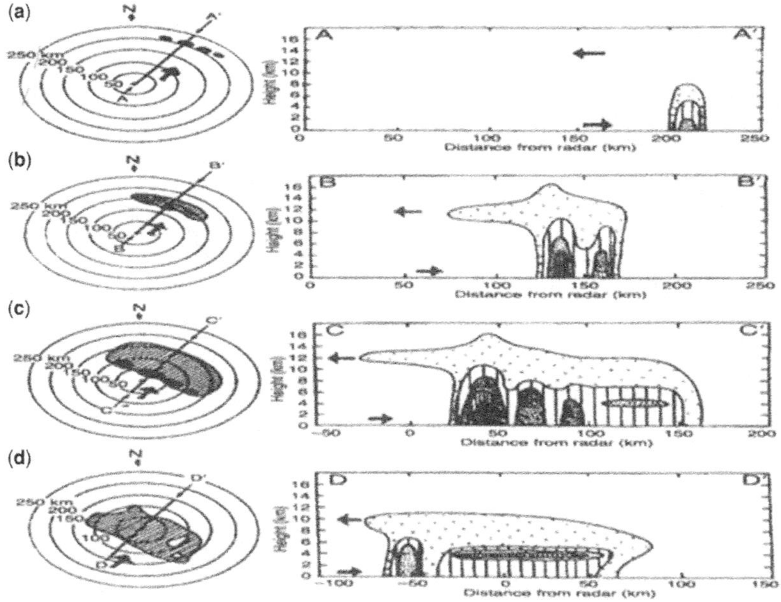

Figure 3.5 Horizontal (PPI) and Vertical (RHI) Structure of Mesoscale Convective System (MCS) from radar.

Genesis of MCSs. The genesis of MCSs requires sustained forcing from the larger-scale environment in order to trigger their formation. Generally, the low-level convergence provides the fuel necessary to support the initiation of a number of cumulonimbus clouds in close proximity. Some sources of low-level convergence are land- and sea-breeze circulations over tropical islands and peninsulas, mountain slope circulations, especially over large mountain ridges, convergence along large-scale, frontal boundaries in middle latitudes, and convergence along the intertropical convergence zone (ITCZ) near the equator and in tropical easterly waves. One of the first signs of the initiation of an MCS is the merger of the anvil clouds emitted by the neighboring thunderstorms. At the same time, a cold pool of air forms near the surface due to cooling of the air by evaporation and melting of precipitation. Both processes appear to be important in the genesis of MCSs.

Cloud Clusters. Cloud clusters are mesoscale systems, where the convection is organized rather randomly. The larger cloud clusters may have squall lines and other thunderstorm units as part of the overall composition of the system. In some tropical clusters, lines of convection may be prevalent, but the orientation of the lines may be parallel to the wind shear vector rather than perpendicular to it as in most of lines. Cloud clusters are quite common in the tropics and are largely responsible for producing the bulk of tropical rainfall. They are particularly common in the ITCZ near the equator, which shifts north and south of the equator depending on season. Some tropical clusters eventually develop into tropical cyclones or hurricanes, which are essentially indistinguishable from the more common clusters that have a much shorter life cycle.

Mesoscale Convective Complex (MCC). Sometimes MCSs grow to such a size that they develop large-scale internal dynamic circulations. An MCC is a unique subset of MCSs, which is defined in terms of the scale and duration of the storm as identified by characteristics observed in infrared satellite imagery (Maddox, 1980). These large-scale MCCs may exist in a semi-balanced state for many hours, or even days. By definition, these storms are large, i.e. greater than 50,000 km^2 in area, and long lasting, i.e. greater than six hours at mature size, with temperature less than or equal to –52°C. These storms contain convective part, often a squall line, which can produce heavy rainfall rates, and a large region of more modest mesoscale ascent, which may produce moderate rain rates over many hours. MCCs tend to occur nocturnally, as late afternoon severe-weather-producing systems move into areas having air masses less conducive to violent weather. Its maximum extent is defined as when the cloud shield, or the overall cloud formation, reaches its maximum area. Its eccentricity (minor axis/major axis) is greater than or equal to 0.7 at maximum extent, so they are fairly round. Like squall lines, an MCC is composed of both convective updrafts and slowly ascending motion. The life cycle of MCCs is such that early in their development, thunderstorm-scale motions prevail. The rainfall is quite showery and there is a great likelihood that the system will produce severe weather, such as tornadoes, hail and flash floods. As the system reaches maturity and produces a stratiform-anvil cloud of maximum horizontal extent, the rainfall becomes light and steady, or more stratiform in character.

Mediterranean Tropical Like Cyclones (TLC) known as Medicanes are meso-scale extreme low pressure systems, resembling the structure of tropical cyclones, as captured by satellites. Their intensity appears normally much weaker than tropical hurricanes. Their genesis is triggered when an upper-level cut-off low is advected over an area, resulting in air mass lifting and cooling causing convective instability (Emanuel, 2005). These meso-scale systems with diameter usually less than 300 km have a rounded structure and a warm core, as well as intense low sea level pressure. Strong winds, heavy precipitation and thunderstorms are associated with the incidence of medicanes, causing occasional severe damages, or resulting in flooding of populated areas, posing a risk to human life (Nastos *et al.* 2015).

3.2.3 Tropical and extratropical cyclones

3.2.3.1 Tropical cyclones

Tropical cyclones are a special class of large-scale rotating wind systems, which occur over sizable portions of the tropical and subtropical oceans. Tropical systems derive their energy from clusters of thunderstorms. Tropical systems include loosely organized thunderstorm systems, namely MCSs, and tightly structured features, such as tropical depressions, tropical storms and hurricanes. Important basic features include: (1) a central warm core; (2) they develop in environments, where only small changes of wind direction and wind speed (shear) occur through the vertical extent of the troposphere and the tropical cyclone's center has comparatively little slope with height (Figures 3.7a and 3.7b).

On a broad scale, the Earth's atmosphere has a region near the equator, namely ITCZ and also the so-called equatorial trough, since it is a region of relatively low sea-level pressure. At around 20–30 degrees of latitude with this location tending to move poleward in the summer, and closer to the equator in the winter is a region of persistent high pressure that is referred to as the subtropical high. This tropical and subtropical region of wind circulation is called the Hadley circulation. It is a form of a direct heat engine in that warm air rising in the ITCZ drives the flow. It is the main mechanism by which heat, moisture and the Earth's rotational momentum are transferred poleward to the latitudes of the subtropical ridge. Each year about eighty-five tropical cyclones form around the globe with low-level sustained winds of 20 m/s or greater. About two-thirds of these tropical cyclones attain hurricane intensity of greater than 33 m/s, about two-thirds form in the Northern Hemisphere and about two-thirds occur in the Eastern Hemisphere (Figure 3.6).

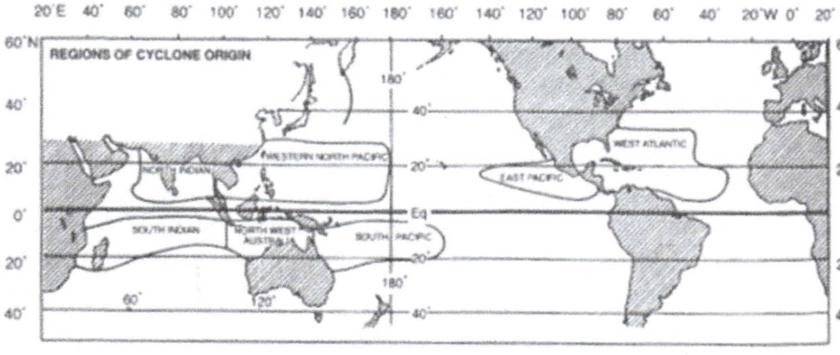

Figure 3.6 Listing of the seven global tropical cyclone basins.

3.2.3.2 Extratropical cyclones

In contrast, middle-latitude cyclones have a cold central region in the troposphere, slope strongly with height and develop in conditions of large change of horizontal wind in the vertical (Figures 3.7a and 3.7b). The stratosphere is a fundamental component of the structure of the middle-latitude cyclone, but is not a fundamental feature to the structure of a tropical cyclone. Specifically, Figure 3.7a presents the differences between the east-west cross-sections of the vertical variation of horizontal wind for the conditions in which middle-latitude cyclones form (left diagram of Fig. 3.7a) and the conditions in which typical tropical cyclones form (right diagram of Fig. 3.7a). In these diagrams the heavy line entitled TROP depicts the tropopause at about 15 km height, which is known to be the level, where temperature begins to rise with increasing height. Similarly, Figure 3.7b presents the differences between the east-west cross-sections of the typical vertical slope of pressure surfaces and associated temperature anomaly conditions, which are related to warm (W) and cold (C) layers for middle-latitude cyclones (left) and tropical cyclones (right), respectively.

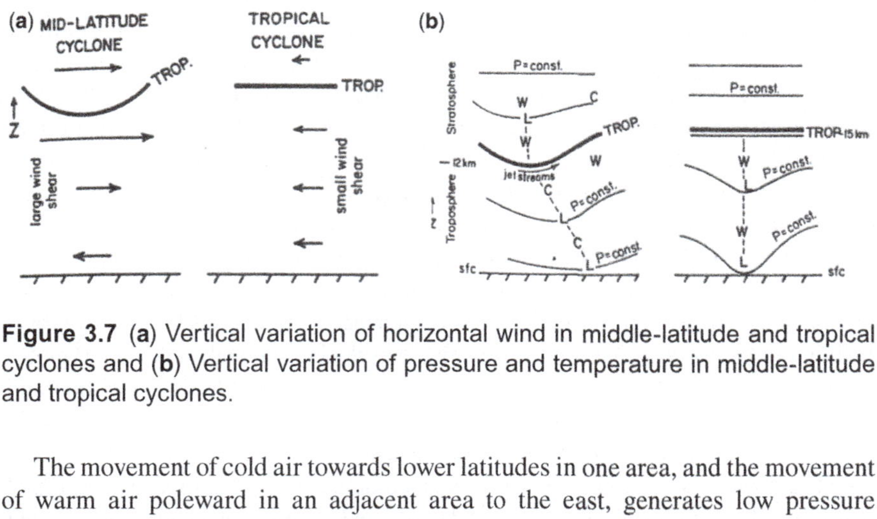

Figure 3.7 (a) Vertical variation of horizontal wind in middle-latitude and tropical cyclones and (b) Vertical variation of pressure and temperature in middle-latitude and tropical cyclones.

The movement of cold air towards lower latitudes in one area, and the movement of warm air poleward in an adjacent area to the east, generates low pressure systems along the polar front. These low pressure systems become extratropical storms if they strengthen as the equatorial movement of cold air and poleward movement of warm air continues. The term "extratropical" is used because the polar front on which they form almost always develops at latitudes poleward of around 30 degrees and is a synoptic-scale, migratory atmospheric disturbance. Its horizontal dimensions are on the order of 10^3 km. The most generally recognized manifestation of the extratropical cyclone occurs near the Earth's surface in the form of a large region of surface low pressure in which air spirals cyclonically and inward to try to fill the depressed region. Large horizontal temperature gradients provide the primary energy source for the disturbance converted to kinetic energy. The extratropical cyclone is always associated with an upper trough. In a well-developed extratropical cyclone, there is a three-dimensional balance, such that air rushing into the system at lower levels is forced upward to high altitudes, where it diverges out and away from the cyclone's center. Indeed, cyclones are locations, where the atmospheric pressure is lower than in its immediate surroundings. This straightforward and simplest possible definition has many ramifications to the structure and dynamics of these features. Since surface pressure is the total weight of the atmosphere in a column directly above it, this definition implies that the air surrounding the cyclone must be denser, thicker, or both.

3.2.4 Features of storms

The development of an MCS begins with the initiation of deep, moist convection, and is influenced by the subsequent organization of isolated convection and by potential interactions between convective cells. The lower atmosphere often contains a warm layer that requires an amount of energy, the convective inhibition (CIN), to force air upward toward the level at which the parcel and environmental temperatures are equal, i.e. the level of free convection (LFC). With further lift to heights greater than the LFC, the air parcel becomes warmer than its environment and accelerates upward by its own buoyancy. Since a major component of the movement of individual storm cells is often oriented with the horizontal winds, the general convective region thus conforms to and maintains the curvilinear shape of the initiating boundary. In some MCCs, a concentrated, closed mid-level mesoscale circulation or mesoscale convective vortex (MCV) may form in the area of warming and persist long after the MCC decays, suggesting that airflow perturbations in MCCs might be predictable over periods of several days.

The release of condensation heating from deep cumulonimbus-type rain clouds is a necessary ingredient to the formation and maintenance of tropical cyclones. In contrast, the rain-producing processes in middle-latitude cyclones, although very important for various aspects of its structure, are not an essential ingredient for its formation and structure. Moreover, the distribution of convective activity in tropical cyclones is usually organized into long quasi-circular rainbands around its center. By contrast, the middle-latitude cyclone has most of its convection cloudiness and rainfall to its eastward and poleward sides. Tropical cyclones do not develop and intensify unless they are located over tropical or subtropical oceans, where surface water temperatures are 26.5°C or higher. Typically, tropical cyclones will weaken rapidly on moving over land, because land surfaces cannot provide sufficient surface to-air transfer of sensible temperature and water vapor, which the tropical cyclone requires. By comparison, middle-latitude cyclones are able to develop and maintain themselves over both land and ocean without the need for similar underlying surface-to-air energy flux requirements.

3.2.4.1 Vertical structure of storms

The relationship between temperature and air pressure in a vertical cross-section through a cyclone is examined. Pressure always declines with height, because there is less atmospheric mass above a point as moving upward through the atmospheric column. But the thickness of the air, which is the distance between constant pressure levels, depends on the temperature of the air between those levels. Moreover, warm air is less dense than cold air, so it expands to occupy a greater volume. This thickness principle requires that there be low pressure aloft above the surface low. This is the structure of a cold-core low (Figure 3.8a). Compared to cold-core lows, which must intensify with height above the surface (Figure 3.8a), warm-core lows actually weaken with height (Figure 3.8b). Similarly, in a warm-core low, the high-pressure area aloft provides a means for evaluating the converging surface air. But this is not the case for a cold-core cyclone, since the low at the surface actually intensifies with height and stronger lows have stronger convergence.

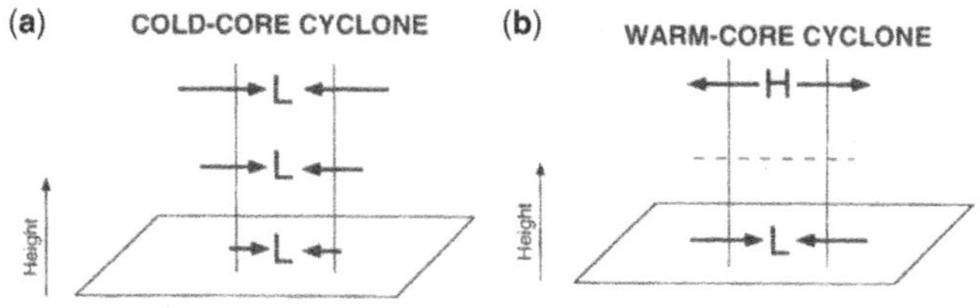

Figure 3.8 Convergence and divergence patterns above (**a**) cold-core and (**b**) warm-core cyclones.

On the other hand, the most common examples of warm-core lows are tropical storms and hurricanes. These systems form over the warm tropical and subtropical oceans from which they derive their energy. Although a simplification, hurricanes can be considered as "bottom-up systems", i.e. they derive their energy from below. Another example of warm-core lows is "heat lows" that form over deserts in the summer. Nevertheless, the basic concept is the "top-down systems". Their birth and sustainance are provided by the jet stream and changing patterns in the upper atmosphere. To understand the nature of these storms, there is a need to understand how they are connected vertically to changes in upper-level flow patterns. Moreover, for a region of surface low pressure to form, there must be less air, or less-dense air, in a vertical column above the low. Surface cyclones are regions of convergence air spirals into the low cyclonically, which are forced to rise. The upward motion generates expansion of the air, its subsequent cooling, and the formation of clouds and possibly precipitation. But this surface convergence poses a problem for the viability of the cyclone, since the added mass of air above the low raises the surface pressure. The solution to this problem is divergence in the upper atmosphere (Figure 3.9a). If the mass of air aloft exiting an imaginary vertical column above the low is equal to or greater than the mass converging at the surface, then the low-pressure system will sustain itself or intensify, respectively, called cyclogenesis. It should now be clear that the key ingredient to the formation of a surface cyclone is upper-level divergence. There are two ways to produce divergence over a given region, by causing the air to accelerate (speed divergence) or by causing the air trajectories to split or spread apart (difluence) (Figure 3.9b). Typically, speed convergence occurs in conjunction with difluence, and speed divergence with confluence.

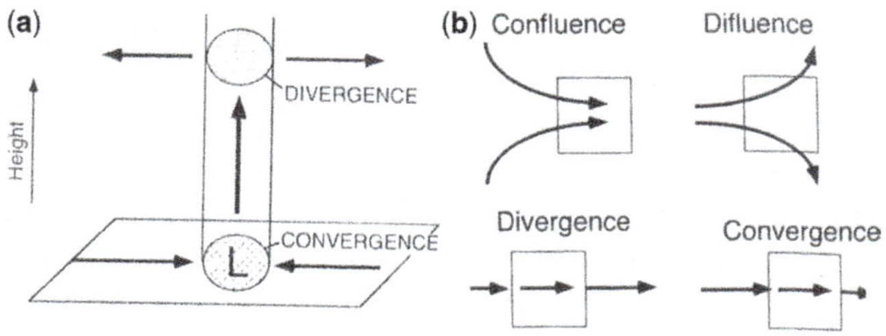

Figure 3.9 (**a**) surface convergence and upper atmosphere divergence and (**b**) confluence and difluence for divergence and convergence.

3.2.4.2 *Seasonality of storms.*

Seasonal tropical cyclone formation frequency can be related directly to a combination of the following six physical parameters, which are referred to as primary climatological formation parameters: (1) Large values of low-level relative vorticity. (2) Large values of Coriolis parameter. (3) A minimum of vertical wind shear or of the horizontal wind difference between the lower and upper troposphere. (4) Ocean energy or an ocean layer of sea temperatures in excess of 26°C to a depth of 50–60 m. (5) Large values of moist lapse-rate instability or positive values of the vertical gradient of equivalent potential temperature between the surface and 500 mb. (6) Large values of middle troposphere (700 to 500 mb) relative humidity (RH). When these six parameters are multiplied together using appropriate scaling factors, a specification of the long period number and location of seasonal tropical cyclone formations is obtained. Moreover, the spatial and seasonal patterns of cyclones are directly linked to the mechanisms responsible for cyclone formation. Cyclones preferentially form when some or all of the following conditions are present: (1) Strong baroclinic zones. (2) Unstable air, i.e. warm surface and cold air aloft. (3) Warm, humid air (southern continental margins). (4) Positive vorticity advection, i.e. lee side of north-south-oriented mountain ranges.

3.2.5 Tornadoes-lightning-flash floods

3.2.5.1 *Tornadoes*

The tornado remains one of the most feared and destructive atmospheric phenomenon on the planet. It is characterized by a relatively small-scale (Orlanski, 1975) columnar vortex circulation, which is often, but not always, associated with a funnel cloud. Tornadoes can range in size from a few meters to 3 km in diameter, as defined by maximum funnel diameter just below cloud base or damage-path width.

Most tornadoes and waterspouts rotate counterclockwise (cyclonically), although a few intense anticyclonic tornadoes have been documented. Figure 3.10 presents a conceptual model of a multi-vortex tornado, where a large tornado undergoes a vortex breakdown into a few smaller suction vortices, which rapidly rotate above the parent vortex core.

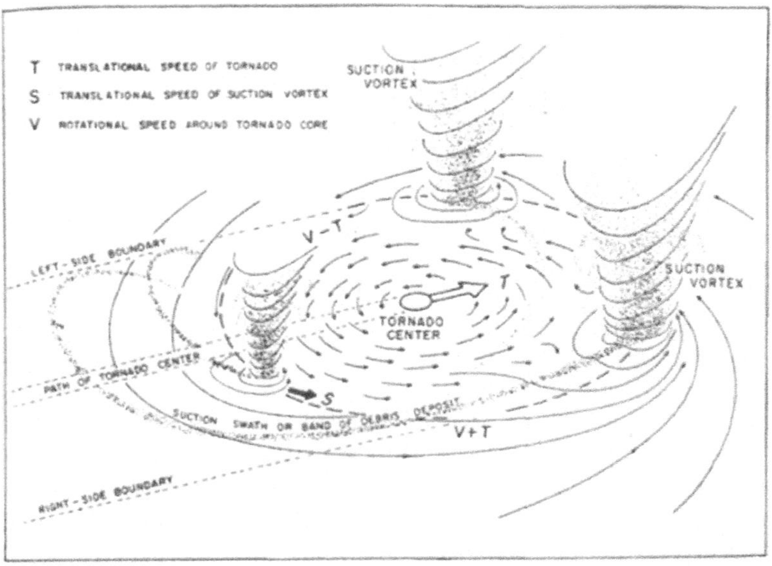

Figure 3.10 Conceptual model of suction vortices in tornado.

Tornadoes occur primarily during the warm season over land in association with intense or severe convective storms. Tornadoes are the most intensely-concentrated of the convective atmospheric vortices. These also include dust-devils on the small end, waterspouts, and then the hurricane, which ranges from 200–2,000 km in diameter. Indeed, the waterspout and tornado are fundamentally the same, but differ only in certain quantitative measures, such as the vortex size and intensity and the parent cloud updrafts. Moreover, multicellular thunderstorm complexes, referred to as MCCs, can also produce tornadoes, hail and flash floods. Tornado prediction has improved dramatically over the past decades. The average lead time for those events that have a tornado warning is 11 minutes. This is attributable to the continuous, high-resolution coverage of geosynchronous satellites, their spectral imagery, water vapor maps, improved use of radar, improved forecaster training and improved local storm-spotter networks and awareness campaigns. Other factors include forecasters' use of lightning location data and interactive models through computer workstations, along with application of new theoretical concepts. Figure 3.11 presents a global map of tornado occurrences.

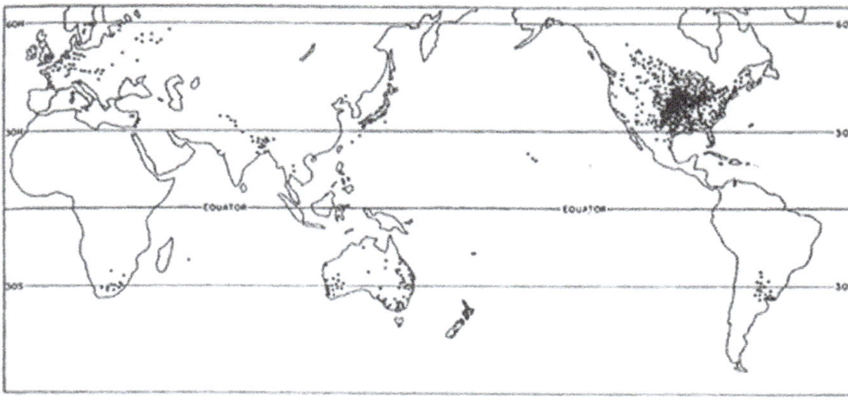

Figure 3.11 Global map of tornado occurrences.

3.2.5.2 Lightning

Lightning is a transient, high-current discharge whose path length is measured in kilometers. Several terminologies are employed when discussing lightning. Discharges occurring within the cloud are called intracloud (IC) events. A cloud-to-ground event is called a CG. In most storms ICs greatly outnumber CGs, the ratio approaching 100 or more in some severe storms. There are occasional discharges from cloud into the clear air (CA). CGs typically lower negative charge to earth (–CG). CG discharges are sometimes initiated by leaders that are positively charged, and the resulting return stroke essentially lowers positive charge to ground (+CG). The +CG now constitutes about 10 per cent of all CGs. The apparent flickering of many lightning CG flashes results from the fact that they are often composed of multiple events, called strokes. While flashes can consist of a single stroke, the average stroke multiplicity for –CGs is about four. Flashes with greater than ten strokes are not uncommon and there have been up to forty-seven strokes documented within a single flash.

Lightning generally is associated with thunderstorms. Conditions conducive to lightning can occasionally occur within tropical storms and hurricanes and winter snowstorms in which convective activity becomes locally intense. Thus, an association between the number of thunderstorms and lightning occurrences was sought. Satellite cloud-top climatologies have also been used as a surrogate for lightning inputs in global climatological models. A major advance in global lightning monitoring from space occurred with the launch of NASA's Optical Transient Detector (OTD) in April 1995. The near-polar, 740 km, orbit allows OTD to scan 1300 km^2 patches of the Earth as it passes overhead twice daily. Lightning is usually found over land areas. Two distinct approaches have been developed to detect and locate CGs. One is based on magnetic direction finding (MDF) and the second on time-of-arrival (TOA) approaches. The current US National Lightning Detection Network (NLDN) consists of a hybrid system using the best of both approaches. The mean flash location accuracy approaches 500 meters and the flash detection efficiency ranges between 80 and 90 per cent, varying slightly by region. It can provide coverage for several hundred kilometers, although the detection efficiency drops off with range.

3.2.5.3 Flash floods

Many flash floods have been associated with convective storms and MCSs. For extreme rainfall events, relatively high moisture contents are often present through a deep tropospheric layer. In addition, weak to moderate vertical wind shear is present through the cloud depth, causing slow moving storms, thereby increasing precipitation accumulation locally. A weak, mid-tropospheric, trough helps to trigger and focus the storms. However, severe flooding events have often been located very near the mid-tropospheric, large-scale ridge position and occurred within normally weak surface pressure patterns. These elusive features complicate a difficult forecast problem, since many of these features often exist and intense rainfall and flooding often do not occur. Forecasting flash flooding involves a consideration of antecedent soil wetness along with the joint conditional probabilities that widespread, persistent storm with high local rainfall rates of long duration develop. Flash flooding often occurs with convection embedded within MCSs. A methodology for flash flood forecasting consists of basic ingredients for deep moist convective heavy precipitation, and development of MCSs conductive to producing flash floods. A critical element in flash flood forecasting is the ability to operationally predict the movement and duration of local heavy rain-producing convective cells in MCCs.

3.2.6 Precipitation efficiency

The precipitation efficiency of a storm is defined as the ratio of the precipitation reaching the ground to the total amount of water vapor entering the storm. The precipitation efficiency can vary from zero for clouds that do not produce precipitation on the ground, to 90 per cent or higher in some flash flood events in humid environments (Maddox, 1980). The precipitation efficiency is usually less in dry environments and for storms with extremely strong updrafts. It is also less for clouds with cold-cloud-base temperatures. Dry environments enhance evaporation of cloud drops and rain. Storms with extremely strong updrafts may tend to carry condensed water high into the anvil before it can grow to precipitation size, where it gets carried downwind and eventually evaporates, never reaching the ground. Clouds with high, cold bases produce precipitation through processes, which tend to be less efficient and allow the frozen particles to be carried higher in the clouds. Warm-cloud-based systems permit cloud droplets growth to raindrops by coalescing and may rapidly convert the inflow moisture to rain, increasing the precipitation efficiency. Under such conditions, radar detection of heavy precipitation may be difficult, especially in rough terrain. Environments with large vertical shear in the horizontal wind tend to be associated with strong updrafts and sometimes with notably drier air, then there is a general tendency for precipitation efficiency to decrease with increasing shear. However, there are high precipitation supercells, which occur under conditions of strong shear and do produce disastrous flash floods. Moreover, airmasses with high precipitable water values shown on synoptic charts may also cause disastrous flash floods. The least efficient storms tend to be severe hailstorms, although the highly efficient storms tend not to produce hail at all. Therefore, it logically follows that the production of large, damaging hail is largely a result of natural storm inefficiency.

3.3 STORM DETECTION

The detection of severe weather and monitoring of precipitation can be achieved by the use of weather radar systems. Radar is an acronym for **Ra**dio **D**etection **a**nd **R**anging. Radar has been instrumental, among others, in the study of severe thunderstorms, observations of variables in the hydrological cycle, identification of rotation associated with mesocyclones and tornadoes, detection of severe hail and damaging winds, and estimation of heavy rainfall associated with flash floods. Indeed, in the last 30 years, ground-based scanning weather radars have revolutionized quantitative rainfall estimation around the world.

3.3.1 Conventional radar

3.3.1.1 Non-coherent radar

Radar basics. The wavelength used by weather radars is determined by the typical size of the different types of falling hydrometeors, such as roughly from 0.1 up to 10 mm for raindrops, up to 25 mm for snowflakes and up to 50 mm for hailstones. Since radar measures volumetric reflectivity of hydrometeors aloft rather than rainfall near the ground, radar rainfall estimation is inherently subject to various sources of error (Battan, 1973). Conventional or non-polarimetric Doppler radars normally transmit horizontally polarized electromagnetic waves, which only give a measure of the horizontal dimension of the cloud particles, such as snow, ice pellets, hail and rain. This one-dimensional picture makes it difficult to tell the difference between precipitation types, since ice particles and larger rain drops are generally not spherical.

Radar Components. The basic radar components are shown in Figure 3.12. The transmitter generates an electromagnetic (EM) wave, which travels through the waveguide connecting the transmitter with the radar antenna. The antenna is composed of three components: the pedestal, the dish (or reflector) and the feedhorn. The reflector is mechanically rotated by the pedestal in the azimuth direction and vertically in elevation. The electromagnetic wave, after the waveguide, is conveyed to the feedhorn, which converts the alternative current wave in a radio wave. This last is then reflected by the reflector to form the radar beam, which is send in the free atmosphere. A part of this signal is back-scattered by hydrometeors, detected and focused by the reflector and sent back to the feedhorn and this energy is called the backscatter. At this time, the radio wave is converted in volts and this energy travels down to the receiver, where it is amplified and processed. The reflected power depends on the hydrometeor and radar characteristics, such as frequency, antenna and distance to the radar.

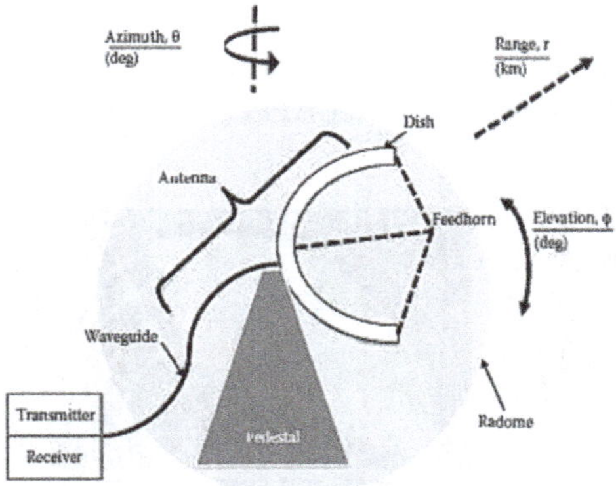

Figure 3.12 Basic components of a weather radar system.

The Radar Beam. Weather radars transmit EM energy in the microwave spectrum that travel at the speed of light in a vacuum at 3×10^8 m s^{-1}. The relationship between radio frequency (f), wavelength (λ), and velocity at the speed of light (c) is the following:

$$c = f\lambda \tag{3.1}$$

where c is 3×10^8 m s^{-1}, f is in cycles per second, or Hertz (Hz), and λ is in m. Figure 3.13 illustrates how the beam volume (or distance between the upper and lower part of the beam) increases with range and is referred to as **beam broadening** or **beam spreading**. This characteristic is one of the major limiting factors of radar measurements taken at far range.

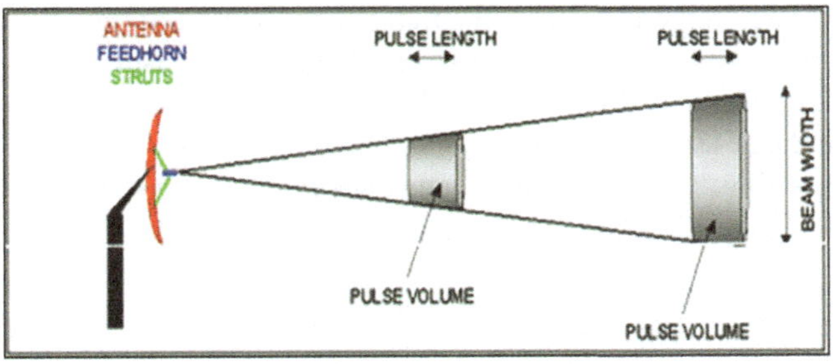

Figure 3.13 Radar beam propagation through the atmosphere.

There are two basic modes of operating a rotating pedestal: **plan position indicator (PPI)** mode and **range height indicator (RHI)** mode. PPIs are obtained by spinning the antenna in the azimuthal direction while keeping it fixed at a constant elevation angle. A full 360 deg rotation in PPI mode constitutes a **surveillance scan** and yields a **tilt** of data. An RHI, on the other hand, keeps the azimuth fixed and varies the elevation angle. RHI mode is useful for interrogating specific storms, whereas a surveillance scan in PPI mode is more practical for operational precipitation estimation. For illustrative purposes, Figure 3.14 shows a PPI mode (a), an RHI mode (b) and a CAPPI (Constant Altitude PPI) mode (c).

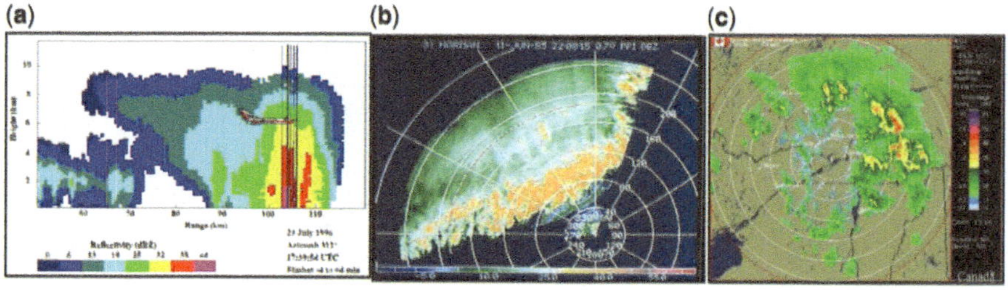

Figure 3.14 Radar images (**a**) RHI, (**b**) PPI, (**c**) CAPPI (from Dalezios, 2015).

The Radar Pulse. The range r from the radar to the target is determined as follows, considering the two-way travel of the pulse to the target and back to the receiver:

$$r = cT/2 \tag{3.2}$$

where c is the speed of light (3×10^8 m sec^{-1}) and T in sec is the elapsed time between a transmitted pulse and the reception of the backscattered energy from the same pulse. The **pulse repetition frequency (PRF)** in sec^{-1} is the number of pulses the radar transmits per second. The reciprocal of the PRF is the **pulse repetition time (PRT)**, which is the elapsed time from the beginning of one pulse to the next one. The **pulse duration** (τ) in sec is how long it takes to transmit a single pulse of energy. Indeed, the radial resolution depends on the pulse duration and typically ranges from 100 to 1000 m. For usual antenna sizes, such as from 2 to 8 m depending on the frequency, the radar beam width is typically about 1 deg, resulting into a radar sampling volume that linearly increases with the distance from the radar. Thus, final Cartesian radar products are commonly provided at a resolution in the order of 1×1 km^2 and 5 min.

Radar equation. The average received power Pr (W) is related to the transmitter power Pt (W) via the following radar equation (Battan, 1973):

$$P_r = [(\pi^3 c)/(1024 \ln 2)][(P_t \tau G^2 \Theta^2)/\lambda^2][|K|^2(Z/r^2)] \tag{3.3}$$

where G denotes the antenna gain (dimensionless), Θ denotes the half power beam width (rad), τ denotes the pulse duration (sec), (where $h = c\,\tau$ and h denotes the pulse width (m)), K denotes the complex dielectric factor of the targets (dimensionless), Z denotes the reflectivity factor (mm^6 m^{-3}), λ denotes the wavelength of the radar (cm), and r denotes the distance to the target (km). In arriving at equation (3.3), it is assumed that the beam is uniformly filled with targets, multiple scattering may be ignored, the total average power equals the sum of the scattered power from individual particles, the beam is Gaussian-shaped, and the targets are Rayleigh scatterers. The radar reflectivity factor Z depends only on the hydrometeor characteristics, namely drop size distribution (DSD) and shape. If these assumptions are met and the targets are known (water, ice, etc.) so as to specify accurately the complex dielectric factor, then Z can be accurately evaluated, via equation (3.3).

3.3.1.2 Coherent or doppler radar

Doppler capability allows radar systems to provide information on the range of the target and the radial velocity of the backscatters. This is derived from the frequency shift induced by the Doppler effect. Doppler velocities are more commonly used for severe weather detection, such as rotation with supercell thunderstorms, and less so for quantitative precipitation estimation. There are two measured variables, the mean Doppler radial velocity v (m s^{-1}), which is the mean of the distribution of the velocity of the hydrometeors within the sampling volume along the radar beam direction and the Doppler spectrum width σ_v (m s^{-1}), which is the standard deviation of the distribution of the velocity of the hydrometeors being sensitive to process, such as atmospheric turbulence that can significantly modify the velocity distribution. Examples of typical Z-R relationships are the classical Marshall-Palmer Z-R relationship (equation 3.4) for non-coherent radars (Marshall & Palmer, 1948) and a commonly used for coherent Doppler and polarimetric radars Z-R relation (Cifelli & Chandrasekar, 2010) (equation 3.5)

$$R_{\mathrm{MP}}(Z_h) = 0.0365Z^{0.625}(\mathrm{mm\,h^{-1}}) \tag{3.4}$$

$$R_{\mathrm{WSR}}(Z_h) = 0.017\,Z^{0.714}(\mathrm{mm\,h^{-1}}) \tag{3.5}$$

where R is rain rate and Z_h is the radar reflectivity factor at horizontal polarization in mm^6 m^{-3}. Radars have been extensively deployed in operational networks, such as the Weather Surveillance Radar–1988 Doppler S-band (WSR-88D).

3.3.2 Polarimetric radar

Dual-polarization radars allow for dual-channel reception of horizontally and vertically polarized electromagnetic waves (Bringi & Chandrasekar, 2004; Seliga *et al.* 1990). Polarimetric radar systems, besides the dual-polarization of the transmitted signal, also have Doppler capability. Hail can be better detected with dual-polarization than with standard radar and helps elliminate any need for vertical integrated liquid (VIL) or other similar products.

3.3.2.1 Principles and variables

Principles. Polarimetric radar systems provide additional information about the scatterer characteristics (Cifelli & Chandrasekar, 2010). For precipitation, these characteristics are related to the size, shape, phase state (liquid or ice), and orientation of hydrometeors in the radar resolution volume, as well as the polarization of the transmitted wave (Hong & Gourley, 2015). Thus, dual-polarization radar can significantly improve the accuracy of precipitation estimation, can be used to more accurately retrieve characteristics of the DSD, can distinguish between very heavy rain and hail, attenuation effects, partial beam blocking and can reduce the effects of non-weather scatterers on radar displays (Bringi *et al.* 1984; Bringi *et al.* 1986). Although dual-polarization techniques have been applied to *S* band and *C* band radar systems for several decades, polarization diversity at higher frequencies, including *X* band, are now widely available to the radar community due to lower cost and the improvements in attenuation correction with polarimetry.

Variables. The DSD can be illustrated in $R(Z_h)$ relations. The reflectivity factor $Z_{h,v}$ is given by

$$Z_{h,v} = [\lambda^4/(\pi^5\ |K|^2)]\int_0^{D_{\max}} \sigma_{h,v}(D)N(D)\mathrm{d}D(\mathrm{mm}^6\ \mathrm{m}^{-3}) \tag{3.6}$$

where K is the dielectric factor of water, $\sigma_{h,v}$ is the backscattering cross-sections at horizontal (h) and vertical (v) polarization, λ is the wavelength, $N(D)$ is the DSD, and D is the particle equivalent diameter. If the Rayleigh approximation is valid, the reflectivity factor $Z_{h,v}$ can be expressed as for non-coherent radars. The assumption of gamma distribution for DSD can be represented with three parameters as

$$N(D) = N_0 D^{\mu}\exp(-\Lambda D) \tag{3.7}$$

where N_0 (mm $^{[-1-\mu]}$ m^{-3}) is a scaling parameter, μ is a distribution shape parameter, Λ (mm^{-1}) is a slope term and D (mm) is the spherical equivalent volume diameter.

Raindrops with an equivolume diameter larger than about 1mm are not spherical, but oblate with their horizontal dimension larger than their vertical one. Their shape is therefore described as an oblate spheroid with semimajor and semiminor axes a and b, respectively, where oblateness is usually expressed by the ratio (b/a). This oblateness means that radar signal linearly polarized in the horizontal plane, interacts with and propagates through hydrometeors not the same way as that one, which is polarized vertically. These differential observations constitute the basis of weather radar polarimetry. The simplest form is a linear relation:

$$(b/a) = 1.03 - \beta D, 1 \leq D \leq 9\,\mathrm{mm} \tag{3.8}$$

where β is the slope of the shape-size relation and takes the value $\beta = 0.062$.

3.3.2.2 Additional variables

A polarimetric radar provides additional variables, as follows:

(1) **Differential reflectivity (ZDR).** The differential reflectivity (ZDR) is defined as

$$ZDR = 10\log(Z_h/Z_v) = Z_h - Z_v(\text{dB}) \quad (3.9)$$

where Z_h and Z_v are defined previously. ZDR is independent of the hydrometeor concentration and strongly depends on the shape of the reflectors, i.e. it is close to 0 for nearly spherical shapes, positive for oblate shapes ($Z_h > Z_v$) with horizontal return, and negative for prolate shapes ($Z_h < Z_v$) with vertical return. ZDR ranges from −8 to 8, though values rarely get that extreme on either side. In rain, as ZDR increases, drop size increases. For raindrops bigger than ~0.5 mm, ZDR ranges from ~0.3 to 3dB. ZDR over 3 is indicative of big drops. Although hail is usually not spherical in shape, ZDR values drop to near zero. However, if hail is small and/or water coated, ZDR gets usually very high, 5 or 6, but this hail is almost never severe in size by the time it reaches the surface. Conversely, negative ZDR can be indicative of very large hail. For snow/ice, ZDR can vary significantly. In general, dry snow is closer to 0 ZDR, with wet snow increasing in amount. Non-weather returns can have a ZDR of literally almost anything. Hail is typically characterized as having high reflectivity (>55dBZ), low ZDR (<1dB), and a local reduction in copolar correlation coefficient CC (0.95 < CC < 0.98). However, when hail diameters become significantly large (>5cm), Mie scattering begins to affect the dual-pol base products. In particular, CC becomes significantly low (<0.9, to as low as 0.7).

(2) **Specific differential phase.** The differential phase shift (Φ_{dp}) [deg] is the difference in phase between the horizontally (*h*) and vertically (*v*) polarized signals. In rain, Φ_{dp} increases with distance and drop oblateness depending on both the shape and concentration of the hydrometeors.

For raindrops with an oblate spheroid shape, *h* polarized waves propagate more slowly than do *v* polarized waves. Φ_{dp} is an important component for rainfall estimation and its spatial derivative is called the specific differential phase shift K_{dp}, [deg km^{-1}], given by

$$K_{dp} = [\Phi_{dp}(r_2) - \Phi_{dp}(r_1)]/[2(r_2 - r_1)][\text{deg km}^{-1}] \quad (3.10)$$

K_{dp} is closely related to the DSD through the rainwater content and is relatively insensitive to ice in the radar volume. For Rayleigh conditions:

$$K_{dp} = (180/\lambda)10^{-3}\,CW\,(1 - r_m)[\text{deg km}^{-1}] \quad (3.11)$$

where C is a constant with magnitude ~3.75, λ is the radar wavelength in meters, W is the water content in g m^{-3}, and r_m is the mass weighted mean axis ratio.

(3) **Linear depolarization ratio (LDR) and copolar correlation coefficient (ρ_{hv}).** LDR and copolar correlation coefficient (ρ_{hv}) are not directly related to rainfall estimation. LDR (dB) is defined as the logarithm of the ratio of cross-polar and copolar signal returns, that is,

$$\mathrm{LDR} = 10\log_{10}(Z_{ex}/Z_h)(\mathrm{dB}) \tag{3.12}$$

where Z_{ex} is the signal received at the vertical polarization (cross-polar return) for a transmitted horizontally polarized wave. The copolar correlation coefficient ρ_{hv} is the coefficient of correlation between Z_h and Z_v of the individual pulses. This parameter is a product that measures how similarly the two radar pulses (horizontal and vertical) are behaving and is sensitive to the distribution of particle sizes, axis ratios, and shapes. Theoretical values are approximately 0.99 for raindrops and ice crystals. Indeed, for high values, confidence is pretty high that whatever is occurring in that area, it's probably uniform. In other words, if its rain, it's probably all rain. For melting aggregates, ρ_{hv} can be less than 0.90. Because ρ_{hv} and LDR are both sensitive to the presence of large wet particles, their melting-layer responses are similar.

The additional variables of the polarimetric radar enable the improvement of analysis and retrieval methods for different applications. Some key advantages are mentioned:

Echo classification. A better distinction can be achieved between meteorological and non-meteorological echoes, such as birds, airplanes, insects, terrain or buildings, as well as a better classification of precipitation echoes in different hydrometeor types, such as drizzle, rain, hail, snow or graupel. Classification algorithms have been developed for different radar frequencies, such as S-band, C-band or X-band.

Microphysical information. The improved hydrometeor classification combined with Doppler capability constitutes an important source of information to better identify and characterize the different microphysical processes in clouds and in precipitation.

DSD retrieval. The DSD can be retrieved from polarimetric radar measurements by assuming a functional form for the DSD, such as the Gamma distribution. Non-parametric methods based on the full Doppler spectrum have also been proposed.

Improved rain rate estimation. Several rain rate estimators based on polarimetric radar variables contribute to the improvement of radar rain-rate estimation for different applications (Gourley *et al.* 2010). ZDR and K_{dp} are commonly used to complement Z_h in multiple power laws used in rain rate estimation. Moreover, polarimetric estimators, combined with more accurate hydrometeor identification, can improve rain rate estimation and help to distinguish hail and rain in convective storms (Giangrande & Ryzhkov, 2008).

Based on the above, it can be stated that acceptable echo classification, artifact-free rainfall products, better identification of microphysical processes and the potential to estimate space-time variability of the DSD, are among the promising achievements of dual polarization. Nevertheless, there remain several sources of uncertainty that can significantly affect the quality of rain rate estimates. In summary, dual-polarization radar is a critical tool for weather research applications, including rainfall estimation, and is at the verge of being a key instrument in operational meteorology.

3.4 STORM MODELLING AND FORECASTING

3.4.1 Storm modelling

Cloud-scale models typically resolve individual storms or small groups of storms, while including simple treatments of the larger-scale environment and of physical processes that act on those larger scales. On the other hand, mesoscale models provide detailed treatment of the evolution of the larger-scale environment, usually initiating the numerical forecasts using analyses of actual, observed weather conditions, while approximating the effects of storms, which act on those scales that are too small to be explicitly represented. Additionally, some models are "dynamic", being based on solutions of the equations of motion governing atmospheric winds and pressure, whereas other models are "kinematic" (non-forecast) driven by specified observed input wind fields. Two salient issues need to be further addressed: (1) the improved representation of convection in existing mesoscale models at grid spacing of 20 km or greater, the proper treatment of surface and boundary layer fluxes, moist downdrafts, and the treatment of form and phase changes of water substance (termed "cloud microphysics") on the scales that can be resolved on the model grid mesh; (2) development of new scheme to represent convection in the range of grid spacing from the order of 1 to; 20 km, where the ratio of grid size to storm diameter approaches unity and conventional approaches and closure assumptions do not apply.

3.4.1.1 Cloud models

Cloud models have proven useful for quantifying the influence of numerical schemes on the evolution of simulated MCSs by exploiting their ability to explicitly represent relevant scales of motion including convection. An one-dimensional cloud model, namely the Convective-Stratiform Technique (CST), has been originally developed for the calculation of rainfall in Florida by means of IR data of the GOES-6 VISSR radiometer (Bendix, 1997). The entire procedure for the calculation of rainfall maps by means of the CST, as well as the adjustment of the CST to the viewing geometry of the Meteosat-3 VISSR sensor, is presented in figure 3.15 and involves the following steps: (1) Estimation of the relation between the average rain rate (Rmean), the convective rain area (Ar) and cloud top temperature (Tc) by computing the cloud model for different model updraft areas. (2) The search for local temperature minima (Tmin) within every Meteosat-3 IR image. (3) The application of a discriminant function to detect non-raining clouds and a slope test to distinguish between young and mature storms. For young storms, the rain rate is calculated. (4) A threshold test based on the anvil-stratiform threshold temperature (Ts) to separate convective core areas and stratiform anvil areas within mature storms. A constant rain rate of 2 mm h^{-1} is assigned to stratiform anvil areas, whereas for core areas the convective rain rate is calculated. (5) A field-of-view correction of the cloud top temperature (Tmin), recorded by the VISSR sensor on board of Meteosat-3, to the 1 km^2 cloud model temperature (Tc). (6) The computation of a rainfall map using time series of consecutive IR images (0.5 h interval).

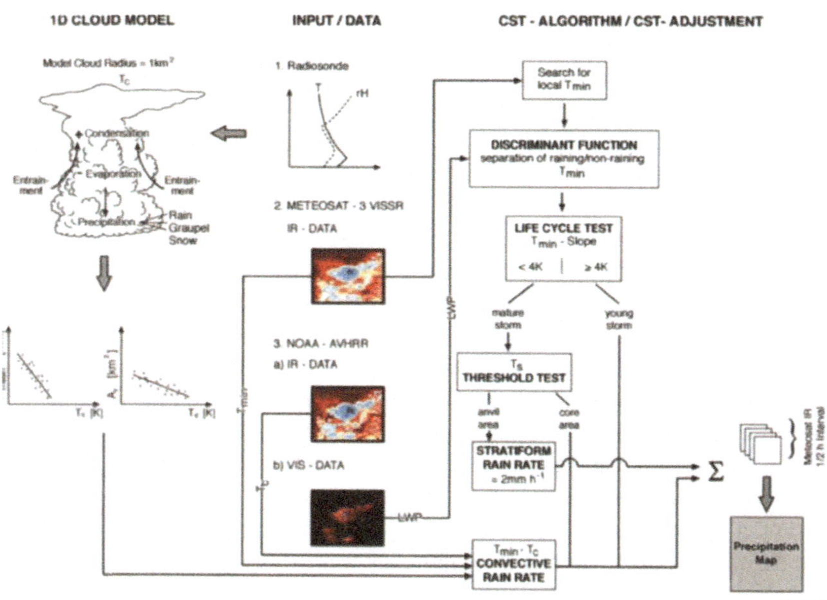

Figure 3.15 CST Cloud model (from Bendix, 1997).

3.4.1.2 Mesoscale models

Carefully initialized mesoscale models can produce improved precipitation and storm forecasts. Moreover, regional climate models (RCM) are used to forecast the fields of environmental precursors of MCSs in a variety of synoptic weather regimes. It appears that among the best methodologies for detecting seasonal to multi-seasonal predictive relationships is based on the simple assumption that changes in future atmosphere-ocean conditions are expected to closely mimic those atmospheric changes, which have been observed in the past. The likely superiority of this approach over the initial-value numerical modeling approach is due to the complex nature of the atmosphere-ocean system and also to the inability of the numerical models to realistically represent and skillfully forecast all of these complicated non-linear processes for thousands and thousands of time steps into the future.

Weather Research and Forecasting (WRF) Model. The WRF model is a state-of-the-art, next-generation mesoscale numerical weather prediction system designed to serve both operational forecasting and atmospheric research needs. It is a non-hydrostatic model, with several available dynamic cores, as well as many different choices for physical parameterizations suitable for a broad spectrum of applications across scales ranging from meters to thousands of kilometers. The dynamic cores in WRF include a fully mass- and scalar conserving flux form mass coordinate version. The physics package includes microphysics, cumulus parameterization, planetary boundary layer (PBL), land surface models (LSM), longwave and shortwave radiation (Skamarock *et al.* 2006). The WRF model has been specifically designed for high resolution applications, and provides an ideal tool for dynamically downscaling GCM simulations. WRF model can be set up by using multiple nests at 30 km and 10 km horizontal grid spacing. The use of a large domain ensures that synoptic weather systems approaching the region of interest are well represented by the time they reach the region. In the vertical, 31 levels can be used with the highest resolution (~20) in the boundary layer.

Eta Model. The Eta dynamics features that deserve the most mention are the following: (1) The step-mountain ("eta") vertical coordinate (Mesinger *et al.* 1988); (2) The Janjic (1994) Arakawa horizontal momentum advection scheme, conserving C-grid defined enstrophy for horizontal nondivergent flow on the model's E-grid, and a number of other quantities; (3) Gravity-wave coupling scheme, used in a two-time level, split-explicit framework; (4) Energy conservation in transformations between the kinetic and the potential energy in space differencing (Mesinger *et al.* 1988); (5) Lateral boundary conditions prescribed or extrapolated along a single outer boundary line, followed by a "buffer" row of points of four-point averaging. The four-point averaging achieves coupling of the boundary conditions of the two C-subgrids.

All of these features are aimed at avoiding or reducing some kind of a physical (as opposed to a mathematical) error (e.g. the eta coordinate), and/or at avoiding various computational modes. A comprehensive "physics package" is in place, with turbulence kinetic energy and cloud water/ice as additional physics prognostic variables (Zhao *et al.* 1997). Recently, at NCEP increasingly comprehensive land-surface parameterizations have been included (Janjic, 1994; Mesinger, 2000). The Eta data assimilation system at NCEP includes the most recent information on the current 3D-Var system available (Rogers *et al.* 2000). Upgrades of an open source version of the Eta model are posted at the CPTEC web site http://etamodel.cptec.inpe.br/. The version originates from the NCEP's Workstation Eta code posted on the NCEP web site http://www.emc.ncep.noaa.gov/mmb/wrkstn_eta, which differs from the NCEP's latest operational Eta by having the WRF-NMM non-hydrostatic option included. In 1993 the United States National Meteorological Center (NMC) began producing operational forecasts with the regional "Eta" model, soon followed by a mesoscale version operating on a finer spatial grid mesh. Results to date have demonstrated the promise of the Eta model to improve prediction of heavy convective precipitation with the inclusion of better sub-grid convective parameterization schemes and more accurate initial conditions.

3.4.2 Storm forecasting

3.4.2.1 Predictive signals

The atmosphere appears to reflect more about the future state of the combined atmosphere-ocean system than does the ocean surface by itself. There are strong multi-season predictive signals for the Atlantic seasonal hurricane activity contained in prior-year West African rainfall, the strength of the Azores anticyclone, as well as the southern oscillation index (SOI). Combined atmospheric precursor signals allow an explanation as much as 40–60 per cent of the year-to-year variance of Atlantic seasonal hurricane activity from one to four seasons in advance. Three types of low-level convergence boundaries that often initiate storms are: (1) cold fronts, (2) cold thunderstorm outflows, such as cold fronts but with cold air mass originating from the storm, (3) drylines, i.e. change between warm, moist air and hot, dry air. A banded area of concentrated temperature contrast or gradient, i.e. a baroclinic zone, often coincides with strong air-flow convergence along these low-level boundaries.

Predicting an MCS may be achieved by anticipating three sequential events: (1) the outbreak of strong (often widespread) convection, (2) the subsequent upscale growth and organization, (3) the ultimate evolution and decay. In addition to successfully forecast an MCS, it is necessary to anticipate the evolution of the following attributes: (1) the type or "mode" of the component convective cells (including modal evolution); (2) the nature and degree or mode of organization of the convective system as a whole; (3) the total area and dimensions, location, movement, and duration of the convective system. There are new operational regional and smaller-scale numerical prediction models providing severe storm forecasters with higher-resolution and more accurate fields of the critical atmospheric parameters that lead to severe thunderstorm and tornado development. The new technology seems to assist in moving from the era of "detected" warnings to the era of "predictive" warnings for tornadoes. This combined with improvements in warning coordination and communication, has contributed to the continued reduction in morbidity and mortality for tornadoes.

3.4.2.2 Strom tracking and nowcasting

Experimental forecasters are able to produce successful daily forecasts of deep convection. A largely unsolved problem in MCS forecasting involves anticipating the timing and location of convective initiation. Since storms often form within a range of 10–20 km of mesoscale boundaries, which may themselves be meso-γ-scale or narrower in width, variability in the boundary layer is probably critical for the storm initiation process along with formation, structure and movement. A brief review of storm tracking and forecasting methods follows (Dalezios, 2012). These methods include extrapolation, knowledge-based nowcasting, numerical models, neural network models and further approaches, such as probability forecasts and modified turning band (MTB) models.

1. **Extrapolation.** Extrapolation methods include steady-state assumption, which involves cross-correlation methods and feature tracking, as well as echo size and intensity trending.

   1.1. **Steady-state assumption.** Forecasting radar echo motion requires determination of their velocity and direction of displacement. Many radar-based methods use tracking algorithms based on pattern recognition, either determining the echo motion through cross correlation (Austin & Bellon, 1974) or tracking particular features of the radar echo. Problems with tracking algorithms are mainly caused by small-scale and short-term variations of the radar echo pattern, e.g. orographical impacts and radar technical limitations. A further approach to advect radar echoes is using winds from the steering level, which can be provided by outputs of Doppler radar measurements.

   1.1.1. **Cross-correlation.** Cross correlation method (Figure 3.16) is used to determine the overall motion of radar echoes. Similar patterns of radar echoes are detected by comparing tracking areas in consecutive scans. The best fit between the tracking areas is found by optimizing the correlation coefficient. The distance between the tracking area and the time lag of the scans determine the displacement vector.

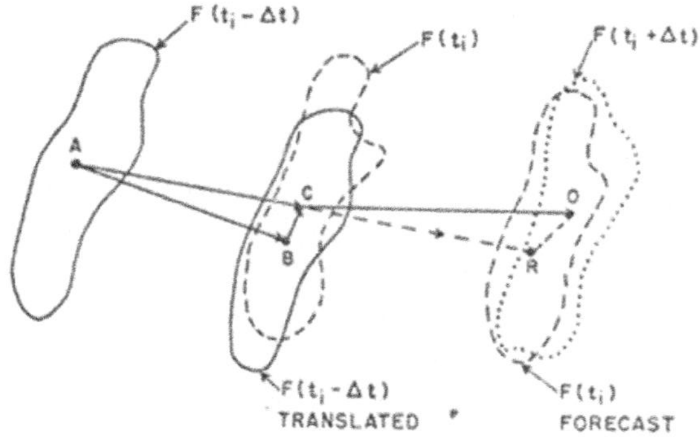

Figure 3.16 Cross-correlation storm tracking method (from Austin & Bellon, 1974).

   1.1.2. **Feature tracking.** It is the mass centroid method, which derives the displacement vector between consecutive radar scans from the distance of the mass centers of two corresponding radar echoes.

The centers are assumed to be representative for individual convective cells of storms. Thus, the mass centroid method provides detailed information about cell tracks and characteristics. A "structured" approach attempts to determine the echo motion, consisting of echo definition, characterization, matching and the forecast step. The variety of features that are identified and recognized include mass, size and centroid, as well as orientation, elongation, intensity distribution, previous size, moment vectors and their root mean square variation and the number of previous recognition of the echo. Other approaches to describe the echo shape use Fourier transformation of the maximal echo diameter and the moments of inertia on the X and Y axes.

1.2. **Echo size and intensity trending.** The main limiting factor to intensity forecasts are initiation and dissipation of precipitation pattern, as well as local effects like orography. A common approach to forecast intensity and echo size trends are weighted or unweighted fits of past developments of the radar echo. Indeed, the individual development of cells is included into a nowcasting technique to forecast the motion of precipitation areas by applying a nonlinear extrapolation to the intensity and echo size forecast, e.g. increase echo size for 15 minutes then decrease. Linear extrapolation has achieved the best results. 30 minutes are supposedly the most appropriate time interval for fitting the extrapolation scheme on the basis that a trend has to be established for a developing cell.

2. **Knowledge-based nowcasting.** Knowledge-based nowcasting systems are observationally based approaches. They combine a variety of data sources and conceptual models to gain information about the initiation, development and dissipation of precipitation. The relation between convergence lines and enhanced convection is the basis for many conceptual models presently used in knowledge-based nowcasting. Investigations of various reasons for initiation of thunderstorms along a dryline, including also investigations concerning vegetation, often causing thunderstorm initiation when its thermal circulation intersects a dryline. The strong impact of local differences in the surface moisture and the moisture profile in the planetary boundary layer (PBL) on storm initiation along convergence lines has also been investigated.
3. **Numerical models.** Numerical models for thunderstorm either include or exclude the use of observational information about convection to define initial conditions. The first reliable forecast for convective events derived by models, which are not initialized with observational data, can be obtained only after six hours, because low level convergences have to be derived by the model itself from large-scale simulation. This implies that the forecast can be more successful if a convective event is more large-scale driven. To overcome these model limitations, observed storm data, e.g. surface rainfall observations, should be used to initialize the numerical model. The rainfall data have to

be converted into initial fields of moisture, divergence, latent heat and similar aspects by applying implicit assumptions on the structure of the convection.

4. **Neural network models.** Artificial neural networks (NN) are a suitable tool for solving complex problems that might be difficult to express in mathematical terms as is rainfall forecasting. The most common neural network is a three-layer network, which consists of the input, the hidden and the output layer. Neurons from different layers are connected using weights. The network is trained on learning sets, consisting of corresponding input and output fields for a variety of variables. The training is carried out by adjusting the connection weights, i.e. the magnitude of the error between real and forecasted output is optimized until a desired accuracy of time limit is reached, known as back propagation. Then, the NN is tested on independent inputs for verification. After successful training and testing, the network is finally able to predict output fields on any newly provided input. To train the NN, various approaches have been chosen. Thunderstorms are forecasted based on surface meteorological data, or data from a numerical weather prediction model and upper air soundings to predict precipitation. Moreover, mathematical rainfall simulation model has been used to provide the input to the NN.
5. **Further approaches.** Other approaches to forecast rainfall using radar data as input include providing probabilities of accumulated spot precipitation, namely probability forecasts (Anderson & Ivarsson, 1991) or statistical approaches, such as the Modified Turning Bands (MTB) Model (Mellor *et al.* 2000).

3.5 HAIL

3.5.1 Hail formation

Hailstones are lumps of ice (often an ice/water/air mixture) with sizes greater than 5mm that are generally opaque with a layered structure. The formation of hail is associated with an unstable atmosphere that produces thunderstorms and a complex chain of physical processes. Hail is formed when small ice particles known as hail embryos are held aloft by strong thunderstorm updrafts within regions of unfrozen super-cooled cloud water, where the temperature is between 0°C and –40°C. Hail embryos grow into hailstones by collecting unfrozen, supercooled liquid cloud droplets through collisions. The collected cloud droplets freeze to the already-frozen embryo, increasing the size, weight, and fall speed of the hailstone. Studies of the internal structure of large hailstones have shown that hailstones can have either a graupel (snow pellet) embryo or a frozen drop embryo. The different hail embryos indicate different growth histories and trajectories and illustrate the complexity within a single hailstorm. The super-cooled cloud water is collected by the hail embryos and freezes to them, resulting in growth to hail (greater than 5 mm diameter) sizes. Growth continues until (1) the supporting updraft weakens, (2) the in-storm motion of the growing hailstone moves it to the downdraft side from where it can fall, or (3) the hailstone grows so large that the updraft can no longer support it.

A theory of hail growth must describe at least the following elements (Knight & Knight, 2001):

(1) The hail embryo formation processes including the microphysics of particle growth and the region or regions in the storm where such growth occurs;
(2) The transport of embryos (either spatially or temporally) to regions where further growth is possible;
(3) The growth trajectories of hailstones passing through the storm updraft and their ability to collect and deplete the super-cooled water.

The above processes have to be understood in terms of the time evolution of the storm's updraft and the storm's ongoing cellular development and structure. Such things as ice nucleation, dominant rain formation process (coalescence vs. ice-graupel process), cloud base temperature, environmental wind hodograph, and updraft strength, width, and evolution are all essential elements of the hail growth process. Hail stone embryo formation generally requires growth times of approximately 15 minutes.

Sources of Hail Embryos. Direct observational evidence from the instrumented aircraft penetrations of Colorado and Alberta storms in the 1970s and early 1980s indicates that hail embryos grow within the evolving main updraft of single cell storms and within the updrafts of developing feeder clouds (the cumulus towers) that flank mature multi-cell and supercell storms (see e.g. Foote 1984; Krauss & Marwitz 1984). In order to be hail embryos, initial ice particle size is important. Embryos must be large enough that they do not rise too quickly in the storm updraft but not so large that they fall from the updraft. Some of this complexity is reviewed by Foote (1985) who classifies a broad spectrum of storm types according to both dynamic and microphysical processes thought to be critical to hail production. Hail embryo sources identified by Foote (1985) include the following: first-ice in a time-developing updraft, first-ice in the core of a long-lived updraft, flanking cumulus congestus, merging mature cell, mature cell positioned upwind, embryos created by melting hail and shedding of the melted water, entrainment of stratiform cloud, or recirculation of embryos that have made a first pass through the updraft core and re-entered the updraft at lower levels during their descent.

The Growth of Large Hail. Large hailstone growth requires large vertical velocities within the cloud. The growth involves combinations of wet and/or dry growth associated with the riming and accretion of super-cooled cloud droplets. Particle growth from 1 mm to 20 mm diameter requires approximately 30 minutes. Hailstones smaller than 20 mm density range from 50 to 890 kg m^{-3}. Larger hailstone densities range from 810 to 915 kg m^{-3}(density of solid ice is 917 kg m^{-3}). For example, hail of 2 cm diameter has a fall speed of 40 m/sec and requires an updraft speed in the super-cooled liquid zone of at least 20 m/sec. In addition, the super-cooled liquid values need to be about 5 g/m^3. For comparison, 4 cm diameter hail has a fall speed of 60 m/sec.

The growth of large hail is theorized to occur primarily along the edges of the main storm updraft where the merging feeder clouds interact with the main storm updraft (WMO 1995). The largest hailstones remain in close proximity to the updraft maximum and grow during most of their journey through the storm. Factors that favour the hail growth are those that increase the growth time of the particle/embryo in the updraft, such as large updraft widths, small differences in the updraft speed and particle terminal velocity, and weak horizontal flow. However, the mature hailstorm often consists of complicated airflow patterns and particle trajectories.

The hailstone growth rate is maximized near the –12°C level, and negligible below –30°C as super-cooled water droplet concentrations become less (Figure 3.17). Hail is most common in mid-latitudes during spring and early summer when surface temperatures are warm enough to create instability for strong thunderstorms, but the upper atmosphere is still cold enough to form ice and large vertical extents of super-cooled liquid water. In most situations the subcloud layer is relatively warm (warmer than 0°C) so hailstones begin to melt during the final portion of their fall to earth, but in many cases the hailstones are too large for melting to be complete, and hail reaches the ground.

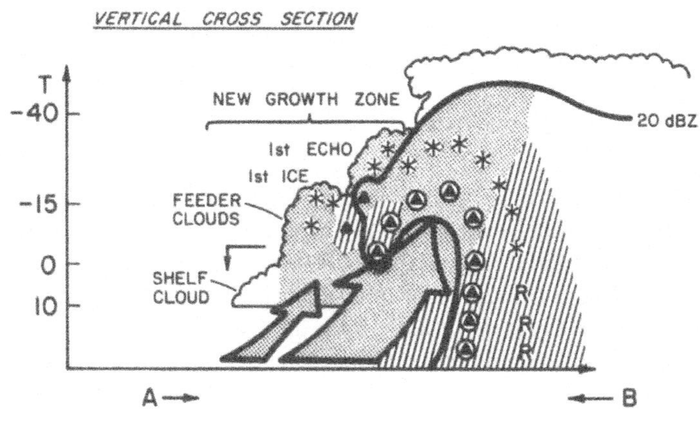

Figure 3.17 Vertical cross-section of hailstorm (from Krauss & Marwitz, 1984).

Hail Climatology. Figure 3.18 illustrates the worldwide distribution of hail days in a year (Court Griffiths, 1986). A superimposed topographic map suggests a strong correlation with the close proxity and distribution of mountains. Factors that play a role include (a) higher thunderstorm frequencies near mountains, probably enhanced by orographic lifting and convergence effects; (b) generally higher terrain, where a falling (and melting) hailstone will spend less time falling from the 0°C level to the ground; (c) generally higher (and colder) cloud bases, providing less opportunity for the falling hailstone to be warmed by cloud water encountered below the 0°C level; and (d) a generally drier subcloud air, providing more opportunity for evaporative cooling to slow the melting process as the stone falls.

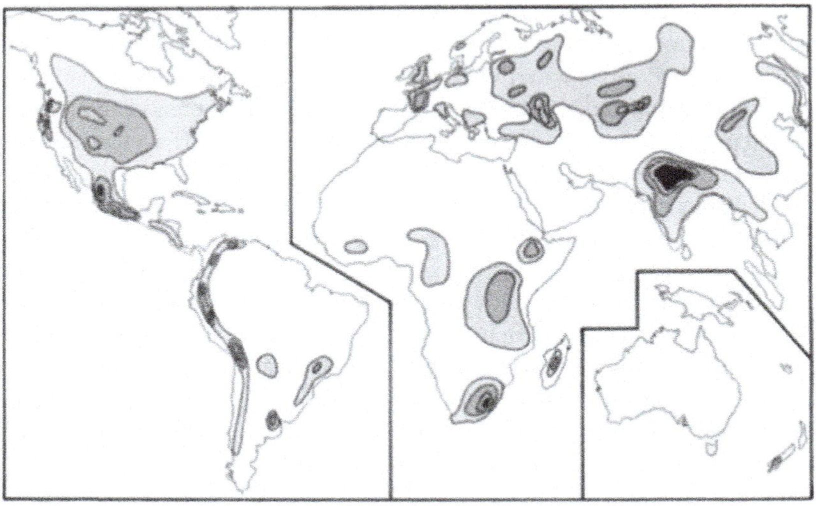

Figure 3.18 Mean annual number of hail days, with contours of 1 Day, 3 Days, 6 Days, and 9 Days (from Court & Griffiths, 1986).

3.5.2 Hail detection

Storm detection, including radar systems, is covered in a previous section. However, in this section some terms may be repeated, although the emphasis is on the identification and detection of hail based on different radar systems. The height of the 45 dBZ contour (a radar echo-intensity level) was a criterion tested in a Swiss hail research program (Waldvogel *et al.* 1979). The Swiss research found that all hailstorms had 45 dBZ contours above the altitude of the –5°C temperature level. There was a False Alarm Rate (FAR) of 50%, largely because some strong rainstorms also met the criterion. However, it is much preferable to make an error and assume that a heavy rainstorm is going to produce hail than to mistakenly believe that a hailstorm is only going to produce heavy rain. The criteria used by the National Hail Research Experiment in the USA (1972–1974) for a declared hail day was defined by radar maximum reflectivity greater than 45 dBZ above the –5°C level. A radar reflectivity (≥45 dBZ) implies that significant supercooled liquid water exists at temperatures cold enough for large hail growth. Waldvogel *et al.* (1979) made tests using the following simple criterion:

$$H_{45} > Ho + 1.4 \text{ Km} \tag{3.13}$$

where H_{45} is the height of the 45 dBZ radar echo contour and Ho is the zero degree level height.

Vertical Integrated Liquid (VIL) is a radar derived estimate of the total mass of precipitation within a vertical column in the clouds. VIL is a function of reflectivity, and converts reflectivity data into an equivalent liquid water content value based on studies of drop-size distribution and empirical studies of reflectivity factor and liquid water content (Amburn, 1996). VIL has units of mass divided by area (kg/m^2). The mass is estimated by the radar reflectivity factor. VIL is the integration of reflectivity within a column of air. A higher VIL means there is more precipitation mass in the column of air. At one time it was thought a certain VIL value could be associated with a certain hail size. However, it was found that the relationship between the VIL value and hail size is much more complex. VIL values are often seasonally and regionally dependent. The VIL versus hail size relationship for one geographic region may not apply to another location with different meteorological conditions. The value is also dependent upon the cloud physics and synoptic considerations for that particular day. VIL values at small radar ranges (e.g. within 20 km of radar) will be underestimated if part of the storm is not observed by radar. VIL values at the edge of the radar display are also underestimated if the lower parts of the storm are not being sampled. VIL values will be higher where all tilt angles are able to sample the entire precipitation/hail core of the storm. In addition, VIL measurements are less accurate for highly tilted storms, and VILs are higher for wet hail than dry hail. VIL can be used as a "guide" for assessing hail size and intensity of precipitation but as mentioned has limitations. The VIL value correlation to hail size can depend on season, synoptic environment, elevation, storm speed, storm structure, or hail reflectivity characteristics. The relationship is generally too complex to issue a severe thunderstorm warning with respect to a certain hail size from the VIL value alone, however, the VIL value can be used as a warning alert if the VIL is very high. Hail mass flux or vertically integrated hail mass can be calculated in the same way as VIL calculations but using a reflectivity conversion to hail mass rather than a rain relationship.

3.5.3 Hail forecasting

Forecasting of hail tends to include general forecasting of severe thunderstorms coupled with estimates of maximum hailstone size (Dalezios & Papamanolis, 1991) or could be based on radar data for storm tracking (Rigo & Llasat, 2016). To forecast hail, deep moist convection is required. In addition an adequate updraft must exist to keep the growing hailstone aloft for an appropriate amount of time; sufficient supercooled water near the hailstone to promote growth as it travels through an updraft, and an initial piece (embryo) of ice or dust for it to grow upon. There is no clear distinction between storms that do and do not produce hailstones. Nearly all severe thunderstorms probably produce hail aloft, though it may melt before reaching the ground. Multi-cell thunderstorms produce many hailstones, but not usually the largest hailstones. In the life cycle of the multi-cell thunderstorm, the mature stage is relatively short so there is not much time for growth of the hailstone. Supercell thunderstorms have sustained updrafts that support large hail formation by repeatedly lifting the hailstones into the very cold air at the top of the thunderstorm cloud. In general, hail larger than 5 cm diameter is associated with supercells. However, non-supercell storms are capable of producing golf ball size hail. In all cases, the hail falls when the thunderstorm's updraft can no longer support the weight of the ice. The stronger the updraft the larger the hailstone can grow. Below are several important factors to consider when forecasting the probability and size of hail.

Convective Available Potential Energy (CAPE): This is the most important factor in determining hail size. CAPE less than 1000 J/kg generally produce borderline severe hail (near 2 cm or less) while CAPE > 2000 J/kg can produce very large hailstones. High CAPE leads to high upward vertical velocities within a thunderstorm. High vertical velocities can suspend hailstones for a longer time and add layers of ice onto already developed hailstones.

Wind Shear: Strong upper level winds allow CAPE to be maximized to its fullest potential. Strong upper level winds tilt the updraft of developing thunderstorms. This allows the updraft and downdraft to be separated from each other. This allows high vertical velocities to persist for longer periods of time without being affected by the weight of the developing precipitation. Directional and speed wind shear also allows the transfer of hail embryos from adjacent, developing cloud turrets to enter a region of greater updraft allowing for the growth of larger sizes.

Precipitable water: The mass of moisture and water will influence the strength of the updraft. High moisture soundings result in water loading. CAPE is reduced with water loading since the force of gravity pushes down on the liquid water drops. Precipitable water values of less than 2.5 cm will not be nearly as influenced by water loading than if precipitable water values are above 3.5 cm. Lower precipitable water values have the potential to produce large hailstones when significant CAPE is present. Low precipitation supercells are notorious for producing large hail.

Freezing level: The freezing level determines the depth of the atmosphere that is warmer than the freezing temperature. If the freezing level is high in the atmosphere, hailstones will have more time to melt than if the freezing level is close to the surface. A high freezing level also decreases the vertical depth in which hailstone formation and growth is possible. The freezing level depends on elevation, the season, and the temperature profile of the atmosphere. High elevation areas will have relatively low freezing levels in all seasons. For low elevation areas, if the freezing level is closer to the surface, strong thunderstorms have a higher probability of producing hail that will reach the surface.

Wet bulb zero level: The wet bulb zero level is defined as the freezing level that will result due to evaporative cooling. The freezing level will lower if there is dry air in the mid-levels of the atmosphere. This occurs due to evaporative cooling of environmental air that entrains into a thunderstorm. This same entrainment can also produce strong and gusty surface winds. This has been shown to be another factor that is associated with many hail days in the mid-western and high plains region of the United States.

Elevation: Higher elevation areas are generally closer to the cold layers of the upper atmosphere. When a hail stone falls, it rapidly begins to melt when the environmental temperature rises above freezing. If the hailstone has to fall through a deep layer of warm air, it will melt and decrease significantly in size. Mountainous regions and the High Plains of the United States have the highest number of hail days per year. Small hail which would normally melt before reaching the surface in a low elevation area reaches the surface in high elevation area. Storms do not need to be as severe in mountainous regions as in lower elevation areas for hail to reach the surface.

Hailstone size is maximized by high CAPE, large wind shear, low freezing levels, low PW, and dry mid-level air.

3.5.4 Hail suppression

The damage to property and crops from a single storm can be in the hundreds of millions of dollars or even exceed $1 Billion. The cost of hail damage provides the motivation for hail suppression operations. Hail suppression operations are presently conducted and have a long history in several countries, including the USA, Canada, Argentina, Greece, France, Spain, Germany, China, and Russia (among others). Delivery of the seeding material varies among the countries and can be delivered using aircraft, rockets, or ground-based generators. The most common hail suppression conceptual model utilized in the current operational hail suppression programs in the World involves the use of glaciogenic (ice-forming) materials (e.g. silver iodide) to seed the developing feeder clouds (near the −10°C level when using aircraft) in the upshear, new growth "propagation" region of hailstorms (Dalezios & Spanos, 1995; Dalezios *et al.* 1996). The glaciogenic reagents initiate the rapid development of small ice particles through the condensation-freezing nucleation process, and thus produce enhanced concentrations of ice crystals that compete for the available, supercooled liquid water in storms. This hail suppression theory is generally referred to as "Beneficial Competition" (World Meteorological Oganization, 1995).

The premise of beneficial competition is that the well-documented natural deficiency of ice nuclei (ice-forming particles) in the atmosphere can be enhanced by the release of additional ice nuclei (glaciogenic seeding material) into developing storm clouds. This is done by the combustion of small amounts of reagent and/or solutions containing silver iodide (AgI), either as pyrotechnics (flares) or from wing-borne solution-burning ice nucleus generators. With either method, from 10^{13} to 10^{14} ice nuclei are produced for each gram of silver iodide burned. This potentially increases greatly the number of precipitation embryos in the cloud. These natural and human-made ice crystals, many of which become precipitation, then "compete" for the available supercooled liquid cloud water within the storm. Because the total amount of supercooled liquid remains essentially unchanged, that same mass is divided among the increased number of embryos, meaning the final maximum size of each individual ice particle is decreased. Hence, the hailstones that form within seeded clouds should be smaller and produce less damage if they survive the fall to the surface. If they are sufficiently small, they will melt completely in the warmer subcloud layer and reach the ground as rain.

Cloud seeding alters the microphysics of the treated clouds, assuming that the existing precipitation process is inefficient due to a lack of natural ice nuclei. Cloud seeding does not alter directly the energy or dynamics of the storm. Any alteration of the storm dynamics that does occur results as a consequence of the increased ice crystal concentrations and the development of additional precipitation-sized ice particles earlier in the cloud's lifetime. Because the mature hailstorm consists of complex airflows and precipitation trajectories, cloud seeding does not affect all hail embryo sources. However, the intent is to modify the primary hail formation process. Therefore, the cloud seeding may not eliminate all of the hail, but attempts to reduce the size and amount of hail. This is meant to prevent the growth of large, damaging hail. The seeding also stimulates the precipitation process by speeding the growth of ice-phase hydrometeors, initially into snow pellets (also called graupel) which fall from the cloud earlier, melt, and reach the ground as rain, instead of continuing to grow into large ice particles that reach the ground as damaging hail. This theory is generally referred to as "Early Rainout". The seeding methodology modifies the initial graupel embryo hail development process. Frozen drop hail embryos may originate from secondary sources (shedding from large existing hailstones, or via a recirculation process at the edge of the main updraft). Cloud seeding can only reduce the hail that grows from frozen drop embryos if the available liquid water can be reduced to limit their growth, or if the dynamics of the storm can be affected to eliminate the recirculation processes that formed the drop embryo in the first place. It should be noted that developing cloud regions are seeded by aircraft flying at cloud tops near the –10°C (typically), rockets fired in the developing cloud turret zone (typically –10 to –20°C), or surface based generators emitting silver iodide smoke particles that are carried aloft by the storm inflow and updrafts. A schematic diagram of the conceptual storm model showing the hail origins and growth processes within a hailstorm is shown in Figure 3.19 (WMO, 1995). The cloud seeding methodology applied to the new growth zone of the storm is illustrated.

The introduction of more precipitation embryos through seeding earlier in a clouds lifetime is generally considered to be highly advantageous, reducing the amount and size of any hail, and making the cloud more efficient as a rain producer in the process. Hail growth and suppression issues are a complicated topic. Large hail is an extreme event involving many non-linear, interacting processes, which is currently not possible to predict accurately with confidence. Supercell storms have been recognized as a particular problem. Extensively practiced glaciogenic seeding technologies have been used operationally in many parts of the world to reduce hail damage. However, scientific evidence to date is still considered by many to be inconclusive and evaluation of the results has proved difficult and the effectiveness remains controversial (Dalezios *et al.* 1991). Generally, evaluation consists of comparing unseeded storms with similar hailstorms that are seeded, and comparing the observed amounts and types of precipitation at the surface, as well as the comparisons of radar characteristics of the storm before, during, and after seeding.

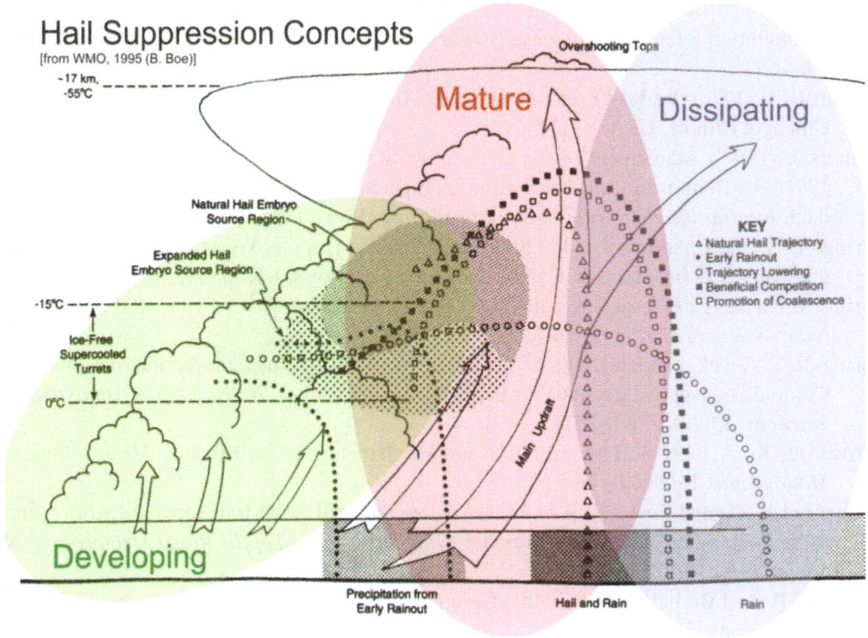

Figure 3.19 Hail suppression concepts (WMO, 1995).

3.6 SUMMARY

Initially there has been a presentation of storm concepts. This is followed by a classification of storms, which has included single-cell, multi-cell and supercell storms, squall lines, mesoscale convective systems (MCS), tropical and extratropical cyclones, description of storm features, tornadoes, lightning and flash floods, as well as precipitation efficiency. Then storm detection has followed, which is conducted through non-coherent radars, coherent or Doppler radars, as well as polarimetric radars. Storm modelling follows, involving cloud and mesoscale models, and storm forecasting, involving storm tracking and nowcasting methods. Finally, the subject of hail is considered, which includes hail formation, detection, forecasting and suppression.

REFERENCES

Adler R. F. and Negri A. J. (1988). A satellite infrared technique to estimate tropical convective and stratiform rainfall. *Journal of Applied Meteorology*, **27**, 30–51.

Amburn S. and Wolf P. (1996). VIL density as a hail indicator. In: *18th Conf. on Severe Local Storms*. Amer. Meteor. Soc., San Francisco, CA.

Anderson T. and Ivarsson K. (1991). A model for probability nowcasts of accumulated precipitation using radar. *Journal of Applied Meteorology*, **30**, 135–141.

Austin G. L. and Bellon A. (1974). The use of digital weather radar recordsf for short-term precipitation forecasting. *Quarterly Journal of the Royal Meteorological Society*, **100**, 658–664.

Battan L. J. (1973). Radar Observation of the Atmosphere. University of Chicago Press, Chicago, Illinois, USA, p. 324.

Bendix J. (1997). Adjustment of the Convective-Stratiform Technique (CST) to estimate 1991/93 El Nino rainfall distribution in Equador and Peru by means of Meteo-3 IR data. *International Journal of Remote Sensing*, **18**(6), 1387–1394.

Bringi V. N. and Chandrasekar V. (2004). Polarimetric Doppler Weather Radar. Principles and Applications, 2nd edn. Cambridge University Press, UK, p. 635.

Bringi V. N., Seliga T. A. and Aydin K. (1984). Hail detection with a differential reflectivity radar. *Science*, **225**, 1145–1147.

Bringi V. N., Vivekanandan J. and Tuttle J. D. (1986). Multiparameter radar measurements in Colorado convective storms. Part II: hail detection studies. *Journal of the Atmospheric Sciences*, **43**(22), 2564–2577.

Browning K. A. (1977). The structure and mechanisms of hailstorms. *Meteorological Monographs* **16**(38), 1–43.

Browning K. A. and Foote G. B. (1976). Airflow and hail growth in supercell storms and some implications for hail suppression. *Quarterly Journal of the Royal Meteorological Society*, **102**, 499–534.

Byers H. R. and Braham R. R. (1948). Thunderstorm structure and circulation. *Journal of Meteorology*, **5**(3), 71–86.

Chisholm A. J. and Renick J. H. (1972). The Kinematics of Multicell and Supercell Alberta Hailstorms. Alberta Hail Studies, 1972, Alberta Research Council Report 72-2. 24–31.

Cifelli R. and Chandrasekar V. (2010). Dual-polarization radar rainfall estimation, pp. 105–125. In: Rainfall: State of the Science, F. Y. Testik and M. Gebremichael (eds), Geophys. Monograph 191, AGU, Washington, DC, p. 287.

Court A. and Griffiths J. F. (1986). Thunderstorm climatology. In: Thunderstorm Morphology and Dynamics, E. Kessler (ed.), University of Oklahoma Press, Norman, USA, pp. 9–40.

Dalezios N. R. (2012). Storms. Lecture Notes. University of Thessaly, Volos, Athens, Greece, p. 480.

Dalezios N. R. (2015). Agrometeorology: Analysis and Simulation (in Greek). KALLIPOS: Libraries of Hellenic Universities (also e-book), ISBN: 978-960-603-134-2, p. 481.

Dalezios N. R. and Papamanolis N. K. (1991). Objective assessment of instability indices for operational hail forecasting in Greece. *Meteorology and Atmospheric Physics*, **45**, 87–100.

Dalezios N. R. and Spanos S. I. (1995). Operational efficiency assessment of hail suppression for agriculture in Greece. *Journal of Weather Modification*, **27**(1), 21–35.

Dalezios N. R., Sioutas M. V. and Karacostas T. S. (1991). A systematic hailpad calibration procedure for operational hail suppression in Greece. *Meteorology and Atmospheric Physics*, **45**, 101–111.

Dalezios N. R., Spanos S. I., Papamanolis N. K. and Krauss T. W. (1996). Climatological and microphysical cloud features towards rain enhancement for agriculture in Greece. *Journal of Weather Modification*, **28**(1), 59–74.

Dennis A. S., Schock M. A. and Koscielski A. (1970). Characteristics of hailstorms of Western South Dakota. *Journal of Applied Meteorology*, **9**, 127–135.

Emanuel K. (2005). Genesis and maintenance of "Mediterranean hurricanes". *Advances in Geosciences*, **2**, 217–220.

Foote G. B. (1984). The study of hail growth utilizing observed storm conditions. *Journal of Climateology Applied Meteorology*, **23**, 84–101.

Foote G. B. (1985). Aspects of cumulonimbus classification relevant to the hail problem. *Journal de Recherches Atmospheriques*, **19**, 61–74.

Foote G. B. and Fankhauser J. C. (1973). Airflow and moisture budget beneath a northeast Colorado hailstorm. *Journal of Applied Meteorology*, **12**, 1330–1353.

Giangrande S. E. and Ryzhkov A. V. (2008). Estimation of rainfall based on the results of polarimetric echo classification. *Journal of Applied Meteorology and Climatology*, **47**(9), 2445–2462.

Gourley J. J., Maddox R. A., Howard K. W. and Burgess D. W. (2010). Impacts of polarimetric radar observations on hydrologic simulation. *Journal of Hydrometeorology*, **11**(3), 781–796.

Hong Y. and Gourley J. J. (2015). Radar Hydrology: Principles, Models and Applications. Taylor and Francis Group, Boca Raton, FL, USA, p. 174.

Janjic Z. I. (1984). Nonlinear advection schemes and energy cascade on semi-staggered grids. *Monthly Weather Review*, **112**, 1234–1245.

Janjic Z. I. (1994). The step-mountain eta coordinate model: further developments of the convection, viscous sublayer, and turbulence closure schemes. *Monthly Weather Review* **122**, 927–945.

Knight C. A. and Knight N. C. (2001). Hailstorms. In: Severe Convective Storms, C. A. Doswell (ed.), AMS Met. Mono. Ch 6, American Meteorological Society, Boston, Massachusetts, USA, pp. 223–254.

Krauss T. W. and Marwitz J. D. (1984). Precipitation processes within an Alberta supercell hailstorm. *Journal of Atmospheric Sciences*, **41**, 1025–1034.

Maddox R. A. (1980). Mesoscale convective complexes. *Bulletin of the American Meteorological Society*, **61**(11), 1374–1387.

Marshall J. S. and Palmer W. Mc. K. (1948). The distribution of raindrops with size. *Journal of Meteorology*, **5**, 165–166.

Marwitz J. D. (1972a). The structure and motion of severe hailstorms, Part I: supercell storms. *Journal of Applied Meteorology*, **11**, 166–179.

Marwitz J. D. (1972b). The structure and motion of severe hailstorms, Part II: multicell storms. *Journal of Applied Meteorology*, **11**, 180–188.

Marwitz J. D. (1972c). The structure and motion of severe hailstorms, Part III: severely sheared storms. *Journal of Applied Meteorology*, **11**, 189–201.

Marwitz J. D. (1972d). Precipitation efficiencies of thunderstorms on the high plains. *Journal of Atmospheric Research*, **6**, 367–370.

Mellor D., Sheffield J., O'Connell P. E. and Metcalfe A. V. (2000). A stochastic space time rainfall forecasting system for real time flow forecasting I: development of MTB conditional rainfall scenario genarator. *Hydrology and Earth System Sciences*, **4**, 603–615.

Mesinger F. (2000). Numerical methods: the Arakawa approach, horizontal grid, global, and limited area modeling. In: General Circulation Model Development: Past, Present and Future. International Geophysics Series, Vol. 70, D. A. Randall (ed.), Academic Press, New York, pp. 373–419.

Mesinger F., Janjic Z. I., Nickovic S., Gavrilov D. and Deaven D. G. (1988). The step mountain coordinate: model description and performance for cases of Alpine lee cyclogenesis and for a case of an Appalachian redevelopment. *Monthly Weather Review* **116**, 1493–1518.

Nastos P. T., Karavana-Papadimou K. and Matzangouras I. T. (2015). Tropical-like cyclones in the Mediterranean: Impacts and composite daily means and anomalies of synoptic conditions. In: Proceedings of the 14th International Conference on Environmental Science and Technology, 3–5 September 2015, Rhodes, Greece, CEST2015_00407.

Rigo T. and Carmen Llasat M. (2016). Forecasting hailfall using parameters for convective cells identified by radar. *Atmospheric Research*, **169**, 366–376.

Rogers E., Black T., Collins W., Manikin G., Mesinger F., Parrish D. and Dimego G. (2000). Changes to the NCEP Meso Eta Analysis and Forecast System: Assimilation of Satellite Radiances and Increase in Resolution. http://sgi62.wwb.noaa.gov:8080/ETA22TPB/

Seliga T. A., Humphries R. G. and Metcalf J. I. (1990). Polarization diversity in radar meteorology; early developments. In: Radar in Meteorology, D. Atlas (ed.), American Meteorological Society, Boston, MA, pp. 109–114.

Skamarock W. C., Klemp J. B., Dudhia J., Gill D. O., Barker D. M., Wang W. and Powers J. G. (2006). A Description of the Advanced Research WRF Version2, NCAR/TN-468+STR, p. 88.

Waldvogel A., Schmid W. and Federer B. (1978). The kinetic energy of hailfalls. Part I: Hailstone spectra. *Journal of Applied Meteorology*, **17**, 515–520.

Waldvogel A., Federer B. and Grimm P. (1979). Criteria for the detection of hail cells. *Journal of Applied Meteorology*, **25**, 1521–1525.

World Meteorological Organization (1995). WMO Meeting of Experts to Review the Present Status of Hail Suppression. Golden Gate National Park, South Africa, 6–10 November. WMP Report No. 26, WMO Technical Document No. 764, R. List, Editor. p. 39.

Zhao Q., Black T. L. and Baldwin M. E. (1997). Implementation of the cloud prediction scheme in the Eta Model at NCEP. *Weather and Forecasting*, **12**, 697–712.

Chapter 4

Floods

Elisa Destro, Efthymios I. Nikolopoulos, Jean-Dominique Creutin and Marco Borga

4.1 FLOOD AND FLOOD RISK

Floods are defined as: "the overflowing of the normal confines of a stream or other body of water or the accumulation of water over areas that are not normally submerged. Floods include river (fluvial) floods, flash floods, pluvial floods, sewer floods, coastal floods, and glacial lake outburst floods" (Field *et al.* 2012). These various flood types are generated by different mechanisms and are associated to different temporal and spatial scales, ranging from localized flash floods that develop in less than one day to those that evolve over days or weeks (riverine floods) and impacts very large basins.

Flooding impacts on where people choose to live, their ability to travel between places uninterrupted and stay connected, their livelihood and security, and from time to time causes enormous damage to property and infrastructure. When societies are not prepared or protected, flooding can have devastating results with widespread destruction of property and a great number of fatalities and population displacement. Floodings have been, and will always be, a persistent problem facing humanity. It cannot be eliminated or even avoided, but with proper understanding, preparedness and appropriate infrastructure development it can be protected against, or its impacts can be mitigated. This is what makes flood prediction a fundamental element of catchment hydrology.

This chapter provides a basic understanding of floods and of the flood risk system, with a focus on small to intermediate size basin (with area less than 1000–3000 km^2) which may be subject to flash floods. In this chapter we will not consider sewer floods, coastal floods and glacial lake outburst floods.

4.1.1 The flood hydrograph and its shape

A typical flood hydrograph is represented in Figure 4.1. Total streamflow during a precipitation event includes the baseflow existing in the basin prior to the storm and the runoff due to the given storm precipitation. Total streamflow hydrographs are usually conceptualized as being composed of:

(i) direct runoff, which is composed of contributions from surface runoff and quick interflow;
(ii) baseflow, which is composed of contributions from delayed interflow and groundwater runoff.

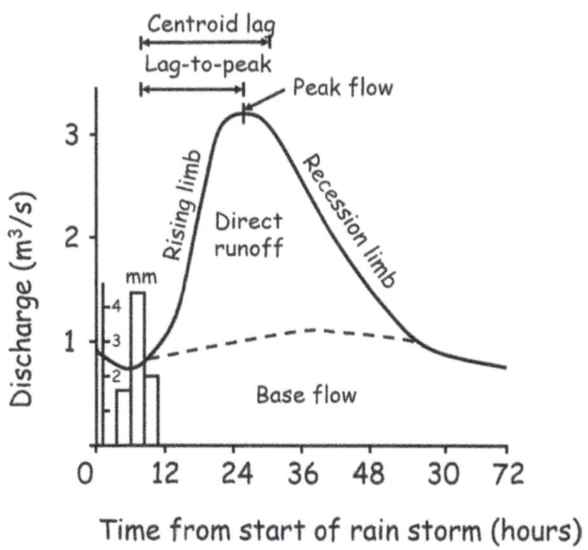

Figure 4.1 Schematic description of a flood hydrograph.

Surface runoff includes all overland flow as well as all precipitation falling directly onto stream channels. Surface runoff in many situations is the main contributor to the peak discharge. Interflow is the portion of the streamflow contributed by infiltrated water that moves laterally in the subsurface until it reaches a channel. Interflow is a slower process than surface runoff. Components of interflow are quick interflow, which contributes to direct runoff, and delayed interflow, which contributes to baseflow (e.g., Bras, 1990).

Groundwater runoff is the flow component contributed to the channel by groundwater. This process is extremely slow as compared to surface runoff.

Schematically in Figure 4.1, the streamflow hydrograph is subdivided into a) Rising Limb: rising portion of the hydrograph, composed mostly of surface runoff. b) Crest: zone of the hydrograph around peak discharge. c) Recession Limb:

portion of the hydrograph after the peak discharge, composed mostly of water released from storage in the basin. The lower part of this recession corresponds to groundwater flow contributions.

The main factors affecting hydrograph shape are (Bras, 1990):

(i) *Catchment characteristics and state*: basin area, basin shape, basin slope, drainage density, drainage network topology, soil type and land use, soil moisture. Most changes in land use tend to modify the amount of runoff for a given storm.
(ii) *Rainfall characteristics*: rainfall intensity, duration, and their spatial and temporal distribution, as well as storm motion, as storms moving in the general downstream direction tend to produce larger peak flows than storms moving upstream.

Hydrographs are also described in terms of the following time characteristics (see Figure 4.1):

(i) *Lag to peak*: time from the effective rainfall centroid to the flood peak time. The lag to peak is largely determined by drainage characteristics such as drainage density, slope, channel roughness, and soil infiltration characteristics. Rainfall distribution in space also affects the lag to peak.
(ii) *Centroid Lag*: time between the center of mass of the effective rainfall hyetograph and the center of mass of the direct runoff hydrograph. The centroid lag is an important concept in the modeling of basin response. Many empirical equations have been proposed in the literature. The simplest of these equations computes the basin lag as a power function of the basin area.
(iii) *Time of Concentration*: time required for water to travel from the most hydraulically remote point in the contributing area to the basin outlet. Time of concentration is often treated as an essentially constant characteristic time for a given watershed, a funcion of the extent and configuration of the contributing area.

4.1.2 Floods as a natural disaster

Floods are the most common natural disaster and the leading cause of natural disaster fatalities worldwide (Doocy *et al.* 2013) and as such represent one of the most burning societal issues hydrology has to face. Flood-related economic losses have increased dramatically over the past decades in most part of the world, and the number of flood fatalities has increased in some continents (Di Baldassarre *et al.* 2010). Data from the global coverage loss databases such as EM-DAT (Centre for the Epidemiology of Disasters, http://www.emdat.be/), NatCatSERVICE (Munich Re, https://www.munichre.com/touch/naturalhazards/en/natcatservice/default.aspx) and Sigma (Swiss Re, http://www.swissre.com/sigma/) show an increase in reported flood disasters and flood losses through time (Kundzewicz *et al.* 2014). However, part of this increase may be attributed to improvements in reporting (Peduzzi *et al.* 2009, 2012), population increase and urbanization in flood-prone areas, increase of property value and reduced awareness about natural risks. For the period 1980–2011, reported flood losses (adjusted for inflation) have increased from an average of US$7 billion per year in the 1980s to some US$24 billion per year in the period 2001–2011, of which an average of 9% was insured (data from Munich Re, NatCatSERVICE, September 2012).

Doocy *et al.* (2013) report that in the same period 1980–2011 approximately 2.8 billion people have been affected by floods with 4.5 million left homeless, at approximately 540,000 deaths and 360,000 injuries. The mortality estimate presented in their study is consistent with the range of estimates presented in other studies (Jonkman & Kelman, 2005). The majority of flood related deaths are concentrated in less developed and heavily populated countries, with Southeast Asia and the Western Pacific region experiencing the highest risk of flood related deaths. For the US, Ashley and Ashley (2008) report 4,585 fatalities over a 47 year period (97.6 deaths/year). No significant increase in flood mortality over time is reported in their study. The majority of flood related deaths were associated to flash floods and were in relation to people in motor-vehicles (63%). The number of flood-related casualties is close to that reported for Europe over the 1970–2005 period by Barredo (2007).

Flash floods, characterized by high velocity flows and short warning times have the highest average mortality rates per event and are responsible for the majority of flood deaths in developed countries. In contrast, riverine floods, which are caused by gradual accumulation of heavy rainfall are less likely to cause mortality because of sufficient time for warning and evacuation. Occasionally floods are associated with secondary hazards such as mudslides in mountainous areas.

The map in Figure 4.2 indicates the location of 3713 flood events during 1985–2010 recorded in the public global database maintained by G.R. Brakenridge at the Dartmouth Flood Observatory (http://floodobservatory.colorado.edu). The map illustrates a clear link between flood reporting and concentrations of population. In addition, regions with few reported floods include the great deserts and high mountains of western North America, southern South America, northern Africa, central Asia and Australia. It is not known if areas of Amazonia actually lack large floods or if instead the online-accessible reporting is sparse; the same holds true for portions of Russia.

4.1.3 The flood risk system: terms and concepts

When discussing about flood risk it is important to remember that 'risk' is entirely a human concern; flood from rivers are predominantly natural events; they are random. The risk arises because the human use and value of the river plains conflicts with their natural function of storage and movement of water during a flood. This is shown in Figure 4.3, which provides both the absolute and relative exposures to river flood risk across countries. In some countries, population and assets are concentrated in flood-prone areas. The highest relative share of population and percentage of economy exposed to floods is found in Cambodia, Bangladesh and Vietnam (Peduzzi *et al.* 2009). Bangladesh is the country with the highest number of people exposed to floods, both in absolute and in relative terms, while it holds ranks four and two in total assets exposed, in absolute and in relative terms, respectively.

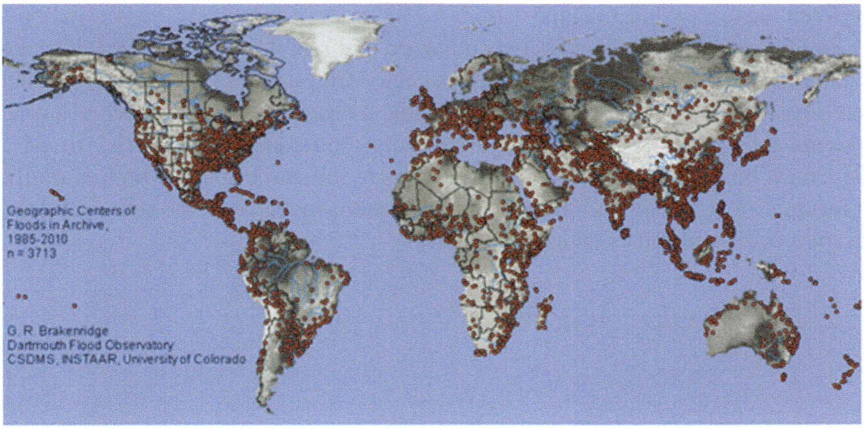

Figure 4.2 Geographic centres of large floods in the dartmouth flood observatory archive. The map illustrates 3713 events over the period 1985–2010.

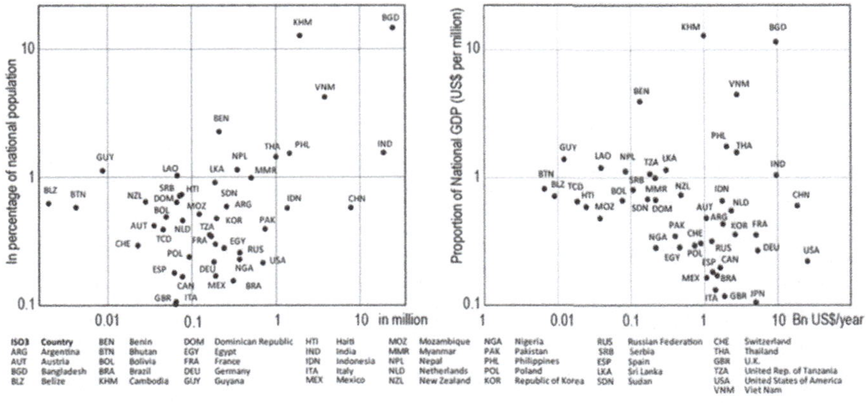

Figure 4.3 Exposure to floods: (left) number of people exposed to floods (per year) in terms of absolute numbers and relative proportions; (right) total assets and GDP exposed to floods (per year), absolute and relative. Source: Peduzzi *et al.* (2009. Peduzzi, P., *et al.* 2011. Preview Global Risk Data Platform, [online], UNEP/GRID Geneva, UNISDR, World Bank. Available from: http://preview.grid.unep.ch/index.php?preview=tools&cat=1&lang=eng).

This shows the importance of the 'flood risk system' concept as the combination of physical and socio-economic characteristics of a region exposed to floods. A difficulty with the terminology of "risk" is that it has been developed across a wide range of disciplines and activities, there is therefore potential for misunderstanding in technical terminology associated with risk assessment, since technical distinctions are made between words which in common usage are normally treated as synonyms. Most important is the distinction that is drawn between the terms "hazard" and "risk".

To understand the linkage between hazard and risk it is useful to consider the commonly adopted Source-Pathway-Receptor-Consequence (S-P-R-C) model (see Figure 4.4). This is, essentially, a simple conceptual model for representing systems and processes that lead to a particular consequence. For a risk to arise there must be hazard that consists of a 'source' or initiator event (i.e. high rainfall); a 'receptor' (e.g. flood plain properties); and a pathway between the source and the receptor (i.e. flood routes including defences).

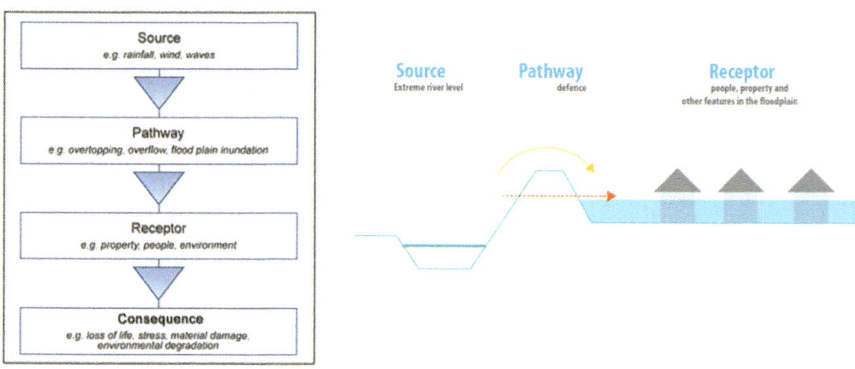

Figure 4.4 Source – Pathway – Receptor-Consequence conceptual model.

A hazard does not automatically lead to a harmful outcome, but identification of a hazard does mean that there is a possibility of harm occurring, with the actual harm depending upon the exposure to the hazard and the characteristics of the receptor.

Thus, to evaluate the risk and take decisions in view of risk management, consideration needs to be made of a number of components:

(i) the nature and probability of the hazard (*p*);
(ii) the degree of exposure of the Receptors (numbers of people and property) to the hazard (*e*);
(iii) the susceptibility of the Receptors to the hazard (*s*);
(iv) the value of the Receptors (*v*).

The value of a receptor is used to translated the degree to which a receptor is impacted into the corresponding loss or damage to society. In its most straightforward form the receptor value is expressed in monetary terms. An alternative could be a sort of (semi) quantitative or qualitative scale. Such a scale is used when monetisation is either not possible or desirable.

Risk is therefore a function of the four variables: p, e, s, v, through the 'risk equation'. The process of flood risk management considers all the four variables included in the risk equation.

4.2 FLOOD TYPOLOGIES AND SCALES

Floods can be generated by a range of processes related to the heavy precipitation properties and catchment state. Floods can also be produced by snowmelt in cold regions and rain on snow events (Merz & Blöschl, 2003). The flood regime is the combined effect of rainfall variability at many time scales and the interactions with landscape dynamics, in particular soil moisture and snow processes (e.g., Robinson & Sivapalan, 1997; Sivapalan *et al.* 2005). Flood processes are often seasonal, i.e. the floods do not occur in all months with the same probability.

The consideration of the interplay between atmospheric input, catchment state (soil moisture, snow) and catchment dynamics provide grounds for typologies of flood events. On this basis, Merz and Blöschl (2003) identified five flood process types: long-rain floods; short-rain floods; flash floods; rain-on-snow floods; and snowmelt floods. These flood types are shortly described below.

Long-Rain Floods: Rainfall over several days or possibly weeks, including low-intensity rainfall, can saturate the catchment and cause high flow conditions. The storage capacity of the catchment is finally exceeded and any additional rain generates a flood event. The rainfall events are synoptic or frontal type storms and often cover a large area up to several thousands of square kilometers.

Short-Rain Floods: Rainfall of short duration and high intensity occurs and can saturate parts of the catchment. Flood runoff results from a combination of runoff from saturated areas, runoff from parts of the catchment where the rainfall intensities exceed infiltration capacity and from fast subsurface flow. The event rainfall depth is moderate to large, and wet antecedent conditions enhance the magnitude of this type of event. Depending on the rainfall patterns, the event can be of a local or regional scale.

Flash Floods: Short, high intensity rainfalls, mainly of convective origin, can trigger floods even if the catchment is in a relatively dry condition. Locally, the rainfall intensities exceed infiltration capacity. Runoff rates often far exceed those of other flood types due to the rapid response of the catchments to intense rainfall, modulated by soil moisture and soil hydraulic properties. Owing to the convective nature of the generating rainfall, the spatial extent of this event is limited (up to 2000–3000 km^2, depending on the climatic conditions) (Figure 4.5). The catchment response tends to be very fast (generally less than one day), partly because of limited spatial coverage of the catchment by the rainstorm, and partly because of most of the flow paths being on the surface.

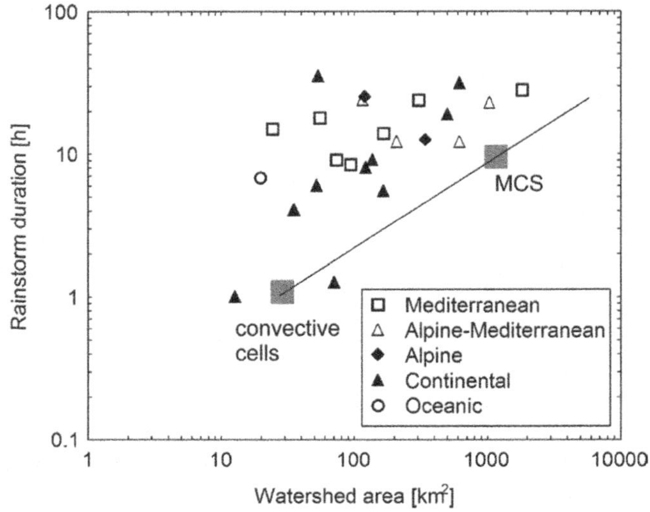

Figure 4.5 Spatial and temporal scales for a number of extreme flash floods analysed by Marchi *et al.* (2010). For each flood, the largest impacted watershed size and the corresponding rainstorm duration were considered. Adapted from Marchi *et al.* (2010).

Rain-on-Snow Floods: Rain falls on an existing snow cover. Moderate rainfall depths can cause enormous runoff depths as a result of a number of mechanisms. During the rainfall, significant long wave radiation and latent heat inputs enhance snowmelt as compared to dry spells. Antecedent snowmelt may saturate large parts of the catchment facilitating overland flow once rain starts. In Alpine catchments, large rain-on-snow floods tend to occur at the end of the winter period when river flows are high due to prior snowmelt. In the lowlands, large rain-on-snow floods can also occur in early winter when snowfall events, snowmelt events and rain events alternate.

Snowmelt Floods: Snowmelt occurs during fair weather periods often associated with a rapid increase in air temperature. The melt energy is mainly global radiation in higher altitudes and turbulent heat exchange in lower altitudes. Snowmelt usually occurs over a period of one or two weeks in sequence, saturating the soils, continuously raising the flows and finally causing a flood. Rainfall may occur but is of relatively minor importance. As there is an upper limit of energy available for melt, these floods are never very extreme. Snowmelt floods can only occur in catchments where a large amount of water is stored in the snowpack. They mainly occur in early winter or in spring, depending on the altitude.

Figure 4.6 shows the interplay between flood seasonality (date of occurrence of the maximum annual flood), catchment elevation and flood types for a case study in Austria (Merz & Blöschl, 2003).

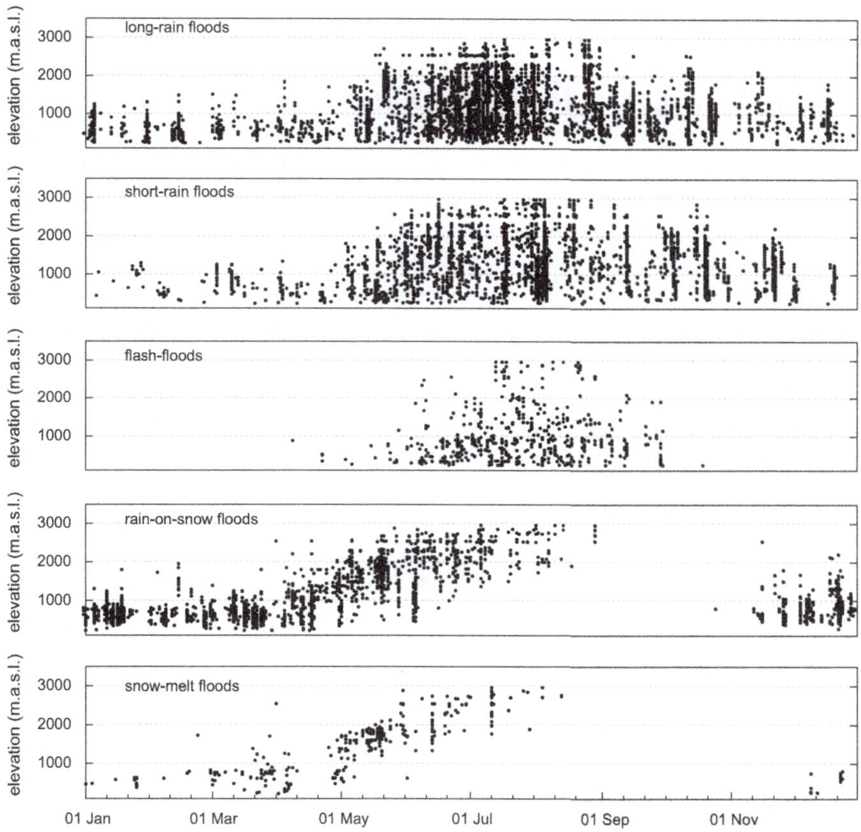

Figure 4.6 Average catchment elevation plotted versus the date of occurrence of the maximum annual floods in Austria, stratified by process type. From Merz and Blöschl (2003).

4.2.1 The case of flash floods

Although flash floods generally affect a limited number of persons when compared with other types of floods, they can be considered as the most deadly type of flood (Doocy *et al.* 2013). According to Barredo (2007), flash floods in Europe caused 2764 fatalities over the period 1950–2005, i.e., 49 casualties per year on average. Similar values of flash flood-related fatalities are reported for the United States (U.S.) by Ashley and Ashley (2008). Owing to these specific flood risk features, more observations are reported in the following for flash floods.

The high risk potential of flash floods is related to their rapid occurrence and to the spatial dispersion of the areas which may be impacted by these floods. Both characteristics limit the ability to issue timely flood warnings. Indeed, the sudden nature of the response is a characterizing feature of flash floods. In the USA flash floods are regarded as having a time to peak up to 6 hours for catchments up to 400 km^2 (Georgakakos & Hudlow, 1984). Marchi *et al.* (2010) showed that this definition may apply to the Mediterranean and Continental areas of Europe as well. The small spatial and temporal scales of flash floods, relative to the sampling characteristics of conventional rain and discharge measurement networks, make also these events particularly difficult to observe and to predict (Borga *et al.* 2008).

Building upon the investigation by Gaume *et al.* (2009), Marchi *et al.* (2010) examined in more detail the characteristics of a number of extreme flash flood occurred in Europe in the last decades. Among other observations, these authors reported the relationship between the catchment area and the unit peak discharge (i.e., the ratio between the peak discharge and the upstream catchment area), known as the flood envelope curve. The envelope curve from Marchi *et al.* (2010) is reported in Figure 4.7 is as follows:

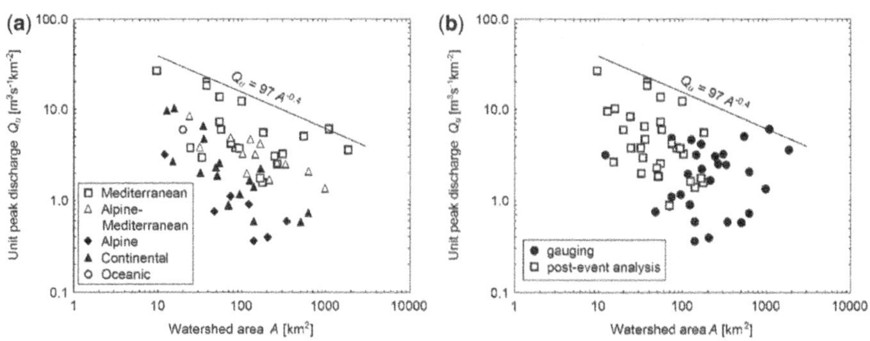

Figure 4.7 Unit peak discharges versus drainage areas; the envelope curve derived from Gaume *et al.* (2009) is also reported. a) climatic regions; b) discharge assessment method. Adapted from Marchi *et al.* (2010).

$$Q_u = 97.0 \cdot A^{-0.4} \tag{4.1}$$

where Q_u is is the unit peak discharge in m^3 s^{-1} km^{-2} and A is the upstream area in km^2. Figure 4.7 shows that the curve is generated by flood events occurring in the Mediterranean region. However, for small basin areas (less than 100 km^2), the flash floods observed under Continental climate (Central Europe) also attain high values of unit discharge, even though these unit peaks seem to decrease with upstream area in a faster way than for the Mediterranean events. This behaviour points out once more the different space and time scales of the generating storm events.

The exponent in the power-law relationship in Eq. 4.1 is smaller than that reported in the literature for extreme floods in the world (Herschy & Fairbridge, 1998) and for Europe (Herschy, 2002). Herschy and Fairbridge (1998) considered 41 maximum floods from the world to develop the following envelope equations:

$$Q_u = 850 \cdot A^{-0.643} \quad A \geq 100\,\text{km}^2 \tag{4.2}$$

For the maximum floods in Europe, Herschy (2002) proposed the following equation:

$$Q_u = 164 \cdot A^{-0.64} \tag{4.3}$$

which provides considerably smaller unit peak discharges than those corresponding to Eq. 4.1 for basin areas larger than 10 km^2. Interestingly, the archive used by Herschy included relatively few flash flood events, reflecting more accurately the behaviour of riverine floods.

Marchi *et al.* (2010) reported that data for more than half of the catchments, and around 80% of the data for basins less than 100 km^2, were provided by post-flood surveys following the methodology described in Borga *et al.* (2008) (Figure 4.7b). These proportions identify the observational problem which characterizes flash foods, which is especially severe for the events characterized by smaller spatial extent. As documented by Ruin et al. (2008), most of the casualties related to flash floods are concentrated in catchments with area less than 100 km^2. So, lack of data at these scales provides a severe limit to the development of understanding flash flood risk.

4.2.2 Forensic analysis of flood peaks

Observations of traces left by water and sediments during a flood provide an opportunity for developing spatially detailed estimates of peak discharges along the stream network (Figure 4.8). This is particularly helpful for better understanding the role of rainfall accumulation and rates and of soil and land use properties in runoff generation for flood events characterised by sharp gradients in runoff response properties and lack of streamgauge data (such as flash floods).

Figure 4.8 a) Example of a flood mark. The vegetation removed from the rocky bank and the moss drenched with silt on the downstream side of the tree show the highest level reached by floodwater (red line). b) The arrow shows the tree with the flood mark and a phase of the topographic survey of the river section. c) Surveying the stream bed.

Indirect methods for flood peak estimation include the slope-area, contracted opening, flow-over-dam, or flow-through-culvert. However, the important thing to note here is that the survey needs to capture not only the maxima of peak discharges: less intense responses within the flood-impacted region are important as well. These can be contrasted with the corresponding generating rainfall intensities and depths obtained by weather radar re-analysis, thus permitting identification of the catchment properties controlling the rate-limiting processes (Zanon *et al.* 2010). Clearly, not all the river sections may be suitable for indirect peak discharge estimation. However, Borga *et al.* (2008) have shown that, provided that a careful logistical planning and properly staffed infrastructure is ensured, post-event surveys may deliver a spatially consistent analysis of the flood response. Surveying the geomorphic response, through mapping of landsliding/debris flow initiation and deposition areas, is important as well. This may help to properly identify the flow processes that occurred in the basin and hence to avoid questionable peak discharge estimates due to incorrect identification and documentation of debris flows.

Marchi *et al.* (2010) has shown the advantages of integrating the indirect peak discharge estimates with spatially distributed model-based flood simulations. With this integration (Figure 4.9), simulated flood hydrographs are first compared with indirect peak discharge estimates, then model-based analysis is carried out where modelled flood hydrographs are consistent with field-derived peak observations. The first step permits to assess the consistency between the various sources of rainfall and discharge observations, whereas the second allows one to evaluate the water balance at the event scale and to provide estimates of the runoff coefficients (ratio of event runoff to event rainfall). Analysis of event runoff coefficients may provide essential insight on how different landscapes 'filter' rainfall to generate runoff and how the observed differences can be explained by catchment characteristics (Blume *et al.* 2007; Norbiato *et al.* 2009). A critical step in the integrated flash flood analysis is to quantify how field- and model-based peak estimates are close enough, taking into account the relevant uncertainties.

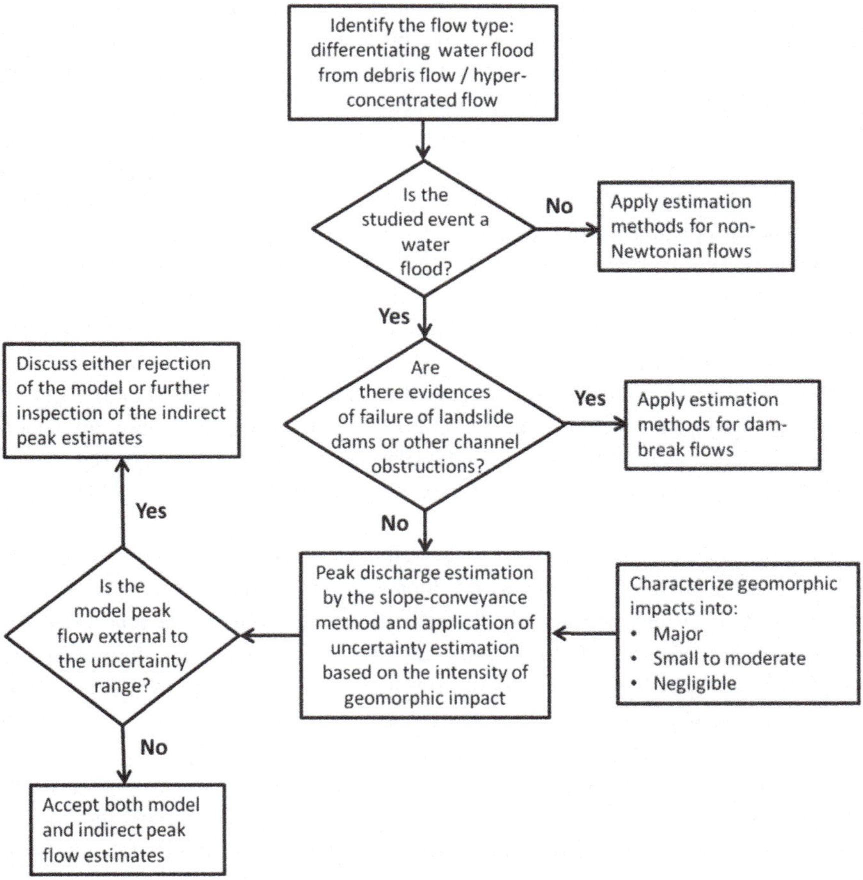

Figure 4.9 Flowchart for integrated use of indirect flood peak estimates and flood modelling: indirect estimate of peak discharge, uncertainty assessment and comparison with model-based peak flows.

4.3 FLOOD RISK MANAGEMENT

Floods have always offered benefits and presented challenges, enriching the land for agriculture and habitat creation by spreading sediment-laden waters across the floodplain, but making the creation of permanent river crossings difficult if not impossible. Nomadic communities learned to live with the episodic nature of floods, but as permanent settlements were established to take advantage of the floodplain, floods began to impact negatively on the lives of those living there. Societies therefore began to take steps to lessen the impact of flooding. At first, these efforts were minimal, consisting of little more than minor adjustments in living style. As populations increased and the economic importance of the floodplain land grew, societies began to take structural actions to keep flood waters away from important areas. Such measures were often difficult to sustain, and invariably were overwhelmed by the next great flood. Today, millennia after these first efforts, the challenges remain. From the earliest recorded attempts of society to deal with flooding until late in the twentieth century, the principal means of mitigating the impacts of floods was flood control. Levees, dykes, diversion channels, dams and related structures were all constructed in an effort to control the natural and periodic rise of rivers. In the middle of the twentieth century, there was a shift to an approach that sought to use structural and nonstructural measures both to prevent flooding and reduce the damages when it occurs (Sayers *et al.* 2013).

As populations and development grew, flood losses continued to increase, and the need to prioritize investment became increasingly acute. A new approach was needed, one that could not only identify the hazards and the consequences faced by society, but was also able to assess the relative significance of the risks faced.

In the 1980s the United Nations put forward the concept of sustainable development (UN, 1987). The ideas of sustainable development supported the increasing concern associated with the environmental consequences of development in general and in floodplains in particular, and the critical role of maintaining ecosystem goods and services. This supported some national governments in moving away from flood management solutions based solely on structural approaches, towards providing a mix of nonstructural and structural responses.

In the late 1970s and early 1980s, risk management began to feature more strongly in the governance of manufacturing industry, air travel and power generation (both hydro-electric and nuclear), covering all industrial activities that placed either the sector or society in general at risk. A seminal paper by the UK Health and Safety Executive (HSE) (HSE, 2001) set out a framework within which both the risk to individuals and society as a whole could be considered and traded against the benefits secured. The HSE introduced the concept that risks (which cannot be eliminated) should be managed to a level that is as low as reasonably practicable (ALARP). In the ALARP methodology, 'practicability' is assessed through consideration of both costs (described as all costs, monetary and nonmonetary) and benefits (described as all benefits, both monetary and nonmonetary). The HSE also introduced the concept of 'unacceptable' risks. In this case, efforts must be made to reduce the residual risk unless the costs of doing so can be demonstrated to be disproportionate to the risk reduction achieved (Figure 4.10).

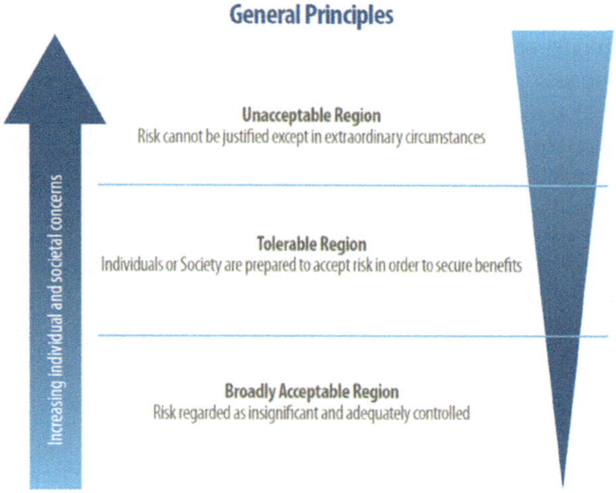

Figure 4.10 The framework of tolerable risk introduced by the HSE in the early 1990s.

The establishment of the Intergovernmental Panel on Climate Change (IPCC) in the late 1980s focused even more attention on the challenge of dealing with an uncertain future.

Major floods on the Rhine River in 1993, again on the Rhine in 1995 and 1997 and in the United Kingdom in 1998 brought increased attention to the growing challenge of flooding. The Rhine flood of 1993 threatened to inundate much of the Netherlands. As a result there was considerably increased activity on the flood risk topic, as both academic and governmental organizations focused their efforts on dealing with growing flood losses across the European Community. During that period the process of risk management start to be used more formally and routinely in flood management, to form the flood risk management (FRM) process. Communicating the residual risk became a central component of FRM, since only through knowledge can individuals and organizations take their own measures and make their own choices regarding the acceptability of the residual risk.

In 2000, the European Union issued a Water Framework Directive addressing the steps necessary to reduce pollution in European rivers and establish river basin management as the framework for cooperative efforts to accomplish the objectives of the Directive. Following additional flood events in Europe during the first decade of the twenty-first century, the European Parliament and Council issued a directive on the 'assessment and management of flood risks' (EC, 2007). A Floods Directive established a framework for this assessment and management, with the goal of reducing adverse consequences of flooding to human health, environment and cultural-economic activity in the European Community. As a first step, the Directive requires that Member States conduct preliminary flood risk assessment of the river basins in their territories, including the assessment of the potential impacts of climate change. It also directed that Member States prepare flood hazard and flood risk maps, and FRM plans for their river basins.

Modern FRM recognizes that there is seldom a single solution to managing flood challenges. Instead, portfolios of FRM measures and instruments are utilized. Such portfolios assembled a range of actions in such a way to reduce risk in an efficient and sustainable manner, and draw upon:

(i) 'hard' structural measures (such as construction of dykes, levees and dams);
(ii) 'soft' structural measures (such as wetland storage)
(iii) nonstructural measures (such as improved flood forecasts and warnings)
(iv) policy instruments (such as land use planning, insurance and other funding incentives, such as homeowner grants for flood proofing).

The criteria for assessing FRM strategies are no longer solely economic, but involve consideration of a much broader set of outcomes, including social justice and ecosystem health. Equally, an increasing recognition of nonstationarity within the flood system (that is, climate, geomorphologic and socioeconomic change) forces an explicit consideration of a full range of plausible ways in which flood risk may shift in the future. This continuous process of adaptation is distinct from the 'implement and maintain' philosophy of a traditional flood defence approach.

Implementing FRM places a high demand on its stakeholders. It involves the collective action of a range of different government authorities and those outside government, including the public and business. This places an increasing emphasis on effective communication and mechanisms to reach consensus without succumbing to the short-termism that may be present in the many competing views. Increasingly, the move towards FRM is becoming embedded in national government policy.

4.3.1 The components of risk

Risk has two components – the chance (or probability) of a potentially harmful event occurring and the impact (or consequence) associated with that event, therefore:

Risk = f (probability of inundation and the associated consequences).

Based on this approach, flood risk assessments start with an assessment of the flood hazard, which indicates the probability and intensity of a possible harmful event (hazard assessment; e.g. Pappenberger *et al.* 2012; Alfieri *et al.* 2013). This reflects both the probability of the occurrence of the initiating event and the probability that flood waters will reach a particular location in the floodplain, taking account of the performance of the intervening system of wetlands, channels, dams, levees, floodwalls and other structures (the pathway of the flood water) (de Moel *et al.* 2015).

This hazard information can be overlaid with socio-economic information, such as land use data, building datasets, information on population, regional gross domestic product (GDP), etc. Doing so will give an indication of what is actually exposed to flooding (i.e. exposure assessment; e.g. Jongman *et al.* 2012). Information on the exposure can be combined with information on the vulnerability of such assets (for instance through damage curves) and the hazard characteristics to estimate the potential damage (damage assessment; e.g. Kreibich & Thieken, 2008; de Moel & Aerts, 2011).

A generalised procedure to estimate direct monetary damage consists of the following: (1) Classification of elements at risk by pooling them into homogeneous classes. (2) Exposure analysis and asset assessment by describing the number and type of elements at risk and by estimating their asset value. (3) Susceptibility (or vulnerability) analysis by relating the relative damage of the elements at risk to the flood impact (Merz *et al.* 2010a, b).

When total damages are calculated for several events with different probabilities, an expected annual damage can be calculated, usually referred to as risk (risk assessment; e.g. Aerts *et al.* 2013; Falter *et al.* 2015) (Figure 4.11). Note that there are various other factors that influence flood risk, which may or may not be included in flood risk assessment as they are sometimes difficult to include in a quantitative analysis. Such factors include warning time, preparedness, social and economic resistance, health, etc.

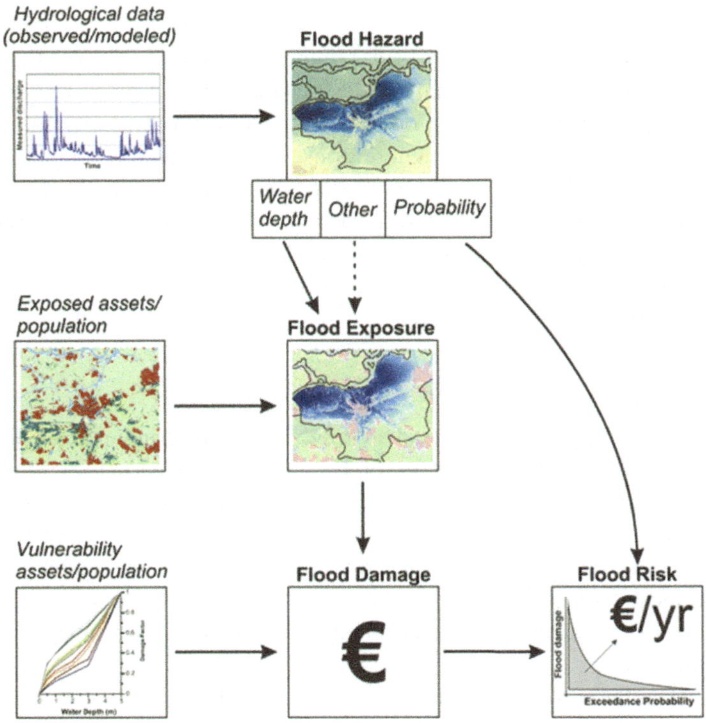

Figure 4.11 Conceptual overview of general flood assessment. From Merz and Thieken (2004).

The following Sections are devoted to the description of the methods used in FRM to assess the hazard and the vulnerability components of risk.

4.4 FLOOD HAZARD ASSESSMENT

The flood hazard assessment aims to relate the hydraulic characteristics of the flood (depth, velocity, duration and extent of flooding) with its return time *T* (a T-year flood is the flood intensity that has a probability of 1/T of being exceeded in any given year). The methodology for such an assessment depends very much on the scale and specific characteristics, such as the geographic setting and the presence of defensive structures, the purpose of the assessment and the available data. In general terms, however, the main hydrological quantity of interest for the flood hazard assessment is the flood hydrograph (or certain parameters, such as the flood peak and the flood volume) associated to the given return time *T* (also called average recurrence interval) (flood frequency estimation).

The methods used for flood estimation can be classified into two groups:

(i) Flood frequency analysis (statistical methods);
(ii) Rainfall-runoff modelling

Flood frequency analysis is based on the analysis of observed historical flood events and estimates the magnitudes of floods with a given return period. Rainfall-runoff modelling, on the other hand, converts a rainfall into a surface runoff using a model of the catchment response based on model parameters which are either calibrated based on observed data or are estimated from catchment characteristics. Flood estimation using the rainfall-runoff method is frequently based on a frequency analysis of extreme rainfall events to establish a precipitation sequence, which is then used in an event-based rainfall-runoff model to simulate the corresponding runoff hydrograph.

4.4.1 Flood frequency analysis

Flood frequency analysis is a statistical approach used to determine the magnitude of a flood event with a certain occurrence probability or return period (Figure 4.12). In contrast to rainfall-runoff modelling, statistical flood frequency analysis is based on observed flood data only, either at the site of interest (*at-site flood frequency analysis*) or from one or several comparable gauged basins within the same region in the case of limited local data availability *(regional flood frequency analysis*). A short data record may also be extended by model simulation and a frequency analysis can then be performed.

A statistical flood frequency analysis is based on the assumption that all events in the observed flood series represent a process that can be described by one single flood frequency distribution. A mathematical function is used to describe the distribution of events, and this function is then extrapolated to give values corresponding to return periods beyond the length of the observed record. Flood frequency analysis can be straightforward if the return period of interest does not significantly exceed the period of observation and all of the observed events are generated by the same flood generating mechanism. Uncertainty, however, increases with increasing return period and, therefore, extrapolation should be avoided as far as possible. If one has to extrapolate, this should be done only as far as necessary and preferably only up to twice the record length.

Figure 4.12 Photograph representative of a flood and flood frequency curve in a mountainous catchment in Austria (river Trisanna at Kapp Nederle, Austria), area 385 km^2, median elevation 2300 m a.s.l. The Generalised Extreme Value distribution has been fitted to the maximum annual peaks with the Bayesian method. Grey shading represents the confidence intervals from 50 to 99.9%.

Additional information to provide independent support to the extrapolated values is valuable, but one should always be aware of the uncertainty. Unfortunately, flood records are frequently of insufficient length, and this introduces significant uncertainty into the flood estimates. In general, a regional flood frequency analysis using data from several stations can be performed to reduce the uncertainty and to "limit unreliable extrapolation when available data record lengths are short as compared to the recurrence interval of interest, or for predicting the flooding potential at locations where no observed data are available" (Castellarin *et al.* 2011).

Current methods for flood frequency analysis assume very often that the data sample consists of independent, identically distributed (iid) events. The criterion of independence implies that there are no autocorrelations, trends or shifts in the sample. For example, the magnitude of one event should not depend on the magnitude of the previous event, and there should be no systematic or abrupt changes over time due to, for example, climate change or anthropogenic influences in the catchment. An example of an abrupt change brought by the construction of a dam is reported in Figure 4.13 for the Zemmbach at Sausteinaste (225 km^2 catchment area). The upstream Schlegeis reservoir was built in 1972. Operation of the dam resulted in a very significant reduction in the flood peaks, as shown in Figure 4.13 left. This change is clearly visible in the time series of the floods, but it is not visible in the flood frequency curve. The reservoir does not only affect the mean/median flood but higher moments as well. For extreme floods, depending on the reservoir size and the management, the free reservoir capacity may be exhausted leading to very little peak reduction. Because the flood peak depends on the free volume of the reservoir, which in turn depends on the reservoir management, a flood of a given return period downstream of the dam is not uniquely defined. This is a very sensitive issue if one considers the number of dams operating around the world, e.g., Nilsson *et al.* (2005) concluded from a global overview that 172 out of 292 investigated rivers are regulated by dams.

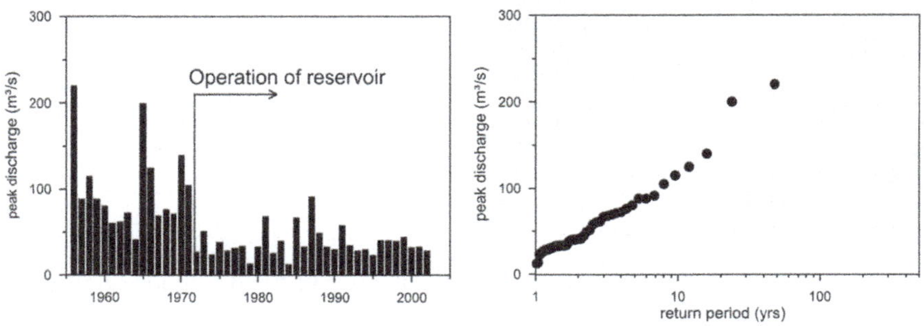

Figure 4.13 (Left) Time series of observed maximum annual peak discharges of the Zemmbach at Sausteinaste (225 km^2 catchment area). The upstream Schlegeis reservoir was built in 1972. (Right) Flood frequency plots of the Zemmbach at Sausteinaste. From Merz and Blöschl (2008).

The requirement of independent events needs to be considered in the general selection methodology for a series of extreme events, which is commonly either an Annual Maximum Series (AMS) or a Partial Duration Series (PDS). The assumption of identical distribution may be violated if the floods are caused by different generating processes.

4.4.1.1 At-site flood frequency analysis

Where sufficiently long data records are available for the site of interest and with reference to the study return time, flood frequency analyses can be relatively straightforward and involves fitting a theoretical statistical distribution to the observed flood data. Several national guidelines recommend performing an at-site analysis where long records (> 30 years) are available. However, for records of 30–50 years, only 2-parameter distributions are recommended for this procedure. For stations with more than 50 years of data, 3-parameter distributions can also be applied, if suitable. Many different statistical distributions are available and commonly applied for flood frequency analysis. For many of them there is no underlying justification for their use other than their flexibility in mimicking the shape of an observed statistical distribution (Coles, 2001). The theoretically correct limit distributions are the Generalized Extreme Value (GEV) distribution in case of AMS and the Generalized Pareto (GP) distribution for PDS.

4.4.1.2 Regional flood frequency analysis

Regional flood frequency analysis is applied when data at a site are insufficient for the reliable estimation of flood quantiles (Cunderlink & Burn, 2001). In this case, flood data from one or several alternative stations with observation records in a "region" are used to improve flood frequency estimation at the site of interest. One could say that space replaces time to increase the sample size and reduce sample uncertainty. A region can be defined geographically or it may comprise stations with similar flow or catchment characteristics. Although a data set constructed with data from different sites may be more heterogeneous, research has shown that more accurate flood frequency estimates are obtained using regional analysis, compared to at-site analysis (Lettenmaier *et al.* 1987). One has, however, to be careful not to overestimate the reduction in uncertainty, as correlation between the stations reduces the size of the sample comprising independent data.

A common approach for a regional flood frequency analysis is the so-called index flood method (Stedinger *et al.* 1993; Hosking *et al.* 1997). This approach estimates the T-year flood q_T as the product of a scale factor, which is called the index flood (often defined as the mean or median annual maximum flood), and a growth factor, which describes the relationship between the dimensionless flood and the recurrence interval, T (the so-called growth curve). In general

$$q_T = g(T) \cdot f(A, P, \ldots) \tag{4.4}$$

where $g(T)$ is the growth curve and the index flood is expressed as a function of climatic and catchment characteristics such as area (A), mean annual precipitation (P) etc.

In equation (4.4) the index flood is site-dependent, while the growth factor is assumed to be the same for the entire homogeneous pooling group. The index flood method thus assumes that the distribution of flood peaks at different sites within a pooling group is the same except for a scale parameter, which is the index flood for a site. The procedure has been developed to improve the estimation of the growth curve due to the exploitation of information from the entire homogenous group as opposed to one-site samples.

A regional flood frequency analysis based on the index flood method hence comprises three steps: (1) identification of regions or similar sites, (2) calculation of the index flood and (3) calculation of the growth curve.

When flood data are available at the site, the mean or median flood is usually used as the index flood. For an ungauged site, or a site with limited data, it can either be estimated using flow data from nearby or similar sites or it can be derived using regional regression formulas. In the first case, the index flood can be scaled to the site of interest based on the catchment area. However, other catchment characteristics can also play a significant role, which can lead to either an under- or over-estimation of the flood frequency at the site of interest. Therefore, regional formulas for the calculation of the index flood can be used. This is a very practical approach when no data or only limited data exist. However, in general, flood estimates derived from catchment descriptors are grossly inferior to estimates made from flood peak data, even those estimated from short records (Reed & Robson, 1999). Brath *et al.* (2001) compared different methods for estimating the index flood at ungauged sites in Northern Italy and found a regression model linking the index flood to a set of catchment descriptors to be the most efficient approach. Kjeldsen and Jones (2010) recently revised the Flood Estimaton Handbook (FEH) procedure for the derivation of the index flood at an ungauged site in the UK using catchment descriptors. They found that local factors are probably not sufficiently represented in the FEH regression models (a single model is used for the whole of the UK), and that flood statistics may benefit from the adjustment of estimates using local data from neighbouring catchments. Their results showed geographical proximity to be the most important factor when identifying a good potential donor site, with little benefit gained by identifying donor sites based on hydrological similarity.

4.4.1.3 Extension of the regional flood frequency analysis

The standardization of methods and techniques for post-flood survey is also instrumental in creating a cohesive European-wide catalogue of flash floods. These data set may prove invaluable to extend flood frequency estimation at scales that are usually ungauged, as shown by Gaume *et al.* (2010) for flash floods and by Merz and Blöschl (2008) for extreme floods. This type of information expansion is particularly important in small catchments, both because fewer and shorter records tend to be available than in larger catchments and because the flood processes are more amenable to analysis than in larger catchments where the regional combination of controls can be relatively more important. A key point made by the end users concern the need to extend the post flood data gathering methodology to collect observations concerning damage and vulnerability characteristics.

4.4.2 Rainfall-runoff modelling

An alternative to the statistical methods of regional flood frequency that rely entirely on regional flood data is to start with local or regional information on rainfall and convert it to a flood using rainfall-runoff models. The transformation from rainfall to flood peak or flood hydrograph involves models ranging from very simple to very complex.

There are advantages and disdvantages to rainfall-runoff modelling over statistical methods. The main advantages include:

(i) data series of precipitation are often longer than runoff series and the climate station network is in some locations more dense than the gauging station network;
(ii) rainfall-runoff models are designed to mimic the most important flood generating processes, including local effects from, say, hydraulic structures. This advantage becomes relevant if high return periods are of interest, as the more extreme floods may be the results of altogether different processe than are accounted for in the limited flood data.

On the other hand, rainfall-runoff models suffer from the fact that the model parameters need to be specified for typically ungauged basins by using some forms of transfering or regionalisation procedure.

The classic example of rainfall-runoff models is the so-called rational method, introduced about 150 years ago and credited either to Mulvaney (1851) or to Kuichling (1889) (Chow, 1964, Ch. 14, p. 6) which is widely used all around the world for flood peak estimation in small ungauged basins and for the design of urban storm drainage systems. An important reason for this method's attractiveness to hydrologists and to authorities alike is certainly its simplicity: the model consists of the single equation

$$q_T = C_T I_{T,tc} A \tag{4.5}$$

where A is the area of the catchment, $I_{T,tc}$ is the annual maximum rainfall intensity with a return period T and averaged over a duration t_r chosen to be equal to the mean residence time t_c of the catchment (concentration time), and C_T is an apparent runoff coefficient. In reality C_T is really a conversion factor that transforms the average rainfall intensity $I_{T,tc}$ to the flood peak q_T of return period T. So, in a sense the rational formula stands for the entirety of the rainfall-runoff-flood peak model used here (combining the runoff generation part and the runoff routing part). The choice of $t_r = t_c$ reflects the nature of the simplest possible runoff routing model used, which assumes that the highest flood peak is produced under resonance conditions.

The rational method's elementary concept and the simplicity of Eq. (4.5) often misleadingly prompt towards a recipe-like application. In reality, however, all steps of the procedure described above involve critical assumptions and estimations for which a number of open research questions exist (Efstratiadis *et al.* 2014).

More complex procedures which may overcome the limitations of the rational method, based either on use of derived distribution methods or on continuous rainfall-runoff models, are available (Rosbjerg *et al.* 2007). Continuous models for estimating a flood of a given return period are usually driven by rainfall generated by stochastic rainfall model (Viglione *et al.* 2013).

4.4.3 Hydraulic modelling

The flood parameters for a given return period, estimated based on methods described in Section 4.4.1 and 4.4.2 above, provide the input for detailed 1D/2D hydraulic modelling (e.g. Ernst *et al.* 2010), for which a large variety of models exists. This can consider hydraulically important features like streets, buildings and channels (Aronica *et al.* 1998). Such models are based on solving physical equations for one/two dimensional flow of water, such as the Saint Venant equations (Aronica & Lanza, 2005). These hydraulic models do not only yield inundation extents and depths but can also provide additional flood indicators like flow velocity and the rate at which the water level rises. These can be important factors in determining the consequences of floods. For instance, flow velocity is an important parameter when assessing erosion and the collapse of buildings and the rate of rise is important when assessing potential fatalities. Modelling of the inundation area can be done using a full 2D model or by coupling 1D and 2D calculations in various degrees. For instance, the LISFLOOD-FP model links separate 1D channel flow calculations with 2D flood propagation calculations using uniform flow formulae (Bates & de Roo, 2000).

4.5 THE CONSEQUENCES OF FLOOD: VULNERABILITY ASSESSMENT

The assessment of the consequences should flooding occur reflects both the vulnerability of the receptors and the chance that a given receptor will be exposed to the flood. In the FRM terminology, the terms exposure and vulnerability are used to quantify the consequences of flood.

Exposure quantifies the number of properties or people, area of habitats, and so on that may be exposed to a given flood event should it occur. This can take various forms, but often datasets on population, land use, buildings, etc. are used in this respect (Wünsch *et al.* 2009), or remote sensing data is utilised (Gerl *et al.* 2014). Such exposure assessments can be binary (i.e. affected or not affected by the flood – Jongman *et al.* 2012a), or can involve various gradations, for instance, by taking various water depths into account (de Moel *et al.* 2011). Such exposure estimates usually relate to physical assets and people. However, also other exposure information can be incorporated, such as cultural values (using datasets on monuments and/or cultural heritage sites), parameters associated with indirect effects (using for instance exposure of infrastructure networks or GDP production) or long-term changes in exposure (related to population growth and economic development; e.g. Elmer *et al.* 2012; Tapia-Silva *et al.* 2011).

Assessment of exposure may need to account for dynamic analysis. Some receptors, such as residential properties, can be considered as static, but other receptors such as people, cars and much wildlife may be dynamic – that is, they are liable to move – and they may or may not be present in the area at the time of a flood. The degree of exposure will influence the risk: for example it will differ depending on the time of day the flood occurs (rush hour, night time and so on).

Vulnerability describes the potential for a given receptor to experience harm during a given flood event. Many flood assessments estimate the direct consequences in monetary terms (Merz *et al.* 2010a). By doing so, different types of consequences are expressed in a single figure (e.g. euros/dollars/£), which allows for easy comparisons, for instance, between areas, over time, with other hazards or comparative evaluations of the effect of different measures (in terms of how much damage is reduced). Many flood damage assessments rely on water depth as main indicator of the hazard, though sometimes distinctions are made based on, for instance, the duration of flood (e.g. the 'Multi-Coloured Manual' in the United Kingdom (UK), Penning-Rowsell *et al.* 2013) or the flow velocity (e.g. HAZUS in the USA, Scawthorn *et al.* 2006). This is usually done by differentiating different damage curves or combining different hazard characteristics into a new indicator (e.g. estimating the stability of buildings using velocity-stage-damage curves; Middelmann-Fernandes, 2010). This is different, however, for damage to agricultural crops, for which the time of flooding (i.e. season) and the duration are decisive as opposed to water depth (Tapia-Silva *et al.* 2011).

The most commonly used approach to assess direct damages is based on depth-damage curves (Smith, 1994; Merz *et al.* 2007). Such curves denote the vulnerability to flooding by relating the water depth to the damage this will cause to a specific type of asset, economic sector or land use category (see e.g. Merz *et al.* 2004; Freni *et al.* 2010; de Moel & Aerts, 2011). However, not all approaches are based on damage curves. For example, Booysen *et al.* (1999) used questionnaires to estimate damages to industries in a case study area. Such an approach is particularly useful for categories that are very heterogeneous and need specific details (such as industrial land use). In addition, some multi-parameter models have been developed including: a conceptual model in the UK (Nicholas *et al.* 2001), a multi-variate regression model to estimate losses in private households in Japan (Zhai *et al.* 2005) and rule based models for loss estimation to companies and private households in Germany (Kreibich *et al.* 2010; Elmer *et al.* 2010). Such multi-parameter methods have the advantage that additional important factors that are often difficult to predict (e.g. warning time, contamination and resilience) can sometimes still be included in the assessment.

The development of probabilistic damage models seems to be an innovative enhancement, since they inherently provide quantitative information about model uncertainty. Probabilistic models using bagging decision trees or Bayesian networks have been successfully validated (Vogel *et al.* 2013; Merz *et al.* 2013). These approaches have the advantage that more variables (than just inundation depth) can be included in the damage assessment and have shown the importance of factors like contamination, precaution and duration of the inundation (Thieken *et al.* 2005; Merz *et al.* 2013).

4.5.1 Social vulnerability to flood and the case of flash floods

Proximity or exposure to a hazard agent, the nature of the hazard itself, built environment characteristics, and engineering solutions are critically important to assess and address population vulnerabilities to hazard impacts. However, toward the end of the 20th century researchers began to question why disaster impacts are unequally distributed in the population, with some localities and population subgroups are afflicted disproportionately by disaster outcomes (Bolin & Bolton, 1986; Bates & Peacock, 1987). These scholars advanced a new dimension of vulnerability that focused on the social and economic forces that shape disaster outcomes, leading to the widespread adoption of the term social vulnerability. Over the past decade, social vulnerability indices have emerged as a leading tool to quantify and map human dimensions of hazards vulnerability (Cutter *et al.* 2003; Dwyer *et al.* 2004; Wu *et al.* 2002).

In their classic work, "At Risk: Natural Hazards, People's Vulnerability, and Disasters", Blaikie *et al.* (1994) define social vulnerablity as "the characteristics of a person or group in terms of their capacity to anticipate, cope with, resist and recover from the impacts of a natural hazard. Social vulnerability involves a combination of factors that determine the degree to which someone's life and livelihood are put at risk." In other words, social vulnerability is defined by the possession of social attributes that increase susceptibility to disasters. Social vulnerability scholars address why types of persons locate in hazardous places, live in inadequate homes, fail to anticipate, resist and/or recover from the aftermath of a disaster, and analyze the economic and social forces that shape and determine these dynamics. These dimensions of Social vulnerability are particularly germane to predict spatial variation in flood casualties.

The research literature on social vulnerability is diverse, addressing a variety of hazards and attributes of social vulnerability including race and ethnicity (Fothergill *et al.* 1999; Lindell & Perry, 2004), and measures of economic status like wealth, income, poverty (Dash *et al.* 1997; Fothergill & Peek, 2004) and factors related to risk perception.

Although discussed in the literature (Adger, 2006), the dynamic character of vulnerability is still not represented in applications to flooding (Wilhelmi & Morss, 2012). This means that vulnerability is considered as a key but static element of the system where the inherent elements are neither interacting with each other, nor evolving during an event or even from one specific event to the next one. Terti *et al.* (2015) were able to show that both social vulnerability and social exposure is subject to variation throughout the day, from 1 day to the other and according to the space and location under concern. The variability of exposure depends on the different occupancy of the same space as a function of time of the day ("quantity" of elements at risk). For example, more people are at work during the working hours, on the road during the rush hours, at home during the rest hours, and often at recreational places during the holidays and weekends (Camarasa Belmonte *et al.* 2010). The variability of sensitivity and coping capacity depends on the different contributions of the individual's characteristics, short-to long-term priorities (e.g., individual socioeconomic status and/or daily constraints), and the way they deal with the natural and social crisis circumstances. For example, there are some classes of workers who are employed by time-sensitive businesses and may not be freely capable to adapt their scheduled activities in case of bad weather conditions. As they may be less adaptable to change their normal work-related journey in fear of losing their job, they could be considered a sensitive population when flooding conditions happen around commuting hours (Ruin, 2010).

Terti *et al.* (2015) developed a conceptual model quantifying the interaction between the flooding circumstances and the variable direct or indirect contributions of the embedded social vulnerability factors (see Figure 4.14). This model integrates the primary factors that need to be considered in the vulnerability analysis to predict the level of human impacts for a specific flood event, with specific focus on flash floods.

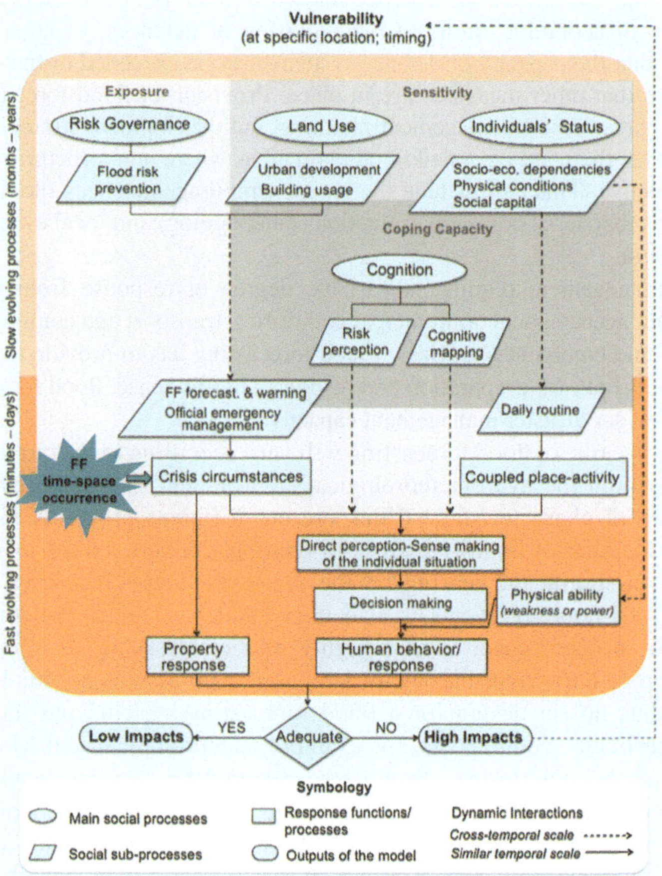

Figure 4.14 Conceptual model of dynamic social vulnerability to flash flood events. from Terti *et al.* (2015).

4.6 FLOOD FORECASTING AND WARNING

A hydrological forecast is an estimate of the future state of some hydrological phenomenon, such as flow rate, cumulative volume, stage level, area of inundation or mean flow velocity, at a particular geographical location or channel section. The lead time of such a forecast is the period from the time of making the forecast (that is, the time origin of the forecast) to the future point in time for which the forecast applies. A flood warning turns a prediction or forecast into information on which many actions are dependent. The fundamental purpose of the warning is to enable individuals and communities to respond appropriately to a major flood threat to reduce the risk of death, injury and property loss.

Flood forecasting and warning is a necessary part of flood risk management, given that no preventive or defence measures can be completely effective. The reality of economic limits to the provision of defences, together with the possibility that the capacity of defence systems may be exceeded or that they may fail, require that other measures are in place. Provision of flood forecasting will also form part of flood management planning and development strategies, which recognize that there are occupied flood plain areas where non-structural measures can be effective. This can include the use of temporary defences (flood gates or demountable barriers), domestic protection (sandbagging) and local evacuation (to flood shelters).

Flood management requires a variable degree of response from the water management agency, local or municipal authorities, transport and communications operations and emergency services. Flood forecasting has to provide information to these users both for preparation and response: in this sense, flood forecasting is part of the wider disaster management capacity.

The precise role of flood forecasting will vary according to the circumstances dictated by both the hydrometeorological environment, the built environment and the social characteristics. Cities present different problems from rural areas: the location of flood-risk areas in relation to rivers, coasts or mountain ranges has a significant bearing on the types of flood forecasting involved. The nature of flooding events is also important, particularly whether floods are regular in occurrence, as in a highly predictable seasonal climate, such as monsoon or hurricane seasons, or irregular, such as violent thunderstorms. There is thus no set design for a flood forecasting system, and the balance between particular components, for example meteorological and hydrological forecasts, scale and timing, have to be adapted to circumstances. Within a given country, a number of different flood types may be encountered and each will require a different forecasting approach. Thus, headwater areas may require a system concentrating on flash floods, whereas flood plain areas may need a system to be focused on the slow build-up of flooding and inundation.

To form an effective real-time flood forecasting system, the basic structures need to be linked in an organized manner. This essentially requires:

(i) Provision of specific forecasts relating to rainfall for both quantity and timing, for which numerical weather-prediction models are necessary;
(ii) Establishment of a network of manual or automatic hydrometric stations, linked to a central control by some form of telemetry;
(iii) Flood forecasting model software, linked to the observing network and operating in real time.

The ultimate end-user of a flood forecastng and warning system is the public, which faces impacts on its housing, possessions, land, livestock and transport. The basic requirement is for people to have time to take necessary actions and make arrangements, which can range from putting protection or proofing arrangements in place, to evacuation, along with basic possessions and livestock, to a place of safety. Social science disaster literature is filled with numerous studies of the many facets of human anticipation of flood hazards. Most of the research focuses on the formal systems designed by governmental organizations to warn the public at risk through ad hoc agencies and services. Key elementary processes have been formalized, documented and modeled for various types of hazards with reference to the temporal coordinate. For floods, Carsell *et al.* (2004) studied the characteristic time duration of different anticipatory processes for the warning-time (maximum potential- and available-warning-times). Lindell (2000) considered the time needed to decide and implement protective actions and Urbanik (2000) assessed the estimation of evacuation times.

For the forecasting and warning system to be useful, the lead time provided by a flood warning has to be sufficiently long to allow response actions to take place. However, the forecasting lead time depends on the basin spatial scale, the associated flood response time and on the space-time scale of the precipitation system which generated the floods.

4.6.1 Flood forecasting, catchment scales and response times

There is a wide range of river basins and systems in the world and to a greater or lesser extent they have their own particularities, reacting quite specifically to heavy rain and storms. In general, however, it is possible to define three main types of basin characterized by their different temporal and spatial responses to a hydrometeorological event:

(i) Urban basins that are very small (a few square kilometres), densely populated and have a high proportion of impermeable surfaces that respond in the order of one or two hours and can overwhelm the capacity of the drainage network;

(ii) Upper watersheds and small to medium catchments, with an area of between 10 and 500 square kilometres, which will respond in a few hours. Those catchments located in upland areas and with steep slopes react quickly;

(iii) Medium-sized rivers with catchment areas of between 500 and 10 000 square kilometres, characterized by long-distance flow propagation with varying contribution of tributaries. For these basins, flood can take days to affect the lower reaches. Basins and continental rivers with catchments in excess of 10000 square kilometres form a subset of large rivers for which flood response is in terms of weeks and reflects major seasonal meteorological conditions.

The basin types and scales outlined above can be monitored and forecasted by a combination of observation and modelling tools, and it is necessary to design a flood forecasting and warning service based on a given combination of conditions. Each type of basin flooding will require different approaches to the necessary monitoring and modelling facilities.

The forecast lead time that it is possible to be achieved in a certain situation is related to the type of catchment and its associated flood response time. However, the basic principle for assessing lead time requirements is the minimum period of advance warning necessary for preparatory action to be taken effectively. This will depend on the needs of the target community or area. Thus, the concept of lead time has to be flexible and the minimum time may be entirely dependent on the catchment structure and the forecasting and warning system facilities. For small and medium scale (less than 500 km^2) catchments, flood response time may be so short that it is extremely difficult to provide an effective warning and response. In these cases, the flood forecasting and warning system should include methods for rainfall forecasting. The extension in lead time which may be made possible by rainfall forecasting depends on the flood-generating precipitation systems. An indication of the relative scales of precipitation systems and the associated lead time for forecasting is shown in Figure 4.15.

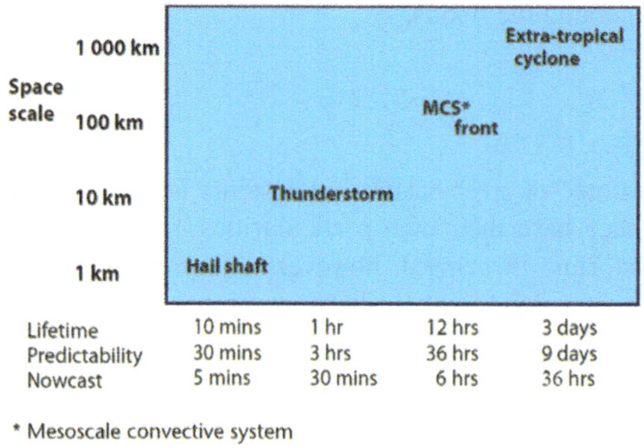

Figure 4.15 Space scales of precipitation systems and their relationship with predictability.

Upland and urban catchments present particular challenges to assuring sufficient lead time for responses to be put in place. Thus, the emphasis must be on effective real-time observation and rapid transmission and processing of data. In these areas it is important to have good quality meteorological observations and forecasts that may be used to produce early, modelled flood estimates. Detailed understanding of the behaviour of localized, high-intensity rainfall and the capacity of drainage structures become important considerations.

In larger catchment areas the lead time associated with rainfall forecast and observation may not be so critical, and over very large basins data sampling may only be necessary at intervals of several hours. More emphasis will, however, need to be placed on identifying the distribution and patterns of rainfall that occur, and on the observation of the hydrological response of contributing basins.

4.6.2 The case of flash floods

Flash flood events are generally very localized in space and time. As such they have a relatively short flood response time and are very difficult to forecast with precision and can subsequently leave people uninformed and subject to surprise in the midst of their daily activities. An examination of flash flood response time has been provided by Creutin et al. (2013), who used a power-law relationship to represent the envelope curve defining the lower limit of lag time (T_L, hours) versus basin area (A, km^2). Examination of the results (Figure 4.16) shows that, the relationship plotted in log-log exhibits a break point in slope at a basin area equal to 350 km^2:

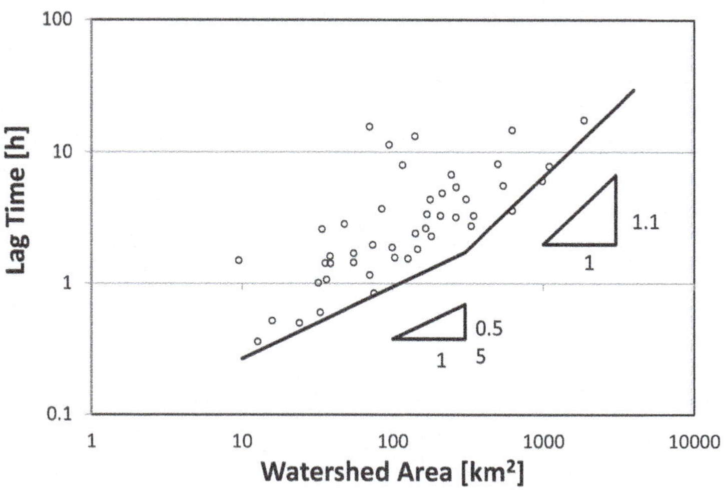

Figure 4.16 Lag time versus watershed area. As discussed in the text, the envelope lines mark the bottom limit of the watershed response time. Adapted from Creutin *et al.* (2013).

$$T_L = 0.08 \cdot A^{0.55} \quad \text{for } A < 350\,\text{km}^2 \tag{4.6}$$

$$T_L = 0.0032 \cdot A^{1.10} \quad \text{for } A \geq 350\,\text{km}^2 \tag{4.7}$$

The exponent in the above equations increases with increasing basin area. This corresponds to a faster increase of response times with increasing basin area, as expected owing to the effect of the transition from channel-dominated flow to a valley bottom-dominated flow when the flow spreads out on the floodplain. The relationships show the dramatic contraction of response time under flash flood conditions. For catchment size up to 100 km^2, which accounts for a significant share of flash flood casualties (Ruin *et al.* 2008), the limit response time is less or equal to 1 hour.

The consequences of the above observations are that forecasting of flash-floods depends critically on meso-scale storm forecasting capable to forecast deep convection events, and requires real time hydrological modeling. The technical requirements for a hydrometeorological flash flood forecasting system include (Collier, 2007): a remote sensing based (radar, satellites) precipitation detection system, a numerical weather prediction (NWP) model, capable to provide short-range Quantitative Precipitation Forecasts (QPF), and a hydrologic/hydraulic model, capable to forecast the stream response from the rain input over a wide range of scales. These requirements are similar to those of more common riverine flood forecasting systems. However, some features characterise flash flood forecasting with respect to riverine flood forecasting and point out to their larger uncertainty (Siccardi *et al.* 2005). These are: (1) the short lead time and the challenge of forecasting convection; (2) the need to provide local forecasts, which means that, on one hand, the rainfall must be monitored and forecasted on a wide range of space/time scales, and, on the other hand, every tributary – generally ungauged – within a wide region can be considered as a potential target for flash flooding.

When dealing with intense, localised and short living events such as flash floods, early warning systems frequently rely on detection indicators, rather than quantitative discharge/level forecasting. These methods aim to replicate the natural threshold behavior of intermittent phenomena whose related state variables and fluxes switch from zero to non-zero values over a short time or space increment (Zehe & Sivapalan, 2009). These methods may be incorporated into diagnostic systems aiming at a broad indication on the spatial distribution of the localized flash flood potential within a relatively large area. Once the monitored or nowcasted precipitation is provided, these systems may be used for early identification of the areas where the triggering threshold will be likely exceeded (as an example see the results reported by the DRIHM EU Project, http://www.drihm.eu/).

Results reported by Norbiato *et al.* (2008) have shown that in humid climates accounting for soil moisture initial conditions is key to evaluate the flash flood potential of storms, particularly in the fringe of the flood/no flood threshold.

Owing to this reason, a number of methods are conditional on the soil moisture regime and are specified with reference to threshold flood peaks at the outlet of the considered catchment, as in the Flash Flood Guidance (FFG) system (Mogil *et al.* 1978; Borga *et al.* 2011) or in the Bayesian method proposed by Martina *et al.* (2006). A number of European research projects, such as FLOODSite (www.floodsite.net/), HYDRATE (www.hydrate.tesaf.unipd.it), and IMPRINTS (www.imprints-fp7.eu), aimed at assessing the advantage for using combined rainfall-soil moisture threshold approached in the case of flash floods (Borga *et al.* 2011; Cabello *et al.* 2011).

The FFG exemplifies the threshold-based method for flash floods (Gourley *et al.* 2012) and represents the depth of rain of a given duration, taken as uniform in space and time in a specific watershed, necessary to cause threshold flooding at the outlet of the watershed. A 2-yr return flow value, often corresponding to the bankfull discharge, is generally used as the threshold peak value. This rainfall depth is computed by running a hydrological model in inverse mode, i.e. by solving iteratively the rainfall input required to obtain the known flood peak provided watershed scale soil moisture conditions. The FFG is compared to either observed or forecasted rainfall of the same duration and in the same watershed. If the nowcasted or forecasted rainfall depth is greater than the FFG, then flooding in the basin is considered likely. As such, the FFG is not a forecast quantity; rather, it is a diagnostic tool. The FFG technique does not predict flash flood timing or location; it is to be used together with monitored or nowcasted precipitation to identify areas where the flood threat may be imminent.

The simplified structure of the threshold-based approaches is beneficial for a number of reasons. They can be applied by non-technical stakeholders at local flood forecasting centres, hence increasing the preparedness of communities exposed to hydro-geomorphic risks (Blöschl, 2008). Moreover, diagnostic methods based on the rainfall thresholds and soil moisture indexes allow processing of local precipitation information and promote close collaboration between hydrologists and meteorologists by simplifying communication about the hydrological status of basins.

4.7 CONCLUSION

In the most simple terms flood risk management is about keeping flood water away from people or people away from flood water. This requires risk management measures and actions both before a flood and during flood emergency. Nevertheless, it is unrealistic to expect that flood risks can be eliminated and so future flood risk management must include a degree of adaptation of society to livng with the flood hazard.

Society is becoming increasingly aware of risks associated with floods. With climate and environmental changes being high on the political agenda it is clear that the links between climate and flood risks have become a major concern for politicians, flood managers and citizens alike. Past research on flood risk management has brought real advances in knowledge and understanding which have been taken up into practice and have shaped changes in policy, as illustrated in this Chapter. However, in front of the environmental and other societal changes, a question arise concerning the conceptually elegant flood risk management (top down) approach: is it possible to quantify future hazards and vulnerabilities to the accuracy needed? The concept of economic risk hinges on the interpretation of the hazard as a probability. However, it may not be possible to specify such probabilities because of the complex nature of the socio-economic processes.

Alternative, bottom-up approaches, may be motivated by a social paradigm, where risk is not defined in monetary but in more qualitative terms. The main goal with these approaches is not to find the optimum, from economic perspective, management strategy but to ensure the wellbeing of people by reducing vulnerability and enhancing resilience (the ability to recover after an event).

An integrated approach to managing floods risk is therefore essential and this should be reflected in the relevant policies. Integration needs to occur across a range of fields: (i) integrating the scale of actions to combine bottom-up and top-down approaches. This will exploit both local expertise and enhance community acceptance in a participatory framework. (ii) Integrating various management strategies including structural measures such as retention basins and nonstructural measures such as land use zoning and flood warnings. A best-mix of such strategies, depending on the local situation, is needed for optimising flood risk management. (iii) Integrating the management of all relevant natural hazards including floods and geo-hydrological hazards (such as landslides and debris flows) in a particular area or region. A holistic approach to emergency planning and management is preferable to a hazard-specific approach. In all types of management strategies, forecasts, early warnings and response play a key role as a primary step to mitigate the social and economic impacts of floods, particularly when dealing with small to moderate spatial scales potentially affected by flash floods.

REFERENCES

Adger W. N. (2006). Vulnerability. *Global Environmental Change*, **16**(3), 268–281.

Aertz J. C. J. H., Botzen W. J. W., de Moel H. and Bowman M. (2013). Cost estimates for flood resilience and protection strategies in New York City. *Annals of the New York Academy of Sciences*, **1294**, 1–104.

Alfieri L., Salamon P., Bianchi A., Neal J., Bates P. and Feyen L. (2013). Advances in pan-European flood hazard mapping. *Hydrological Processes*. doi:10.1002/hyp.9947.

Aronica G. T. and Lanza L. G. (2005). Drainage efficiency in urban areas: a case study. *Hydrological Processes*, **19**, 1105–1119.

Aronica G., Nasello C. and Tucciarelli T. (1998). A 2D multilevel model for flood propagation in flood affected areas. *ASCE Journal of Water Resources and Planning Management*, **124**(4), 210–217.

Ashley S. T. and Ashley W. S. (2008). Flood fatalities in the United States. *Journal of Applied Meteorology and Climatology*, **48**, 805–818.

Barredo J. I. (2007). Major flood disasters in Europe: 1950–2005. *Natural Hazards*, **42**(1), 125–148. doi:10.1007/s11069-006-9065-2.

Bates F. L. and Peacock W. G. (1987). Disasters and social change. In: The Sociology of Disasters, R. R. Dynes, B. Demarchi and C. Pelanda (eds), Franco Angeli Press, Milan, Italy.

Bates P. D. and de Roo A. P. J. (2000). A simple raster-based model for flood inundation simulation. *Journal of Hydrology*, **236**, 54–77.

Blaikie P., Cannon T., Davis I. and Wisner B. (1994). At Risk: Natural Hazards, People's Vulnerability and Disasters. Routledge, London.

Blöschl G. (2008). Flood warning – on the value of local information. *International Journal of River Basin Management*, **6**(1), 41–50.

Blume T., Zehe E. and Bronstert A. (2007). Rainfall-runoff response, event-based runoff coefficients and hydrograph separation. *Hydrological Science Journal*, **52**(5), 843–862.

Bolin R. and Bolton P. (1986). Race, Religion, and Ethnicity in Disaster Recovery. Program on Environment and Behavior, Monograph No. 42, Institute of Behavioral Science, University of Colorado, Colorado.

Booysen H. J., Viljoen M. F. and de Villiers du T. G. (1999). Methodology for the calculation of industrial flood damage and its application to an industry in Vereeniging. *Water SA*, **25**(1), 41–46.

Borga M., Gaume E., Creutin J. D. and Marchi L. (2008). Surveying flash floods: gauging the ungauged extremes. *Hydrological Processes*, **22**(18), 3883–3885. doi:10.1002/hyp.7111.

Borga M., Anagnostou E. N., Blöschl G. and Creutin J.-D. (2011). Flash flood forecasting, warning and risk management: the HYDRATE project. *Environmental Science & Policy*, **14**, 834–844.

Bras R. L. (1990). Hydrology: An Introduction to Hydrologic Science. Addison-Wesley, New York, 643 pp.

Brath A., Castellarin A., Franchini M. and Galeati G. (2001). Estimating the index flood using indirect methods. *Hydrological Sciences Journal*, **46**(3), 399–418.

Burton I., Kates R. and White G. (1978). The Environment as Hazard. Oxford University Press, New York, USA.

Cabello A., Velasco M., Barredo J. I., Hurkmans R. T. W. L., Barrera-Escoda A., Sempere-Torres D. and Velasco D. (2011). Assessment of future scenarios of climate and land-use changes in the IMPRINTS test-bed areas. *Environmental Science and Policy*, **14**(7), 884–897.

Camarasa Belmonte A. M., Lopez-Garcia M. J. and Soriano-Garcia J. (2010). Mapping temporally-variable exposure to flooding in small Mediterranean basins using land-use indicators. *Applied Geography*, **31**(1), 136–145.

Carsell K. M., Pingel N. D. and Ford D. T. (2004). Quantifying the benefit of a flood warning system. *Natural Hazards Review*, **5**(3), 131–140.

Castellarin A., Kohnová S., Gaál L., Fleig A. K., Salinas J. L., Toumazis A., Kjeldsen T. R. and Macdonald N. Eds. (2011). Review of applied statistical methods for flood frequency analysis in Europe. COST Action ES0901 "European procedures for flood frequency estimation – FloodFreq", Report Working group 2.

Chow V. T. Ed.-in-chief. (1964). Handbook of Applied Hydrology, Section 8-1. McGraw Hill, New York.

Coles, S. G. (2001). An Introduction to Statistical Modeling of Extreme Values. SpringerVerlag, London.

Collier C. (2007). Flash flood forecasting: what are the limits of predictability? *Quarterly Journal of the Royal Meteorological Society*, **133**(622A), 3–23.

Creutin J. D., Borga M., Gruntfest E., Lutoff C., Zoccatelli D. and Ruin I. (2013). A space and time framework for analyzing human anticipation of flash floods. *Journal of Hydrology*, **482**, 14–24.

Cunderlink J. M. and Burn D. H. (2001). The use of flood regime information in regional flood frequency analysis. *Hydrological Sciences Journal*, **47**(1), 77–92.

Cutter S. L., Boruff B. J. and Shirley W. L. (2003). Social vulnerability to environmental hazards. *Social Science Quarterly*, **84**(1), 242–261.

Dash N., Peacock W. G. and Morrow B. (1997). And the poor get poorer: a neglected black community. In: Hurricane Andrew: Ethnicity, Gender and the Sociology of Disaster, W. G. Peacock, B. H. Morrow and H. Gladwin (eds), Routledge, London, pp. 206–225.

De Moel H. and Aerts J. C. J. H. (2011). Effect of uncertainty in land use, damage models and inundation depth on flood damage estimates. *Natural Hazards*, **58**(1), 407–425.

De Moel H., Aerts J. C. J. H. and Koomen E. (2011). Development of flood exposure in the Netherlands during the 20th and 21st century. *Global Environmental Changes*, **21**(2), 620–627.

de Moel H., Jongman B., Kreibich H. *et al.* (2015). Flood risk assessments at different spatial scales. *Mitigation and Adaptation Strategies for Global Change*, **20**, 865. doi:10.1007/s11027-015-9654-z

Di Baldassare G. *et al.* (2010). Flood fatalities in Africa: from diagnosis to mitigation. *Geophysical Research Letters*, **37**, L22402.

Doocy S., Daniels A., Murray S., Kirsch T. D. (2013). The human impact of floods: a historical review of events 1980–2009 and systematic literature review. *PLOS Currents Disasters*, 2013 Apr 16. Edition 1. doi: 10.1371/currents.dis.f4deb45790493 6b07c09daa98ee8171a.

Dwyer A., Zoppou C., Nielsen O., Day S. and Roberts S. (2004). Quantifying Social Vulnerability: A Methodology for Identifying Those at Risk to Natural Hazards. Geoscience Australia, Canberra.

Efstratiadis E., Koussis A. D., Koutsoyiannis D., and Mamassis N. (2014). Flood design recipes vs. reality: can predictions for ungauged basins be trusted?, *Natural Hazards and Earth System Sciences*, **14**, 1417–1428.

Elmer F., Thieken A. H., Pech I. and Kreibich H. (2010). Influence of flood frequency on residential building losses. *Natural Hazards and Earth System Sciences*, **10**, 2145–2159.

Elmer F., Hoymann J., Duethmann D., Vorogushyn S. and Kreibich H. (2012). Drivers of flood risk change in residential areas. *Natural Hazards and Earth System Sciences*, **12**(5), 1641–1657.

Ernst J., Dewals B. J., Detrembleur S., Archambeau P., Erpicum S. and Pirotton M. (2010). Micro-scale flood risk analysis based on detailed 2D hydraulic modelling and high resolution geographic data. *Natural Hazards*, **55**, 1818–209.

European Commission (EC) (2007). Directive 2007/60/EC of the European Parliament and of the Council of 23 October 2007 on the Assessment and Management of Flood Risks.

Falter D., Schröter K., Nguyen D., Vorogushyn S., Kreibich H., Hundecha Y., Apel H. and Merz B. (2015a). Spatially coherent flood risk assessment based on long-term continuous simulation with a coupled model chain. *Journal of Hydrology*. doi:10.1016/j.jhydrol.2015.02.021.

Falter D., Dung V. N., Vorogushyn S., Schröter K., Hundecha Y., Kreibich H., Apel H., Theisselmann F. and Merz B. (2015b). Continuous, large-scale simulation model for flood risk assessments: proof-of-concept. *Journal of Flood Risk Management*. doi:10.1111/jfr3.12105.

Field C. B. *et al.* (ed.) (2012). Managing the Risks of Extreme Events and Disasters to Advance Climate Change Adaptation. Special Report of Working Groups I and II of the Intergovernmental Panel on Climate Change (IPCC), Cambridge University Press, Cambridge.

Fothergill A. and Peek L. A. (2004). 'Poverty and disasters in the United States: a review of recent sociological findings'. *Natural Hazards*, **32**, 89–110.

Fothergill A., Maestas E. G. M. and Darlington J. D. (1999). Race, ethnicity, and disasters in the United States: a review of the literature. *Disasters*, **23**(3), 156–173.

Freni G., La Loggia G. and Notaro V. (2010). Uncertainty in urban flood damage assessment due to urban drainage modelling and depth-damage curve estimation. *Water Science Technology*, **61**, 2979–2993. doi:10.2166/wst.2010.177.

Gaume E., Valerie B., Pietro B., Newinger O., Barbuc M., Bateman A., Blaškovicová L., Blöschl G., Borga M., Dumitrescu A., Daliakopoulos J., Garcia J., Irimescu A., Kohnova S., Koutroulis A., Marchi L., Matreata S., Medina V., Preciso E., Sempere-Torres D., Stancalie G., Szolgay J., Tsanis I., Velasco D. and Viglione A. (2009). A compilation of data on European flash floods. *Journal of Hydrology*, **367**(1–2), 70–78. doi:10.1016/j.jhydrol.2008.12.028.

Gaume E., Gaál L., Viglione A., Szolgay J., Kohnová S. and Blöschl G. (2010). Bayesian MCMC approach to regional flood frequency analyses involving extraordinary flood events at ungauged sites. *Journal of Hydrology*, **394**, 101–117. doi:10.1016/j.jhydrol.2010.01.008.

Georgakakos K. P. and Hudlow M. D. (1984). Design of national realtime warning systems with capability for site-specific flash flood forecasts. *Bulletin of American Meteorological Society*, **67**, 1233–1239.

Gerl T. M., Bochow M. and Kreibich H. (2014). Flood damage modeling on the basis of urban structure mapping using high-resolution remote sensing data. *Water*, **6**(8), 2367–2393.

Gourley J. J., Erlingis J. M., Hong Y. and Wells E. B. (2012). Evaluation of tools used for monitoring and forecasting flash floods in the United States. *Weather and Forecasting*, **27**(1), 158–173.

Health and Safety Executive (HSE) (UK) (2001). Reducing Risks, Protecting People – HSE's Decision-Making Process. HMSO, London.

Herschy R. W. (2002). The world's maximum observed floods. *Flow Measurement and Instrumentation*, **13**(5–6), 231–235.

Herschy R. W. and Fairbridge R. W. (Eds), (1998). Encyclopedia of Hydrology and Water Resources. Kluwer, Dordrecht, The Netherlands.

Hosking J. R. M. and Wallis J. R. (1997). Regional frequency analysis. *An Approach Based on L-Moments*. Cambridge University Press, Cambridge, UK, 224 pp.

Jongman B., Ward P. J. and Aerts J. C. J. H. (2012). Global exposure to river and coastal flooding—long term trends and changes. *Global Environmental Change*, **22**, 823–835.

Jonkman S. N. and Kelman I. (2005). An analysis of the causes and circumstances of flood disaster deaths. *Disasters*, **29**(1), 7597.

Kjeldsen T. R. and Jones D. A. (2010). Predicting the index flood in ungauged UK catchments: on the link between data-transfer and spatial model error structure. *Journal of Hydrology*, **387**, 1–9.

Kreibich H. and Thieken A. H. (2009). Coping with floods in the city of Dresden, Germany. *Natural Hazards*, **51**(3), 423–436.

Kreibich H and Thieken A. H. (2008). Assessment of damage caused by high groundwater inundation. *Water Resources Research*, **44**(9) 10.1029/2007WR006621.

Kundzewicz Z. W. *et al.* (2014). Flood risk and climate change: global and regional perspectives. *Hydrological Sciences Journal*, **59**(1), 1–28. doi:10.1080/02626667.2013.857411.

Lettenmaier D. P., Wallis J. R. and Wood E. F. (1987). Effect of regional heterogeneity on flood frequency estimation. *Water Resources Research*, **23**, 313–323.

Lindell M. K. (2000). An overview of protective action decision-making for a nuclear power plant emergency. *Journal of Hazardous Materials*, **75**, 113–129.

Lindell M. K. and Perry R. W. (2004). Communicating Environmental Risk in Multiethnic Communities. Sage, Thousand Oaks.

Marchi L., Borga M., Preciso E. and Gaume E. (2010). Characterisation of selected extreme flash floods in Europe and implications for flood risk management. *Journal of Hydrology*, **394**, 118–133, doi:10.1016/j.jhydrol.2010.07.017.

Martina M. L. V., Todini E., Libralon A. (2006). A Bayesian decision approach to rainfall thresholds based flood warning. *Hydrology and Earth System Sciences*, **10**, 413–426.

Merz B. and Thieken A. H. (2004). Flood risk analysis: concepts and challenges. *Österreichische Wasser-und Abfallwirtschaft*, **56**(3–4), 27–34.

Merz R. and Blöschl G. (2003). A process typology of regional floods. *Water Resources Research*, **39**(12), 1340. doi:10.1029/2002WR001952.

Merz R. and Blöschl G. (2008). Flood frequency hydrology 1: temporal, spatial and causal expansion of information. *Water Resources Research*, **44**, W08432, doi:10.1029/2007WR006744.

Merz B., Kreibich H., Thieken A. H. and Schmidtke R. (2004). Estimation uncertainty of direct monetary flood damage to buildings. *Natural Hazards and Earth Systems Science*, **4**, 153–163.

Merz B., Thieken A. H. and Gocht M. (2007). Flood risk mapping at the local scale: concepts and challenges. In: Advances in Natural and Technological Hazards Research, S. Begum, M. Stive and J. Hall (eds), Springer, Dordrecht, pp. 231–251.

Merz B., Kreibich H., Schwarze R. and Thieken A. (2010a). Review article 'assessment of economic flood damage'. *Natural Hazards and Earth Systems Science*, **10**, 1697–1724. doi:10.5194/nhess-10-1697-2010.

Merz B., Hall J., Disse M. and Schumann A. (2010b). Fluvial flood risk management in a changing world. *Natural Hazards and Earth Systems Science*, **10**, 509–527.

Merz B., Kreibich H. and Lall U. (2013). Multi-variate flood damage assessment: a tree-based data-mining approach. *Natural Hazards and Earth Systems Science*, **13**(1), 53–64.

Middelmann-Fernandes M. H. (2010). Flood damage estimation beyond stage-damage functions: an Australian example. *Journal of Flood Risk Management*, **3**(1), 88–96.

Mogil H. M., Monro J. C. and Groper H. S. (1978). NWS's flash flood warning and disaster preparedness program. *Bulletin of the American Meteorological Society*, **59**, 690–699.

Nicholas J., Holt G. D. and Proverbs D. (2001). Towards standardizing the assessment of flood damaged properties in the UK. *Structural Survey*, **19**, 163–172.

Nilsson C., Reidy C., Dynesius M. and Revenga C. (2005). Fragmentation and flow regulation of the world's large river system. *Science*, **308**, 405–408.

Norbiato D., Borga M., Degli Esposti S., Gaume E. and Anquetin S. (2008). Flash flood warning based on rainfall depth-duration thresholds and soil moisture conditions: an assessment for gauged and ungauged basins. *Journal of Hydrology*, **362**(3–4), 274–290, doi:10.1016/j.jhydrol.2008.08.023.

Norbiato D., Borga M., Merz R., Blöschl G. and Carton A. (2009). Controls on event runoff coefficients in the eastern Italian Alps. *Journal of Hydrology*, **375**, 312–325.

Pappenberger E., Dutra E., Wetterhall F. and Cloke H. L. (2012). Deriving global flood hazard maps of fluvial floods through a physical model cascade. *Natural Hazards and Earth Systems Science*, **16**, 4143–4156.

Peduzzi P. *et al.* (2009). Assessing global exposure and vulnerability towards natural hazards: the disaster risk index. *Natural Hazards and Earth System Sciences*, **9**, 1149–1159.

Peduzzi P. *et al.* (2011). Preview Global Risk Data Platform, [online], UNEP/GRID Geneva, UNISDR, World Bank. http://preview.grid.unep.ch/index.php?preview=tools&cat=1&lang=eng (accessed 18 October 2012).

Peduzzi P. *et al.* (2012). Global trends in tropical cyclones risk. *Nature Climate Change*, **2**(4), 289–294.

Penning-Rowsell E. C. (2013). A 'realist' approach to the extent of flood risk in England and Wales. In: Comprehensive Flood Risk Management: Research for Policy and Practice, F. Klijn and T. Schweckendiek (eds), Taylor and Francis, London, p. 450.

Reed D. W. and Robson A. J. (1999). Flood Estimation Handbook 3: Statistical Procedures for Flood Frequency Estimation. Institute of Hydrology, Wallingford.

Robinson J. and Sivapalan M. (1997). An investigation into the physical causes of scaling and heterogeneity of regional flood frequency. *Water Resources Research*, **33**(5), 1045–1059.

Rosbjerg D. (2007). Regional flood frequency analysis, In: Extreme Hydrological Events: New Concepts for Security, NATO Science Series IV: Earth and Environmental Sciences, Vasiliev, O. F., van Gelder, P. H. A. J. M., Plate, E. J., Bolgov, M. V. (eds), **78**, 151–171.

Ruin I. (2010). Conduite à contre-courant et crues rapides, le conflit du quotidien et de l'exceptionnel. *Annales de Géographie*, **674**, 419–432.

Ruin I., Creutin J. D., Anquetin S. and Lutoff C. (2008). Human exposure to flash floods–relation between flood parameters and human vulnerability during a storm of September 2002 in Southern France. *Journal of Hydrology*, **361**(1), 199–213.

Sayers P., Li Y., Galloway G., Penning-Rowsell E., Shen F., Wen K., Chen Y. and Le Quesne T. (2013). *Flood Risk Management: A Strategic Approach.* UNESCO, Paris.

Scawthorn C., Flores P., Blais N., Seligson H., Tate E., Chang S., Mifflin E., Thomas W., Murphy J., Jones C. and Lawrence M. (2006). HAZUS-MH flood loss estimation methodology. II: damage and loss assessment. *Natural Hazards Review*, **7**(2), 72–81.

Siccardi F., Boni G., Ferraris L. and Rudari R. (2005). A hydrometeorological approach for probabilistic flood forecast. *Journal of Geophysical Research*, **110**, 1–9.

Sivapalan M., Blöschl G., Merz R. and Gutknecht D. (2005). Linking flood frequency to long-term water balance: incorporating effects of seasonality. *Water Resources Research*, **41**, W0612, doi:10.1029/2004WR003439.

Smith D. I. (1994). Flood damage estimation—a review of urban stage-damage curves and loss functions. *Water SA*, **20**, 231–238.

Stedinger J. R., Vogel R. M. and Foufoula-Georgiou E. (1993). Frequency analysis of extreme events. In: Handbook of Hydrology, D. R. Maidment (ed.), McGraw-Hill, Inc., New York, NY, 18.1–18.66.

Tapia-Silva F.-O., Itzerott S., Förster S., Kuhlmann B. and Kreibich H. (2011). Estimation of flood losses to agricultural crops using remote sensing. *Physical and Chemistry of Earth*, **36**(7–8), 253–265.

Terti G., Ruin I., Anquetin S. and Gourley J. J. (2015). Dynamic vulnerability factors for impact-based flash flood prediction. *Natural Hazards*, **79**(3), 1481–1497.

Thieken A. H., Müller M., Kreibich H. and Merz B. (2005). Flood damage and influencing factors: new insights from the August 2002 flood in Germany. *Water Resources Research*, **41**(12), W12430.

United Nations (1987). Report of the World Commission on Environment and Development. General Assembly Resolution 42/187, 11 December.

Urbanik T. (2000). Evacuation time estimates for nuclear power plants. *Journal of Hazardous Materials*, **75**(2–3), 165–180.

Viglione A., Merz R., Salinas J. L., and Blöschl G. (2013). Flood frequency hydrology 3. A Bayesian analysis. *Water Resource Research*, **49**(2), 675–692.

Vogel K., Riggelsen C., Scherbaum F., Schröter K., Kreibich H. and Merz B. (2013). Challenges for Bayesian network learning in a flood damage assessment application. In: ICOSSAR 2013, New York.

Wilhelmi O. V. and Morss R. E. (2012). Integrated analysis of societal vulnerability in an extreme precipitation event: a Fort Collins case study. *Environmental Science and Policy*. doi:10.1016/j.envsci.2012.07.005.

Wu S. Y., Yarnal B. and Fisher A. (2002). Vulnerability of coastal communities to sea level rise: a case study of Cape May County, New Jersey, USA. *Climate Research*, **22**, 255–270.

Wünsch A., Hermann U., Kreibich H. and Thieken A. H. (2009). The role of disaggregation of asset values in flood loss estimation: a comparison of different modeling approaches at the Mulde River, Germany. *Journal of Environmental Management*, **44**(3), 524–541.

Zanon F., Borga M., Zoccatelli D., Marchi L., Gaume E., Bonnifait L. and Delrieu G. (2010). Hydrological analysis of a flash flood across a climatic and geologic gradient: the September 18, 2007 event in Western Slovenia. *Journal of Hydrology*, **394**, 182–197, doi:10.1016/j.jhydrol.2010.08.020.

Zehe E. and Sivapalan M. (2009). Threshold behaviour in hydrological systems as (human) geo-ecosystems: manifestations, controls, implications. *Hydrology and Earth Systems Science*, **13**, 1273–1297.

Zhai G., Fukuzono T. and Ikeda S. (2005). Modeling flood damage: case of Tokai Flood 2000. *Journal of American Water Resources Association*, **41**, 77–92.

Chapter 5

Droughts

Nicolas R. Dalezios, Ana M. Tarquis and Saeid Eslamian

5.1 DROUGHT CONCEPTS

5.1.1 Drought definitions and types

Drought constitutes part of natural climatic variability. It is recognized that drought is considered a natural phenomenon occurring over a region periodically. The main cause of drought is the deficiency or lack of precipitation in a region over an extended period of time. This is the reason that droughts are also referred to as 'non-events'. Indeed, it is very important to establish the context in which the drought phenomenon and its associated impacts are being described, which leads to a better definition. Specifically, if drought is considered a phenomenon it is certainly an atmospheric phenomenon. Studies in several areas around the world have shown that drought periods are often characterized by a large decrease in the amount of rainfall per rainy day, by an increase in the continentality of clouds and by lack of rain-producing clouds (Dalezios *et al.* 2009). In general, there is evidence that droughts are associated with persistence of ridges or centers of high pressure systems at middle-level in the troposphere. Moreover, the corresponding reduced cloud cover results into positive temperature anomalies in the lower atmosphere, which produces the middle-level pressure anomaly and favors subsidence at high level keeping the atmosphere significantly drier and more stable than normal (Dalezios *et al.* 2016d).

It is recognized that there is absence of a universally accepted and precise definition of drought, since there is a wide variety of sectors affected by drought, its diverse temporal and spatial distribution and the water demand for different uses (Heim, 2002). Definitions of drought are usually region and application or impact specific. In particular, droughts are regional in extent and each region maintains specific climatic characteristics. In addition, definitions of drought should be application specific, since drought impacts vary between different sectors. Indeed, by considering drought a hazard, there is a tendency to define and classify droughts into different types. Specifically, drought is considered one of the major natural hazards with significant impact to several sectors of the economy, society and environment (Sivakumar *et al.* 2005). Moreover, there are several unique characteristics, which differentiate droughts from other environmental hazards, namely its slow onset, often characterized as a creeping phenomenon, its non-structural impacts, which can be regional or local lasting for a long time or a very short time, as well as the absence of a universal definition leading to inaction (Wilhite *et al.* 2000). Nevertheless, the impacts of droughts may be severe and are neither immediate nor easily measured. All the above may accumulate difficulties in drought assessment and response, which may result into slow progress on drought preparedness plans and mitigation actions.

Definitions of drought can be classified as either conceptual or operational. Conceptual definitions are general and help the public to understand the concept of drought. On the other hand, operational definitions contribute to the identification of the severity and duration of drought and are more useful in drought assessment and management. In the international literature, three operational definitions are normally considered, namely meteorological or climatological, agricultural or agrometeorological and hydrological drought (Wilhite *et al.* 2000). In addition, as a fourth type of drought, the socioeconomic impacts of drought can also be considered. With the exception of meteorological drought, the other types of drought, such as agricultural and hydrological, focus on the human or social aspects of drought, in terms of the interaction between the natural characteristics of meteorological drought and human activities that depend on precipitation, to provide adequate water supplies to meet societal and environmental demands (Keyantash & Dracup, 2002). These diverse drought types impact different sectors, although in most cases the impacts related to each sector overlap both temporally and spatially. Drought concepts refer to conditions of precipitation deficit, soil moisture, streamflow, plant wilting, wild fires, famine, as well as other components. Indeed, drought is multi-faceted issue and drought quantification requires a multi-faceted assessment. Moreover, drought monitoring involves climate data, soil moisture, streamflow, groundwater, reservoir and lake levels, snow pack, short-, medium- and long-range forecasts, as well as vegetation health/stress and fire danger. Needless to say, the relationship between the different drought types is complex. A brief description of the above mentioned drought types follows.

Meteorological or climatological drought is a region-specific natural event, due to the regional nature of atmospheric phenomena, originating from multiple causes. It is defined as the degree of dryness specified by deficiencies of precipitation and the dry period duration. Meteorological drought is generally characterized by a precipitation anomaly being lower than average in a region for some period of time and by prolonged and abnormal moisture deficiency.

Agricultural or agrometeorological drought refers to the agricultural impacts resulting from deficiencies in the water availability for agricultural use. Indeed, agricultural drought is defined in terms of crop failure and exists when soil moisture is depleted so that crop yield is reduced considerably. Specifically, agricultural drought is described by the availability of soil water to support crop and forage growth and there is no direct relationship between precipitation and infiltration of precipitation into the soil. In fact, infiltration depends on antecedent moisture conditions, precipitation intensity, soil type and slope. Soils with low water holding capacity are typical of drought-prone areas, which are more vulnerable to agricultural drought.

Hydrological drought is normally defined by the departure of surface and/ or subsurface water from some average conditions over a long time period resulting from meteorological drought. Hydrological drought is considered to be a period during which the actual water supply, either surface water or groundwater, is less than the minimum water supply necessary for normal operations in a particular region (watershed). Such as agricultural drought, there is no direct relationship between precipitation amounts and the surface and subsurface water supplies. There is also significant time lag between departures of precipitation and the appearance of these deficiencies in surface and subsurface components of the hydrological system.

Finally, **socioeconomic** impacts of drought are defined in terms of loss from an average or expected return and can be measured by both social and economic indicators (McVicar & Jupp, 1998). Indeed, socioeconomic drought refers to the gap between supply and demand of economic goods brought on by the three other types of drought described before, such as water, food, raw materials, transportation, hydroelectric power, as a result of a weather-related shortage in water supply. Socioeconomic drought is different from other types of drought, since its occurrence depends on the spatiotemporal processes of supply and demand. All these drought types along with the temporal sequence of procedures are shown schematically in Figure 5.1 (from Wilhite *et al.* 2000).

5.1.2 Factors and features of drought

Several factors may be implicated as potential causes of drought. Among these factors are mentioned: ENSO (El Nino Southern Oscillation), abnormal sea surface temperature patterns in areas other than the equatorial eastern Pacific, soil moisture desiccation, as well as nonlinear behavior of the climate system. It is also suggested that climate is changing and that droughts are becoming more frequent and/or more severe (IPCC, 2012). However, there have always been droughts and records show that drought events appear to be within the range of statistical expectations. Examination of the paleoclimatic record for the Great Plains suggests that the droughts of the 1930s, 1950s and 1980s were eclipsed several times by droughts within the last 2000 years, and that more severe droughts could occur in the future (Woodhouse & Overpeck, 1998). Indeed, the massive Australian drought of 1895–1903, which followed rapid growth of rural enterprises in the 1870s and 1880s, as well as the great American drought of the 1930s, are good examples of lack of assessment of the environment conditions leading to inappropriate land use. Specifically, the number of sheep in Australia fell from an estimated 106 million in 1891 to 54 million in 1902, cattle numbers were almost halved to about 7 million, and dust storms were common as a result of the vast expansion of land plowed for wheat. In addition, in areas, such as the Sahel, where nomadism and intermittent grazing have prevailed and are more or less in balance with environmental conditions, more intensive exploitation has resulted into disastrous impacts for social systems and ecosystems during drought events.

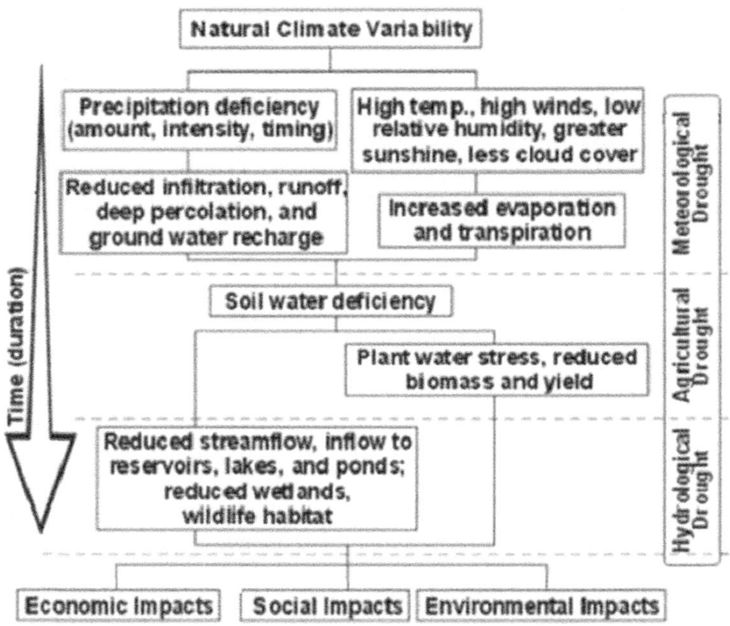

Figure 5.1 Drought types and temporal sequence (from Wilhite *et al.* 2000).

The frequency of droughts around the world, as well as their possible links with phenomena such as El Nino, constitute hazard evidences even for the general population. Specifically, the 1975–1976 drought in Western Europe had widespread impacts on agriculture, domestic and industrial water supplies, as well as on river and canal traffic. Moreover, the drought in Britain, particularly southern England, has been the worst for almost 250 years. In particular, in some areas, water supplies to domestic consumers were cut for up to 17 hours per day, and production of root and vegetable crops was down by as much as 40%. Similarly, droughts of the early to mid- 1980s in Africa affected more than 40 million people. Furthermore, the Canadian Prairie Provinces are particularly sensitive to rainfall shortages and associated soil losses due to wind erosion (Wheaton & Chakravarti, 1990). Specifically, the 1984 conditions are considered equivalent to those of the "Dust Bowl" years of 1936–1937, with farmers losing up to half of their grain crop to the value of about $2.5 billion. In addition, the 1988 drought in the USA was rated as one of the worst in 100 years, with an estimated impact on the economy of $40 billion. An unexpected impact in this drought event was on barge traffic on the lower Mississippi river, where the industry suffered an income loss of about 20%. Moreover, the El Nino event of 1997–1998 was linked to drought in Central America and southeast Asia, with major impacts on vegetable quantity and quality in Central America and on coffee and palm oil in southeast Asia, both leading to reduced exports and increased prices on global markets. It is assessed that worldwide disasters triggered by droughts are twice as frequent during year two of ENSO warm events as during other years, particularly in southern Africa and southeast Asia.

For the assessment and monitoring of droughts, several drought features are usually detected. Indeed, conventional and/or remote sensing data and methods can be employed to show the temporal and spatial variability of several drought features quantitatively (Dalezios *et al.* 2012; Dalezios *et al.* 2014). Some key features are described as follows.

Severity or intensity of drought is defined as the escalation of the phenomenon into classes ranging from mild, moderate, severe and extreme. The severity is usually determined through drought indicators and indices based on the above mentioned classes. The regions affected by severe drought evolve gradually and there is a seasonal and annual shift of the so-called epicenter, which is the area of maximum severity.

Periodicity is considered the recurrence interval of drought or the frequency of occurrence.

Duration of a drought episode is defined as the time interval from the start and end time usually in months. Since drought is a complex phenomenon, the assessment of start and end time is a complicated technical subject.

Onset is the beginning of a drought and is determined by the occurrence of a drought episode, which is assessed through indicators or indices reaching certain threshold value.

End time of a drought episode signifies the termination of drought based again on threshold values of indicators or indices. It is often difficult to determine the onset and the ending of a drought and on what criteria these determinations should be made.

Areal extent of drought is considered the spatial coverage of the phenomenon as is quantified in classes by indicators or indices. Areal extent varies in time and remote sensing has contributed significantly in the delineation of this parameter by counting the number of pixels in each class. From a planning perspective, the spatial characteristics of drought may have serious implications for several sectors of the economy, such as agriculture, energy, transportation, health, recreating and tourism.

5.2 DROUGHT RISK IDENTIFICATION

Risk analysis attempts to investigate and better understand drought hazards with the objective to develop pro-active measures and procedures before a crisis. The components of risk analysis are incorporated within the drought risk management framework. Indeed, the objective of risk management is to reduce the threats of hazards, whereas at the same time attempts to maximize any related benefits (Smith, 2013). Indeed, drought risks are usually not clearly understood and remain less visible, as compared, for example, to risks associated with floods. Moreover, drought impacts and damages are not systematically recorded, international standards are not yet adopted, and data bases are still very rare. This justifies the lack of comprehensive and credible assessments of drought risks internationally, at the present time.

Risk identification involves drought quantification, modelling and event monitoring through the development of early warning systems. Quantification of drought is usually accomplished through drought indicators, which are variables presenting drought characteristics, such as severity, duration, magnitude, periodicity, areal extent, onset and end time. Basic data for meteorological, agricultural or hydrological drought indicators are climatic and environmental variables, such as temperature and precipitation, stream flows, soil moisture, reservoir storage, groundwater levels, snow pack and vegetation. Data analysis, interpretation and aggregation leads to drought indicators and/or indices. There is already considerable number of reviews on the use of drought indicators and indices based on conventional and/or remotely sensed data (Heim, 2002; Keyantash & Dracup, 2002). The scientific and operational validity of an index, i.e., how each indicator is combined and weighted in the index and how an index value is related to geophysical and statistical characteristics of drought, is still under research. For the assessment of indices, a set of weighted decision criteria is usually assigned to each index, which are based on expected featrures of each index, namely robustness, tractability, transparency, sophistication, expendability and dimensionality. It is clear that the above criteria weights correspond to the relative importance of the evaluation criteria, thus, are difficult to be precisely justified. The list may be expanded or condensed, however, the above criteria provide a reasonable framework for the evaluation of drought indices.

Employing drought indices is a pragmatic way to process and convert large amounts of data into quantitative drought assessment for a number of applications, such as drought severity classification, drought forecasting, preparedness planning or impact assessment. Indeed, precipitation is the essential parameter for a number of drought indicators, however, several other indicators are also significant in the assessment and monitoring of drought severity. In addition, it is recognized that remote sensing has gradually been established as an important multi-scaling tool for the detection of the spatial and temporal distribution and features of drought. In summary, it is best to consider multiple indicators to assess the existence and severity of drought.

The classification of drought indices is based on the type of impacts they are referred to. Indeed, the taxonomy of drought indices is based on the variables that are used or the use of disciplinary data (Niemeyer, 2008). As already mentioned, there are three widely used categories, namely meteorological, agricultural and hydrological drought indices. However, three more categories are also considered, namely comprehensive, combined and remotely sensed drought indices. Specifically, comprehensive drought indices use a number of meteorological, agricultural and hydrological variables to draw a comprehensive picture of drought, such as the Palmer Drought Severity Index (PDSI) (Palmer, 1965). Moreover, combined drought indices, which are also termed hybrid or aggregate or composite, are derived by incorporating existing drought indicators and indices into a single measure, such as the US Drought Monitor (USDM) (Dalezios *et al.* 2016b). Finally, remotely sensed drought indices use information from remote-sensing sensors to map the condition of the land, such as the NDVI. In summary, although some drought indices, such as SPI (Standardized Precipitation Index) and NDVI (Normalized Difference Vegetation Index), are widely used, the large spectrum of drought indices reflects a fundamental lack of universal definition and concepts, as well as different operational requirements. In addition to the variability in the types and applications of drought indices (e.g., meteorological versus hydrological), the interrelationship of drought indices with drought impacts has led to aggregate drought indices to cover additional aspects and applications.

5.2.1 Global composite drought indices

During the last decades, a web-services-based environment is being developed for integration of regional and continental drought monitors, for computation and display of spatially consistent drought indicators on a global scale, such as in situ SPI, satellite-based indices and modeled soil moisture, and for drill-down capacity to regional, national and local drought indicators. Each continental drought monitor is developed and functions according to the unique conditions of the specific continent. At the present time, there are several regional/continental drought monitor systems, with the objective to be integrated into a Global Drought Monitor (GDM), in order to coordinate and exchange information towards a Global Drought Information System (GDIS) (Figure 5.2). Specifically, the four major regional/continental models, which are developed, are the North American Drought Monitor (NADM), which consists of USDM, Canada and Mexico, the European Drought Observatory (EDO) model, the African Drought Monitor

(ADM) and the Australian Drought Monitor model (Dalezios *et al.* 2016b). It is recognized that there is an international need to continue working toward newer and potentially better drought indices that can also account for a changing climate in which there may be a shift in both temperature and precipitation regimes. The USDM system and the EDO model are briefly presented.

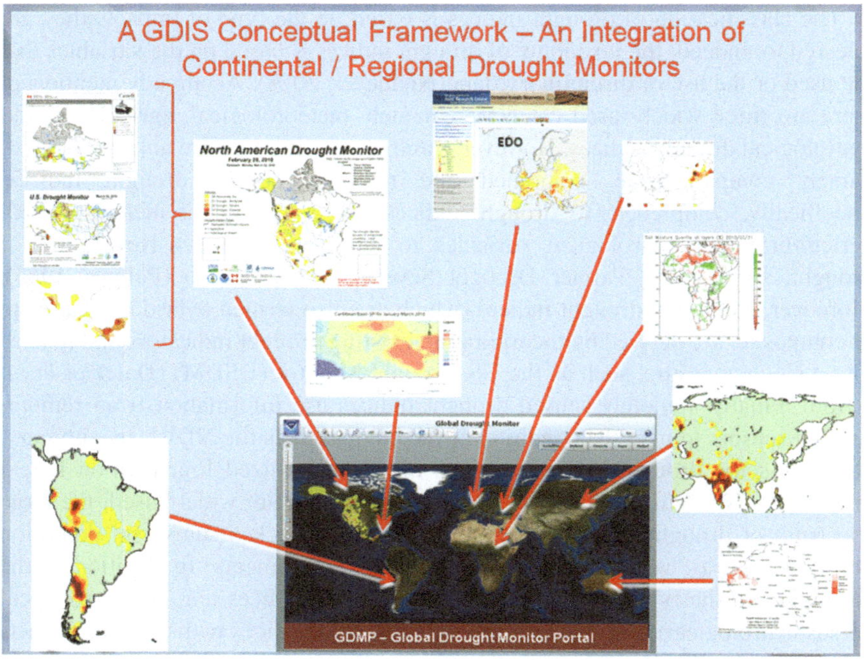

Figure 5.2 Global drought information system (GDIS).

5.2.1.1 US drought monitor (USDM)

The USDM (Wilhite *et al.* 2000) is a composite drought index. The USDM integrates multiple indices, such as SPI and PDSI, as well as indicators, such as vegetation and hydrological conditions, into a weekly map of drought. This information is then subjected to expert interpretation for refinement. Because of its composite nature, USDM can respond to the needs of various water users including water planners and the agriculture industry. The USDM system is based on a composite of multiple indicators within short- and long-term time frames, to develop a ranking drought methodology leading to a single index. The USDM is an operational product issued weekly since 1999 and provides a general up-to-date summary of current drought conditions across the Lower 48 States, Hawaii, Alaska and Puerto Rico. The USDM uses a Fujita-like drought classification as shown in Table 5.1 with four drought categories (D1-D4) and an abnormally dry category (D0) based on percentile rank. The USDM system is flexible enough to integrate new tools and data, as well as additional available information, in order to improve the accuracy level. Moreover, it may be possible to utilize several existing drought indices, and determine the most suitable for any particular area or season for drought monitoring and DEWS. Furthermore, the system also uses hybrid types of drought indices, where satellite data are being merged with surface data, such as the Vegetation Drought Response Index (VegDRI) (Brown *et al.* 2008).

Table 5.1 Drought severity ranking (Fujita scale).

Percentile Rank	Drought Severity
21–30	D0 Abnormally Dry
11–20	D1 Drought – Moderate
6–10	D2 Drought – Severe
3–5	D3 Drought – Extreme
0–2	D4 Drought – Exceptional

The USDM functions in two drought impact assessment schemes. Specifically, the short-term scheme approximates drought-related impacts, which primarily respond to precipitation, but also to other secondary factors on time scales ranging from a few days to a few months, such as topsoil moisture, non-irrigated agriculture, unregulated stream flows, wildfire danger, as well as range and pasture conditions. Specifically, the **short-term** composite drought index consists of 35% Palmer Z-Index, 25% 3-Month Precipitation, 20% 1-Month Precipitation, 13% Climate Prediction Center Soil Moisture Model and 7% Palmer (Modified) Drought Index. Similarly, the **long-term** scheme approximates drought-related impacts that respond to precipitation on time scales ranging from several months to a few years, such as reservoir stores, irrigated agriculture, groundwater levels, and well water depth. Indeed, the long-term composite drought index consists of 25% Palmer Hydrologic Drought Index, 20% 12-Month Precipitation, 20% 24-Month Precipitation, 15% 6-Month Precipitation, 10% 60-Month Precipitation and 10% Climate Prediction Center Soil Moisture Model.

5.2.1.2 European drought observatory (EDO) model

It is recognized that the deficit or lack of average precipitation is the major factor and driver of drought. When this precipitation deficit leads to a decrease of soil moisture at levels, which cannot satisfy the water demand of the plants with impacts on the physiological processes, then agricultural drought starts. The EDO consists of a methodology that combines different drought indices, namely SPI, soil moisture anomalies and the Fraction of Absorbed Photosynthetically Active Radiation (fAPAR) anomalies, in order to identify not only areas affected by agricultural drought, but also areas with the potential to be affected. Specifically, the SPI-n (Mckee *et al.* 1993) is a statistical indicator, which compares the total precipitation recorded at a particular location during a period of n months with the long-term rainfall distribution for the same period at that location. Needless to say, SPI is one of the most common drought indicators, which has been selected by WMO as a key meteorological drought indicator to be produced operationally by meteorological services. Moreover, droughts can affect and reduce the vegetation growth rate, which is influenced by changes either in the solar interception of the plant or in the light use efficiency. The EDO model also uses soil moisture (pF), which is a significant variable in hydrological, climatological, biological and ecological processes, since it contributes to the interactions between the atmosphere and land surface. Specifically, the fAPAR represents the fraction of the solar energy, which is absorbed by vegetation and it is used as drought indicator due to its sensitivity to vegetation stress. The EDO methodology leads to a classification scheme, which is based on three drought impact levels that reflect the different stages of the typical agricultural drought cause–effect relationship. In particular, the three levels are called: "Watch", "Warning" and "Alert". Two additional levels, namely "Partial recovery" and "Recovery", are also considered to identify the stages of the vegetation recovery process. In general, several conventional or remotely sensed drought indices are employed for drought assessment (Dalezios *et al.* 2016b). It is recognized that the development of a composite drought index by combining and integrating meteorological and remote sensing indicators can also help to reduce false alarms, such as in the case of vegetation indices, where a biomass reduction can be produced by causes different from a drought induced water stress.

5.2.2 Composite indices of different drought types

The selection of an appropriate drought index is based on several factors and criteria, such as drought type, implementation cost, data requirements, performance consistency, clarity and practicality, scientific validity, temporal and spatial variability and sensitivity, as well as specific horizontal and duration scale, e.g. watershed versus climatic classification. Moreover, there should be well-defined quantitative thresholds for drought onset and end time, statistical consistency within drought levels, as well as with other indices.

5.2.2.1 Meteorological drought indices

The primary requirement for the development and implementation of a drought index is data availability. Such data include just precipitation indices, such as

Rainfall Anomaly Index (RAI) (Van-Rooy, 1965), Bhalme-Mooley Drought Index (BMDI) (Bhalme & Mooley, 1980), Drought Severity Index (DSI) (Bryant *et al.* 1992), National Rainfall Index (NRI) (Gommes & Petrassi, 1994), Effective Drought Index (EDI) (Byun & Wilhite, 1999) and Drought Frequency Index (DFI) (Gonzalez & Valdez, 2006). Additional meteorological variables have been considered in order to achieve high correlation with drought impacts and accounting for temporal temperature trends. These indices include modifications to SPI (McKee *et al.* 1993) to develop more comprehensive indices, such as the Reconnaissance Drought Index (RDI) (Tsakiris & Vagelis, 2005) that incorporates evapotranspiration resulting in better association with impacts from agricultural and hydrological droughts. Indeed, the Standardized Precipitation Evapotranspiration Index (SPEI) (Vicente-Serrano *et al.* 2010) is a similar index, which is sensitive to long-term trends in temperature change and SPEI performs similarly to SPI when there are no trends. Moreover, Keetch-Byram Drought Index (KBDI) (Keetch & Byram, 1968) is based on temperature with extensive applications to wildfire monitoring. In addition, Palfai Aridity Index (PAI) (Palfai, 1990) has considered groundwater on top of these two indicators with applications mainly to basins. Besides temperature and evapotranspiration, PDSI (Palmer, 1965) also considers stream flow and soil moisture in order to present a more comprehensive approach of the water balance. Indeed, PDSI is identified as a "comprehensive" drought index (Niemeyer, 2008) and remains one of the most widely used index. There are improvements of the PDSI, which include self-calibration capacity (Wilhite *et al.* 2000) and modifications to the evapotranspiration methods by replacing the original Thornthwaite method (Thornthwaite, 1948) with other formulations. Moreover, Drought Early Warning Systems (DEWS) can be based on meteorological drought indices in order to provide timely information on drought for decision making. Indeed, meteorological drought indices are essentially considered far more useful than raw data, especially for drought monitoring for near-real time decision making.

Other uses of meteorological drought indices involve the assessment of thresholds for a number of drought features, such as onset, severity, magnitude, duration, periodicity and end time. Furthermore, a meteorological drought index can also be used as ground-truthing information for modelling efforts or remotely sensed detection of several drought features. A classification and grouping of drought indices is considered, which is based on recently conducted studies using similarity characteristics. These classes include atmospheric drought indices, indices of precipitation anomaly, aridity indices, soil moisture indices, combined or recursive indices, and indices based on remotely sensed information (Dalezios *et al.* 2016d). Needless to say, this classification refers to indices from all types of drought, however, the current presentation is focused on meteorological drought indices. Table 5.2 presents an indicative list of available and commonly used meteorological drought indices in different classes. A brief description of the classes of meteorological drought indices follows.

Table 5.2 Indicative list of meteorological drought classes and indices (from Dalezios *et al.* 2016d).

Classification of Drought Indices	
1. ***Atmospheric Drought Indices***	4. ***Recursive Indices***
1.1 Saturation Deficit	4.1 Fooley Anomaly Index (FAI)
2. ***Precipitation Anomaly Indices***	4.2 Bhalme-Mooley Drought Index (BMDI)
2.1 Precipitation Index	4.3 Palmer Drought Severity Index (PDSI)
2.2 Relative Precipitation Sum	4.4 Standardized Precipitation Index
2.3 Relative Anomaly	4.5 Surface Water Supply Index (SWSI)
2.4 Standardized Anomaly Index (SAI)	4.6 Reclamation Drought Index (RDI)
2.5 Average Standard Anomaly	4.7 Palmer Drought Index (PDI)
3. ***Aridity Indices***	4.8 Palmer Crop Moisture Index (CMI)
3.1 Lang's Rainfall Index	4.9 Keetch-Byram Drought Index (KBDI)
3.2 De Martone Aridity Index	4.10 Effective Drought Index
3.3 Ped's Drought Index (PDI1)	4.11 Reconnaissance Drought Index (RDI)
3.4 Selyaninov's Hydrothermal Coefficient	5. ***Remotely Sensed Information***
3.5 Thornthwaite Index	5.1 Crop Water Stress Index (CWSI)
3.6 Potential Water Deficit	5.2 Vegetation Index
3.7 Potential Evaporation Ratio	5.3 Normalized Difference Vegetation Index
3.8 Aridity Index: Moisture Available Index	5.4 Stress Degree Days
3.9 Relative Evaporation	
3.10 Surface Energy Balance	
3.11 Bowen Ratio	

1. **Indices of atmospheric drought** Low humidity is the typical characteristic of dry spell. The water vapor saturation deficit is usually considered for the identification of atmospheric drought, although in temporal scale shorter than a month, or even a few days. Nevertheless, the impact of these days could be disastrous for certain species. These indices are not widely used indices, although they could be occasionally used for water use efficiency in irrigation scheduling. Indeed, the corresponding meteorological drought index takes the form of a typical meteorological element, namely the saturation deficit.

2. **Indices of precipitation anomaly** There is a relationship between some antecedent precipitation amounts and different types of drought indices. It is estimated that the duration of this previous period of time could range from 3 or 4 weeks up to years. Indeed, it is assessed that most of the time the onset of drought occurs after an abnormal rainy season or period. As a result, the deviation from normal or average precipitation amount signifies the first class of drought indices. Nevertheless, there are obviously certain combinations, or normalization or generalization of deviation for possibly better delineation of the phenomenon. For example, there have to be long-term comparisons between a drought index and yield data before the establishment of threshold values for agricultural drought. Needless to say, there are several existing precipitation anomaly indices (Table 5.2).
3. **Aridity indices** Aridity is a climatic feature, associated with the shortage and inability of precipitation to maintain the vegetation of plants. Indeed, the theoretical base for the aridity index is the evapotranspiration/precipitation ratio. The simplest approach is to use empirical models to estimate evapotranspiration based only on temperature or degree-days, since the difference in the aridity indices is based on the approximation of the evapotranspiration. There are several types of aridity indices (Table 5.2).
4. **Recursive drought indices** Recursive drought indices describe and delineate the cumulative effect of prolonged moisture deficiency usually on a month basis. Indeed, these indices are widely used for the identification of meteorological droughts or dry spells, since these indices maintain a kind of memory, where current values depend on previous values of meteorological variables. The term recursive indices originates from the method of computation. There are several recursive indices including the family of PDSI (Table 5.2).
5. **Remotely sensed drought indices** The majority of remotely sensed drought indices, which are currently and widely used, are based on spectral reflectance of vegetation, also known as vegetation indices. This justifies their main use as indices of agricultural drought, which is covered in another section. Nevertheless, there has been recently significant progress and development in the field of remote sensing due to technological and scientific advancements with increasing reliability in data and methods. In general, it is recognized that any meteorological drought index can be adjusted to be a remotely sensed index, if the computation of precipitation and/or temperature, which are normally used in the index, is based on remotely sensed algorithms or methods. It is anticipated that the development and use of remotely sensed meteorological drought indices is expected to increase drastically in the forthcoming years due to the availability of new precipitation and temperature data platforms at a global scale.

5.2.2.2 Agricultural drought indices

Approaches to characterize agricultural drought mainly involve monitoring of soil water balance and the subsequent deficit when drought occurs. This applies to several non-remote-sensing agricultural drought indices, such as Relative Soil Moisture (RSM) (Thornthwaite & Mather, 1955), Crop Moisture Index (CMI) (Palmer, 1965), which is similar to PDSI, however, models short-term agricultural drought by considering moisture deficit only in the top 5ft of soil column (Narasimhan & Srinivasan, 2005), and Crop Specific Drought Index (CSDI) originally designed for corn and its variant for soybean (Meyer *et al.* 1993). In addition, Agricultural Drought Index (DTx) (Matera *et al.* 2007) calculates the daily transpiration deficit (DT) for x days. Indeed, DTx uses the CRITeRIA soil moisture balance model with inputs including soil, crop, and weather conditions in addition to temperature anomalies, which affect evapotranspiration. Moreover, increased spatial and temporal resolutions are considered in developing Soil Moisture Deficit Index (SMDI) and Evapotranspiration Deficit Index (ETDI) (Narasimhan & Srinivasan, 2005). This approach considers the soil component of the SWAT hydrological model that has a resolution of 16 km^2, as compared to the 7000 to 160000 km^2 resolutions of SPI and PDSI, respectively. Specifically, within the top 2m of the soil component, "soil profile", SMDI detects soil moisture deficit at varying depths: top 2ft (SMDI2), 4ft (SMDI4), and 6ft (SMDI6). Furthermore, SMDI2 and ETDI, which considers the evapotranspiration deficit, have been recommended for monitoring short-term drought conditions and SMDI6 for long-term monitoring, respectively. In addition, remotely sensed vegetation indices, such as NDVI, Temperature Vegetation Dryness Index (TVDI) (Domenikiotis *et al.* 2008), Enhanced Vegetation Index (EVI) (Liu & Huete, 1995), Vegetation Drought Response Index (VegDRI) (Brown *et al.* 2008), Temperature Condition Index (TCI) (Kogan, 1995) and Normalized Difference Water Index (NDWI) (Foley, 1957) are also used to monitor general vegetation state and health. Table 5.3 presents an indicative list of conventional and remotely sensed agricultural drought indices.

5.2.2.3 Hydrological drought indices

Hydrological drought indices aim to provide a comprehensive presentation of delayed hydrological impacts of drought. Indeed, the advanced Palmer Hydrological Drought Index (PHDI) (Palmer, 1965) model considers precipitation, evapotranspiration, runoff, recharge and soil moisture. Nevertheless, the PDSI family of indices show ever lacked the snow component accumulation, which lead to the development of Surface Water Supply Index (SWSI) (Shafer & Dezman, 1982), probably the most popular of this group. Furthermore, the Reclamation Drought Index (RDI) (Wells *et al.* 2004) has improved SWSI by incorporating temperature and hence calculated a variable water demand as input. Moreover, the Remote Sensing Drought Index (RSDI) (Stahl, 2001) bases its model on homogeneous drought-stricken regions that comprise several neighboring low-flow gauging stations. Indeed, the RSDI first calculates the deficiency in streamflow compared with historical records and then uses cluster analysis to delineate the drought-stricken regions. In addition, there is extensive diachronic research on low-flow indices, such as a recent review on low-flow hydrology, which is of paramount importance for the development and implementation of sustainable water resource management practices (Vafakhah *et al.* 2014). Moreover, there have been numerous attempts to estimate low-flow indices, such as the estimation of the low flow index (7Q10), the 7-day, 10-year low-flow, using principal component regression (PCR) based on physiographic and hydrological variables (Eslamian *et al.* 2010). Other indices consider water balance modelling, namely Groundwater Resource Index (GRI) (Mendicino *et al.* 2008) and Water Balance Derived Drought Index (WBDDI) (Vasiliades *et al.* 2011). The GRI focuses on groundwater resources and uses geo-lithological conditions information in a distributed water balance model, whereas the WBDDI artificially simulates runoff for ungauged and low-data watersheds. There is a recent review on hierarchical drought analysis including surface and groundwater hydrological drought modelling (Dalezios *et al.* 2016b).

Table 5.3 Conventional and satellite-based agricultural drought indices (from Dalezios *et al.* 2016c).

Conventional Drought Indices	Satellite-based Drought Indices
1. Agricultural Drought Index (DTx)	1. Normalized Difference Vegetation Index
2. Bhalme-Mooley Drought Index (BMDI)	2. Deviation NDVI index
3. Corn Drought Index	3. Enhanced Vegetation Index (EVI)
4. Crop Moisture Index (CMI)	4. Vegetation Condition Index (VCI)
5. Crop Specific Drought Index	5. Monthly Vegetation Condition Index
6. Evapotranspiration Deficit Index (ETDI)	6. Temperature Condition Index (TCI)
7. Global Vegetation Water Moisture Index	7. Vegetation Health Index (VHI)
8. Leaf Water Content Index (LWCI)	8. Normalised Difference Temperature Index
9. Moisture Availability Index (MAI)	9. Crop Water Stress Index (CWSI)
10. Reclamation Drought Index (RDI)	10. Drought Severity Index (DSI)
11. Soil Moisture Anomaly Index (SMAI)	11. Temperature- Vegetation Dryness Index
12. Soil Moisture Deficit Index (SMDI)	12. Normalized Difference Water Index
13. Soil Moisture Drought Index (SMDI)	13. Remote Sensing Drought Risk Index
14. Standardized Vegetation Index (SVI)	14. Vegetation Drought Response Index
15. Computed Soil Moisture	
16. Agro-Hydro Potential	

5.2.2.4 Remotely sensed drought indices

Drought assessment based on remote sensing data and methods is growing rapidly, mainly due to the increasing number of relevant satellite systems and their gradually improving capabilities. It is stated that remote sensing data and methods can delineate the spatial and temporal variability of several drought features in quantitative terms. Satellite data are consistently available and can detect several drought features. Nevertheless, remote sensing methodologies and data can be incorporated into several aspects of drought, such as vulnerability, damage assessment and warning. Indeed, the contribution of remote sensing could involve relief and, possibly, preparedness or warning, as well as disaster prevention, where satellite temporal resolution is not a crucial limitation. Moreover, a major consideration of remote sensing for drought assessment and disaster reduction is the continuous availability of data for operational uses. There are essentially two types of remote sensing systems for drought assessment and operational monitoring, namely meteorological and environmental (or resource) satellites. Specifically, meteorological satellites are also two types, namely geostationary such as METEOSAT and geosynchronous such as NOAA/AVHRR, and can contribute to operational monitoring and assessment of drought. Similarly, environmental satellites such as LANDSAT, SPOT and recently IKONOS, or WV2 with high to very high spatial resolution, but low frequency of occurrence, can contribute to land-use classification and qualitative features of drought and less to quantitative assessments. Tables 5.2 and 5.3 present indicative lists of internationally used conventional and remotely sensed drought indices (Dalezios *et al.* 2016c, 2016d), respectively. Needless to say, most of the conventional drought indices can be adjusted to incorporate remotely sensed data.

Remote sensing capabilities provide an accepted approach to compensate any loss of conventional information. However, remote sensing techniques have to improve the ability to delineate and simulate the spatial features of environmental hazards and, in particular, droughts. Furthermore, there are dissimilarities in temporal and spatial averages as computed by modelling efforts, as exist in the real world and as observed by remote sensing systems. As a result, there is a need for compatibility between remotely sensed data and mathematical modelling of the corresponding quantification schemes in order to be useful for monitoring and assessing droughts. Moreover, new sensors and algorithms have steadily enabled the incorporation of improved remotely sensed information in drought identification and assessment. Indeed, new sensors have higher spatial resolution, which is expected to overcome a current shortcoming in drought indices products. Specifically, novel noise reduction algorithms and other atmospheric correction algorithms improve the thematic accuracy of remote sensing datasets. Remotely sensed indices are diverse and new indices are frequently developed. Besides NDVI, which is still very effective and widely used (Dalezios *et al.* 2016b) other indices, such as VegDRI, Vegetation Condition Index (VCI) (Kogan, 1995), Temperature Condition index (TCI) and Vegetation Health Index (VHI) (Kogan, 1995) are currently operationally used

(NDMC, 2011; NOAA, 2011). Traditionally used bands include near-infrared (NIR), red and short-wavelength infrared (SWIR). The Land Surface Temperature (LST) has also been used along with NDVI to improve drought identification accuracy.

Remote sensing data and methods can be employed in several aspects of drought. Indeed, satellite systems provide temporally and spatially continuous data over the globe and, thus, they are potentially better and relatively inexpensive tools for regional applications, such as monitoring vegetation conditions, agricultural drought and crop yield assessment than conventional weather data. For these types of applications, appropriate remote sensing systems are those that provide low spatial and high temporal resolution data, since daily coverage and data acquisition are required. In particular, the series of geosynchronous, polar-orbiting meteorological satellites, e.g. NOAA/AVHRR and/or MODIS, fulfil the above requirements and there are already long series of databases. In addition, remote sensing is a useful tool to analyze the vegetation dynamics for agricultural drought on local, regional or global scales (Keyantash & Dracup, 2002), as well as to assess the vegetative stress and to determine the impact of climate on vegetation. Satellite-derived vegetation indices have been extensively used for identifying periods of vegetative stress in crops, which represent an indication of agricultural drought, or vegetation status, in general. Again, in Table 5.3, an indicative list of remotely sensed agricultural drought indices is presented.

5.2.2.5 Aggregation of drought indices

There is an increasing trend in the development of composite drought indices, which constitutes a means to incorporate and more effectively exploit information that is readily available and proven to be useful in field-specific drought indices (Niemeyer, 2008). Indeed, non-hybrid drought indices are not just comprehensive, but mainly useful for particular areas and specific goals or applications. Specifically, the Lincoln Declaration (WMO, 2009) has recommended the creation of a new composite hydrological drought index that would cover stream flow, precipitation, reservoir levels, snowpack, and groundwater levels. In general, hybrid drought indices can provide a higher correlation with actual impacts sustained in the ground. In fact, most hybrid drought indices are comparatively recent, including the USDM and VegDRI. Specifically, VegDRI combines SPI and PDSI in addition to two NDVI-based indicators, namely Percent Average Seasonal Greenness (PASG) and Start of Season Anomaly (SOSA). Moreover, a combination of SPI, SWSI, and PDSI has been conducted to develop the integrated HDI (Karamouz *et al.* 2009). The predicted non-stationarity in future climates (IPCC, 2012) has instigated research for including future temporal patterns in drought assessment. In particular, the SPEI accounts for the increase in the duration and magnitude of droughts resultant from higher temperatures. Additional research has been conducted for specific regions including Australia and the Czech Republic (Dalezios *et al.* 2016b).

5.2.3 Description of representative drought indices

As already mentioned, indices are also indicators and constitute an attempt to quantify drought features, such as magnitude and severity. It is also known that drought indices mainly originate from the lack or deficit of precipitation. However, several other drought indicators are also important for understanding drought behavior in a region and the assessment or monitoring of drought severity through multiple indicators. New types of data and platforms, such as remote sensing, are also employed for assessing and monitoring drought. Indeed, the advantage to using remotely sensed data is that they allow for high spatial resolution and frequency of occurrence, which allows for near real time data analyses. However, the main drawback of remote sensing data is the short period of available records. Moreover, drought indices also account for temperatures and their impacts to the water balance of a basin. Specifically, the Standardized Precipitation Evapotranspiration Index (SPEI) (Vicente-Serrano *et al.* 2010) is based on the difference between precipitation and potential evapotranspiration, and the Reconnaissance Drought Index (RDI) (Tsakiris & Vagelis, 2005; Dalezios *et al.* 2012) utilizes the ratio of precipitation over potential evapotranspiration. For the illustration of drought quantification, three indices are presented, namely the Standardized Precipitation Index (SPI) for meteorological drought, the Reconnaissance Drought Index (RDI) also for meteorological drought and the Vegetation Health Index (VHI) for agricultural drought, respectively.

Standardized Precipitation Index (SPI). The SPI (McKee *et al.* 1993) attempts a quantification of the precipitation deficit for multiple time scales, namely for 3, 6, 9, and 12 month periods, respectively. It is stated that at least 20–30 years of serially complete monthly values are required, although 50–60 years (or more) are preferred and considered optimal. The time series of rainfall records at a station are fitted to a gamma distribution. This is conducted by using the maximum likelihood estimation of the gamma distribution parameters, β and γ (equation 5.1)

$$P(x) = \frac{x^{\gamma-1}\exp(-x/\beta)}{\beta^{\gamma}\Gamma(\gamma)} \quad {}_{\gamma>0} \tag{5.1}$$

where $P(x)$ is the probability density frequency (pdf) equation and x is the variable. A classification system is used to define drought severities (or intensities) resulting from the SPI (Table 5.4). The SPI is computed by dividing the difference between the normalized seasonal precipitation and its long-term seasonal mean by the standard deviation (equation 5.2). Thus,

$$SPI = \frac{X_{ij} - X_{im}}{\sigma} \tag{5.2}$$

Table 5.4 Standardized precipitation index (SPI) classification scale (from McKee *et al.* 1993).

Standardized Precipitation Index Value	Moisture Level
+2.0 and greater	Extremely wet
+1.5 to 1.99	Very wet
+1.0 to 1.49	Moderately wet
–.99 to .99	Near normal–mild dry
–1.0 to –1.49	Moderately dry
–1.5 to –1.99	Severely dry
–2.0 and less	Extremely dry

where, X_{ij} is the seasonal precipitation at the *i*th raingauge station and *j*th observation, X_{im} the long-term seasonal mean and σ is its standard deviation. A drought event exists when the SPI is continuously negative and reaches an intensity of –1.0 or less. The drought event ends when the SPI becomes positive. Indeed, each drought event has a duration defined by its beginning and end, and intensity (or severity) for each month as long as the drought event continues. The positive sum of the SPI for all the months during a drought event can be termed the "magnitude" of drought. In analogy to PDSI, SPI may be used for monitoring both dry and wet conditions. Seven classes of SPI are shown in Table 5.4. Moreover, during an international expert workshop at the University of Nebraska (8–11 Nov 2009) it was announced via the "Lincoln Declaration on Drought Indices" that the SPI is adopted by WMO and be used to characterize meteorological droughts around the world and the National Meteorological and Hydrological Services (NMHSs) are encouraged to use the SPI and provide this information on their websites (WMO, 2009).

Reconnaissance Drought Index (RDI). The RDI is based on the ratio of precipitation over potential evapotranspiration. The initial value of the index for a certain period, indicated by a certain month (k) in a year, is calculated by equation (5.3):

$$a_k = \frac{\sum_{j=1}^{j=k} P_j}{\sum_{j=1}^{j=k} PET_j} \tag{5.3}$$

where, P_j and PET_j are the precipitation and the potential evapotranspiration of the *j* month of the hydrological year, respectively. For real world applications, if a_k is calculated as a general indicator of meteorological drought, it is suggested to use periods of 3, 6, 9 and 12 months, respectively. In case of selecting a 12-month period, the result could be directly compared with the Aridity Index, which is produced for the area under study. If a_{k12} for a certain year is lower than the computed Aridity Index, then the area is affected by drought during this year. Two expressions are used for the RDI, namely Normalized RDI (RDI_n) and Standardized RDI (RDI_s). The Normalized RDI (RDI_n) (equation 5.4) and the Standardized RDI (equation 5.5), which is usually preferred, are computed by:

$$RDI_n(k) = \frac{a_k}{\bar{a}_k} - 1 \tag{5.4}$$

$$RDI_{st}(k) = \frac{y_k - \bar{y}_k}{\hat{\sigma}_k} \tag{5.5}$$

where a_k is the initial value for each month, $\bar{a}_k$ is the average value of a_k, y_k is the $\ln a_k$, $\bar{y}_k$ is its arithmetic mean and $\hat{\sigma}_k$ is its standard deviation. Drought classes are shown in Table 5.5.

Table 5.5 RDI drought classification scheme (from Tsakiris & Vagelis, 2005).

Drought Categories	RDI Values
Extremely Wet	>2.00
Very Wet	1.50 to 1.99
Moderately Wet	1.00 to 1.49
Near Normal	0.99 to –0.99
Moderately Dry	–1.00 to –1.49
Severely Dry	–1.50 to –1.99
Extremely Dry	<–2.00

Vegetation Health Index (VHI). The VHI is a combination of the Vegetation Condition Index (VCI) and the Temperature Condition Index (TCI), both derived from NOAA/AVHRR satellite data. The Vegetation Health Index (VHI) represents the overall vegetation health and is used for drought mapping and crop yield assessment. Indeed, the computation of VHI on a monthly basis uses satellite data of brightness temperature (BT) and the Normalized Difference Vegetation Index (NDVI). The VCI is based on NDVI and is an extension of NDVI and is expressed by the following equation (5.6):

$$VCI = 100 * \frac{NDVI - NDVI_{min}}{NDVI_{max} - NDVI_{min}} \tag{5.6}$$

where NDVI, $NDVI_{max}$ and $NDVI_{min}$ are the smoothed ten-day NDVI, its multi-year maximum and its multi-year minimum, respectively, for each pixel, in a given area. Similarly, TCI is based on the same concept as VCI and is expressed by the following equation (5.7):

$$TCI = 100 * \frac{BT_{max} - BT}{BT_{max} - BT_{min}} \tag{5.7}$$

where BT, BT_{max} and BT_{min} are the smoothed ten-day radiant temperature, its multi-year maximum and its multi-year minimum, respectively, for each pixel, in a given area. The five classes of VHI that represent agricultural drought, as well as no drought conditions are presented in Table 5.6. Specifically, from Table 5.6 it is evident that drought severity is decreasing with increasing VHI values. VHI is expressed by the equation (5.8):

$$VHI = 0.5 * (VCI) + 0.5 * (TCI) \tag{5.8}$$

Table 5.6 VHI drought classification scheme (from Kogan, 1995).

VHI Values	Vegetation Drought Class	Drought Classes Numbers
≤10	Extreme drought	1
≤20	Severe drought	2
≤30	Moderate drought	3
≤40	Mild drought	4
>40	No drought	5

Moreover, VCI and TCI vary from zero, for extremely unfavorable conditions, to 100, for optimal conditions. VCI and TCI characterize the moisture and thermal conditions of vegetation, respectively. Thermal conditions are especially important when moisture shortage is accompanied by high temperature, increasing the severity of agricultural drought and having direct impact on vegetation's health. In several parts of the world, TCI and VCI have proven to be useful tools for the estimation of agricultural drought (Kogan, 1995; Dalezios *et al.* 2014).

5.2.4 Drought early warning systems (DEWS)

DEWS emphasize on monitoring drought conditions (Wilhite *et al.* 2000) based on the use of drought indicators and indices. Specifically, the United States Drought Monitor (USDM) system is based on a composite of multiple indicators covering various short- and long-term time frames, with the objective to develop a ranking methodology for drought analysis leading to a single product. The USDM system has also the flexibility to integrate new tools and data and additional information, if available, in order to improve the level of accuracy. Moreover, there are hybrid types of drought indices, where satellite data are combined with surface data, such as the Vegetation Drought Response Index (VegDRI). Based on the availability and quality of data for any particular area, it is possible to employ several available drought indices, in order to determine the most suitable for any particular area or season for drought monitoring and DEWS. Using an approach, such as the USDM that is looking at all available indicators, it would also allow for the flexibility to implement more temperature-based indicators for drought monitoring and early warning systems. Furthermore, drought indices can also account for a changing climate in which there may be a shift in both temperature and precipitation regimes. For illustrative purposes, two case studies using empirical models and leading to DEWS, one based on RDI and the other based on VHI, respectively, are briefly presented.

Meteorological DEWS: RDI. The cumulative monthly areal extent values of the extreme RDI drought class, i.e. class 4 of Table 5.5 with values lower than −2, for all the drought episodes is plotted. Then, two figures are produced, namely Figure 5.3 for droughts of large areal extent and Figure 5.4 for droughts of small areal extent, respectively. Moreover, curve fitting is conducted for each of these figures leading to the following polynomials, namely equation (5.9) for droughts of large areal extent and equation (5.10) for droughts of small areal extent, respectively, both with high coefficient of determination.

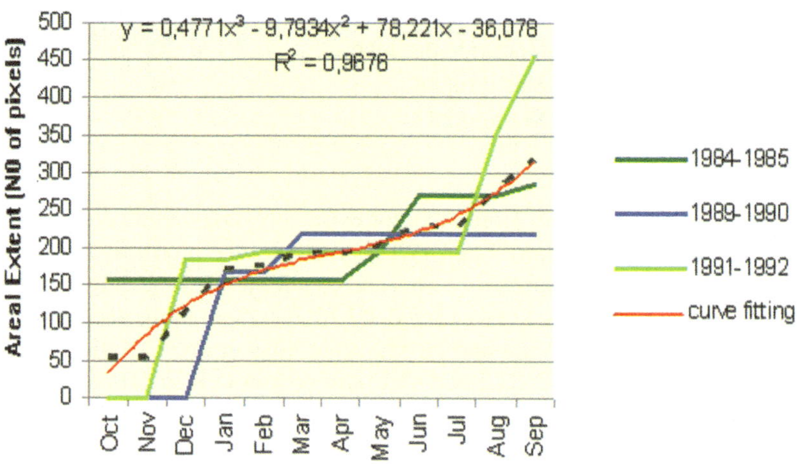

Figure 5.3 Cumulative large areal extent (no of pixels 8 × 8 km²) of extreme drought (<−2.0) during drought years based on RDI (from Dalezios *et al.* 2012).

$$y = 0.4771x^3 - 9.7934x^2 + 78.221x - 36.078 \quad (R^2 = 0.9676) \tag{5.9}$$

$$y = 0.4868x^2 - 3.3415x + 4.781 \quad (R^2 = 0.9618) \tag{5.10}$$

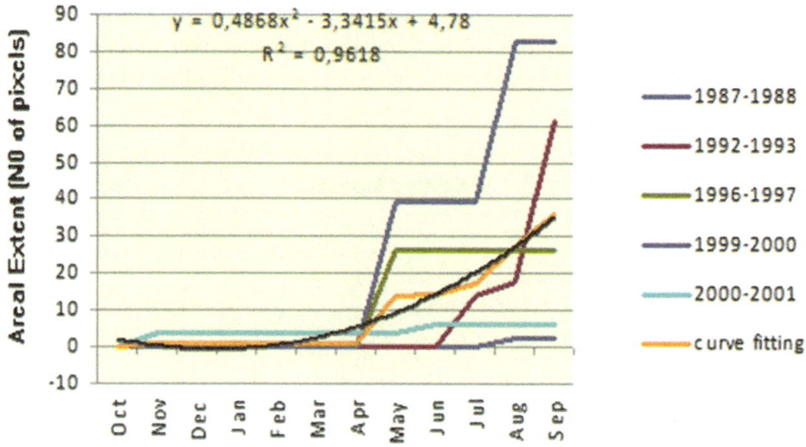

Figure 5.4 Cumulative small areal extent (no of pixels 8 × 8 km^2) of extreme drought (<–2.0) during drought years based on RDI (from Dalezios *et al.* 2012).

Indeed, for droughts of large areal extent (Figure 5.3), drought starts during the first three months of the hydrological year, whereas for droughts of small areal extent (Figure 5.4), drought starts in spring (April). This finding signifies the possibility of using the fitted curves for monitoring and early warning drought assessment in a region. This finding justifies the use of the fitted curves of Figures 5.3 and 5.4, along with the corresponding equations (5.9) and (5.10), respectively, for drought prognostic assessment or DEWS.

Agricultural DEWS: VHI. The cumulative monthly areal extent curves of the two merged classes is presented in Figure 5.5, which correspond to the four VHI severity classes of agricultural drought (Table 5.6). Moreover, curve fitting is conducted for each of these curves leading to the following polynomials, namely equation (5.11) for high severity areal extent drought and equation (5.12) for low severity areal extent drought, respectively, both with high coefficient of determination.

$$y = 0.0905x^2 + 4.3574x \qquad (R^2 = 0.9168) \tag{5.11}$$

$$y = -3.7413x^3 + 34.977x^2 - 6.8352x \quad (R^2 = 0.9998) \tag{5.12}$$

The two curves of Figure 5.5, namely the cumulative monthly areal extent curve of high severity classes (equation 5.11) and the corresponding curve of low severity classes (equation 5.12), respectively, present the range of values that can be assigned to agricultural drought every year during the warm season. Indeed, these findings signify the possibility of using the fitted curves for monitoring and early warning drought assessment in any region, although the actual VHI values may be different for different regions. Consequently, the fitted curves of Figure 5.5, along with the corresponding equations (5.11) and (5.12), respectively, can be used for first-guess drought prognostic and monitoring assessment leading to DEWS.

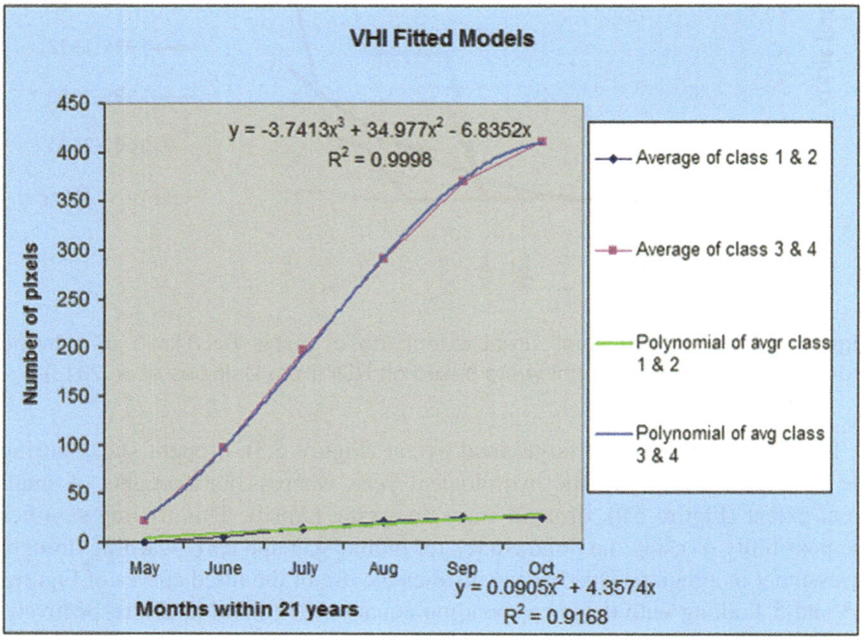

Figure 5.5 Fitted models of cumulative areal extent (number of pixels) of average monthly drought VHI values for the two sums of severity classes (from Dalezios *et al.* 2014).

5.3 DROUGHT RISK ASSESSMENT

Drought risk assessment includes risk estimation, meaning the risk of such drought events, i.e., event probabilities, as well as magnitude-duration-frequency and areal extent relationships for hazard assessment. Moreover, risk estimation also involves vulnerability assessment and its uncertainty. Indeed, the combination of all the particular risks leads to the total drought risk for all the severities (intensities), return periods and elements at risk resulting into the quantitative risk assessment (QRA).

5.3.1 Drought severity-duration-frequency (SDF) relationships

The theory of extreme events (Gumbel, 1958) has been a research subject for several decades in meteorology, hydrology and, in general, in natural sciences.

Moreover, statistical frequency analysis of climatic extremes has been extensively used internationally with very satisfactory results for several hazards. However, frequency analysis of droughts is not easily accomplished, since droughts are not universally quantified phenomena. Indeed, the frequency of drought occurrence cannot sufficiently and fully cover drought hazards, unless it is quantitatively related to other features and characteristics of drought, such as severity and duration. This has led to the development of drought Severity-Duration-Frequency (SDF) relationships.

Data for high-intensity extreme events are, in general, specified by three variables, namely frequency, duration and either depth or mean intensity. Moreover, the areal extent of particular common drought episodes is another significant drought feature, which can be considered through remote sensing data and methods. The frequency of an extreme event is usually expressed by its return period or recurrence interval, which is defined as the average interval of time within which the magnitude of the event is equated or exceeded once. In addition, the magnitude of an extreme event is provided by the total depth occurring for a specific duration and data for extreme events can be usually presented by depth - duration - frequency graphs for several sites throughout the region or the watershed of interest. For the estimation of extreme events, such as droughts, in which the return periods are needed when the depths and duration are provided, it is necessary to assume a specific mathematical form of the frequency distribution. There are several theoretical distributions that have been tested against the cumulative severities, i.e. the ΣZ values of various durations, of the Palmer drought method. These include the Extreme Value I (EVI, Gumbel), the Generalized Extreme Value (GEV), the three parameter Log-Normal (LN3) and the Log-Pearson (LP3) distributions (Dalezios *et al.* 2000). Application of the non-parametric Kolmogorov-Smirnov two-sample test, at 95% confidence level, and fitting of the above theoretical frequency distributions to the ΣZ values show that the EVI provides a reasonable and acceptable approximation of the frequency of the calculated ΣZ values (Figure 5.6). Indeed, the EVI has been applied in numerous extreme drought analyses and it is easier to be used than the LN3 and LP3 distributions, respectively, since it is a two-parameter frequency distribution. For the above reasons the EVI has been used for drought SDF analysis.

It is recognized that the quantitative assessment of drought severity and duration appears to be a difficult subject. The severity of drought is defined as the cumulative sum of successive negative values of the Z-index (ΣZ) and the duration of drought is defined as the corresponding number of successive months with continuous negative Z-index values (Palmer,1965). Moreover, drought frequency is defined as the return period of a specific cumulative Z-index (ΣZ) value for successive drought months. In the current example, the data base for the development of the SDF relationships for droughts consists of the computed monthly Z-index time series from 28 stations over Greece (Dalezios *et al.* 2000). It is also mentioned that, for the development of the SDF relationships, a common period from 1957 to 1983 is used, for the 28 stations. The analysis has been conducted for each station for the drought periods, i.e. successive negative Z-index values. A brief description of the steps, which are followed to develop the drought SDF relationships, is presented.

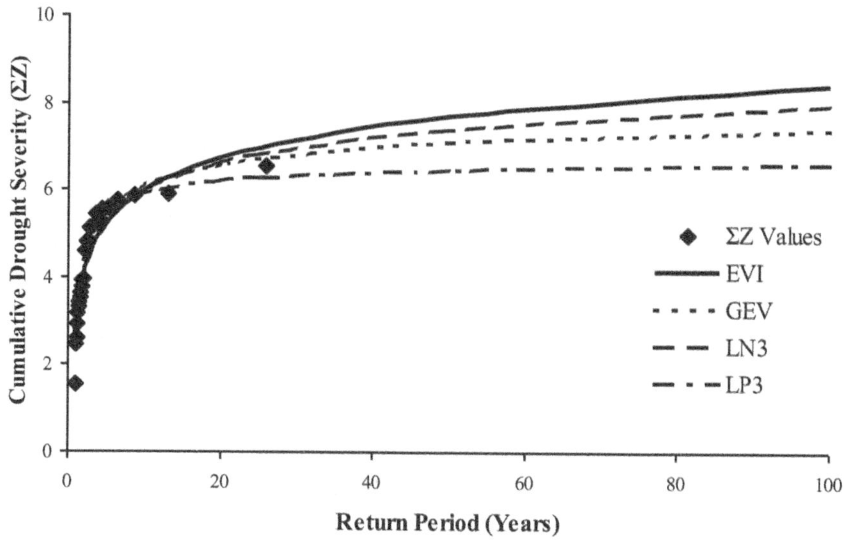

Figure 5.6 Comparison of EVI (Gumbel) and other theoretical frequency distributions for cumulative drought (negative ΣZ values) of three-month duration at Volos (BL) station (from Dalezios *et al.* 2000).

Step 1: Probability Tables. Based on the computed monthly Z-index time series, the drought episodes for each station are identified and tabulated, i.e., the cumulative drought severity using successive (negative) values of Z-index (ΣZ) time series, along with the corresponding duration in months. As a result, multiple events per year are also accounted in the SDF analysis. A typical example is presented in Table 5.7 for Volos (BL) station of three-month drought duration. In particular, in Table 5.7 column 1 shows the ranking numbers, column 2 shows the cumulative (negative) Z-index (ΣZ) values in ascending order, in column 3 the corresponding probability (P) of occurrence is presented using the Weibull plotting position equation $P = [m/(n + 1)]$, where m is the current ranking number and n is the total number of data points and column 4 shows the corresponding drought return period T of three-month duration using the equation $T = [1/P]$, where P has been previously defined. In addition, columns 5 to 7 show the ΣZ estimates of EVI, as well as the upper and lower 95% confidence limits, respectively. The final number of produced tables is equivalent to the number of identified classes-durations for each of the 28 stations in Greece.

Table 5.7 Probability table of cumulative drought severities (negative ΣZ values) of three-month duration for volos (BL) station (from Dalezios *et al.* 2000).

	Cumulative Drought Severity (ΣZ)	Exceedance Probability	Return Period (Years)	EVI (Gumbel)	95% Upper Confidence Limit	95% Lower Confidence Limit
(1)	(2)	(3)	(4)	(5)	(6)	(7)
1	6.54	0.04	26.00	6.97	8.47	5.47
2	5.91	0.08	13.00	6.24	7.45	5.03
3	5.85	0.12	8.67	5.80	6.85	4.76
4	5.75	0.15	6.50	5.49	6.41	4.56
5	5.57	0.19	5.20	5.23	6.07	4.40
6	5.55	0.23	4.33	5.02	5.78	4.26
7	5.45	0.27	3.71	4.84	5.53	4.14
8	5.15	0.31	3.25	4.68	5.32	4.03
9	5.12	0.35	2.89	4.53	5.12	3.93
10	4.82	0.38	2.60	4.39	4.95	3.83
11	4.8	0.42	2.36	4.26	4.78	3.74
12	4.61	0.46	2.17	4.14	4.63	3.65
13	3.94	0.50	2.00	4.03	4.49	3.56
14	3.91	0.54	1.86	3.91	4.36	3.47
15	3.77	0.58	1.73	3.80	4.23	3.38
16	3.65	0.62	1.63	3.70	4.11	3.28
17	3.53	0.65	1.53	3.59	4.00	3.18
18	3.44	0.69	1.44	3.48	3.89	3.08
19	3.33	0.73	1.37	3.37	3.78	2.96
20	3.17	0.77	1.30	3.26	3.67	2.84
21	2.94	0.81	1.24	3.14	3.56	2.71
22	2.61	0.85	1.18	3.01	3.45	2.56
23	2.6	0.88	1.13	2.86	3.33	2.38
24	2.48	0.92	1.08	2.68	3.20	2.17
25	1.55	0.96	1.04	2.44	3.02	1.85
		0.50	2	4.03	4.49	3.56
		0.20	5	5.19	6.00	4.37
		0.10	10	5.96	7.06	4.86
		0.05	20	6.70	8.09	5.31
		0.04	25	6.93	8.41	5.45
		0.02	50	7.65	9.42	5.88
		0.01	100	8.37	10.43	6.31

Step 2: Fitting Gumbel Distribution. For each duration in months for drought and wet periods, the identified cumulative severities for drought and wet periods are plotted versus the corresponding return period and a statistical distribution is fitted to the plotted data points. In particular, column 2 (Table 5.7) is plotted versus column 4. The extreme value law is implemented to drought severities by fitting the EVI distribution, which has the following Cumulative Distribution Function (CDF), $F(x)$:

$$F(x) = \exp[-\exp(-A(x - U))] \tag{5.13}$$

where A, U are the fitted parameters, which are computed for each duration from data and are equal to:

$$A = \frac{1.283}{\sigma} \tag{5.14}$$

$$\mathrm{U} = \bar{x} + 0.45\sigma \tag{5.15}$$

where $\bar{x}, \sigma$ are the mean and the standard deviation of the data, respectively. The fitted EVI distribution to the data points of Table 5.7 is shown in Figure 5.7 for Volos (BL) station. The procedure of fitting EVI distribution is applied to all the identified drought durations for each station. As expected, each curve is plotted on top of the previous, since it corresponds to ascending drought duration. Moreover, the data could be occasionally grouped in high durations in order to enlarge the data base and achieve an acceptable fit.

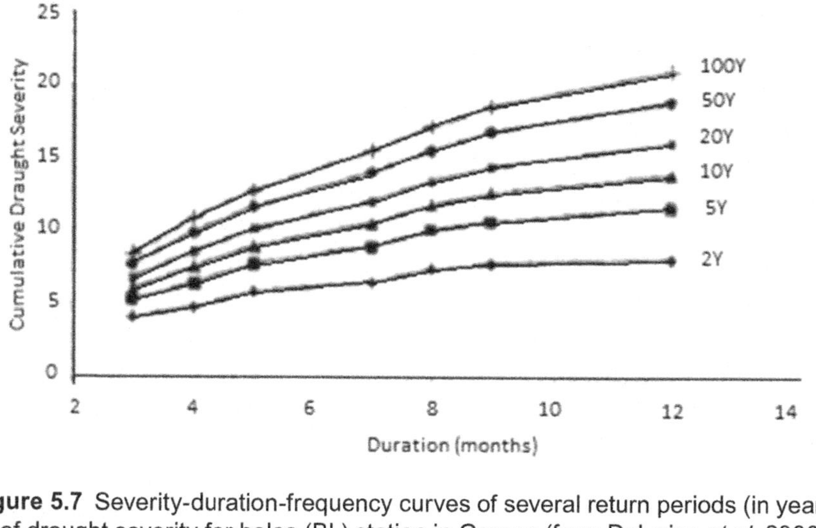

Figure 5.7 Severity-duration-frequency curves of several return periods (in years, Y) of drought severity for bolos (BL) station in Greece (from Dalezios *et al.* 2000).

Step 3: Severity – Duration – Frequency (SDF) Curves. Using equation (5.13) drought severities (ΣZ) are computed, which correspond to return periods of 2, 5, 10, 20, 30, 50 and 100 years, respectively, for each identified drought duration. It should be mentioned that the low frequency drought events, i.e. events with return periods of 50 and 100 years, respectively, have a much larger uncertainty than more frequent drought events. This is mainly the result of the small sample, which is what usually happens, and the extrapolation of the fitted theoretical frequency distribution to the most extreme events. However, the sample used in this example is reasonably long (27 years), thus minimizing the uncertainty of extreme events as it is shown by the 95% confidence limits indicated in Table 5.4 for the station of Volos (BL). Moreover, the SDF relationships are developed, using the ΣZ values with the above return periods, and the corresponding diagrams are produced for each station. Then, from these diagrams, a series of maps of SDF relationships can be produced over a region by combining severities, durations and frequencies. For illustrative purposes, Figure 5.7 shows the diagram of the developed SDF relationships of drought and wet periods for Volos (BL) station, where each curve refers to one of the above return periods.

5.4 DROUGHT RISK MANAGEMENT

The principle of drought policy focuses on risk management through the implementation of preparedness plans and mitigation measures. Specifically, preparedness refers to pre-disaster activities designed to improve the level of readiness or increase the operational and institutional efficiency for responding to a drought episode. Similarly, mitigation includes short- and long-term actions, programs, or policies applied during and in advance of drought that reduce the degree of risk to human life, productive capacity and property. There are also alternative drought mitigation measures and programs dealing with the so-called marginal waters in the framework of water scarcity management in a region (Dalezios *et al.* 2016a). Moreover, emergency response is always expected to be part of any drought management framework, since it is unlikely that policy makers can anticipate, avoid or reduce all potential impacts through mitigation programs. Indeed, there are several components of a drought risk reduction plan and strategy. For instance, such components include the availability of reliable information for decision making policies that encourage assessment and use the above information, a framework for risk management measures, as well as consistent and effective actions. Nevertheless, the main goal of national drought plans remains the improvement of preparedness and response efficiency by reinforcing early warning and monitoring, impact and risk assessment, as well as mitigation and response. Until recently, droughts used to focus on response efforts, whereas, at the present time, the trend is on mitigation as the fundamental element of a drought plan. Thus, many plans adopt a more risk management approach to drought management, becoming thus more pro-active (Dalezios *et al.* 2009).

5.4.1 Drought impacts

Drought impacts refer to a number of drivers that may convert physical drought causes, such as reduced average precipitation, deficient soil moisture and low water levels, into disaster events for vulnerable societies and economies. Indeed, some obvious disaster trends are attributed to socio-economic factors rather than to the frequency and magnitude of geophysical processes. There are certainly several causes and drivers, which justify the increasing trend in disaster impact, despite the fact that the frequency of geophysical events remains almost the same and there are many positive efforts to reduce disasters. Specifically, these types of drought risk causes and drivers may include land pressure, climate variability and change, economic growth, technological innovation, population growth, urbanization, inequalities, political change, social expectations, risk governance capacities and global interdependence, among others.

Nevertheless, droughts lead to several significant impacts, which can be divided into direct and indirect impacts. Specifically, direct drought impacts may include reduced cropland, reduced water levels, agricultural production losses, food security problems, increased fire hazards, forest and rangeland productivity, livestock and wildlife mortality rates, damage to wildlife and fish habitat, rural livelihoods, as well as impacts in the urban and economic sectors, among others. In addition, droughts may contribute to ecosystem decline, migration and conflict. On the other hand, the consequences of these direct impacts are usually considered indirect impacts. Moreover, drought impacts can also be classified based on the affected sector, leading to environmental, economic or social types of impact. Specifically, environmental impacts refer to the losses resulted as a direct consequence of drought or indirectly, such as wildfire damage to plant and animal species. Furthermore, several economic impacts affect agriculture and related sectors. Finally, social impacts refer to public safety, health, quality of life issues, water use conflicts and regional inequities in relief and impacts distribution.

5.4.2 Drought mitigation

Drought mitigation plans are based mainly on three fundamental components applied to provincial, national or regional level, respectively. First, early warning systems serve as the basis for decision making during a drought period and have to be associated with a dissemination system that distributes reliable and timely information. Second, it is important to undertake risk assessment in order to determine the subject and the causes of risk, which is accomplished through impact studies of drought events. Third, it is necessary to specify appropriate mitigation actions in order to reduce the risk of each impact for future drought events (Wilhite *et al.* 2000). In addition, response actions are also considered, although the main goal of drought mitigation plans remains to reduce the vulnerability to drought events. Several mitigation actions are briefly presented.

There are assessment programs, which include the development of criteria or triggers for specific mitigation actions in response to drought, the establishment of new data collection networks, the development of early warning and monitoring systems, as well as monitoring climatic and water supply conditions. Moreover, legislative actions include measures to protect environmental flows, to examine water rights for possible modification during water shortages, to guarantee low-interest loans to farmers and to impose limits on urban development. Similarly, there is a need for augmentation of water supplies during droughts, which includes reviewing reservoir operation plans and rehabilitating reservoirs to operate at design capacity. In addition, public awareness of drought severity involves periodic reports and pamphlets, information meetings and workshops, as well as focal points for support and technical assistance on water conservation and other water related activities, such as evaluation of water quantity and quality. There are additional mitigation actions, which involve water conservation programs, such as water use efficiency, metering and leak detection programs, emergency response programs, resolution of water use conflicts and drought contingency plans.

Besides the above mentioned conventional drought mitigation measures, there exist significant non-conventional measures, which make use of the so-called marginal waters and could be applied as a supplement for additional support of the efficiency of water management schemes in arid and semi-arid regions. In particular, marginal waters refer to waste waters, saline, sea and atmospheric waters. Non-conventional measures, which could have an important impact on arid areas, include the consideration of water dams in small streams in conjunction with small hydroelectric works, the provision of desalination plans to convert brackish or sea water to drinking water, the reuse of wastewater effluents and water streams of inferior quality in agricultural applications or for the replenishment of the aquifers. At the present time, there are efforts to reduce the cost of desalination for municipal, domestic and industrial use, whereas wastewater reuse is currently limited to a small percentage of irrigation water use, but is expected to become more significant in the near future. Special reference is given to atmospheric water as a non-conventional measure for drought mitigation. Specifically, atmospheric water can assist to combat drought in terms of weather modification programs for rain enhancement through cloud seeding (Dalezios *et al.* 2009; Dalezios *et al.* 2016a). It is stated that rain enhancement in selected arid areas is expected to improve the quantity of water in the hydrological system. Eventually, the incorporation of non-conventional measures to combat drought may assist the overall water management strategy conceived and be applied to arid regions in several ways.

5.5 SUMMARY

Drought has been considered not just a phenomenon, but a hazard. At first, drought concepts have been described, including drought definitions and types, factors and features. This is followed by a comprehensive presentation of the components of drought risk with emphasis on drought issues. Specifically, drought risk identification has been analyzed, including specific drought indices, with emphasis on composite indices for different drought types and drought early warning systems (DEWS). Then, drought risk assessment has been considered, involving the development of drought severity-duration-frequency relationships. In the final section of drought risk management, drought impacts and drought mitigation measures have been discussed.

REFERENCES

Bhalme H. N. and Mooley D. A. (1980). Large-scale drought / floods and monsoon circulation. *Monthly Weather Review*, **108**(8), 1197–1211.

Brown J. F., Wardlow B. D., Tadesse T., Hayes J. J. and Reed B. C. (2008). The vegetation drought response index (VegDri): an integrated approach for monitoring drought stress in vegetation. *GIScience and Remote Sensing*, **45**(1), 16–46.

Bryant S., Arnell N. W. and Law F. M. (1992). The long-term context for the current hydrological drought. Institute of Water and Environmental Management (IWEM), Conference on the Management of Scarce Water Resources, 13–14 October 1992.

Byun H. R. and Wilhite D. A. (1999). Objective quantification of drought severity and duration. *Journal of Climate*, **12**(9), 2747–2756.

Dalezios N. R., Loukas A., Vasiliades L. and Liakopoulos H. (2000). Severity-duration-frequency analysis of droughts and wet periods in Greece. *Hydrological Science Journal*, Vol. **45**(5), 751–770.

Dalezios N. R., Bampzelis D. and Domenikiotis C. (2009). An integrated methodological procedure for alternative drought mitigation in Greece. *European Water*, **27/28**, 53–73.

Dalezios N. R., Blanta A. and Spyropoulos N. V. (2012). Assessment of remotely sensed drought features in vulnerable agriculture. *NHESS*, **12**, 3139–3150.

Dalezios N. R., Blanta A., Spyropoulos N. V. and Tarquis A. M. (2014). Risk Identification of Agricultural Drought for Sustainable Agroecosystems. *NHESS*, **14**, 2435–2448.

Dalezios N. R. and Eslamian S. (2016). Regional design storm for Greece within the flood risk management framework. *International Journal of Hydrology Science and Technology (I.J.H.S.T)*, 6(1), 82–102.

Dalezios N. R., Angelakis A. N. and Eslamian S. (2016a). Water Scarcity Management: Part 1: Methodological Framework. *International Journal of Hydrology Science and Technology* (accepted, in press).

Dalezios N. R., Dercas N. and Eslamian S. (2016b). Water Scarcity Management: Part 2: Satellite-based Composite Drought Analysis. *International Journal of Hydrology Science and Technology* (accepted, in press).

Dalezios N. R., Gobin A., Tarquis A. M. and Eslamian S. (2016c). Agricultural Drought Indices: Combining Crop, Climate and Soil Factors. Book chapter 5 in Vol. 1 of 3-Volume Handbook of Drought and Water Scarcity (HDWS). S. Eslamian (ed.). Taylor and Francis (accepted, in press).

Dalezios N. R., Dunkel Z. and Eslamian S. (2016d). Meteorological Drought Indices: Definitions. Book chapter 3 in Vol. 1 of 3-Volume Handbook of Drought and Water Scarcity (HDWS). S. Eslamian (ed.). Taylor and Francis (accepted, in press).

Domenikiotis C., Dimakis E., Kanellou E. and Dalezios N. R. (2008). Use of the Remotely Sensed Derived TVDI for Drought Monitoring in Thessaly Watersheds. AgEng2008: Agricultural and Biosystems Engineering for a Sustainable World, Crete, 23–25 June, (CD).

Eslamian S., Ghasemizadeh M., Biabanaki M. and Talebizadeh M. (2010). A principal component regression method for estimating low flow index. *Water Resource Management*, **24**, 2553–2566.

Foley J. C. (1957). *Droughts in Australia: Review of Records from Earliest Years of Settlement to 1955.* Melbourne: Commonwealth of Australia, Bureau of Meteorology.

Gommes R. A. and Petrassi F. (1994). Rainfall variability and drought in sub-Saharan Africa since 1960. FAO, Rome, Italy. 100p.

González J. and Valdés J. (2006). New drought frequency index: definition and comparative performance analysis. *Water Resource Research*, **42**(11): W11421. doi:10.1029/2005WR004308.

Heim R. R. Jr. (2002). A review of twentieth-century drought indices used in the United States. *Bulletin of the American Meteorological Society*, **83**(8), 1149–1165.

IPCC (2012). Managing the Risks of Extreme Events and Disasters to Advance Climate Change Adaptation, Special Report of IPCC, 582p.

Karamouz M., Rasouli K. and Nazif S. (2009). Development of a hybrid index for drought prediction: case study. *Journal of Hydrologic Engineering*, **14**(6): 617–627. doi:10.1061/(ASCE)HE.1943-5584.0000022.

Keetch J. J. and Byram G. M. (1968). *A Drought Index For Forest Fire Control.* Asheville, NC, 32p.

Keyantash J. and Dracup J. A. (2002). The quantification of drought: an evaluation of drought indices. *Bulletin of American Meteorological Society*, **83**(8), 1167–1180.

Kogan F. N. (1995). Application of vegetation index and brightness temperature for drought detection, *Advances in Space Research*, **15**, 91–100.

McKee T. B., Doesken N. J. and Kleist J. (1993). The relationship of drought frequency and duration to timescales. *Preprints, Eighth Conf. on Applied Climatology*, Anaheim, CA, Amer. Meteor. Soc., 179–184.

Liu H. Q. and Huete A. (1995). A feedback based modification of the NDVI to minimize canopy background and atmospheric noise. *IEEE Transactions on Geoscience and Remote Sensing*, **33**(2), 457–465. doi:10.1109/36.377946.

McVicar T. R. and Jupp D. L. B. (1998). The current and potential operational uses of remote sensing to aid decisions on drought exceptional circumstances in Australia: a Review. *Agricultural Systems*, **57**(3), 399–468.

Mendicino G., Senatore A. and Versace P. (2008). A groundwater resource index (GRI) for drought monitoring and forecasting in a mediterranean climate. *Journal of Hydrology (Amst)*, **357**(3–4), 282–302. doi:10.1016/j.jhydrol.2008.05.005.

Narasimhan B. and Srinivasan R. (2005). Development and evaluation of soil moisture deficit index (smdi) and evapotranspiration deficit index (ETDI) for agricultural drought monitoring. *Agricultural and Forest Meteorology* **133**(1–4), 69–88. doi:10.1016/j.agrformet.2005.07.012.

NDMC (2011). Vegetation Drought Response Index. National Climatic Data Center. http://drought.unl.edu/vegdri/VegDRI_Main.htm

Niemeyer S. (2008). New drought indices. *Options Méditerranéennes. Série A: Séminaires Méditerranéens*, Options Méditerranéennes. Série A: Séminaires Méditerranéens, 267–274.

NOAA (2011). STAR—Global Vegetation Health Products. NOAA. http://www. star. nesdis.noaa.gov/smcd/emb/vci/VH/vh_browse.php.

Pálfai I. (1991). Az 1990 évi aszály Magyarországon. *Vízügyi Közlemények*, **2**, 117–132.

Palmer W. C. (1965). Meteorological drought. Weather Bureau Research Paper No. 45, US Department of Commerce, Washington, DC. 58 pp.

Shafer B. and Dezman L. (1982). Development of a surface water supply index (SWSI) to assess the severity of drought conditions in snowpack runoff areas. Proceedings of the Western Snow Conference, pp. 164–175.

Sivakumar M. V. K., Motha R.P. and Das H.P. (Eds), (2005). Natural Disaster and Extreme Events in Agriculture. Springer, ISBN-10 3-540-22490-4, 367p.

Smith K. (2013). Environmental Hazards: Assessing Risk and Reducing Disaster. Routledge, 6th Edition, ISBN0-415-22463-2, 478p.

Stahl K. (2001). Hydrological drought-a study across Europe. Universitätsbibliothek Freiburg.

Thornthwaite C. W. (1948). An approach toward a rational classification of climate. *Geography Review*, **38**(1), 55–94.

Thornthwaite C. W. and Mather J. R. (1955). The water balance. *Climatology*, **8**, 1–104.

Tsakiris G. and Vangelis H. (2005). Establishing a Drought Index Incorporating Evapotraspiration. European Water 9/10: 3-11.

Vafakhah M., Eslamian S. and Bozchaloei S.K. (2014). Low-flow hydrology. Book chapter 20, in Handbook of Engineering Hydrology: Fundamentals and Applications. S. Eslamian (ed.), Taylor and Francis, 433–453.

Van-Rooy M. P. (1965). A rainfall anomaly index (RAI) independent of time and space. *Notos*, 14: 43–48.

Vasiliades L., Loukas A. and Liberis N. (2011). A water balance derived drought index for pinios river basin, Greece. *Water Resources Management*, **25**(4), 1087–1101.

Vicente-Serrano S. M., Begueria S. and Lopez-Moreno J. I. (2010). A multiscalar drought index sensitive to global warming: The standardized precipitation evaportranspiration index. *Journal of Climate* **23**, 1696–1718.

Wilhite D. A., Hayes M.J., Kinutson C. and Smith K. H. (2000). Planning for drought: moving from crisis to risk management. *Journal of the American Water Resources Association*, **36**(4), 697–710.

Wheaton E. E. and Chakravarti A. K. (1990). Duststorms in the canadian prairies. *International Journal of Climatology*, **10**(8), 829–837.

Woodhouse C. A. and Overpeck J. T. as before (1998). 2000 years of drought variability in the central United States. *American Meteorological Society Bulletin*, **79**(12): 2693–2714.

WMO (2009). Lincoln declaration on drought indices. World Meteorological Organization (WMO) http://www.wmo.int/pages/prog/wcp/agm/meetings/wies09/documents/ Lincoln_Declaration_ Drought_Indices.pdf (accessed 22 April 2011).

Chapter 6

Land desertification

Costas Kosmas and Orestis Kairis

6.1 DESERTIFICATION CONCEPTS AND CHARACTERISTICS

The concept of "desertification" has been introduced by the scientific community in the early 20th century (Lavauden, 1927) referring to the low land productivity of rangelands in Tunisia, although the term is mainly attributed to Aubréville (1949), who connected the deforestation in West Africa with erosion and desertification. The United Nations Convention to Combat Desertification (UNCCD) has defined desertification as "the land degradation in arid, semi-arid and dry sub-humid areas, resulting from various factors, including climatic variations and human activities" (UNEP, 1994) considering climate change as an additional factor of desertification. Desertification has been and still is a controversial issue. In the previous decades, this was largely due to the lack of a common understanding of "what to measure" and "how to measure it". During the 1980's the need for a general and flexible approach to combat desertification became more keenly felt.

Although desertification processes are frequently grouped into physical, chemical, biological and anthropogenic (Perez-Trejo, 1992), the mechanism of desertification is characterized by the reduction of available soil water to the growing plants resulting into critical low plant productivity. Desertification is the consequence of a series of important degradation processes in semi-arid and arid regions, where water is the main limiting factor of land use performance on ecosystems. The Millennium Ecosystem Assessment (2005) considered desertification as "human induced land degradation" that reduces the capacity of the land to produce the ecosystem goods and services from which our society it depends on (Figure 6.1). Hill *et al.* (2008) provide another interesting aspect of the relationship between degradation and desertification outlining land degradation as a process in which disturbances have exceeded land resilience and have led to an irreversible loss of the land's biological production capacity. Desertification of an area will proceed if certain land components are degraded beyond specific thresholds, above which further change is irreversible. Undoubtedly desertification is associated with degradation processes with the distinction that desertification is defined as land degradation in arid areas, while land degradation not only refers to the arid regions (UNCCD, 2009).

Figure 6.1 Grazing land highly degraded due land to soil erosion and desertified (southern Greece left photo, central Morocco right photo).

A holistic view of land degradation issue defines it as the deterioration in physical and chemical properties of the soils that occur due to changes in environmental state (Imeson & Emmer, 1992). Soil degradation can be divided into natural, which refers mainly to soil formation factors (pan formation, calcification, leaching, etc.) and anthropogenic comprising concepts related to cultivation practices (compaction, runoff, crusting, etc.), land management (decline in soil organic matter, nutrient depletion, nutrient toxicity, etc.) and non-rational management of natural resources (pollution, salinisation, etc.) (Lal, 1998). Irrespective of the scientific classifications and approaches, the main causes of land degradation and of subsequent desertification in human biological time are human-induced (Salvati *et al.* 2015). This perspective has been accepted at EU level since 2001, when a significant turn in research priorities was noted indicating the important role of local and national stakeholders in conducting research programs that explore the incentives of land use decisions leading many times to unsustainable practices, which damage the environment (Geeson *et al.* 2015).

An assessment carried out by the International Centre for Arid and Semi-Arid Land Studies (ICASALS) revealed that approximately 70% of arid lands show more or less intense signs of desertification (Dregne *et al.* 1991). Cropland is experiencing the highest risk of land degradation and desertification. Land degradation and desertification affect over one billion people (Rubio & Recatala, 2006).

6.2 CAUSES AND PROCESS OF LAND DESERTIFICATION

For many years the scientific community has raised the debate over the nature and the causes of desertification. The following two opposing scientific theories have been developed: (a) the single-factor and (b) multiple-factor causation of desertification. The theory of single factor causation uses various root causes of the phenomenon, such as demographic changes or improper land management that leads to human induced desertification. The other theory suggests that desertification is the result of several factors that are strongly linked to the local conditions that is impossible to draw general patterns of desertification (Geist *et al.* 2004).

D'Odorico *et al.* (2013) categorized the most important feedback mechanisms that have been proposed as responsible for land degradation by several desertification theories. These mechanisms alter the properties of soils, transform the relationship between vegetation and climate and modify the composition of plant communities. According to this classification scheme the following three major processes of land degradation are distinguished: (a) the loss of nutrient-rich topsoil due to wind and water erosion, (b) the decrease in soil water storage capacity induced by various causes including unsustainable agricultural practices or overgrazing, and (c) the accumulation of salts or other toxic substances in the soil. Other authors (Eswaran *et al.* 2001) particularly focused on the important distinction between factors and causes of land degradation. Based on this theory, factors are the biophysical processes and attributes that define the type of degradation processes, such as soil erosion, salinization, soil sealing, etc. while causes are considered as the biophysical, socioeconomic, and political agents affecting the rate of land degradation. A comprehensive study carried out by the EU research project DESIRE[1] have shown that in 17 study sites located in the Mediterranean and Eastern Europe, Latin America, Africa and Asia, the main processes or causes of land degradation and desertification identified in these study sites were: (a) soil erosion including water and tillage erosion, (b) soil salinization, (c) water stress, (d) forest fires, and (e) overgrazing.

6.2.1 Soil erosion

Soil erosion is a significant process of land degradation and, consequently, desertification. The European Mediterranean region, characterized by high rainfall erosivity and high soil erodibility, is generally susceptible to soil erosion. Several studies have reported the important role of vegetation cover and land-use management to the variation in runoff and sediment yields. Francis and Thornes (1990) identified the key limit of plant cover (40%), under which accelerated soil erosion occurs. Kosmas *et al.* (1997) found that the lowest rates of annual runoff

[1]http://www.desire-project.eu/

and sediment loss occurred in hilly areas cultivated with olives and highlighted the significant role of the annual plants understory vegetation in limiting soil loss in moderately steep olive orchards. Recent efforts for quantifying the phenomenon have been carried out (Cerdan *et al.* 2010) illustrating that soils with permanent vegetation cover generate soil sediment losses by at least of one order of magnitude lower than those on arable land.

Although scientists have extensively dealt with runoff erosion, it is very important to point out that soil losses related to tillage may be more relevant than those associated with water erosion over most of the hilly cropland (Govers *et al.* 1994). Tillage erosion is a progressive down slope translocation of soil caused mechanically by tillage implements, such as mouldboard, chisel, duck foot chisel, harrow, etc. Displacement of soil materials mainly occurs in the upper convex parts (summit, shoulder, backslope) of a hillslope and deposition in the concave parts (footslope, toeslope). Tillage erosion is affected by tillage implement plough depth, tillage direction. A tandem disk may be more erosive than a mouldboard plough operation. By ploughing the soil at shallow depth in any direction of tillage operation, tillage erosion is significantly reduced. When the movement of tractor is along the contour lines, then soil displacement is highly restricted. Any other direction of tillage operation increases tillage erosion.

Studies carried out during the executions of the EU research project DESIRE have shown that among the most important factors related to human activities and affecting soil erosion in areas prone to desertification were land terracing, rate of land abandonment, and land use intensity (Kairis *et al.* 2013). Sloping areas with high percentage of terracing had higher soil erosion rates possibly due to poor design or maintenance of terraces, or because terraces tend to be constructed in terrain sensitive to erosion, where even higher rates occurred before terrace construction. Sloping areas with high rate of land abandonment (higher than 25 ha per 10 years per 10 km^2) were subjected to higher soil erosion. Furthermore, sloping land subjected to ploughing, disking or harrowing are usually subjected to higher degree of water and tillage erosion. In cases that the existing regulations for land protection were implemented, the degree of soil erosion was significantly reduced. Furthermore, areas with high (higher than 10%) elderly index were subjected to moderate or severe erosion due to unwillingness or inability of old-aged farmers to apply measures for soil erosion protection.

6.2.2 Soil salinization

According to the United Nations Environment Program (UNEP, 1992) soil salinization is one of the main degradation processes in arid zones, which can lead to land desertification. Salinization is included among the eight major threats for soil quality and sustainability in Europe (Montanarella, 2007). The concentration of salts in the root zone increases the osmotic potential of soil solution (Singh, 2015) and consequently the normal absorption of water and nutrients.

Soil salinization affects mainly plain areas with poor drainage conditions and especially coastal areas. Except the important climate characteristics of low annual rainfall (lower than 650 mm) and high aridity indices, human actions, related to water resources exploitation, such as ground water exploitation, water consumption/water demands, irrigation percentage of arable land, and water scarcity are greatly affecting soil salinization in areas prone to desertification. Areas of low water availability accompanied by over-exploitation of ground and surface water resources are more vulnerable to soil salinization. Furthermore, under high rates of water consumption/ water demands, soil salinization is more likely to occur. Areas characterized by high water scarcity (water consumption per sector) are more vulnerable to soil salinization and land desertification (Kairis *et al.* 2013). Furthermore, soil salinization and desertification can be affected by soil water storage capacity. Soils of high water storage capacity are more vulnerable to soil salinization.

6.2.3 Water stress

Water stress can be considered as an important process of desertification in areas with high rainfall seasonality. An irregular distribution of rainfall, such as this observed under semi-arid climatic conditions with most of the rainfall occurring in the period from October to May, enhance water stress. Human actions, such as increasing rates of ground water exploitation, promotes water stress in the growing plants. Furthermore, dry areas subjected to high tourism changes or to high population density are more vulnerable to desertification, since urban water consumption increases at the expense of water used for plant growth.

Studies carried out during the execution of the EU research project DESIRE have shown that indicators related to climate, soil, water and water use, vegetation, land use, fires, water runoff, land management, tourism, social and institutional characteristics are affecting water stress. The most important indicator related to soils affecting water stress is slope gradient. Under high slope gradients (greater than 25%), the growing plants are subjected to higher water stress, since surface water runoff is expected to be higher. In addition, under high rates of deforestation or fire frequency the growing plants are mainly removed reducing water demands and water stress risk. Water stress is negatively related to soil erosion control measures. Field sites in which no or low soil erosion control measures are undertaken, water stress and desertification risk is high. Tourism change is positively related to water stress. Areas under high tourism change (higher than 5% – number of overnight stays in a specific destination in one year averaged by overnight stays in the last 10 years) are more vulnerable to water stress, since urban water consumption increases in charge of water used for plant growth.

6.2.4 Forest fires

Extensive forest fires occurring in dry areas and induced by man are causes promoting desertification. The main land degradation process leading to desertification under these causes is soil erosion. The socio-economic transformation observed in the last decades, especially the migration of rural people to urban areas, have enhanced frequency of forest fires. The reduction of grazing intensity in some areas resulted in a dramatic increase in the availability of vegetation fuel favouring forest fires. However, in other areas, such socio-economic transformations resulted in overgrazing the land, in which farmers deliberately put fires to eradicate unpalatable vegetation favouring high erosion rates leading to desertification (Papanastasis, 1998).

Detailed studies carried out by the EU research project DESIRE in areas prone to forest fires, located in Portugal, Mexico, Botswana, have shown that desertification risk due to soil erosion caused by forest fires are related to climate, soil, vegetation, husbandry, land use, and social characteristics (Figure 6.2). Areas characterized by relatively low annual rainfall (less than 650 mm) have been subjected to frequent fires characterized by moderate or severe erosion and desertification risk.

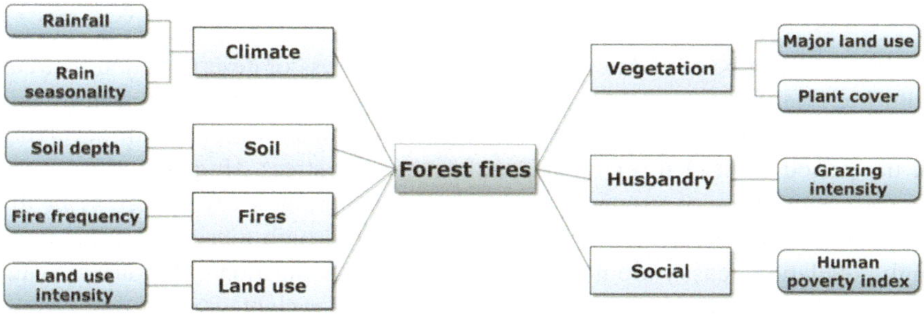

Figure 6.2 Important indicators related to soil erosion and desertification risk in areas subjected to forest fires (Kairis *et al.* 2013).

The type of vegetation, which is related with the frequency of fires, is related to desertification due to soil erosion. Shrublands used as pasture are affected by higher soil erosion rates and desertification risk. In the opposite, areas covered with pine or deciduous forests or areas used for recreation are characterized by slight degree of soil erosion (Kairis *et al.* 2013). Areas with soil depth less than 30 cm are usually characterized by severe or very severe erosion, since such soils have limited soil water storage capacity generating higher surface water runoff and sediment losses. Furthermore, land use intensity is negatively related to degree of soil erosion and desertification. Forested areas under low land use intensity are usually covered with high amount of flammable biomass favouring forest fires and therefore soil erosion and desertification. In addition, areas subjected to low grazing intensity (lower than the grazing capacity) are mainly characterized by moderate to severe soil erosion and desertification risk. The same studies have shown that areas with high human poverty index are subjected to moderate or slight soil erosion and desertification risk (Kairis *et al.* 2013).

6.2.5 Overgrazing

Overgrazing is an important cause of land desertification affecting biodiversity, soil degradation, and soil erosion. Animals overgraze the land by eating parts of plants or whole plants removing selectively the most palatable species in favour of the less palatable species (Figure 6.3). Under such conditions, land is dominated by certain plant species (non-palatable species) and other species are disappeared (palatable species) greatly affecting biodiversity. Furthermore, as large number of animals walks over the land exert high pressure on the soil with their hooves causing soil compaction, and reduction of soil infiltration rate favouring higher surface water runoff. Furthermore, overgrazing is usually accompanied by fires, which are deliberately ignited by farmers to eliminate unpalatable plant species. Under such conditions, soils remain unprotected to raindrop splashing impacts resulting in accelerated soil erosion rates, causing severe problems of land degradation and desertification.

Figure 6.3 Grazing lands under various intensities of land use (photo by C. Kosmas).

Studies carried out during the execution of the European research project DESIRE in areas prone to desertification have shown that important indicators affecting land degradation in overgrazing areas are related to climate, soil, agriculture, land use, fires, water use, social, and institutional characteristics. Areas characterized by low annual rainfall (lower than 650 mm per year) are subjected to moderate or severe erosion and desertification risk. Furthermore, areas with high aridity index are subjected to moderate or low soil erosion, since they receive low amounts of rain during the year for generating runoff and sediment loss even though they are overgrazed. South-facing slopes are subjected to higher erosion rates and desertification risk than north-facing slopes under arid and semi-arid climatic conditions.

As appeared from the previous study, overgrazed areas under high rates of burning (greater than 26 ha per year per 10 km^2 territorial surface) are subjected to lower desertification risk, since grazing animals are excluded for a certain period for rehabilitation of the land. Furthermore, areas subjected to high rates of land abandonment (greater than 26 ha per 10 years per 10 km²) are subjected to high desertification risk. It seems under such conditions animals are removed from areas affected from fire to abandoned areas for grazing. In addition, highly fragmented land property (number of parcels greater than 10) is mainly subjected to moderate or severe soil erosion and desertification risk (Kairis *et al.* 2013). Furthermore, field sites belonging to old-aged farmers are also usually characterized by moderate or severe desertification risk, since old people are usually not able to apply measures for controlling grazing and soil erosion.

6.3 FACTORS OF LAND DESERTIFICATION

6.3.1 Climate

Rainfall, amount and distribution, is a major factor affecting biomass production, and soil erosion rates on hilly lands. Arid, semi-arid, and dry sub-humid climatic conditions are characterized by seasonal distribution of rainfall, which makes the existing ecosystems vulnerable to land degradation and desertification. Studies conducted during the execution of the EU research projects MEDALUS[2] (Kosmas *et al.* 1999) have shown that soil erosion in hilly areas of the Mediterranean region with shrubby vegetation increases as annual rainfall decreases from 650 mm to 300 mm (Kosmas *et al.* 1997) (Figure 6.4). The increase in soil erosion is attributed to the decreasing biomass production accompanied with lower vegetation cover.

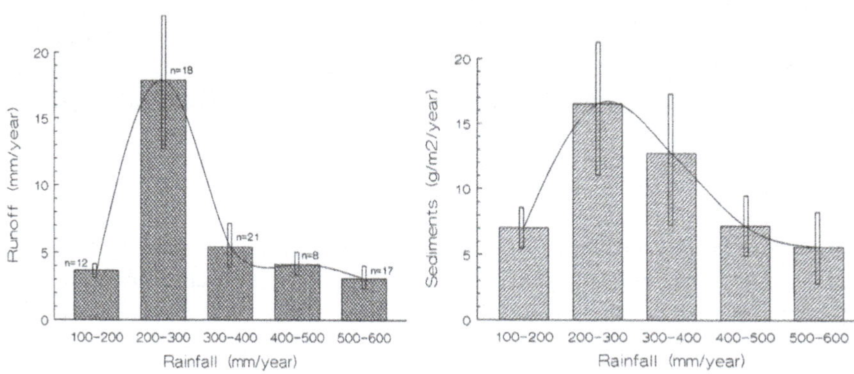

Figure 6.4 Relation of surface water runoff and sediment losses with annual rainfall measured in various study sites located in the Mediterranean Europe (Kosmas *et al.* 1997).

Annual rainfall and rain seasonality greatly affects the rate of burned area in areas affected by desertification (Kosmas *et al.* 2013). Low amounts of rainfall combined with high rainfall seasonality favours extensive fires in drylands. Rainfall

[2] Site.

seasonality affects soil erosion, plant species composition and growth rate. Very high inter-annual rainfall variability causes periods of particularly long drought and sudden rainfall events of high-intensity. Reduced biomass production directly affects the organic matter content of the soil and the aggregation and stability of the surface horizon against erosion. In addition, reduced vegetation growth is followed by a net decrease in carbon dioxide assimilation by the plants, and thus it greatly contributes to climate change at local and regional scale, and to desertification in a broad sense. However, the existing vegetation under arid and semi-arid climatic conditions presents a great capacity of adaptation and resistance to dry conditions. Under those climatic conditions, many species can survive several months through prolonged droughts with soil moisture content below the theoretical wilting point protecting extensive areas from desertification.

6.3.2 Water resources

In the last four decades, the favourable soil and climatic conditions in areas with semi-arid or dry sub-humid climatic conditions, the new technologies of farming, the increasing demands for foods and the availability of ground or surface water has resulted in intensive farming of the lowlands, as well as of the hilly lands. Furthermore, the development of fast transportation means and the availability of cheap holiday-offers have encouraged the expansion of domestic and international mass tourism in coastal areas over the last 50 years. Under such conditions, water quality is under high risk of deterioration. Water resources are under severe physical, social, economic and environmental stresses, compounding to the water uses. Soil salinization, resulting from poor quality of water, is one of the key processes that could lead to desertification especially the plain areas along the coast. It is a growing problem all over the world and affects millions of hectares in Europe.

Increasing salt levels in the top soil layers can affect plant growth and productivity. High concentrations of various salts (e.g. sodium chloride, magnesium and calcium sulphates and bicarbonates) affect plants growth both directly by their toxicity and indirectly by increasing osmotic potential and lowering root water uptake. In dry climates, continuous salt accumulation could lead to desertification, while in humid or sub-humid climates moderate or severe salinization may occur periodically.

Water scarcity, by definition, means diminishing water resources (for example due to climate change) and/or a pressure on the supply of available resources from an increasing demand. The World Health Organization uses the level of 1000–2000 cubic meters per person per year to identify risk on water scarcity. When these values drop below 1000 cubic meters per person per year, then areas are considered as experiencing water scarcity. Water scarcity is especially important indicator for assessing desertification risk in areas where the main cause of land degradation is water stress and soil salinization. Areas under high water scarcity are subjected to higher desertification risk.

6.3.3 Soils

Desertification will proceed, in a certain landscape, when the soil is not able to provide the plants with rooting space, water and nutrients. In the semi-arid and the sub-humid zones, the land becomes irreversibly desertified when the rootable soil depth is not capable to sustain a certain minimum vegetation cover. Extensive studies (DESIRE, DESRTLINKS[3], MEDALUS EU research projects) have shown that soil parameters greatly affecting desertification are parent material, soil depth, slope gradient, slope aspect, soil texture and amount of rock fragments on the soil surface. These parameters are related to water availability to the plants and to soil erosion resistance.

Areas with soils formed on different types of parent materials exhibit various degrees of sensitivity to land degradation and desertification (Kosmas *et al.* 1993). For example, limestone produces shallow soils with a relatively dry moisture regime characterized by moderate erodibility and slow vegetation recovery. Many areas with soils formed on limestone have been greatly degraded and decertified. Areas with soils formed on shale or schist are normally characterized by high productivity that may supply appreciable amounts of previously stored water to the stressed plants protecting areas from desertification (Kosmas *et al.* 2000; Koojman *et al.* 2005).

Slope aspect is defined as the orientation of the land with respect to the sun. In Mediterranean climatic conditions, slopes with southern and western facing aspects are warmer and have higher water evaporation rates than northern and eastern aspects. As a consequence, southern exposed slopes are usually more degraded or are more sensitive to desertification than northern exposed slopes.

Slope gradient greatly affects the amount of surface water runoff and soil sediment loss. Critical ranges of soil slope values significantly affecting erosion have been defined in Mediterranean conditions (Davidson & Theoharopoulos, 1992; Kosmas *et al.* 2000) with severely eroded soils to be mostly observed in slopes greater than 12% or 18%. As the slope becomes steeper, the runoff coefficient increases, the kinetic energy and carrying capacity of surface water flow becomes greater, soil stability and slope stability decreases, and soil sediment loss increases.

Soil depth can be considered as a very important soil parameter affecting land desertification. Soil depth decrease due to soil erosion is a serious threat to the soil quality and productivity in hilly areas. Shallow soils have lower water storage capacity and, therefore, surface water runoff is generating shortly after a moderate rainfall event. Areas with shallow soils are usually abandoned with higher rates than areas with deep soils, since crop production becomes unprofitable in areas with shallow soils. The critical value of 30 cm has proved that under semi-arid conditions is fundamental for supporting perennial natural vegetation with appropriate percentage cover sufficient to protect soils from erosion (Kosmas *et al.* 2000).

Another important factor that determines soil erosion and land desertification is the percentage of soil's rock fragments, especially in areas with poor plant

[3]http://www.kcl.ac.uk/projects/desertlinks/

cover. Studies have shown that rock fragments with dimensions of cobbles (9.0–19.0 cm diameter) offer no protection from erosion in contrast to coarse gravels (2.0–7.0 cm diameter) that effectively reduces soil erosion (Moustakas *et al.* 1995). The presence of rock fragments, especially in the dimension of cobbles, along sloping areas with soil formed on parent materials rich in rock fragments, such as conglomerates, shale-sandstone, etc. despite their normally low productivity, may supply appreciable amounts of previously stored water to the stressed plants and ensure an adequate biomass production in dry years (Kosmas *et al.* 1993).

6.3.4 Vegetation

Vegetation is a crucial factor, which affects soils in all its dynamics including erosion control, water redistribution over and within the soil, and the microbial activity. Surface water runoff is greatly controlled by vegetation, which can be readily altered in hilly areas depending on climatic conditions and the period of the year. Extensive Mediterranean areas cultivated with rainfed crops (Figure 6.5), such as cereals, vines, almonds and olives are mainly confided to hilly lands with shallow soils very sensitive to erosion (Kosmas *et al.* 1997). These areas become vulnerable to erosion and desertification, because of the decreased protection effect of the vegetation cover from raindrop impact during heavy rains. Soil erosion measurements conducted along the Mediterranean Europe and Portugal during the execution of the EU research projects MEDALUS have shown that under the existing land management practices vineyards generate the highest amount of runoff and sediment loss followed by land uses, which covered with eucalyptus, cereals, shrub-land and olive groves under minimum tillage in a decreasing rate. Olives present a particularly high adaptation and resistance to long term droughts and under semi-natural conditions support a remarkable diversity of flora and fauna even higher than some natural ecosystems, protecting hilly areas from erosion and desertification.

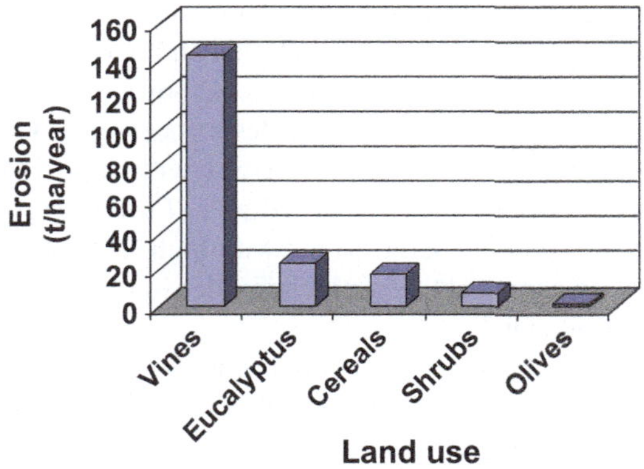

Figure 6.5 Average annual soil erosion rates measured under various land uses along the Mediterranean Europe (Kosmas *et al.* 1997).

Many studies have demonstrated that both runoff and sediment loss decrease exponentially with increasing percentage of plant cover (Elwell & Stocking, 1986). A vegetation cover of 45–50% is considered as critical value, since above this limit soils are adequately protected from raindrop impact and soil erosion is significantly reduced (Francis & Thornes, 1990). Furthermore, plant cover reduces significantly summer soil surface temperatures and soil water conservation. Plant cover is a crucial indicator of land desertification, especially in areas affected by water stress. In addition, climate greatly affects plant cover. High annual air temperatures promote low plant cover due to high evapotranspiration demands. Furthermore, low amounts of annual rainfall negatively affect plant cover. In addition, high rain seasonality indices usually promote low plant cover percentages.

6.3.5 Socio-economics

Soil degradation and desertification is affected by both biophysical and socio-economic factors. Salvati *et al.* (2015) directly correlate the increasing soil degradation and land sensitivity to desertification in Italy with the quite complex relationships among environmental and socio-economic parameters. Changes in the socio-economic structure of an area can be both a driving force and an impact on land degradation.

Changes in the social structure can occur when the local socio-economic context has become incapable of satisfying the population's needs. This may be the result of land degradation caused by man in the landscape over time, or due to low land productivity affecting the income of the population. When such problems arise, degraded areas are abandoned, deprived of the conditions that might mitigate the problem, for example by encouraging the use of good management practices to avoid destructive phenomena.

Studies carried out by the EU research project DESIRE in areas prone to desertification have shown that population density, population growth rate, old age index, allocated subsidies, land fragmentation and land ownership can greatly affect vulnerability to desertification. Areas affected by soil erosion or water stress are more vulnerable to land degradation and desertification under high population densities (higher than 100 people km^{-2}). Areas with high population density are usually subjected to high rates of burned area. Furthermore, areas with low population density are mainly characterized with high rates of land abandonment affecting positively or negatively desertification depending on the characteristics of the physical environment.

Population growth rate is especially important for forested areas affected by fires. Low population growth rates accompanied with high old age index leads to decreasing animal population grazing the forested land and increasing forest fires due to higher remaining flammable biomass. In the opposite, areas characterized by high population growth rates are mainly characterized as supporting low vegetation cover enhancing water runoff and land degradation.

The old age index is related to the applied land management practices and to the introduction of new technologies. Older people tend to remain under the traditional land management practices, while younger farmers are relatively easily convinced to introduce new technologies and land management practices, considering sustainability of land resources.

Land fragmentation can affect land management practices and decision-making related to the structure of the farm. A highly fragmented land is not easily mechanized and is usually cultivated in a direction in which soil erosion is not restricted. Farm ownership represents a fundamental factor for decision-makers for addressing political measures to individual farmers or to more effective public level. Farm ownership is related to land use intensity. Land use intensity is mainly characterized as low to moderate in areas characterized as owner-farmed, while high land use intensity can be mainly identified in areas characterized as tenant- or state-farmed.

Farm subsidies have ensured an adequate income to farmers for maintaining landscapes especially in less favoured areas, but this support has favoured the intensification and specialization process in agriculture by applying more fertilizers and pesticides, irrigating the land, increasing the number of animals, etc. Under this financial support, unproductive agricultural areas cultivated with cereals, critical to desertification, have remained under cultivation leading to further deterioration of soil conditions. Number of animals has increased in unsustainable numbers, causing severe degradation in vegetation and soils in grazing lands.

6.4 DESERTIFICATION RISK IDENTIFICATION

Land desertification is considered an extreme stage of land degradation generated by various processes or causes of land degradation, such as soil erosion, soil salinization, forest fires, water stress, overgrazing, etc. Therefore, the risk assessment of desertification requires the identification: (a) of the most important degradation process or cause and (b) the present state of the physical environment and socio-economic characteristics of the area. The risk for assessing the impact of the various processes on land degradation and desertification can be achieved by: (a) using a series of indicators, (b) applying models for assessing specific degradation process, and (c) assessing the stage of land degradation by describing certain land characteristics.

6.4.1 Using indicators

Indicators are valuable tools for tracking positive or negative changes in complex systems and monitoring progress on protection of the environment. The UNCCD regards indicators as appropriate tools for supporting a wide range of activities, such as assessing and mapping the extent of desertification, as well as determining the causes, quantifying the impacts, justifying expenditure for mitigation measures and monitoring the efficiency of the measures undertaken. The UNCCD Conference of the Parties (COP9) decided recently to accept a set of eleven impact indicators focused on on-site erosion to monitor desertification.

The European research project DESERTLINKS has set a series of indicators to assist the National Action Plans of the Annex IV countries and their needs for available indicators databases on assessing desertification (Kosmas *et al.* 1999; Brandt *et al.* 2005). The resulting DIS4ME Indicator System (Desertlinks, 2005), provides information on 148 indicators of relevance to Mediterranean desertification and constitutes a valuable tool to enable stakeholders to identify areas prone to desertification, to understand how important the problem is, and to assess measures on combating desertification. It should be pointed out that when a complex process is considered, such as soil erosion, a single indicator is generally not sufficient. A set of indicators would be necessary to effectively analyze soil erosion.

Based on the research carried out during the execution of the EU DESIRE project in various areas in Latin America, China, eastern and Western Europe and Africa, a series of effective land management indicators affecting desertification risk have been identified based on the degradation process or cause. As Figure 6.6 shows, the most effective land management indicators for reducing desertification risk related to the land degradation process of soil erosion in agricultural areas are runoff water storage, sustainable farming, and land terracing. These land management practices become more effective under low land use intensity and implementation of existing regulations for environmental protection. Desertification risk in agricultural areas in which the main process of land degradation is tillage erosion, is greatly affected by tillage operations, tillage direction and mechanization index. Sloping areas under high mechanization index in which soils are tilled by a mouldboard perpendicular to the contour lines in the down-slope direction are subjected to high desertification risk.

In pastures, the most important land management indicators affecting desertification risk due to soil erosion are grazing control, soil water conservation and runoff water storage. Tillage operations become especially important indicators in pastures cultivated for forage production. Of course, reduction of the rate of burned area and land use intensity are crucial indicators for reducing desertification risk in such areas. Desertification risk in forested areas in which the main process of land degradation is soil erosion can be substantially reduced, if fire protection measures are applied and the land is subjected to low grazing intensity. Desertification risk is greatly reduced, if the rate of burned area is minimized.

In the previous mentioned research, it was also found that desertification risk due to soil salinization can be affected by the land management indicators of water quality, rate of ground water exploitation, soil drainage, and irrigation percentage of arable land (Figure 6.6). Areas with good quality of ground water in which exploitation rate is not higher than the recharge rate are subjected to low desertification risk. Furthermore, the expansion of irrigated land using poor quality of water increases the risk of desertification, especially in soils characterized with poor drainage conditions. The construction of drainage ditches in poorly drained soils for lowering ground water level or removing excess of irrigation water in areas with arid, semi-arid climatic or dry sub-humid climatic conditions is expected to reduce desertification risk.

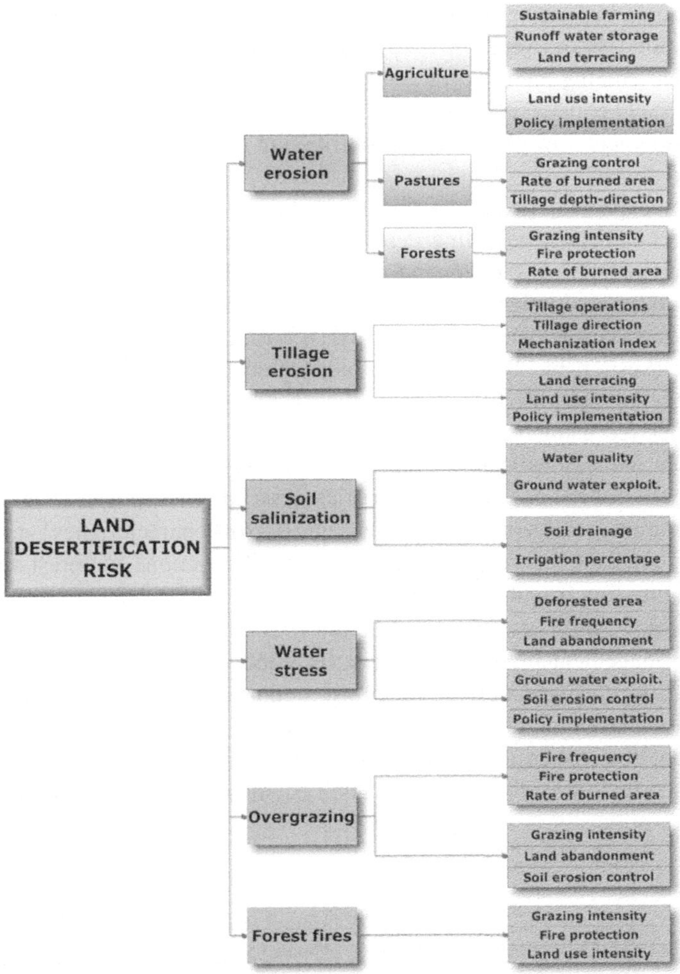

Figure 6.6 Important land management indicators affecting desertification risk in various land uses, processes, and causes of land degradation.

The most important land management indicators affecting desertification risk in areas in which water stress is the most important process of land degradation are: rate of deforested area, rate of land abandonment, fire frequency, rate of ground water exploitation, soil erosion control, and policy implementation (Figure 6.6).

The highest impact on desertification risk can be attributed to the rate of land abandonment and policy implementation. Under high rates of land abandonment, land desertification risk due to water stress is reduced. Of course, the implementation of existing policy on environmental protection and especially those regulations related to water resources management alleviates desertification risk.

The most important indicators related to land management affecting desertification risk in overgrazed areas are: fire frequency, fire protection, rate of burned area, grazing intensity, rate of land abandonment, and soil erosion control (Figure 6.6). The analysis of the data has shown that fire protection and rate of land abandonment have the greatest contribution on combating desertification in overgrazed areas.

As it is expected, fire protection and grazing intensity are two important management indicators affecting desertification risk in forested areas. Adequate measures on forest fire protection in forested areas subjected to low or moderate rate of grazing reduces significantly the desertification risk. Another important indicator related to land management in forested areas is land use intensity. If the rate of wood extraction is sustainable, then the land is adequately protected from desertification.

Based on the research carried out by the EU project DESIRE, an expert system has been derived using a series of indicators and classified according to the land degradation and desertification processes or causes (Figure 6.7). The proposed indicators are either of state characteristics describing the physical environment (air temperature, soil depth, population density, soil characteristics, etc.), which are rather constant, or affected mainly by human actions (deforestation, land abandonment, fire frequency, land terracing, etc.) all varying with time. The developed tool can be used for: (a) the analysis of a wide range of alternatives for land management practices in order to reduce land desertification risk, (b) the evaluation and selection of the most important indicators through which desertification risk may be assessed in a variety of locales worldwide, and (c) the development of a consensus among various groups (politicians, managers, experts, etc.) in assessing desertification risks. The system is loaded in the website: http://www.desire-project.eu.

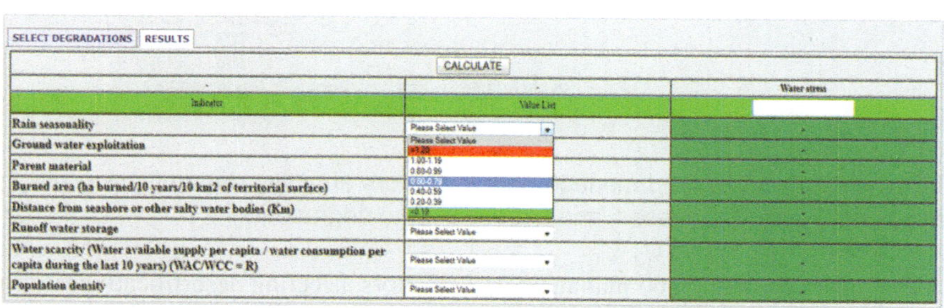

Figure 6.7 The DESIRE expert system showing the related indicators for a specific process or cause of desertification risk.

Based on the research carried out during the execution of the MEDALUS and DESERTLINKS projects (http://www.kcl.ac.uk/projects/desertlinks/indicator_system/introduction.htm), a methodology has been developed using a series of indicators to identify the environmentally sensitive areas (ESAs) and the environmentally sensitivity index (ESI) to desertification in order to understand the factors causing desertification.

In that approach, a set of 15 key indicators describing different desertification factors (climate, soil, vegetation, management) (Figure 6.8) are used to derive the environmentally sensitivity composite index (ESI) of land desertification (Kosmas *et al.* 1999). Each of these factors is grouped into various uniform classes with respect to its behavior on desertification and weighting factors are assigned in each class. Then the following four qualities are evaluated (a) soil quality, (b) climate quality, (c) vegetation quality, and (d) management quality. After the computation of the above mentioned four composite indices corresponding to the four environmental qualities, the ESI is calculated by using their geometric mean.

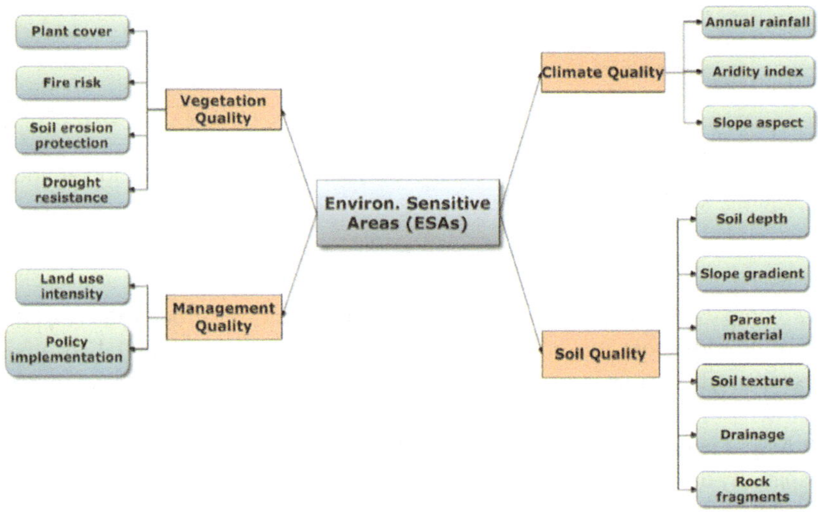

Figure 6.8 The indicators used for identification environmentally sensitive areas (ESAs) to desertification.

The EU funded DESERTLINKS research project has developed the DIS4ME online system (Brandt & Geeson, 2005) (http://www.kcl.ac.uk/projects/desertlinks/accessdis4me.htm). This system, which is available to the public, includes many simple indicators, some key headline indicators (simple indicators integrating several aspects of a more complex system), as well as composite indices. The DIS4ME website includes an interactive tool for calculating the value of ESI at a point. Values for each parameter are selected from pull down lists, and the index is calculated, giving additional feedback on the particular factors to note (Figure 6.9).

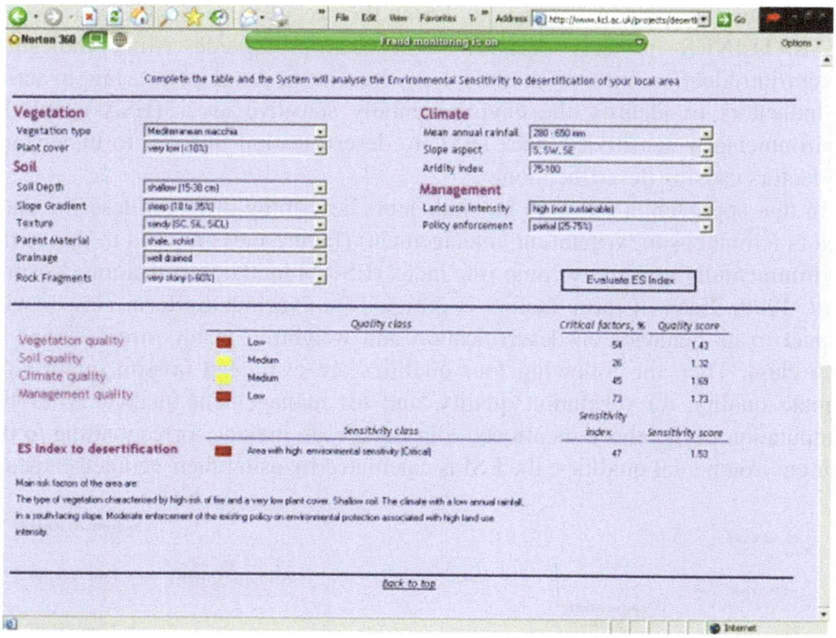

Figure 6.9 Screenshot of the environmentally sensitive index (ESI) tool found on the DIS4ME website.

6.4.2 Applying models

More complex methods than the previously described have been developed by combining indicators into indices. These methods have the potential for giving access to much more sophisticated data sources, and to producing maps of different scenarios over very large areas. The PESERA model (Pan-European Soil Erosion Risk Assessment) (Kirkby, 1999; Kirkby *et al.* 2000) is one of the physically-based and spatially distributed models for quantifying soil erosion at European scale. This methodology combines topography, climate, vegetation, soil and land management characterisitcs into a single regional degradation indicator index (RDI) predicting soil erosion rates and therefore desertification risk of sloping areas.

The PESERA model was mainly developed for use as a regional diagnostic tool (Kirkby *et al.* 2004) in a watershed, for evaluating erosion rates in various types of land use with different soil and topography, when executed in a GIS (Geographical Information System) environment (PESERA-GRID-Fortran90TM-grid based code) or for estimating erosion rates along a catena in the Microsoft Office Excel macro environment (PESERA-VBA-ExcelTM-point based code) (Irvine & Kosmas, 2007). An example of the application of this model in three dimensions is the estimation of soil erosion risk at European Union level at 1-km resolution (Figure 6.10).

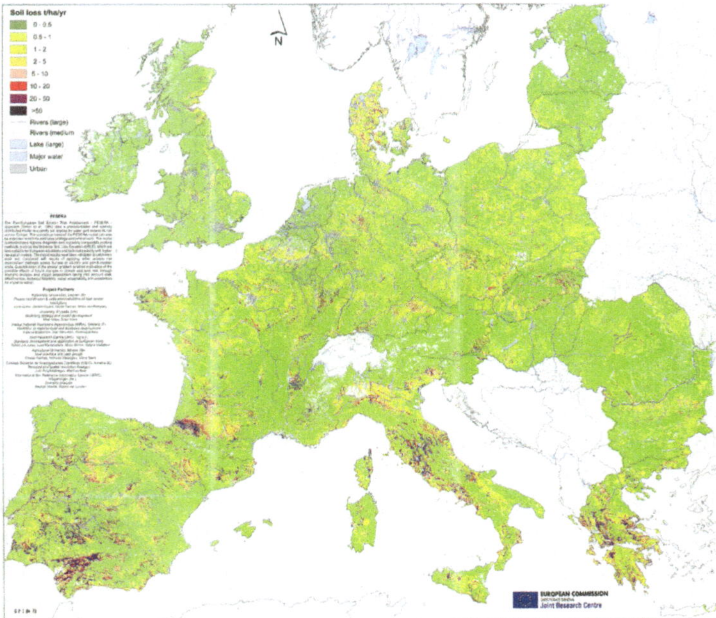

Figure 6.10 Map of Estimated Erosion Risk for Europe at a 1-km Resolution (Source: PESERA Project).

6.4.3 Assessing the state of land degradation

Assessing the state of land degradation by describing certain land characteristics is another methodology for describing desertification risk. Such techniques have been developed using remote sensing methods (Zha & Gao, 1997; Yang *et al.* 2005). Vegetation cover is considered as a direct indicator of land degradation. The assessment of vegetation cover is based only on vegetation growth and observed changes in vegetation with time. Vegetation status can reflect the environmental variation caused by land degradation. Studies have been shown that areas with sparse vegetation are likely to be at risk of desertification. However, this cannot be always the case, especially in mountainous areas, where additional data such as DEM have to be considered. The use of high-resolution satellite imagery for desertification monitoring of large regions is not feasible due to high costs, effort and data unavailability. Therefore, multi-scale remote sensing monitoring approach is proposed in combination with the use of coarse resolution land cover maps to monitor large-scale processes with the scope to identify critical areas to desertification. After the identification, these areas are in detail analyzed using higher resolution satellite data (Sun & Chen, 2000).

Normalized Difference Vegetation Index (NDVI) developed by the National Oceanic Atmospheric Administration (NOAA) satellites, with the advance very high-resolution radiometer (AVHRR), was used. Data from AVHRR have a daily frequency and are relatively cheap or even available free of charge. Many remote sensing studies, at large spatial and temporal scales, have been carried out using AVHRR-NDVI data for studying arid and semi-arid environment (Schmidt & Karnieli, 2000). Feoli *et al.* (2000) suggest that the desertification risk index (DRI) in Mediterranean areas should be high, when the yearly average temperature is high all over the year, the yearly amount of rainfall is concentrated only in some period and the NDVI after the dry season is low. The DRI, in combination with NDVI, provides an indirect estimation of the vegetation cover, which is similar to the regional degradation index suggested by Kirkby *et al.* (2000), which take into consideration soil erosion calculated on the basis of remote sensing data (Sommer, 2001).

6.5 DESERTIFICATION RISK ASSESSMENT

According to UNCCD consultancy report (Grainger, 2009) the establishment of an effective and global monitoring system for desertification is necessary. This need becomes important due to the fact that there is a lack of reliable estimates of desertification extent, as well as of the rate of its change. The functionality of a monitoring system should be based on two pillars. The first concerns the use of a common set of indicators to describe the degradation processes, whereas the second dictates baseline surveys to be conducted in each affected country using the above mentioned indicators. Towards this direction, a well-designed remote sensing system supported by ground data can help in monitoring extended areas at low cost and in a relatively short time. A remote sensing surveillance system, subject to the previous rules, can contribute to early and rapid detection of degradation processes that is necessary to prevent further deterioration of land (Higginbottom & Symeonakis, 2014). Lawley and others (2015) documented that combining remote sensing techniques and capabilities with precise, site-based, detailed and specialized indicators of the state of vegetation (compositional, structural and functional) the scientific community can achieve the best possible result for monitoring.

Kosmas and others (2013) created an empirical field-scale approach in defining desertification risk in order to monitor the phenomenon. This approach was based on the type of environmentally sensitive area (Kosmas *et al.* 1999) and on the main process or cause of degradation determined for each land. The logical framework developed in the previously mentioned work allows the assessment of desertification risk, with the use of a limited number of biophysical and socio-economic indicators in different areas of environmental sensitivity by utilizing data of regional or local level and consequently it can be widely applied (Kairis *et al.* 2013).

Santini *et al.* (2010) proposed a new methodology that combined GIS data with various models simulating environmental processes and resulted in an Integrated Desertification Index (IDI). Since modeling is able to analyze better the spatio-temporal dynamics of environmental processes, they combined several simulation models describing six factors of desertification (overgrazing pressure-CAIA model, water erosion-RUSLE model, wind erosion-WEAM model, sea water intrusion-SHARP model, vegetation productivity-MOD17 model, soil fertility-CENTURY model). The frequency of desertification risk assessment can be carried out once every five or ten years by measuring certain indicators related to the ecosystem performance, such as vegetation cover change, appearance of erosion features, water resources availability, or social poverty, etc.

Cost assessment of the measures for land protection depends on the physical environmental characteristics, the application of land management practices, the stage of land degradation and the effective processes of land degradation. For example, in a cropland in which the main degradation process is soil erosion, measures of reduction of soil erosion, such as minimum tillage and enhancing the vegetative cover, can be effective in minimizing desertification risk. However, definition of the appropriate measures requires actions, such as characterization of the physical environment, description of the applying land management practices, analysis of the socio-economic characteristics, registration of the existing level of technology, etc. Ferrara *et al.* (2012) estimated that the monetary cost for data collection, processing and updating in order to apply the ESA methodology to the study area of Agri basin in southern Italy is 8.05 € per hectare. However, in case that permanent structures (building terraces) or reforestation or reduction of number of animals etc. are necessary actions to protect the land from desertification, then the cost has to be assessed taking into consideration the local economic conditions.

The first worldwide assessment of desertification direct cost was based on reports of yield declines on lands under different grades of desertification (UNCCD, 2013). This first reliable global estimation was $26 billion per annum (UNEP, 1980). After that, there is a lack of similar cost approaches. Instead, there are several estimations of the economic losses caused by desertification at national level in peer-reviewed and non-peer-reviewed journals. These cost estimates are derived as proportion of national income recording among others 2% of Gross Domestic Product (GDP) in India (Reddy, 2003), 0.4% of GDP in the USA (Pimentel *et al.* 1995), 9% of GDP in Burkina Faso (Lallement, 1989), 0.9–12.5% of GDP in Mali (Bishop & Allen, 1989) and 8% of GDP in Niger (Nkonya *et al.* 2011).

6.6 DESERTIFICATION RISK MANAGEMENT

Management of desertification plays a crucial role for securing ecosystem services, such as food production, regulation of climate and enhancing living conditions for local populations. Our society must encourage promoting soil, water and vegetation conservation, protection of biological diversity, as well as of the prevention and mitigation of land degradation and desertification. Land management or responses to land ecosystem degradation and desertification (LEDD) are defined in the European research project LEDDRA (http://leddra.aegean.gr/) as any type of action, planned (formal, institutionalized) or unplanned (informal, non-institutionalized), in LEDD-affected regions that purports either to directly tackle a LEDD problem or to address other socio-economic problems, as well as individual and collective goals. The success of responses in meeting environmental and social goals depends on the degree of their socio-ecological fit; i.e. how they fit and how well adapted they are to the biophysical and societal conditions prevailing in a socio-ecological system. The various land management practices can have various impacts on natural resources affecting positively or negatively desertification risk.

6.6.1 Land management practices

Some of the most important land management parameters identified by the European research project DESIRE in areas affected by desertification were the following: major land use, type of vegetation land cover, percentage of land cover, land use intensity, tillage operations, grazing intensity, rate of deforested area, fire frequency, rate of burned area, ground water exploitation, sustainable farming, soil erosion control, soil water conservation, runoff water storage, grazing control and land abandonment.

Major land use affects land management and therefore desertification risk. For example, in a forested land, desertification is affected by the rate of actual and sustainable yields of wood and the intensity of grazing. In the opposite, in a cropland, desertification is affected by the intensity of cultivation, such as frequency of ploughing, fertilization, irrigation, etc. Major land use is related to the rate of land abandonment. Areas used for agriculture are characterized by a higher rate of land abandonment than those used as pastures.

Land use intensity, related to the intensity in which the land resources are used, can have negative impacts on desertification. Repeated cultivation of soils associated with other management practices (application of herbicides and pesticides) results in decreased organic matter content and aggregate stability favouring soil crusting, overland flow, and erosion. Soil compaction is also another form of soil degradation resulting from heavy machineries used in agricultural areas.

Tillage operations carrying out by the various tillage implements, such as mouldboard plough, chisel plough, duckfoot chisel, harrow, etc. can have detrimental impacts on land degradation and desertification. The availability of heavy powerful machinery favoured deep soil ploughing and high speeds in directions usually perpendicular to the contour lines. This resulted in the displacement of huge amounts of soil materials from the upper convex parts (summit, shoulder, backslope) of a hillslope to the concave parts (footslope, toeslope) and decreased significantly the production of the various crops negatively affecting land degradation and desertification. Tillage erosion affects soil quality for long terms, by changing soil depth and consequently water holding capacity, nutrient availability, organic matter content, and crop yield.

Grazing intensity is related to the pressure imposed on the growing vegetation by the grazing animals. The number of animals is the main factor that affects the productivity of the grazed lands. An overgrazing land is vulnerable to desertification, if the high number of animals remains for long period in the pasture, and the soil, topographic and climatic conditions are adverse. Besides overgrazing, undergrazing can also cause desertification, due to the growth of high amount of flammable biomass affecting ignition of wildfires and leading to high erosion rates and land degradation. Therefore, a sustainable grazing of the pasture land is considered the best land management practice for protecting such areas from desertification.

Large scale deforestation of semiarid and dry sub-humid areas accompanied by intensive cultivation and overgrazing resulted in accelerated erosion and the formation of badlands with very shallow soils. Such areas are very sensitive to land degradation and desertification. If trees are removed in a higher rate than the sustainable number, the stand density decreases gradually, the soil is partially covered with vegetation and therefore is degraded from raindrop impact and soil erosion.

Fire frequency in an area usually has adversely affected regeneration of many plant and animal communities, with some of them experiencing loss of species diversity, followed by site degradation. Forest fires are frequent in pine forested areas. Pastures are frequently subjected to man-induced fires in order to renew the biomass production. Burned ecosystems lack partially or totally vegetation covers for a period ranging from months to years. During this period, the soil is exposed to wind and water erosion followed by flooding of lowland and siltation of water reservoirs. Furthermore, forest fires greatly contribute to climate change by increasing greenhouse gasses.

Sustainable farming is defined as an agricultural system evolving towards greater human utility, increased efficiency of resource use, minimum depletion of non-renewable resources, and protection of land against degradation and desertification. Sustainable farming may include actions such as: (a) land utilization schemes within the land capability limits, (b) minimum tillage, (c) enhancing vegetation cover, (d) tillage of soil in the up-slope direction, (e) minimum depth of ploughing, etc. (Figure 6.11). Minimum tillage may have favourable effects on soil aggregation and reduction of soil crusting and soil erosion. If the soil moisture level is optimal, a minimum tillage is generally favourable, because the implements break up the clods, incorporate the organic matter into the soil, kill weeds, and create a more favourable seed bed. Frequent tillage operations, especially those involving heavy equipments, have detrimental effects on surface soil degradation such as accelerating oxidation of soil organic matter, breaking down stable soil aggregates.

Figure 6.11 Examples of sustainable farming including no tillage (left) and minimum tillage -inducing plant cover (right).

Soil erosion control measures may include: conservation tillage, contour farming, stabilization structures, vegetated waterways, strip cropping, enhancement of understory growing vegetation, adequate plant cover, terracing, small water reservoirs, etc. (Figure 6.12). Conservation tillage is a land management practice, which can significantly reduce rill and interill erosion in hilly areas cultivated with annual crops. The incorporation of plant residues into the soil after harvesting increases soil organic matter content and improves soil aggregate stability. Contour farming is considered a very effective practice for reducing soil erosion, but in many cases farmers face security problems when slope gradient is greater than 12%. Stabilization structures protect soil erosion in channels, where water is concentrated. Such structures reduce runoff water velocity to non-erosive rates. Enhancement of understory annual vegetation is an appropriate land management for reducing soil erosion in hilly areas cultivated with perennial crops, such as olives, vines, almonds, etc. Terracing is an alternative management practice for water conservation and erosion control. Terracing decreases the slope length and reduces the damage by surface water runoff. Strip cropping is a practice in which hilly areas are not cultivated uniformly with the same crop. There are alternative strips with various crops growing in different period. Furthermore, small water reservoirs are constructed at farm level or in collaboration of nearby farmers. Water is temporarily stored into the reservoir reducing down slope water runoff.

Soil water conservation techniques may include: mulching, weed control, management of soil surface for maximum water infiltration and vapour adsorption, cultivation, etc. Weed control usually takes place by means of mechanical tillage or chemical sprays, whereas mulching or grazing is alternatively applied in organic farming. Water vapour adsorption in arid and semi-arid climatic conditions becomes one of the most important sources of available water for the growing annual vegetation during the dry period. Management of the land by: (a) reducing the density of the growing vegetation and increasing the soil-atmosphere interface, (b) using surface mulches, such as rock fragments or plant residues partially covering the soil surface, and (c) ploughing the soil for increasing macro porosity can beneficially and significantly affect water vapour adsorption and soil water conservation. Cultivation of the land using various tillage implements, such as mouldboard, chisel, duck foot chisel, harrow, etc is considered as a management practice for conserving subsoil water. However, ploughing the soils in hilly areas can cause severe problems of land degradation, due to mechanical displacement of soils usually down slope.

Runoff water storage is defined as the volume of runoff rain water stored into the soil or in small ponds by various actions such as construction of small ponds, keeping adequate plant cover on the soil surface, etc. (Figure 6.13). There are many factors and processes affecting water runoff, such as vegetation density, biological activity, soil crusting, micro-topography, soil texture and structure, plant cover, rainfall amount and intensity, slope gradient, land management practices, etc. The degree of crusting of soil surface has a strong influence on infiltration rates and therefore surface water runoff rates. Micro-topography of the soil surface consisting of random roughness on the surface, together with cultivation features such as plough ridges retards water flow and increases amount of water infiltrating into the soil. The presence of terraces increases adsorption of water by the soil during heavy rainfall events.

Figure 6.12 Cultivation of the land in rows parallel to the contour lines (left) and terracing of land for soil erosion control (photos by C. Kosmas).

Grazing control is a management action aiming at the establishment of equilibrium between herbivores and the resource base of rangelands so that sustained production is ensured. Generally, there are two main systems of grazing a land, namely: (a) continuous, and (b) rotational. The continuous grazing system is beneficial to the animals, since they are free to walk around and to select the best plants or plant parts available in the pasture. However, such a system is detrimental for the pasture, since plant species do not have the chance to recover and in many cases soil remains partially covered by vegetation for protecting it from the raindrop impact. The rotational grazing system involves division of the grazing land in pieces allowing the animals graze sequentially over the growing season of the year. Such a grazing control system allows the vegetation to recover and to protect soil from erosion.

Figure 6.13 Typical examples of surface water runoff storage in small ponds (left) or keeping adequate plant cover for reducing surface water runoff (right) (photos by C. Kosmas).

Irrigation percentage of arable land defines the intensity in which arable land and water resources are used. Due to the expected warmer and drier climate accompanied with increase in aridity and drought hazard, the extension of irrigation will be necessary over large areas for efficient agricultural production and protection of the environment. The expansion of irrigated land will probably aggravate the existing problem of brackish water intrusion in aquifers through diminishing groundwater recharge or reduced amount of rainfall. The sustainable use of water resources and the appropriate land management systems and land management practices will be the appropriate tools for combating desertification in those areas. The uncontrolled use of water resources in the last decades for both irrigation and domestic consumption resulted in severe environmental impact on the water quality of the most productive aquifers and on soil degradation due to salinization.

Land abandonment is related to decreasing of land profitability, usually resulting in change of land use from agriculture to grazing land. Land abandonment may have positive or negative impacts on desertification depending on the soils and climatic conditions of the area (Figure 6.14). Soils under favourable climatic conditions that sustain plant cover may improve with time by accumulating organic materials, increasing floral and faunal activity, improving soil structure, increasing infiltration capacity, and therefore decrease erosion potential and desertification risk. In the opposite, areas with soils relatively shallow, of low organic matter content, with thin solum, low resilience capacity and vegetation cover, in steep slopes, with adverse climatic conditions, and overgrazed result in negative impacts of land abandonment. The natural restoration processes in these areas are really slow, almost impossible, and persistent erosion processes can lead to severe soil degradation and desertification.

Figure 6.14 Examples of areas with positive (left) and negative (right) effects of land abandonment on desertification (photos by C. Kosmas).

6.6.2 Assessing land management practices

There are various methods or techniques in assessing the applied land management techniques on combating desertification. One simple technique is to assess the impacts of a certain degradation process caused due to applied land management practice and to recommend the appropriate changes. However, the assessment of a certain land management practice requires an integrated approach among physical environment and socio-economic characteristics in which the links between human and natural systems is non-linear. Based on the approach designed by the European research project LEDDRA, the following techniques has been adapted for assessing a land management practice or a specific response to land degradation and desertification: (a) identification of land ecosystem degradation and desertification (LEDD) problems and their impact on socio-ecological resilience, (b) natural and human drivers or causes of LEDD problems, (c) identification of previous and current responses to LEDD problems, (d) identification of the necessary elements to implement successfully fitting of responses. The necessary elements for fitting of responses include: (a) description of the natural and socio-economic capitals, (b) description of natural and socio-economic functions. Fitting of a response must fulfil certain selected criteria, such as secure capitals, functions, and socio-ecological resilience.

As an example of "fitting" a response is given for the Messara Valley located in south Crete (Greece). The average annual rainfall is 620 mm, while the average air temperature 17.8°C. The area is characterized mainly as gently to moderately sloping (areas with slope gradient lower than 12% cover about 47% of the area). The soils are mainly moderately deep to deep (soil depth greater than 60 cm), cultivated mainly with olive trees, vegetables, orange trees and vines. The area has intensively cultivated after the participation of Greece to the European Union (1981) accompanied by significant land-use transformations, wide use of fertilizers, and over-exploitation of natural resources.

From an ecological point of view, the 'fitting' of the response "intensive cultivation" is assessed using the following criteria, which reflect the degree to which the response affects critical ecological functions of the socio-ecological system: (1) protect soil water storage capacity, (2) secure low salt content in soils, (3) reduce flooding risk in lowland areas, (4) minimize water pollution, and (5) protect land from desertification. Based on existing methods and techniques for assessing the previous criteria, soil water storage capacity is reduced significantly in parts of the slopping land of Messara Valley, due to high rates of soil erosion negatively affected (Figure 6.15). The criterion of securing low salt content in the soils is not satisfied due to over-exploitation of groundwater and intrusion of brackish water in the aquifers of Messara Valley. The criterion of minimizing ground water pollution is moderately satisfied (Figure 6.15). The criterion of protecting land from desertification is partially satisfied.

After aggregating all the indices deriving from the corresponding criteria, "fitting" of response "intensive cultivation" can be considered as high and moderate for 60.5% and 25.9% of the area, respectively (Figure 6.16).

Another response "the integrated land management" is proposed instead of the existing response "intensive cultivation" for securing sustainability of natural resources combined with maximum crop production, promoting biodiversity, and protecting farmer's health is an alternative. The application of this response is controlled by the local authorities and farmers receive extra payment. After aggregating all the indices derived from the corresponding criteria, the "fitting" of this response can be considered as very high and high in the whole area.

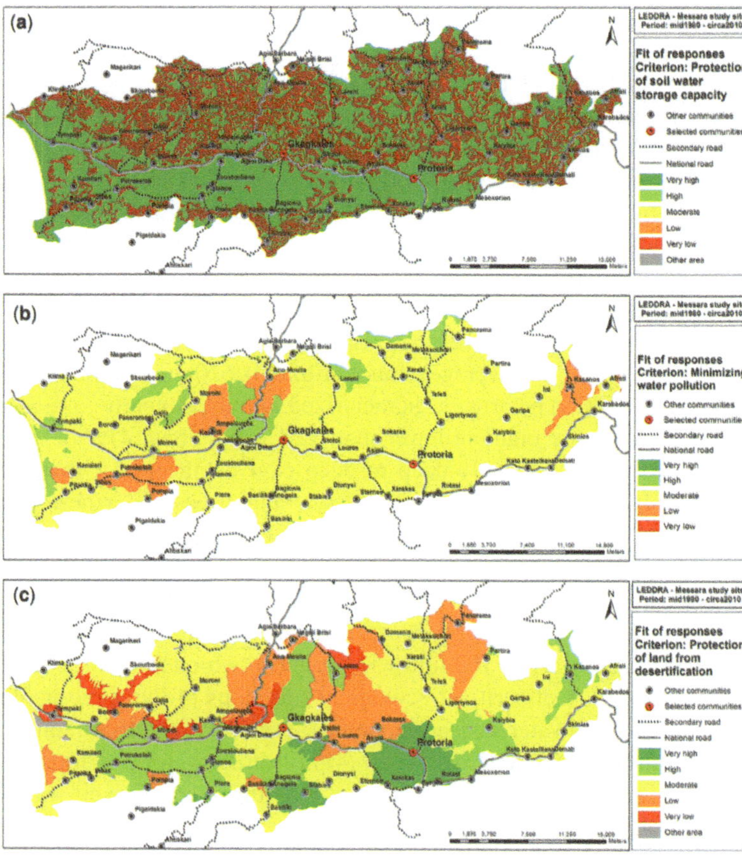

Figure 6.15 Fitting of response "intensive cultivation" based on the criteria: (**a**) protection of soil water storage capacity, (**b**) minimizing water pollution, and (**c**) minimizing desertification risk.

6.7 SUMMARY

Desertification constitutes a globally-acknowledged issue with ecological, social, economic, implications at both regional and local scale. Desertification is the consequence of a series of important degradation processes, such as soil erosion, soil salinization, soil sealing, etc. where water is the main limiting factor of land use performance on ecosystems. Desertification is occurring in both developed and emerging countries and affect arid, dry and even sub-humid areas due to the interplay of several interacting biophysical (e.g. climate, soil, vegetation) and human factors (e.g. land-use, settlement distribution, agricultural concentration, industry, tourism). Risk assessment of desertification requires the identification of the most important degradation process or cause and the present state of the physical environment and socio-economic characteristics of the area. The assessment of various processes or causes on identification of desertification risk can be achieved by using a series of indicators, applying models for assessing specific degradation process, and assessing the stage of land degradation by describing certain land characteristics.

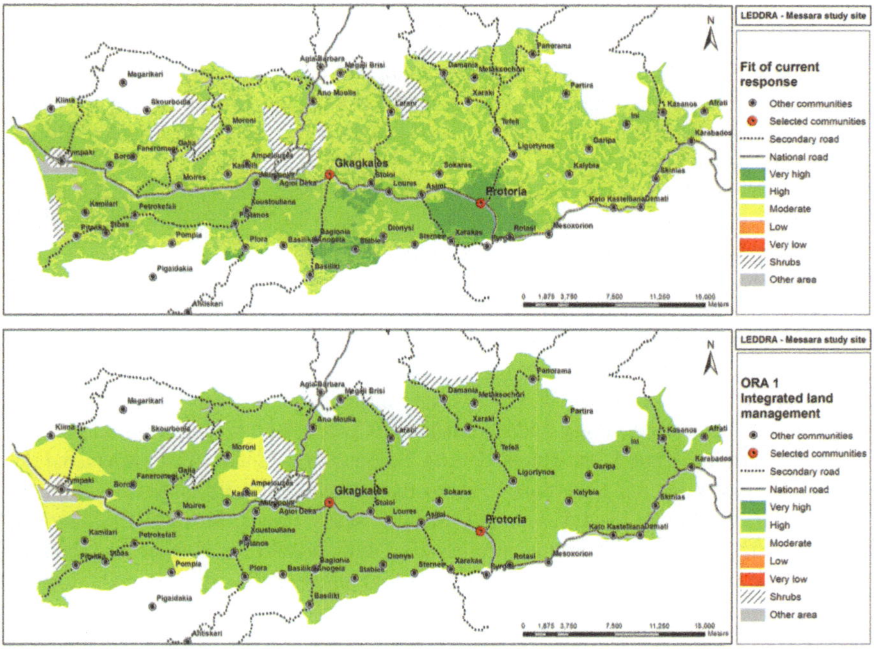

Figure 6.16 Aggregate index maps derived by the criteria presented in figure 6.15 for the response "intensive cultivation" (above) and "integrated land management" (below).

REFERENCES

Aubréville A. (1949). Climats, forêts et désertification de l'Afrique tropicale. Société d'Editions Géographiques, Maritimes et Coloniales, Paris, France.

Barbero-Sierra C., Marques M. J. and Ruíz-Pérez M. (2013). The case of urban sprawl in Spain as an active and irreversible driving force for desertification. *Journal of Arid Environments*, **90**, 95–102. March 2013.

Bishop J. T. and Allen J. (1989). The On-site Costs of Soil Erosion in Mali. Environment Department Working Paper No. 21. World Bank, Washington, DC.

Brandt J. and Geeson N. (2005). Desertification Information System to Support National Action Programmes in the Mediterranean (DISMED). DIS4ME, Desertification Indicator System for Mediterranean Europe. www.unibas.it/desertnet/dis4me/. (accessed 20 February 2012).

Brunner A. C., Park S. J., Ruecker G. R. and Vlek P. L. G. (2008). Erosion modelling approach to simulate the effect of land management options on soil loss by considering catenary soil development and farmers perception. *Land Degradation & Development*, **19**(6), 623.

Cerdan O., Govers G., Le Bissonnais Y., Van Oost K., Poesen J., Saby N., Gobin A., Vacca A., Quinton J., Auerswald K., Klik A., Kwaad F. J. P. M., Raclot D., Ionita I., Rejman J., Rousseva S., Muxart T., Roxo M. J. and Dostal T. (2010). Rates and spatial variations of soil erosion in Europe: a study based on erosion plot data. *Geomorphology,* **122**, 167–177.

D'Odorico P., Bhattachan A., Davis K. F., Ravi S. and Runyan C. W. (2013). Global desertification: drivers and feedbacks. *Advances in Water Resources*, **51**, 326–344. January 2013, ISSN 0309-1708, http://dx.doi.org/10.1016/j.advwatres.2012.01.013.

Davidson D. A. and Theocharopoulos A. P. (1992). A survey of soil erosion in Viotia, Greece. In: Past and Present Soil Erosion, M. Bell, J. Boardman (eds), Oxford Books, Oxford, pp. 149–154.

DESERTLINKS (2005). Combating desertification in Mediterranean Europe. Linking science with stakeholders. Project DESERTLINKS. EVK2-CT-2001-00109. Available at: http://www.kcl.ac.uk/projects/desertlinks/

Dregne H., Kassas M. and Rozanov B. (1991). A new assessment of the world status of desertification, UNEP. *Desertification Control Bulletin*, **20**, 6–18.

Elwell H. A. and Stocking M. A. (1986). Vegetative cover to estimate soil erosion hazard in Rhodesia. *Geoderma*, **15**, 61–70.

Eswaran H., Lal R. and Reich P. F. (2001). Land degradation: an overview. In: Responses to Land Degradation. Proc. 2nd. International Conference on Land Degradation and Desertification, Khon Kaen, Thailand, E. M. Bridges, I. D. Hannam, L.R. Oldeman, F. W. T. Pening de Vries, S. J. Scherr and S. Sompatpanit (eds), Oxford Press, New Delhi, India.

Feoli, E., Giacomich, P., Mignozzi, K., Oztürk, M. and Scimone, M. (2003). Monitoring desertification risk with an index integrating climatic and remotely-sensed data: an example from the coastal area of Turkey. *Management of Environmental Quality: An International Journal*, **14**(1), 10–21.

Ferrara A., Salvati L., Sateriano A. and Nolè A. (2012). Performance evaluation and cost assessment of a key indicator system to monitor desertification vulnerability. *Ecological Indicators*, **23**, 123–129. December 2012.

Francis C. F. and Thornes J. B. (1990). Runoff hydrographs from three Mediterranean vegetation cover types. In: Vegetation and Erosion, Processes and Environments, J. B. Thornes (ed.), J. Wiley & Sons, Chichester, pp. 363–384.

Geeson N., Quaranta Q., Salvia R. and Brandt J. (2015). Long-term involvement of stakeholders in research projects on desertification and land degradation: how has their perception of the issues changed and what strategies have emerged for combating desertification? *Journal of Arid Environments*, **114**, 124–133. March 2015.

Geist M. H. and Lambin E. F. (2004). Dynamic causal patterns of desertification. *Bioscience*, **54**(9), 817–829.

Govers G., Vandaele K., Desmet P., Poesen J. and Bunte K. (1994). The role of tillage in soil redistribution on hillslopes. *European Journal of Soil Science*, **45**, 469.

Grainger A. (2009). Developing a Baseline Survey for Monitoring Biophysical and Socio-economic Trends in Desertification, Land Degradation and Drought. Consultancy report, UNCCD, February, 2009.

Higginbottom Th. and Symeonakis El. (2014). Assessing land degradation and desertification using vegetation index data: current frameworks and future directions. *Remote Sensing*, **6**, 9552–9575. doi:10.3390/rs6109552.

Hill J., Stellmes M., Udelhoven Th., Röder A. and Sommer S. (2008). Mediterranean desertification and land degradation: mapping related land use change syndromes based on satellite observations. *Global and Planetary Change*, **64**(3–4), 146–157. December 2008. ISSN 0921-8181, http://dx.doi.org/10.1016/j.gloplacha.2008.10.005.

Imeson A. and Emmer I. M. (1992). Implications of climatic change on land degradation in the Mediterranean. In: Climatic Change and the Mediterranean, L. Jeftic, J. D. Milliman and G. Sestini (eds), Edward Arnold, London, pp. 95–128.

Irvine B. and Kosmas C. (2007). PESERA User's Manual. PESERA Technical Report Deliverable 15, contract QLK5-CT- 1999-0(1323). Revised for the ENVASSO Project (Contract 022713) by R.J.A. Jones. European Commission FP5 & 6 research projects, p. 133.

Kairis Or., Kosmas C., Karavitis Ch., Ritsema C., Salvati L., Acikalin S., Alcalá M., Alfama P., Atlhopheng J., Barrera J., Belgacem A., Solé-Benet A., Brito J., Chaker M., Chanda R., Coelho C., Darkoh M., Diamantis I., Ermolaeva O., Fassouli V., Fei W., Feng J., Fernandez F., Ferreira A., Gokceoglu C., Gonzalez D., Gungor H., Hessel R., Juying J., Khatteli H., Khitrov N., Kounalaki A., Laouina A., Lollino P., Lopes M., Magole L., Medina L., Mendoza M., Morais P., Mulale K., Ocakoglu F., Ouessar M., Ovalle C., Perez C., Perkins J., Pliakas F., Polemio M., Pozo A., Prat C., Qinke Y., Ramos A., Ramos J., Riquelme J., Romanenkov V., Rui L., Santaloia F., Sebego R., Sghaier M., Silva N., Sizemskaya M., Soares J., Sonmez H., Taamallah H., Tezcan L., Torri D., Ungaro F., Valente S., de Vente J., Zagal E., Zeiliguer A., Zhonging W. and Ziogas A. (2013). Evaluation and Selection of Indicators for Land Degradation and Desertification Monitoring: Types of Degradation, Causes, and Implications for Management. Environmental management. Springer Science + Business Media, New York. Published online: 28 June 2013. DOI: 10.1007/s00267-013-0110-0.

Kirkby M. (1999). Regional desertification indicators (RDIs). In: The Medalus Project, Mediterranean Desertification and Land-use Manual on Key Indicators of Desertification and Mapping Environmentally Sensitive Areas to Desertification, C. Kosmas, M. Kirkby and N. Geeson (eds), European Commission EUR 18882, Luxembourg, pp. 48–65.

Kirkby M. J., Jones R. J. A., Irvine B., Gobin A., Govers G., Cerdan O., Van Rompaey A. J. J., Le Bissonnais Y., Daroussin J., King D., Montanarella L., Grimm M., Vieillefont V., Puigdefabregas J., Boer M., Kosmas C., Yassoglou N., Tsara M., Mantel S., Van Lynden G. J. and Huting J. (2004). Pan-European Soil Erosion Risk Assessment: The PESERA Map, Version 1 October 2003. Explanation of Special Publication Ispra 2004, No.73 (S.P.I.04.73). European Soil Bureau Research Report No.16, EUR 21176 and 1 map in ISO B1 format. Office for Official Publications of the European Communities, Luxembourg, p. 18.

Koojman A. M., Jongejans J. and Sevink J. (2005). Parent material effects on Mediterranean woodland ecosystem in NE Spain. *Catena*, **59**, 55–68.

Kosmas C., Danalatos N. G., Moustakas N. K., Tsatiris B. and Yassoglou N. (1993). The impacts of parent material and landscape position on drought and biomass production of wheat under semi- arid zone of Greece. *Soil Technology*, **6**, 337–349.

Kosmas, C., Danalatos, N., Cammeraat, L. H., Chabart, M., Diamantopoulos, J., Farand, R., Gutierrez, L., Jacob, A., Marques, H., Martinez-Fernandez, J., Mizara, A., Moustakas, N., Nicolau, J. M., Oliveros, C., Pinna, G., Puddu, R., Puigdefabregas, J., Roxo, M., Simao, A., Stamou, G., Tomasi, N., Usai, D. and Vacca, A. (1997). The effect of land use on runoff and soil erosion rates under Mediterranean conditions. *Catena*, **29**, 45–59.

Kosmas C., Kirkby M. and Geeson N. (1999). Manual on: Key Indicators of Desertification and Mapping Environmentally Sensitive Areas to Desertification. European Commission, Energy, Environment and Sustainable Development, EUR 18882, p. 87.

Kosmas C., Danalatos N. G. and Gerontidis St. (2000). The effect of land parameters on vegetation performance and degree of erosion under Mediterranean conditions. *Catena*, (1), 3–17. (i.f. 1.346).

Kosmas C., Kairis Or., Karavitis Ch., Ritsema C., Salvati L., Acikalin S., Alcalá M., Alfama P., Atlhopheng J., Barrera J., Belgacem A., Solé-Benet A., Brito J., Chaker M., Chanda R., Coelho C., Darkoh M., Diamantis I., Ermolaeva O., Fassouli V., Fei W., Feng J., Fernandez F., Ferreira A., Gokceoglu C., Gonzalez D., Gungor H., Hessel R., Juying J., Khatteli H., Khitrov N., Kounalaki A., Laouina A., Lollino P., Lopes M., Magole L., Medina L., Mendoza M., Morais P., Mulale K., Ocakoglu F., Ouessar M., Ovalle C., Perez C., Perkins J., Pliakas F., Polemio M., Pozo A., Prat C., Qinke Y., Ramos A., Ramos J., Riquelme J., Romanenkov V., Rui L., Santaloia F., Sebego R., Sghaier M., Silva N., Sizemskaya M., Soares J., Sonmez H., Taamallah H., Tezcan L., Torri D., Ungaro F., Valente S., de Vente J., Zagal E., Zeiliguer A., Zhonging W. and Ziogas A. (2013). Evaluation and Selection of Indicators for Land Degradation and Desertification Monitoring: Methodological Approach. Environmental management. Springer Science + Business Media, New York. Published online: 25 June 2013. DOI: 10.1007/s00267-013-0109-6.

Lal R. (1998). Soil quality and sustainability. In: Methods for Assessment of Soil Degradation, R. Lal, W. H. Blum, C. Valentine and B. A. Stewart (eds), CRC Press LLC, Boca Raton, pp. 17–30.

Lallement, D. (1989). Environment and the Poor: Development Strategies for a Common Agenda. Transaction Books, New Brunswick.

Lavauden L. (1927). Les forêts du Sahara. *Revue des Eaux et Forêts*, **7**(65), 329–41.

Lawley V., Lewis M., Clarke K. and Ostendorf B. (2015). Site-based and remote sensing methods for monitoring indicators of vegetation condition: An Australian review. Ecological Indicators, Available online 7 April 2015, http://dx.doi.org/10.1016/j.ecolind.2015.03.021.

M.E.A. – Millennium Ecosystem Assessment (2005). Ecosystems and Human Wellbeing: Desertification Synthesis. World Resource Institute, Washington DC.

Montanarella L. (2007). Trends in land degradation in Europe. In: Climate and Land Degradation, M.V. Sivakumar, N. N'diangui (eds), Springer, Berlin.

Moustakas N. C., Kosmas C. S., Danalatos N. G. and Yassoglou, N. (1995). Rock fragments I. Their effect on runoff. erosion and soil properties under field conditions. *Soil Use and Management*, **11**, 115–120.

Nkonya E., Gerber N., Baumgartner P., von Braun J., De Pinto A., Graw V., Kato E., Kloos J. and Walter T. (2011). The Economics of Desertification, Land Degradation, and Drought: Toward an Integrated Global Assessment. ZEF Discussion Papers on Development Policy No. 150. Centre for Development Research, Bonn. NRSA (1995) Report on Area Statistics.

Papanastasis V. P. (1998). Livestock grazing in Mediterranean ecosystems: an historical and policy perspective. In: Ecological Basis of Livestock Grazing in Mediterranean Ecosystems, V. P. Papanastasis and D. Peter (eds), European Commission Science, Research and Development. Proceedings of the International Workshop, October 23–25, 1997, Thessaloniki (Greece). European Communities, 1998, Belgium, pp. 5–9.

Perez-Trejo F. (1992). Desertification and Land Degradation in the European Mediterranean. Eur. Comm. Report EUR 14850, p. 63.

Pimentel D., Harvey C., Resosudarmo P., Sinclair K., Kurz D., McNair M., Crist S., Shpritz L., Fitton L., Saffouri R. and Blair R. (1995) Environmental and economic costs of soil erosion and conservation benefits. *Science*, **267**, 1117–1123.

Reddy R. (2003). Land degradation in India: extent, costs and determinants. *Economic and Political Weekly*, **38**, 4700–4713.

Rubio J. L. and Recatala L. (2006). In Desertification in the Mediterranean Region: a Security Issue, Springer, p. 134. pp. 133–165.

Salvati L. (2014). A socioeconomic profile of vulnerable land to desertification in Italy. *Science of The Total Environment*, **466–467**, 287–299. 1 January 2014.

Salvati L., Mavrakis A., Colantoni A., Mancino G. and Ferrara A. 2015. Complex adaptive systems, soil degradation and land sensitivity to desertification: a multivariate assessment of Italian agro-forest landscape. *Science of The Total Environment*, **521–522**, 235–245. 15 July 2015.

Santini M., Caccamo G., Laurenti A., Noce S. and Valentini R. (2010). A multi-component GIS framework for desertification risk assessment by an integrated index. *Applied Geography*, **30**(3), 394–415. July 2010.

Schmidt, H. and Karnieli, A. (2000). Remote sensing of the seasonal variability of vegetation in a semi-arid environment. *Journal of Arid Environments*, **45**(1), 43–59. May 2000.

Singh A. (2015). Soil salinization and waterlogging: A threat to environment and agricultural sustainability. *Ecological Indicators*, **57**, 128–130. October 2015.

Sun, S., Crepeau, C. and Wang, G. (2000). A pilot research of desertification monitoring by application the SPOT4 vegetation satellite data. In: Desertification Monitoring Center, State Administration of Forestry (ed.), Desertification Monitoring Techniques in China, China Forestry Press, Beijing, p. 264.

UNCCD – United Nations Convention to Combat Desertification (2009). Summary for Decision Makers, Revitalizing the UNCCD, p. 3. 2009.

UNCCD – United Nations Convention to Combat Desertification (2013). White paper I: economic and social impacts of desertification, land degradation and drought. In 2nd Scientific Conference, 9–12 April 2013, Bonn, Germany.

UNEP – United Nations Environment Programme (1980). Study on Financing the United Nations Plan of Action to Combat Desertification. Report to the Secretary General. Document A/35/396. United Nations Environment Programme, Nairobi.

UNEP – United Nations Environment Programme (1992). World Atlas of Desertification. Edward Arnold, United Nations Environment Programme, London.

UNEP – United Nations Environment Programme (1994). Use of terms – cf. United Nations Convention to Combat Desertification, Art.1, p. 4., available at http://www.unccd.int.

Yang, X, Zhang, K., Jia B. and Ci, L. (2005). Desertification assessment in China: an overview. *Journal of Arid Environments*, **63**(2), 517–531.

Zha Y. and Gao J. (1997). Characteristics of desertification and its rehabilitation in China. *Journal of Arid Environments*, **37**, 419–432.

Part 3

Biophysical Hazards Methodologies

Chapter 7

Frost and heatwaves

Nicolas R. Dalezios and Panagiotis T. Nastos

7.1 FROST AND HEATWAVES HAZARDS

Frost is a meteorological phenomenon that occurs when the air temperature near the earth's surface drops to 0°C or below. Indeed, in meteorology, frost may be considered either as a form of mineral or a climatic condition, or even a hazard with several human impacts (Webb and Snyder, 2013). At first, frost as a form of mineral refers to a solid phase of water, since it crystallizes according to the hexagonal system featuring six-sided plates, needles, clusters, and columns. Second, as a climatic condition, the frost of specific interest is hoar frost or white frost that accumulates on surfaces in places with appropriate temperatures and the lack of sunshine. Indeed, the "hoar" implies a gray or grayish tone often seen on objects covered with frost. Hoar frost is indicative of three conditions: (1) surfaces on which the frost forms must be 0°C or below; (2) the surrounding air is saturated at 0°C or slightly below; and (3) nuclei are present so that the process of sublimation can take place. Each of these three conditions is part of the natural environment periodically. Finally, frost as a hazard implies near 0°C temperatures. Thus, a plant may suffer "chilling" injury, but not as a result of frost or freezing. Whatever the process of formation, frost is a symptom of a climatic condition in which temperatures have been reduced through radiation or advection to the freezing stage. This means that frost, particularly hoar frost, is a symptom of or ancestor to plant or crop loss. Frost can freeze plant tissue, which marks the end of growth for the plant, and this can occur at either end of the growing season. Since destruction is resulted, it is evident that frost is a natural hazard along with other climatic hazards.

On the other hand, heat is a feeling of discomfort, which is related to exceptionally high temperatures. In general, heat refers to a thermal discomfort, which activates the thermoregulatory system of the body. Continuous heat increases the discomfort and may cause adverse effects on health particularly combined with high levels of humidity or when exacerbated by a person's physical condition. If heat lasts for several days or longer it is called a heatwave. Although heatwave can be easily understood as a term, there is confusion among meteorologists about the limits or threshold values of the term. Indeed, the heatwave is one of the most special kinds of unusual and extreme weather. Heatwave is characterized by higher than normal temperatures, low winds, and generally good weather. Although unusual temperatures can occur any time of the year and anywhere, extreme heatwaves occur mainly during summer (Bampzelis *et al.* 2006). In the mid and high-latitudes, heatwaves are embedded in the usual course of summer weather, whereas in the tropics they are endemic.

A holistic and integrated approach to environmental hazards, including frost and heatwaves, has been gradually considered using common methodologies, such as risk analysis, which involves risk management and risk assessment (Smith, 2013). Indeed, through risk analysis there is an attempt to investigate and better understand frost and heatwaves hazards with the objective to develop pro-active measures and procedures before a crisis.

7.1.1 Frost concepts

Frost is a natural environmental risk with impacts on various human activities (Webb and Snyder, 2013). Technically, the word "frost" refers to the formation of ice crystals on surfaces, either by freezing of dew or a change of phase from vapor to ice (Blanc *et al.* 1963). Indeed, the word frost is used by the public to describe the condition when plants face injuries due to freezing. There are several definitions of frost, such as the following: (1) when the surface temperature drops below 0°C and the existence of low air temperature that causes damage or death to the plants (Ventskevich, 1958); (2) the occurrence of a temperature less than or equal to 0°C measured in a "Stevenson-screen" shelter at a height between 1.25 and 2.0 m (Hogg, 1971); (3) frost is defined as the condition, which exists when the air temperature near the earth's surface drops below 0°C (Kalma *et al.* 1992). A freeze exists when over a widespread region the air remains below freezing (0°C) for a sufficient period of time, which is at least one or two days. Based on the existing bibliography, Snyder and Paulo de Melo-Abreu (2005) use the following definition: a freeze exists when over a widespread region the air remains below freezing. Indeed, water within plants may or may not freeze during a frost event. Moreover, when plants freeze, intracellular or extracellular freezing takes place. Specifically, intracellular freezing is immediately fatal and extracellular freezing may cause injuries. The injury extent depends on plant resistance to frost. This resistance is associated with a critical temperature below which a malfunction or death of the cell takes place. As a result, the agricultural community recognizes frost as a hazard.

One of the areas directly affected by frost with disastrous results is agriculture. When frost occurs in active growth may cause major damage, which can reach up to the total destruction of production. Moreover, frost could cause injuries to sensitive crops. The extent of injuries depends on frost type, severity, frequency and duration of the phenomenon. Specifically, in the case of spring frosts it is not the severity and frequency as in winter frosts, however, they are the most harmful for agriculture, since during the spring season most plants are at the flowering stage or fruits and horticulture are susceptible even to pathetic frosts. Plant resistance to frost, variation in hardiness between different plant parts, different rates of tissue acclimation and deacclimation may determine the extent of freezing injury of the plant. In subtropical climates, frost crop damages result in economic losses that greatly exceed losses caused by any other weather related phenomenon. Frost is one of the most important threats to farmers, since it accounts nearly 50% of the total losses due to weather related phenomena.

Frost forecasts help farmers to reduce any possible injuries to their crops. Forecasting accuracy may be improved by using local meteorological factors and site specific climate data (Bagdonas *et al.* 1978). Frost protection methods, either active or passive, help farmers combat frost. There is a worldwide need for more and better spatial information on frost occurrence and its associated risk. There is growing recognition of the potential value of meteorological information in decision making. Deciduous fruit trees are very susceptible to frost damage in spring, with loss of fruit yield and even permanent tree damage. Several protective devices are available, such as wind machines, sprinklers and heaters, although these are expensive. Accurate minimum temperature forecasts can help the grower in making decisions whether or not to protect, on a particular occasion, and so may help to avoid economic loss.

Excessive cold may cause death directly through the effects of exposure and is indirectly responsible for deaths from causes, such as fatal heart attacks while clearing snow and asphyxiation in stranded vehicles. The annual average mortality rate attributable to winter storms and cold in the USA up to the mid-1990s was between 130 and 200 (Changnon *et al.* 1996). The number of cold-related fatalities may be increasing, because an aging population becomes probably more sensitive to the effects of cold. The impact of a severe winter extends beyond increased fatalities. Such impacts include production losses in industry, crop losses, transportation losses in revenue and damage to roads and bridges, losses in retail sales, and losses resulting from increased energy consumption during cold winters. Moreover, unemployment, the consumer price index, or the inflation rate may all rise, as well as the number of business failures may increase and the balance of payments may drop.

7.1.2 Heatwaves concepts

The activation of the temperature control mechanism is based on the energy balance of the body. The energy source is food, transformed by the body's metabolism. In hot weather, there are thermal discomfort conditions, leading to imbalances in the energy balance and thermoregulation of the organization's activities (Tetzlaff, 2013). The result is enhanced transfer of energy from the body core to the surface by imposing the increased blood circulation in order to maintain thermal comfort. The surface temperature and surface energy flows, however, are mainly affected by environmental parameters. Quantification of heat is a function of the overall assessment of the body's energy balance. Nevertheless, factors of energy balance are air temperature, humidity, ventilation, radiation, physical activity, clothing, mass and body shape, health status, age, and individual susceptibility. In overheating, the body's temperature regulation system cannot maintain an acceptable temperature range in key parts of the body.

Heatwaves can develop more frequently, when there are more hot days. In recent years, the number of days with heat stress, and of days leading to heatwaves, increased in the mid-latitudes (Jendritzky & Birger, 2009). The climate change forecast points to further increase of heat stress in many parts of the world. Excessive heat possibly contributes to more illness and mortality than any other direct, weather-related cause (Kalkstein *et al.* 1996). Most heat-related deaths occur in mid-latitude cities, in northern India and China, eastern and Midwestern USA and Western Europe, with infrequent, but extreme heatwaves. Subtropical and tropical cities with higher mean summer temperatures seem less vulnerable, partly due to acclimatization of residents and perhaps due to more efficient cooling of houses. The occurrence of heatwaves and their adverse effects is globally wide spread. In the last century, about 150 heatwaves were documented to reach disaster level (EM-DAT, 2012). The frequency of heatwaves is expected to change in the future (IPCC, 2007). Observations for Central Europe already show, that there is a drastic increase of the high (>30°C) and very high temperatures, and during the last decades the number of days with high temperature has been doubled, whereas the increase of the mean temperature was about 1°C. Heat-related deaths, averaging about 1000 per year, appear to be increasing in the USA, exceeding those caused by other hazardous weather conditions (Changnon *et al.* 1996).

7.1.3 Fog hazard

Fog is defined as a cloud at or near the surface. Fog consists of water droplets suspended above or near the surface and/or ice particles suspended above or near the surface (Croft, 2013). Moreover, mist consists of suspended and falling liquid hydrometeors. Finally, rime consists of suspended and falling solid hydrometeors. Fog is observed worldwide at all locations and elevations under substantially variant synoptic, mesoscale, and microscale conditions of the atmosphere. As such, its formation and intensity may occur any time or day of the year, and its coverage may be widespread, regional, or highly localized. Although fog is often very shallow in the vertical layer, it is very often close to the surface. In combination with other atmospheric constituents (natural or anthropogenic), fog also has the potential to damage the health and welfare of human, plant, and animal communities, as well as insect populations. These may occur in the air, within the ground surface layers, or the water cycle components. The occurrence of smog dates back to the burning of coal reserves to at least the thirteenth century. Smog episodes occur around the world including London, Mexico City, Tehran, Los Angeles, and in many portions of Southeast Asia.

Fog is a hazard to land, sea, and air transportation. The production, maintenance, and dissipation of fog constitutes significant risks and hazards for multiple modes of transportation, with impacts to populations. For example, the Bosphorus and Dardanelles Straits experience closures in Turkey due to dense fog, and the closures impact the flow of oil and grain from ports in the Russian Black Sea. Moreover, many multiple-vehicle accidents occur when visibility is severely restricted on highways. Similarly, the aviation industry loses millions of dollars each year when fog causes aircraft diversions and delays, inconveniencing thousands of passengers and incurring increased operating costs. Costly adjustments, such as automatic aids for aircraft, are possible and many fog dispersal techniques have been used. Supercooled fog can be modified, but accounts for only a small proportion of fogs. Warm fog can also be modified using thermal energy, but high costs of installation and operation of dispersal systems and the problem of pollution have discouraged their use (Kunkel, 1980). Visibility is one of the most difficult meteorological phenomena to forecast. Several objective fog forecasting techniques have been tried, but there is not entirely reliable method yet. Publicity of the hazards and driver education remain the surest ways of reducing the dangers of fog on highways.

7.2 FROST AND HEATWAVES CHARACTERISTICS

This section presents different frost classifications schemes, one for each selected criterion, as well as the main characteristics and features of heatwaves.

7.2.1 Frost classification

The frost phenomenon is characterized by varying severity and extent. The scale of the atmospheric mechanisms, which constitute the driving force of frost occurrence, vary considerably. Frost is generally classified into different types (Kalma *et al.* 1992), based on certain criteria, as follows.

(1) **Frost genesis criterion.** Frost is classified based on the genesis criterion into advection frost and radiation frost, both originating from synoptic meteorology factors and principles. Specifically, radiation frost is produced by intense radiation during the night from land surface causing cooling, which is associated with anticyclonic circulation during calm or apnea and cloudless nights (Figure 7.1a). In other words, radiation frosts are the result of intense longwave radiation, which causes cooling of the earths's surface due to energy loss during calm and clear nights. These types of frost have several features, such as an associated temperature inversion, clear and calm nights, as already mentioned, and usually air temperature higher

than 0°C during daytime. In radiation frosts, there is usually temperature inversion, which means that the temperature at an average height of about 15 meters above the ground is usually higher about 4–5°C than the corresponding surface temperature (Figure 7.2). In particular, figure 7.2 shows the formation of a frost pocket in a valley with trees planted on the valley slopes and deep in the valley. During the night, due to radiation the land gets cool. The cool air reaches the bottom of the valley, while forcing the hot air to rise upward. This situation favors thermal inversion and the formation of thermorofis.

(a) (b)

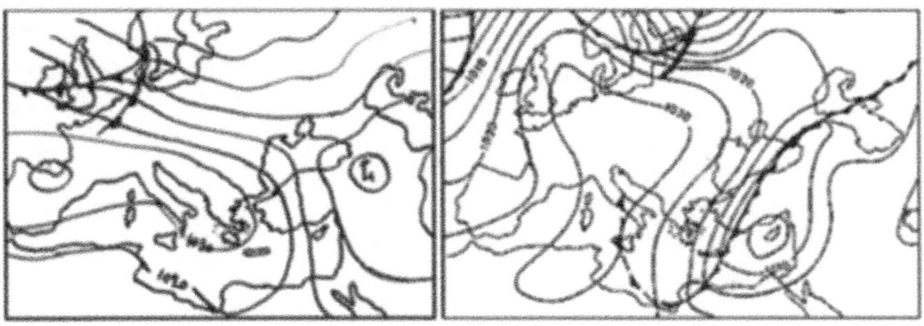

Figure 7.1. (**a**) Synoptic surface map with predominantly anticyclonic circulation over Europe favoring radiation frost over Greece. (**b**) Synoptic surface map showing strong invasion of cold air masses, following the cold front passage and causing advection frost.

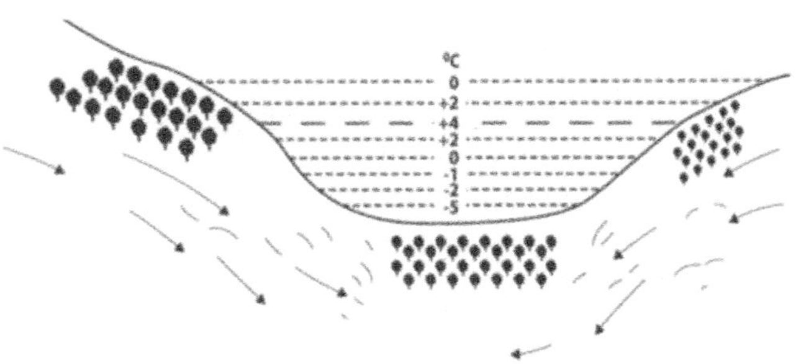

Figure 7.2. Formation of frost pocket in a valley.

On the other hand, advection frost is produced by sudden and strong invasion of cold air masses often from polar region, which usually follows the passage of a cold front from a region (Figure 7.1b). Advection frosts may develop during day or night and they are also associated with windy conditions and a well-mixed atmosphere, as well as air temperature may be less than 0°C during daytime. Moreover, Snyder and Paulo de Melo-Abreu (1995) have indicated cases where the observed conditions are considered as "micro-scale advection frosts". These frosts occur in valleys and surface depressions, when radiation frost conditions exist in a region, but cold air drainage on a local scale leads to rapid loss of temperature within the radiation frost area. The most damaging frosts are the advective frosts followed by radiation frosts. In this case advection frosts drop the temperature to near 0°C or below followed by clearing skies and drop in wind. As a result a severe radiation frost occurs.

In summary, it can be stated that the mechanisms for the formation of radiation frosts are: anticyclone centers, lack of barometric field, apnea or wind speed up to 5m/sec, absence of cloudiness, as well as low to mild relative humidity. Similarly, the mechanisms for the formation of advection frosts are: strong pressure gradient after the cold front passage, high wind speed and usually low relative humidity.

(2) **Frost duration criterion.** Based on the duration of the phenomenon, frost is characterized as total when the maximum daily temperature is less than or equal to 0°C and partial frost when the minimum daily temperature is less than or equal to 0°C. If the total frost lasts for several consecutive days, it can have catastrophic consequences, particularly since in such cases frost penetrates into the soil, resulting in the destruction of seeds and roots, which are embedded therein. In general, the most common type of frost is partial frost, which is equally devastating, especially when it occurs during spring, where the plants are in active growth stage.

(3) **Frost humidity criterion.** Atmospheric humidity is another frost criterion, which may cause condensation of water droplets creating ice crystals (frost), along with temperature drop to negative values. This type of frost is called white frost. On the other hand, when there is no condensation along with the drop in temperature below 0°C, then this type of frost is called black frost.

(4) **Frost severity criterion.** Frost is classified based on the severity criterion into: mild frost when the temperature ranges between 0°C and −4°C, moderate frost when the temperature ranges between −4.1°C and −10°C and severe frost when the temperature ranges from −10.1°C and below.

Climatological freezing characteristics. Frost has several features, such as the recurrence period, the space-time-frequency analysis, as well as the frost persistence. The recurrence period is extremely useful, because it allows the determination of the expected extreme value of a quantity, based on the average of the extreme values of the same magnitude in a given period. As frequency of frost occurrence is defined the likelihood of partial frost occurrence for several consecutive days, whereas frost severity is defined as the absolute minimum (negative) temperature, which occurs on successive days of frost. Moreover, frost duration is defined as the number of consecutive days in which there is partial frost.

Frost has also been related to the average length of frost-free period (or growing season in days) during the year and the frost severity-duration-frequency relationships. Needless to say, the frost-free period is very important for several parts of the world. A frost-free period is defined as the period between the latest spring and the earliest autumn frosts. For example, in Greece, based on data from 86 meteorological stations, the frost-free period reaches more than 240 days per year mainly in the southern regions and decreases from south to north (Dalezios, 2015). There are several climatic zones based on the duration of the mean annual frost-free period. For example, in subtropical regions, such as the Mediterranean region, a zone comprises regions that experience occasional winter frosts. In addition, frost events can cause significant damages to crops, or these zones include agricultural regions with very important farming, such as Mediterranean agriculture with sensitive crops and horticulture.

7.2.2 Heatwaves characteristics

The basic meteorological parameters related to heatwaves are air temperature, humidity, solar radiation, long wave radiation, wind speed and air quality. Other type of parameters are also examined, such as duration of heat, severity and frequency of the phenomenon. Weather and environmental conditions are usually unstable and almost permanently activate the body's thermoregulatory system. Usually the quantitative assessment of heat is conducted through indicators, which provide a simplified description of the body's energy balance. Indicators classify environmental conditions or recommend possible physical activity, or require the cooling of the body so as to prevent health damages caused by overheating. If the cooling mechanisms are not effective enough to maintain the temperature of the core at a constant optimum level, then heat stress conditions result. Taking as a base temperature the body temperature, a simple definition of a heatwave would be when the temperature of the environment is expected to rise above the body temperature of 37°C, however, other meteorological factors, besides temperature, must also be taken into account. The heatwave threshold temperature is also expected to vary from region to region and from season to season. There are some places in the world that temperature never exceeds 37°C, still sometimes the term heatwave can be used for unusual high temperatures observed there. Temperatures in excess of 40°C for a long period during the day result in weakening of the human body. The situation is worsened from the activities of humans and especially labour works. Most vulnerable humans are the elder and the babies.

There are different environmental conditions in every climate zone around the world. As expected, the climatic conditions show that in humid tropical regions, thermal stress occurs in more than 300 days per year, where permanent populations have been acclimatized and prepared in endemic heatwave conditions. On the other hand, in mid-latitudes, the heat stress shows a range of occurrence approximately between 20 and 100 days per year. Moreover, in mid-latitudes, days with heat stress occur mainly during the summer and then grouped into heatwaves, usually exceeding climatic average temperatures and lasting from two to about five days, on the average, dominated by high-pressure systems. The duration is of some relevance, because it takes a few days for the heat to penetrate the walls of most types of standard non-air-conditioned buildings and to raise the indoor temperature to heat stress conditions. The severity of each heatwave depends on the temporal and local particularities of high-pressure systems, which, in rare cases, allow for heatwaves to persist for weeks under very hot conditions.

Two deadly heatwaves have occurred in Europe in 2003 and 2010, where the number of deaths reached more than 66,000 in 2003 and 55,000 in 2010, respectively (EM-DAT, 2012). Also, these two heatwaves have resulted into property damages, mainly in forestry and agriculture. Both heatwaves have been caused by a long-term middle-latitude air flow field, known internationally as a "block" system. However, there are two additional factors, high temperatures and low rainfall. Specifically, the mean monthly maximum temperature in both cases, i.e. July 2010 in Russia and August 2003 in France, was approximately 10°C higher than the climatic mean. Moreover, in both cases the number of days with extremely high daily maximum temperature was extraordinary, where the lowest thresholds were 36°C in Western Europe in 2003, and 30°C in Russia in 2010, respectively.

7.3 FROST AND HEATWAVES RISK IDENTIFICATION

Frost and heatwaves risk identification involves hazard quantification. Table 7.1 presents a list of extreme air temperature indices, for frost or heatwaves assessment and monitoring (http://cccma.seos.uvic.ca/ETCCDMI/list_27_indices.html.).

7.3.1 Frost quantification

The contribution of meteorology to the frost problem can be summarized as in the early detection of the phenomenon, and consideration of the climatic conditions to estimate the frequency of probability of phenomenon occurrence in some area. Then considering the annual temperature variation, forecasting methods and modern frost assessment methods are developed.

Annual temperature variability. The annual temperature variability is approximated with sufficient accuracy by sinusoidal functions of the form:

$$T_i(D) = A_i + B_i \sin(\tfrac{360}{365} D - F_i) \quad (7.1)$$

where the quantities A_i (in °C), B_i (in °C) and Fi (in degrees) are constants for each location. The variable D represents the day of the year ($D = 1, \ldots, 365$) and the index i denotes the mean, i.e. mean maximum (mean-max), mean minimum (mean-min), absolute maximum (max-max) and absolute minimum (min-min) daily temperature.

Table 7.1. Extreme air temperature indices recommended by the ETCCDMI. (See also http://cccma.seos.uvic.ca/ETCCDMI/list_27_indices.html).

ID	Indicator Name	Indicator Definitions	UNITS
TXx	Max Tmax	Let Tx_{kj} be the daily maximum temperatures in month *k*, period *j*. The maximum daily maximum temperature each month is then $TXx_{kj} = \max(TX_{kj})$	°C
TNx	Max Tmin	Let Tn_{kj} be the daily minimum temperatures in month *k*, period *j*. The maximum daily minimum temperature each month is then $TNx_{kj} = \max(Tn_{kj})$	°C
TXn	Min Tmax	Let TX_{kj} be the daily maximum temperatures in month *k*, period *j*. The minimum daily maximum temperature each month is then $TXn_{kj} = \min(Tx_{kj})$	°C
TNn	Min Tmin	Let Tn_{kj} be the daily minimum temperatures in month *k*, period *j*. The minimum daily minimum temperature each month is then $TNn_{kj} = \min(Tn_{kj})$	°C
TN10p	Cold nights	Let Tn_{ij} be the daily minimum temperature on day *i* in period *j* and let $Tn_{in}10$ the calendar day 10th percentile centred on a 5-day window (Zhang *et al.* 2005b). The percentage of time is determined where $Tn_{ij} < Tn_{in}10$	Days
TX10p	Cold days	Let Tx_{ij} be the daily maximum temperature on day *i* in period *j* and let $Tx_{in}10$ be the calendar day 10th percentile centred on a 5-day window (Zhang *et al.* 2005b). The percentage of time is determined where $Tx_{ij} < Tx_{in}10$	Days
TN90p	Warm nights	Let Tn_{ij} be the daily minimum temperature on day *i* in period *j* and let $Tn_{in}90$ be the calendar day 90th percentile centred on a 5-day window (Zhang *et al.* 2005b). The percentage of time is determined where $Tn_{ij} > Tn_{in}90$	Days
TX90p	Warm days	Let Tx_{ij} be the daily maximum temperature on day *i* in period *j* and let $Tx_{in}90$ be the calendar day 90th percentile centred on a 5-day window (Zhang *et al.* 2005b). The percentage of time is determined where $Tx_{ij} > Tx_{in}90$	Days

(*Continued*)

Table 7.1. Extreme air temperature indices recommended by the ETCCDMI. (See also http://cccma.seos.uvic.ca/ETCCDMI/list_27_indices.html) (*Continued*).

ID	Indicator Name	Indicator Definitions	UNITS
DTR	Diurnal temperature range	Let Tx_{ij} and Tn_{ij} be the daily maximum and minimum temperature respectively on day i in period j. If I represents the number of days in j, then $DTR = \frac{\sum_{i=0}^{I}(Tx_{ij} - Tn_{ij})}{I}$	°C
FD0	Frost days	Let Tn_{ij} be the daily minimum temperature on day i in period j. Count the number of days where $Tn_{ij} < 0°C$	Days
SU25	Summer days	Let Tx_{ij} be the daily maximum temperature on day i period j. Count the number of days where $Tx_{ij} > 25°C$	Days
1DO	Ice days	Let Tx_{ij} be the daily maximum temperature oil day i in period j. Count the number of days where $Tx_{ij} < 0°C$	Days
TR20	Tropical nights	Let Tn_{ij} be the daily minimum temperature on day i in period j. Count the number of days where $Tx_{ij} > 20°C$	Days
GSL	Growing season Length	Let T_{ij} be the mean temperature on day i in period j. Count the number of days between the first occurrence of at least 6 consecutive days with $T_{ij} > 5°C$ and the first occurrence after 1st July (1st January in SH) of at least 6 consecutive days with $T_{ij} < 5°C$	Days
WSDI*	Warm spell duration indicator	Let Tx_{ij} be the daily maximum temperature on day i in period j and let $Tx_{in}90$ be the calendar day 90th percentile centred on a 5-day window (Zhang *et al.* 2005b). Then the number of days per period is summed where, in intervals of at least 6 consecutive days:- $Tx_{ij} > Tx_{in}90$	Days

7.3.1.1 Frost modelling and forecasting methods

Frost forecasting is very important especially for the prevention of disasters in agriculture and particularly in crops. Frost forecasting is conducted through the analysis of daily synoptic weather charts, as well as the application of empirical models. Theoretical and applied methods are implemented, which combine air temperature, dew point temperature and land surface temperature through heat transfer. Moreover, formulas are applied, which consider the behavior of the temperature variability within the lower part of the atmospheric boundary layer (ABL), namely between 1.5 m to 24 m. There are two levels of frost forecasting, namely general and local forecast. General forecast consists of the identification of the properties and characteristics of the air mass, which prevails over the area during the night and early morning. In the local forecast, some representative sites of the area are selected, such as a valley, or the slopes or the top of a hill. The drop of temperature during the night depends on the duration of the night, meaning that the longer the duration of the night, the greater the drop in temperature is and the heat lost by the ground. The amount of heat lost by the land surface is a function of soil moisture and temperature, cloud cover, vegetation cover, air temperature and other factors.

Some frost forecasting methods are presented below (Bagdonas *et al.* 1978):

(1) **Rule of maximum – minimum.** It has been shown that the minimum temperature (T_{min}) at the next day ($N + 1$) follows a linear relationship with the maximum temperature (T_{max}) of day (N), when the forecast is issued, namely:

$$T_{min} = aT_{max} - b, \quad \text{where } a, b \text{ constants} \tag{7.2}$$

(2) **Rule of wet bulb temperature.** This method takes into account the effect of atmospheric water vapor, which prevents heat loss from the land surface.

$$T_{min} = aT_w - bT_d - c \tag{7.3}$$

where T_w is the wet bulb temperature at the time of sunset of day N, T_d the dry bulb temperature at the same day and time, and a, b, c, constants, depending on the location. Many times a and b are very small, so equation (7.3) can be simplified to: $T_{min} = T_w - c$ and is called Kammerman formula.

(3) **Craddock formula.** It is an empirical formula for the minimum temperature T_{min} based on the dry bulb temperature T_d at 12:00Z, and dew point temperature T_{dew} also at 12:00Z.

$$T_{min} = aT_d + bT_{dew} + 2.12 + c \tag{7.4}$$

The values of the temperatures are given in degrees °F. The a and b are constants. The constant c is given as a function of the average cloud on prognostic hours 18:00Z, 24:00Z, and 06:00Z and the average wind speed (in knots) at the same hours. The formula is not valid when there is fog at night.

(4) **Gold formula.** This formula computes the minimum temperature T_{min} as follows:

$$T_{min} = a\,T_{1500Z} + b\,T_{dew1500Z} - c \tag{7.5}$$

where T_{1500Z} is the temperature at 15:00Z, $T_{dew1500Z}$ is the dew point temperature at 1500Z and a, b and c are constants.

(5) **Faust formula for soil frost.** Faust has provided an empirical formula in which if the amount of cloud at night is less than 2/8 and the average wind speed is less than 2 knots, then soil frost occurs when the sum $[T + [1/(2T_{dew})]]$ at 14:00 local time less than 79°F.

(6) **Multiple linear regression.** For the forecast of the minimum temperature Y of the day, a multiple linear regression can be used as follows:

$$Y = a_1X_1 + a_2X_2 + a_3X_3 + a_4X_4 + a_5X_5 \tag{7.6}$$

wherein X_1 is the dry bulb temperature, X_2 is a function of the cloud cover: $X_2 = 0.9$ m + 0.5 k, where m = amount of low cloud and k amount of medium clouds into eighths, X_3 is a function of the wind speed and direction, X_4 is the dew point temperature and X_5 is a function dependent on whether frost has occured the previous day, then takes the value 1 and 0 if not. The variables a_1 to a_5 are calculated using the least squares method in a specified area.

(7) **Model ANGELA.** The physical model of ANGELA System (WMO, 2010) refers to the drop of temperature at night. In this model, the land surface temperature is a function of temperature at the sunset time and several hours after the sunset, given by:

$$Tn = Ts - K \times n^{1/2} \tag{7.7}$$

where Tn is the temperature n hours after sunset in °C, Ts is the temperature at the time of sunset in °C, K is the coefficient of temperature drop and n is the number of hours after sunset.

(8) **Remote sensing methods.** Frost assessment using remote sensing methods is based on temperature observations from InfraRed (IR) bands of meteorological satellites, such as NOAA/AVHRR, METEOSAT, or MODIS. From these satellites, brightness temperature is usually observed from thermal IR channels on a pixel basis, from which land surface temperature (LST) can be computed through regression analysis with conventional surface temperature observations, again on a pixel basis (Dalezios, 2015). It is known that meteorological satellites are characterized by high temporal resolution, e.g. for METEOSAT every 30 minutes, but with rather coarse spatial resolution ranging between 1 to 2.5 Km or 1000 m for MODIS. As a result, temperature monitoring can be conducted through meteorological satellites leading to estimation and assessment of frost, especially for radiation frost in cloudless nights

during spring season in frost-prone valleys or land surface patches. For illustrative purposes, figure 7.3 delineates the surface temperatures from a METEOSAT IR image in the area of Katerini in Northern Greece consisting of 26 pixels. Moreover, table 7.2 presents the number of pixels for each corresponding temperature from METEOSAT IR during the night of 31 March 1994 in the same area of Katerini in Northern Greece.

		1	1	2
		1	0	1
		0	0	1
2	1	-1	0	1
1	0	-2	-1	0
0	-1	-2	-1	0
			0	-1

Figure 7.3. Observed temperatures from METEOSAT IR (2.5 × 2.5 Km resolution) at 04.00 am of 31 March 1994 in the area of Katerini in Northern Greece.

Table 7.2. Number of pixels for each corresponding temperature during the night of 31 March 1994 in the area of Katerini (Figure 7.8) from METEOSAT IR.

Time 31.3.94	**Area of Katerini in Northern Greece**					
	3°C	**2°C**	**1°C**	**0°C**	**–1°C**	**–2°C**
3:30	–	5	8	8	4	1
4:00	–	2	8	9	5	2
5:00	–	10	9	6	1	–
6:00	1	12	12	5	1	–
6:30	2	9	9	6	–	–
7:00	6	10	6	4	–	–

7.3.1.2 Frost early warning systems (FEWS)

Frost monitoring is based on the development of FEWS. Frost hazard is based on temperature and its spatiotemporal variability. Quantification of frost hazard uses a methodological approach based on the minimum temperature consideration (Dalezios & Lavrediadou, 1995). The database consists of series of satellite records (e.g. LANDSAT, METEOSAT, NOAA/AVHRR) from which temperature is extracted on a pixel basis. For monitoring frost a so-called phenomenological approach is used based on Kalman filtering, which belongs to estimation and control theory (Dalezios, 1987). Specifically, an one step-ahead forecasting on a pixel basis using 2-D satellite temperature images is considered. In this way, temperature time series are developed for each pixel and then the one step-ahead forecasting is attempted. The adopted approach comes from the optimal estimation theory and in the current application the adaptive Kalman filter is employed (Dalezios, 1987). Specifically, as already mentioned, the system model is the so-called phenomenological temperature model, which is based on the assumption that the daily temporal variability of temperature follows a sinusoidal function. There are also several other FEWS methods based on the frost forecasting methods presented in section 7.3.1.1 above, or based on similar approaches.

7.3.2 Heatwaves quantification

This section of heatwaves quantification covers two groups of heatwaves indices and methods of heatwaves early warning systems (HEWS).

7.3.2.1 Heatwaves indices

This section presents two groups of indices related to heatwaves, namely the thermal indices (Epstein & Moran, 2006) and the joint distribution indices (Beniston, 2009).

(1) **Thermal indices.** These are four indices, which are briefly presented.

The effective temperature (ET) index was proposed by Houghton and Yaglou (1923). This index was established to provide a method for determining the relative effects of air temperature and humidity on comfort. The Effective Temperature (ET) is defined as the temperature of a still, saturated atmosphere which would, in the absence of radiation, produce the same effect as the atmosphere in question. It indicates the combined effects of relative humidity, air velocity, air temperature and clothing.

The **WetBulb Globe Temperature (WBGT)** is a measure of the heat stress in direct sunlight, which takes into account: temperature, humidity, wind speed, sun angle and cloud cover (solar radiation). This differs from the heat index, which takes into consideration temperature and humidity and is calculated for shady areas. The risk of heat disorders depends on wet-bulb globe temperature (WBGT) rather than on dry bulb temperature. WBGT is defined by dry-bulb temperature, wet-bulb temperature, and globe temperature (Yaglou & Minard, 1957).

Physiologically Equivalent Temperature (PET) is defined as the physiological equivalent temperature at a given place (outdoors or indoors). It is equivalent to the air temperature at which – in a typical indoor setting (without wind and solar radiation) – the heat balance of the human body (work metabolic rate 80 W of light activity, that should be added to the basic metabolic rate 86.5 W, Stolwijk and Hardy 1977; heat resistance of clothing 0.9 clo, which is the reference clothing insulation value used for the formulation of PET) is maintained with core and skin temperatures equal to those of the under assessment conditions (Mayer & Höppe, 1987; Höppe, 1999). The following assumptions are made for indoor reference climate: Mean radiant temperature equals air temperature (Tmrt = Ta). Air velocity is set to 0.1 m/s. Water vapour pressure is set to 12 hPa (approximately equivalent to relative humidity of 50% at Ta = 20.0°C). According to Hoppe (1999), the assumption of constant values for clothing and activity in the calculation of PET was made in order to define an index independent of individual behavior.

Universal Thermal Climate Index (UTCI) is defined as the equivalent ambient temperature (°C) of a reference environment providing the same physiological response of a reference person as the actual environment (Weihs *et al.* 2012). The calculation of the physiological response to the meteorological input is based on a multi-node model of human thermoregulation (Fiala *et al.* 2001), associated with a clothing model. Static clothing insulation is adjusted to the ambient temperature considering seasonal clothing adaptation habits of Europeans, which notably affects human perception of the outdoor climate (Havenith *et al.* 2012). In this point it is worth mentioning that while in principle the PET scale represents thermal sensations of thermal environment experienced by specific population, the UTCI scale represents heat/cold stress intensities regardless the population type.

(2) **Joint distribution indices.** The four combined climate indices, based on air temperature and precipitation, concern Cool/Dry days (CD), Cool/Wet days (CW), Warm/Dry days (WD), Warm/Wet days (WW). These indices are defined by the exceedances of the joint quartiles of temperature and precipitation using the 25th and 75th percentile levels in order to capture a larger number of events (Beniston, 2009).

The CD index is defined as the number of days with the daily mean air temperature (T) below the 25th percentile of the daily mean temperature (T25) and simultaneously the daily precipitation (P) below the 25th percentile of the daily precipitation (P25), thus (T < T25 and P < P25).

The CW index is defined as the number of days with the daily mean air temperature (T) below the 25th percentile of the daily mean temperature (T25) and simultaneously the daily precipitation (P) above the 75th percentile of the daily precipitation (P75), thus T < T25 and P > P75.

The WD index is defined as the number of days with the daily mean air temperature (T) above the 75th percentile of the daily mean temperature

(T75) and simultaneously the daily precipitation (P) below the 25th percentile of the daily precipitation (P25), thus T > T75 and P < P25.

The WW index is defined as the number of days with the daily mean air temperature (T) above the 75th percentile of the daily mean temperature (T75) and simultaneously the daily precipitation (P) above the 75th percentile of the daily precipitation (P75), thus T > T75 and P > P75.

7.3.2.2 Heatwaves early warning systems (HEWS)

Heatwaves monitoring is based on the development of HEWS. Heatwaves hazard is based on temperature variability. Quantification of heatwaves hazard is based on the maximum temperature consideration (Bampzelis *et al.* 2006; Tsiros *et al.* 2008). The database consists of series of satellite records (e.g. LANDSAT, METEOSAT, NOAA/AVHRR) from which temperature is extracted on a pixel basis. For monitoring heatwaves a so-called phenomenological approach is used based on Kalman filtering (Dalezios, 1987), which belongs to estimation and control theory. Specifically, an one step-ahead forecasting on a pixel basis using 2-D satellite temperature images is considered. In this way, temperature time series are developed for each pixel and then the one step-ahead forecasting is attempted. The adopted approach comes from the optimal estimation theory and in the current application the adaptive Kalman filter is employed (Dalezios, 1987). Specifically, as already mentioned, the system model is the so-called phenomenological temperature model, which is based on the assumption that the daily variability of temperature follows a sinusoidal function. There are also several other HEWS methods based on either statistical regression analysis or on physically-based energy (heat) balance methods.

7.3.3 Fog modelling and assessment

The advent of a scientific and structured observational (or empirical) study of fog has followed cloud classification schemes and advanced instrumentation for the study of the atmosphere, particularly on more localized scales. As a result, there is an improved understanding of the observed properties and behaviors of fog based on weather observation data and fog's associated characteristics as a function of dynamic processes in the atmosphere. Fog prediction is developed in order to provide opportunity for the avoidance, mitigation, or prevention of its impacts. Fog has been classified by its formative method, e.g. radiative, location, e.g. sea fog, or atmospheric processes, e.g. frontal fog, and has led to many types being recognized internationally, as well as locally, called fogs. Further development of these methods have included statistical and time series analyses to produce simple conditional climatologies of fog occurrence, severity and trends, as well as Markov Chain and other analytical methods to assess fog occurrence and probability distributions for specific locations or regions or even severity. In such cases, the information is limited to the period of record and how representative each location is with regard to its surrounding area. Further specification of fog occurrence is possible when these techniques are linked to the synoptic setting or when statistical or climatic values are mapped across a region in order to infer the frequency of occurrences, severities, and fog coverage or behavior. However, although predictions or diagnosis are developed, these findings are limited due to an observational framework and time scale in terms of hours or days, without full consideration of the process-oriented nature of fog formation, maintenance, evolution, and dissipation. As the observational study of fog has matured, the study of cloud physics and the microphysical processes involved in cloud growth provided distinct thermodynamic information as to the internal mechanisms and factors associated with fog formation and dissipation, as well as its radiative impacts.

7.4 FROST AND HEATWAVES RISK ASSESSMENT

Frost and heatwaves risk assessment includes risk estimation, which involves the risk of such events, i.e. event probabilities, as well as magnitude-duration-frequency and areal extent relationships for hazard assessment.

7.4.1 Frost frequency analysis

The daily minimum temperature below 0°C is used as a threshold and the number of days with such a temperature, or partial frost, is identified for each station. If successive days have minimum temperatures below 0°C, they are identified as one frost episode having as duration the number of successive days and intensity or severity the absolute minimum temperature below 0°C recorded during that period. Several episodes are grouped according to their duration, for the whole period of study, and are ranked according to their severity for each station. Similarly, the frequency of an extreme event is usually expressed by its return period or recurrence interval, which may be defined as the average interval of time within which the magnitude of the event is equated or exceeded once. The analysis of extreme events is usually presented by severity-duration-frequency (SDF) relationships for several stations throughout the region of interest.

For the estimation of extreme events, such as frosts, in which the return periods are required, when the severities and duration are given, it is necessary to assume a particular mathematical form of the frequency distribution. Several theoretical distributions have been tested against the cumulative severities of extreme phenomena of various durations. These include the Extreme Value I (EVI, Gumbel), the Generalized Extreme Value (GEV), the three parameter Log-Normal (LN3) and the Log-Pearson (LP3) distributions (Dalezios *et al.* 2000). Application of the non-parametric Kolmogorov-Smirnov two sample tests at 95% confidence level and visual inspection of the fitting of the above theoretical frequency distributions to cumulative intensity values indicate that the EVI provides overall, a reasonable and acceptable approximation of the frequency of the calculated severity values (Dalezios & Lavrediadou, 1994). Furthermore, the EVI has been used in numerous studies of extreme phenomena (Dalezios *et al.* 2000). Data used in this study include daily series of minimum temperature from 15 meteorological stations in Greece. The threshold is temperature below 0°C. A brief description of the steps, which are followed to develop the SDF relationships, is presented.

Step 1: Probability Tables. The frost episodes for each station are identified, when the minimum temperature for successive days is below 0°C. Absolute minimum temperatures of partial frost below 0°C of each episode are used in order to rank the episode's severity. In this way, multiple episodes for the whole period are calculated for several durations. In table 7.3 column 1 shows the ranking numbers, column 2 shows the absolute minimum temperature values in ascending order, column 3 the corresponding probability (P) of occurrence using the Weibull plotting position equation, where m is the current ranking number and n is the total number of data points and column 4 shows the corresponding return period T duration using the equation, where P was previously defined. Table 7.3 provides an example Probability Table for Agrinio Station and for 2-days duration partial frost episodes.

Table 7.3. Frost absolute minimum temperatures of 2-days duration with the corresponding probabilities (P) and return periods (T) for Agrinio station.

Rank	Abs. Min Temper.	Probability $P=m/(n+1)$	Return Period $T=1/P$	Rank	Abs. Min Temper.	Probability $P=m/(n+1)$	Return Period $T=1/P$
1	0.2	0.025	40.00	21	2.4	0.525	1.90
2	0.5	0.05	20.00	22	2.6	0.55	1.82
3	0.6	0.075	13.33	23	2.6	0.575	1.74
4	0.8	0.1	10.00	24	2.6	0.6	1.67
5	1	0.125	8.00	25	2.8	0.625	1.60
6	1	0.15	6.67	26	2.8	0.65	1.54
7	1.2	0.175	5.71	27	3	0.675	1.48
8	1.3	0.2	5.00	28	3	0.7	1.43
9	1.3	0.225	4.44	29	3	0.725	1.38
10	1.4	0.25	4.00	30	3.2	0.75	1.33
11	1.4	0.275	3.64	31	3.5	0.775	1.29
12	1.6	0.3	3.33	32	3.6	0.8	1.25
13	1.6	0.325	3.08	33	3.6	0.825	1.21
14	1.8	0.35	2.86	34	3.8	0.85	1.18
15	1.8	0.375	2.67	35	4	0.875	1.14
16	1.9	0.4	2.50	36	4.8	0.9	1.11
17	2	0.425	2.35	37	4.8	0.925	1.08
18	2.2	0.45	2.22	38	5	0.95	1.05
19	2.2	0.475	2.11	39	7	0.975	1.03
20	2.3	0.5	2.00				

Step 2: Fitting Gumbel Distribution. For each frost episode the identified absolute minimum temperatures are plotted versus the corresponding return period and the EVI distribution (Dalezios & Lavrediadou, 1994), is fitted to the plotted data points, which has the following Cumulative Distribution Function (equation 7.8):

$$F(x) = \exp\left[-\exp(-A \cdot (x - U))\right] \tag{7.8}$$

where A, U are the fitted parameters, which are computed for each duration from the data. The procedure of fitting EVI distribution is applied to all the identified episodes for each station. An example of fitting extreme value distributions of 2-days duration for the station of Kavala in Northern Greece is shown in Figure 7.4. The best fit is provided by Gumbel EVI curve, which is selected.

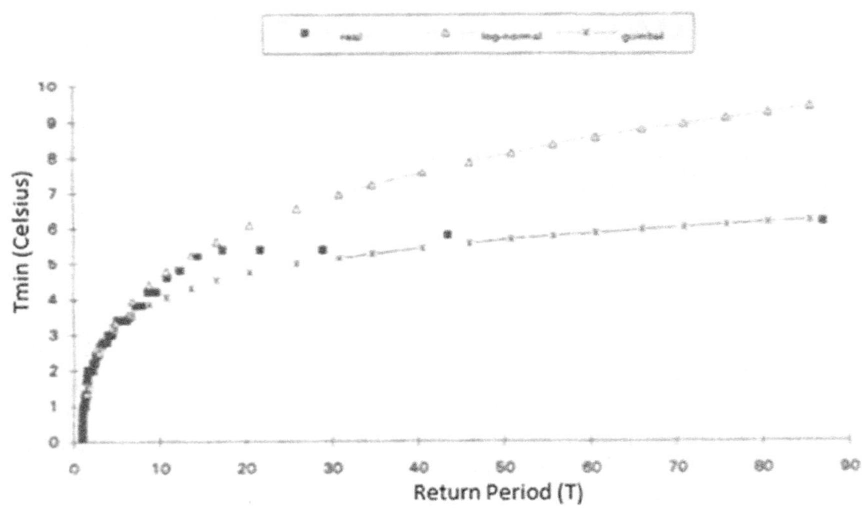

Figure 7.4. Fitting extreme value distributions for partial frost episodes of 2-days duration for Kavala station in Northern Greece (from Dalezios & Lavrediadou, 1994).

Step 3: Severity – Duration – Frequency (SDF) Curves. Finally, using the Gumbel distribution, cumulative periods are computed, which corresponds to return periods of 2, 5, 10, 25, 50 and 100 years, respectively, for each identified heatwave duration as seen in Figure 7.5 for the station of Larissa. The SDF curves appear to be, as expected, since for decreasing frequencies there is a corresponding increase in severities or intensities. There is a corresponding increase in frost severities, respectively, which tend to become asymptotic to the x axis, as seen in Figure 7.5.

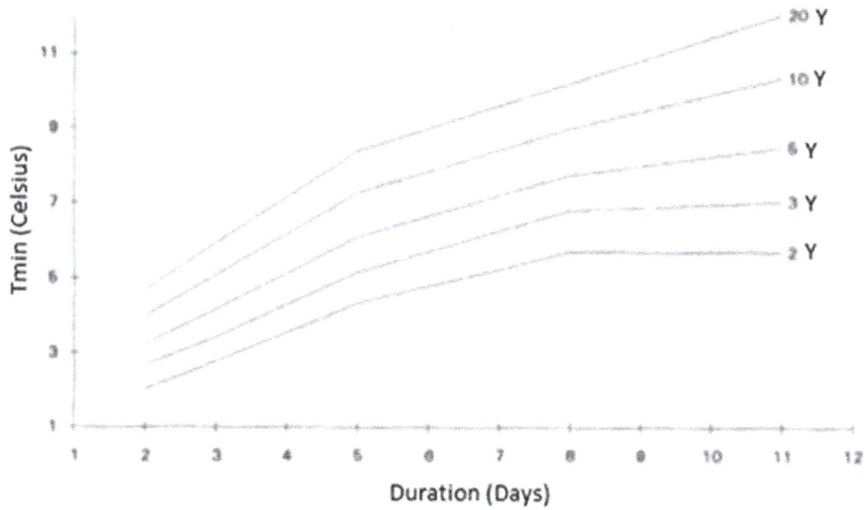

Figure 7.5. Frost Severity-Duration-Frequency Curves for Kavala station in Northern Greece (from Dalezios & Lavrediadou, 1994).

7.4.2 Heatwaves frequency analysis

Taking as threshold the daily maximum temperature of above 35°C, the number of days with such a temperature is identified for each station. If successive days have maximum temperatures above 35°C, they are identified as one heatwave episode having as duration the number of successive days and intensity or severity the cumulative temperatures above 35°C reported during that period. Several episodes are grouped according to their duration, for the whole period of study, and are ranked according to their intensity for each station.

The frequency of an extreme event is usually expressed by its return period or recurrence interval, which may be defined as the average interval of time within which the magnitude of the event is equated or exceeded once. The magnitude of a heatwave, is given by the absolute maximum temperature occurring in a particular duration and data for extreme events can be usually presented by severity – duration – frequency graphs for several points throughout the region of interest. For the estimation of heatwaves, in which the return periods are required, when the severities and duration are given, it is necessary to assume a particular frequency distribution, such as the Extreme Value I (EVI, Gumbel), the Generalized Extreme Value (GEV), the three- parameter Log-Normal (LN3) and the Log-Pearson (LP3) distributions (Bampzelis *et al.*

2006; Dalezios *et al.* 2000). Application of the non-parametric Kolmogorov-Smirnov two-sample tests at 95% confidence level indicate that the EVI provides overall, a reasonable and acceptable approximation of the frequency of the calculated severity values (Bampzelis *et al.* 2006). Furthermore, the EVI has been used in numerous extreme phenomena studies (Dalezios *et al.* 2000). Data used in this study include daily series of maximum temperature from 13 meteorological stations in Greece. The geographical distribution of the stations is delineated in Figure 7.6. The threshold is temperature above 35°C. A brief description of the steps, which are followed to develop the SDF relationships, is presented.

Figure 7.6. The geographical distribution of the stations in Greece.

Step 1: Probability Tables. The heatwaves episodes for each station are identified, when the maximum temperature for successive days exceeds 35°C. Cumulative temperatures over 35°C of each episode are used in order to rank the episode's severity. In this way, multiple episodes for the whole period are calculated for several durations. Again, in table 7.4 column 1 shows the ranking numbers, column 2 shows the maximum temperature values in ascending order, column 3 the corresponding probability (P) of occurrence, where m is the current ranking number and n is the total number of data points and column 4 shows the corresponding return period T ($T = 1/P$), where P was previously defined. Table 7.4 provides an example Probability Table for Larissa Station and for six days duration heatwave episodes.

Table 7.4. Probability table of duration of 6 successive days of cumulative maximum daily temperature above 35°C for Larissa station.

Rank	Cumulative Temperature (in °C)	Probability P	Return Period T (years)	Rank	Cumulative Temperature (in °C)	Probability P	Return Period T (years)
1	30.6	0.059	17	9	12.0	0.529	1.89
2	26.2	0.118	8.5	10	11.8	0.588	1.7
3	25.8	0.177	5.67	11	11.6	0.647	1.55
4	24.2	0.235	4.25	12	10.3	0.706	1.42
5	21.2	0.294	3.4	13	8.2	0.765	1.31
6	18.6	0.353	2.83	14	6.4	0.824	1.21
7	16.2	0.412	2.43	15	5.0	0.882	1.13
8	13.2	0.471	2.13	16	5.0	0.941	1.06

Step 2: Fitting Gumbel Distribution. For each heatwaves episode the identified maximum temperatures are plotted versus the corresponding return period and a statistical distribution is fitted to the plotted data points. Specifically, the extreme value law is used by fitting the EVI distribution (Bampzelis *et al.* 2006), which has, as before, the following Cumulative Distribution Function (equation 7.9):

$$F(x) = \exp\left[-\exp(-A \cdot (x - U))\right] \tag{7.9}$$

where A, U are the fitted parameters, which are computed for each duration from the data. The procedure of fitting EVI distribution is applied to all the identified heatwaves episodes for each station. An example of fitting the EVI curves for the station of Larissa is shown in Figure 7.7 for durations from 1 to 9 days. As expected, each curve is plotted on top of the previous, since it corresponds to ascending duration.

Step 3: Heatwaves Severity – Duration – Frequency (SDF) Curves. Finally, using the Gumbel distribution, cumulative heatwave periods are computed corresponding to return periods of 2, 5, 10, 25, 50 and 100 years, respectively, for each identified heatwave duration as seen in Figure 7.8 for the station of Larissa. In this study a different approach of the term heatwave is used by producing SDF curves for several stations across Greece taking into account only dry bulb temperature. The SDF curves appear to be as expected, since for decreasing frequencies there is a corresponding increase in severities as seen in Figure 7.8. The development of SDF curves for 13 stations in Greece clearly indicates that cumulative temperatures in continental stations, such as Larissa, Agrinio and Tripoli, are much higher than in coastal stations, such as

Chania, Irakleio and Rhodes (Figure 7.9). Thus, continental stations are more vulnerable to heatwaves than coastal stations, as expected. In Figure 7.9 the cumulative temperatures for all 13 stations can be seen for 4-days heatwave duration and 100-years return period. The expected cumulative temperatures for several heatwave durations can be helpful to farmers when planning their crops and for irrigation and water needs.

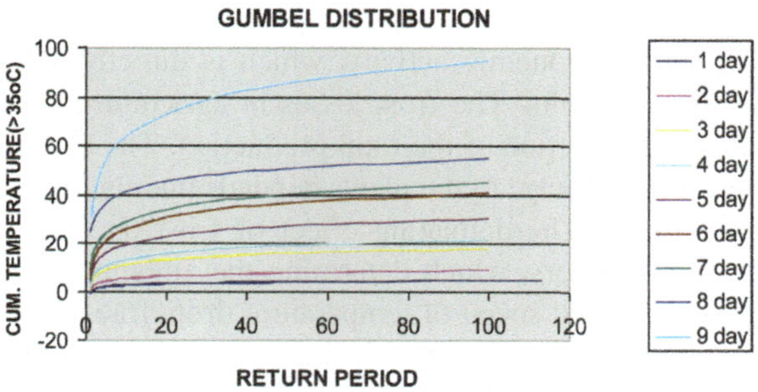

Figure 7.7. Gumbel Distribution for total heatwave episodes for Larissa station.

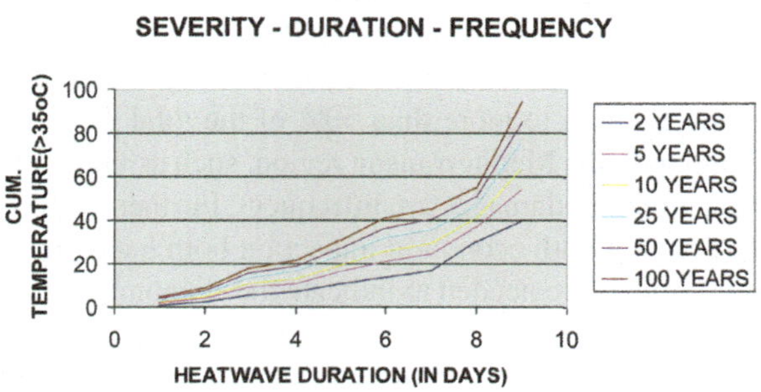

Figure 7.8. Heatwaves Severity-Duration-Frequency Curves for Larissa station.

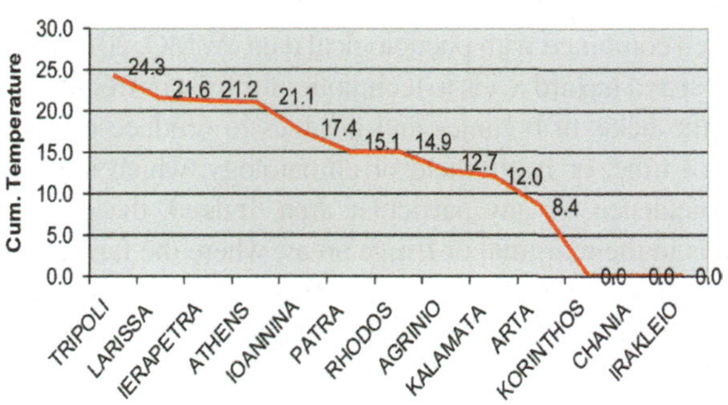

Figure 7.9. Cumulative Temperature Values for all stations (Return Period 100 years, 4-day heatwave duration).

7.5 FROST AND HEATWAVES RISK MANAGEMENT

Frost and heatwaves risk management involves impacts, protection and mitigation measures, as well as the consideration of fog mitigation.

7.5.1 Frost impacts and mitigation

7.5.1.1 Frost impacts and prevention

Agriculture is a field of economic activity, which is directly affected by frost, usually with disastrous results. The frost losses in agriculture are too large and sometimes reach the destruction of the total production. The critical temperature below which there is no damage to the plants depends mainly on the development stage of the plant tissue. The damaging effect of frost on plants is related to cellular scale activity. Factors, which determine the amount of damage, e.g. in horticulture, are the rate and speed of temperature drop, freezing conditions, the rate of temperature increase, the time of freezing, the botanical species and variety, the vegetative part and stage of the plant, as well as the age of the plant. Winter frosts contribute the most to total frost injuries. The most damaging events are the successive frosts, i.e. advection frosts followed by severe radiation frosts. Frost damages account to more than half of the total damages due to weather related phenomena in Greece (Dalezios, 2015). Contribution refers to total frost injuries. Fruit trees damages account to more than 50% of the total frost injuries. Special horticulture plantation in the Mediterranean region, such as vines, may account for 20 % of the losses, although damages are infrequent. Further studies are needed to relate reliable forecasting with economic impact on both local and regional scale. Additionally, more studies are needed to indicate the economic benefits of accurate frost forecasting for different crop categories.

The occurrence of frost has an economic effect on high-value crops, although crops can be protected. Frost-risk maps and dates of first and last frost are simple, but useful applications to agriculture. These maps are made at the macro- to mesoscale and are useful for specifying general planting dates for cereal crops and for the assessment of crop damage when combined with phenological data (WMO, 2010). The accumulated experience of frost as a hazard leads to techniques for frost or freeze prevention. These methods are in the fields of botanics and genetics to produce plants that mature in shorter periods of time, or in the field of climatology, which may define the areal extent of frost incidence for any particular area. Indeed, the greatest emphasis in frost prevention is in the marginal or fringe areas, where the farmer is willing to risk planting in the hope to produce a crop. The frost prevention techniques applied in agriculture include both treatment of the phenomenon, as well as crop resistance. It is observed that the best time to protect an area from frost, is even before planting. This means the need to select a plant species, the area and the time of planting.

Long before agriculture is considered for a particular area or crop, a climatic record or history would reveal the probabilities of frost or freeze incidence and the chance of success in the proper plants and methods are used. Additional terrain analysis helps to identify local microclimatic conditions in basins, alluvial fans, slopes, plains, or other topographic conditions. Hilly or valley sites could and should be evaluated. In the competition for the early harvest–high price crop, farmers for many years have tried to protect young plants with plastic or paper shields. However, the shield protects within limited temperature ranges. In addition, the shield may speed growth and thus "harden" the plant against chill, especially cranberries, to prevent low temperatures from damaging tender berries near the end of the growing season.

7.5.1.2 Frost protection and mitigation methods

To limit damage to agriculture from frosts various protection methods are used around the globe. These methods are classified as either passive or active. Most of the practical techniques listed below to combat the phenomenon of frost, are only effective for radiation frosts, although some can be applied to advection frosts.

(1) **Passive methods.** The passive protection measures include microclimatological research prior to any use of a field and selecting the most suitable crop for each region. Passive methods are utilized prior to the frost event in order to avoid or minimize injuries. Passive methods include the following:
   (1) Proper site selection. It is probably the most important passive protection method. Hilltops and middle portions of hillsides are most volatile to advection frosts where the temperatures observed are usually higher than down-wind sides and low spots that are sheltered from the wind. Exactly the opposite is observed during a radiation frost event.
   (2) Cold air drainage management. Cold air drained downhill can be diverted by using methods such as building a solid wall or using a wooden fence. This diversion can effectively provide protection.
   (3) Plant selection
   (4) Avoiding soil cultivation
   (5) Plant covers
   (6) Canopy trees provide protection from radiation frosts since there is an enhanced long-wave radiation downward from the trees. The effect is that temperatures are higher under these trees than in the open space.

   Passive methods of frost protections are used more widely in agricultural regions basically because are more cost-effective and more beneficial. In Greece, passive protection methods are widely utilized to help farmers combat freezing injuries (Dalezios, 2015). Low prices and overproduction is responsible for the unwillingness of farmers to invest in active protection methods.

(2) **Active methods.** Active frost protection measures focus to modify the microclimate of the field, in order to avoid low temperatures leading to

frost in crops. In other words, they cover the development of physical equipment, such as fans, heaters, brushes, sprinklers, and plant shields to modify temperatures or reduce radiation. Active protection methods are costly, since most of them are fuel dependent. Active methods are deployed during a frost night. Whatever method or combination of methods is chosen to prevent destruction by frost, the choice is usually to modify temperatures a few degrees, usually not more than four or five. The techniques involve reducing radiation, improving wind circulation, discouraging sublimation, or creating a fog or smoke cover. The physical modification of the environment in the immediate condition of a frost or freeze hazard includes the deployment of the following: wind machines, sprinklers, heaters, surface irrigation, or foggers. Specifically, wind machines and micro-sprinklers are used as active protection methods only in a small scale. For example, horizontal (conventional) blowing machines are deployed in the Argolic plain in Greece to protect citrus, namely sweet oranges and mandarins (Dalezios, 2015). Overall wind machines protect less than 10% of the citrus crop in Greece. Over-plant micro-sprinklers are used over kiwifruit. Under-plant and over-plant sprinklers are deployed in citrus regions to protect less approximately 20% of the crop.

7.5.2 Heatwaves protection and mitigation

Excessive heat possibly contributes to more illness and mortality than any other direct, weather-related cause mainly in urban areas. This is almost certainly the result of climatic modification and heat retention due to urbanization, the heat island effect, plus the pollution trapping and concentrating effects of stagnant atmospheric conditions of heatwaves, adding to those of heat stress. Death rates are also higher among the aged, especially as a result of aggravating effects on pre-existing conditions, such as heart disease or cancer.

Increased demands for air-conditioning and refrigeration can produce overloading of power supply systems during heatwaves, leading to power restrictions and breakdowns, tending to aggravate the heat-stress situation. Assessment of the causes of death in the 1995 Chicago heatwave included factors, such as inadequate warning systems, insufficient time to acclimatize, the heat island effect, an aging population, an inadequate ambulance service, and the inability of many residents to properly ventilate homes due to fear of crime or lack of resources for fans or air-conditioning (Changnon *et al.* 1996). During a subsequent heatwave in Chicago in July 1999 the death toll was much lower, because lessons had been learned from the previous event. Heatwave plans included timely warnings, activation of cooling shelters, frequent broadcasting of information and the ready availability of helplines (Palecki *et al.* 2001). In any such event, indeed in almost any hazard event, there are benefits. In this case air-conditioner sales increased, sales of ice creams and ice set records, private ambulance operators were busier than normal, people went to shopping malls and movie theaters to escape the heat in air-conditioned facilities, and merchants at lakeside outlets benefited from record attendances at the height of the heatwave. Properly spaced green areas are the most effective and aesthetically pleasing means of controlling urban temperature excess by improving ventilation and circulation and reducing heat storage capacity.

Building materials with lower heat conductivity and storage properties, and water bodies within or close to cities help to keep maximum temperatures down and encourage mixing and ventilation resulting from enhanced temperature differentials. Adequate surveillance systems are needed, to alert the public and authorities that potentially dangerous weather is imminent. A promising approach is based on the identification of high risk air masses historically associated with increased mortality (Kalkstein *et al.* 1996). A similar approach, using a weather type classification scheme developed for North America, has been used in several heat stress warning systems worldwide.

The combination of high temperatures with drought conditions is very critical for the initiation and spread of forest fires, especially when warm and dry winds prevail, which favor the expansion and dissemination of forest fires. In the middle-latitude and subtropical regions, such as Greece, the initiation of most fires occurs during midday, where the maximum daily temperatures and the minimum relative humidity are recorded (Domenikiotis *et al.* 2002). The destruction of vegetation by forest fires can have effects on the soil surface and the hydrological cycle by increasing albedo, surface runoff, reduce evaporation, increase erosion, incidence of floods, as well as it can contribute to climate change and land degradation, leading possibly to desertification. Furthermore, the gases released by the burning of biomass can contribute to a worsening of the greenhouse effect. Specifically, the summer of 2007 in Greece has been characterized by a combination of exceptionally severe drought conditions with equally severe heatwaves with very high temperatures, resulting into numerous forest and grassland fires throughout Greece for the whole summer (Tsiros *et al.* 2008).

It can be stated that, in general, droughts and heat waves have shown to be associated with the persistence of ridges or centers of high pressure systems at the middle level in the atmosphere (Dalezios, 2015). Moreover, the corresponding reduced cloud cover results in positive temperature anomalies in the lower atmosphere, which produces the middle level pressure anomaly and favors subsidence in the high level keeping the atmosphere significantly drier and more stable than normal. Studies in several areas around the world have shown that drought periods are often characterized by a large decrease in the amount of rainfall per day, by an increase in the continentality of the clouds and by a lack of rain-producing clouds. Moreover, prolonged drought periods usually result into heatwaves, which are equally associated with high pressure systems, subsidence and stability in the atmosphere and extremely high temperatures (Dalezios & Eslamian, 2016).

7.5.3 Fog mitigation

Presently, fog is known to be formed through three basic mechanisms, each of which are found to a lesser or greater extent in the creation of each fog type or classification. The basic processes are radiative, i.e. cooling, advective, i.e. cooling and/or lift, and mixing, i.e. thermodynamic. Making use of the historical empirical evidence, statistical evaluations, and the principles of physical-chemistry, researchers and application-specialists have used this information to enact two basic types of mitigation or prevention strategies with regard to fog hazards and a third option for use in other cases: (1) fog dispersion, (2) air quality management, and (3) special cases. In the first case fog dispersion has focused on relatively costly mechanical means, i.e. bulk mixing, as opposed to thermodynamic and physical methods, i.e. heating and/or seeding, that have been applied predominantly at aerodrome locations. For air quality issues, emphasis has been placed on reduced exposure, e.g. avoidance by remaining indoors or masking through the use of personal filters, reduction of contributing sources, or ventilation (similar to bulk mixing) and is often a function of the population affected, i.e. receptors, and the resources required for cost-effective implementation and the likely effectiveness.

In support of the avoidance, mitigation, or prevention of fog hazards, a variety of operational support and decision making tools and information is available to forecasters, as well as the affected and responding communities, e.g., aerodromes, emergency managers, or similar personnel. These are also linked to numerical or statistical guidance packages. These are particularly effective when tied to GIS databases and decision-support software or artificial intelligence and are more commonly used in a military or disaster-related type of response, e.g. application in catastrophe modelling. The additional deployment of meso- and microscale surface-based observing networks is expected to increase the spatial and temporal acuity of data as related to fog occurrence and its evolution, so that it may be more accurately detected, assessed, and compared with forecasts. In some instances, highway systems have been deployed to provide "instant" warning signs to alert drivers to rapidly changing visibility in fog-prone regions.

Satellite and similar remote sensing platforms offer a variety of products, namely the website of "Nighttime Fog and Low Cloud Images" as produced by NOAA-NESDIS, USA: http://www.orbit.nesdis.noaa.gov/smcd/opdb/aviation/fog.html, and provide a gross estimate of fog occurrence, severity, and coverage by channel differencing, as well as through examination of sounder data to construct vertical and near-surface profiles of temperature and moisture in the atmosphere. This information covers fog after its formation and thus allows tracking its movements and evolution. Although microwave sensors and ground-based radar, lidar, and profiler platforms may offer additional information and operational support, none are presently suited to fog detection or prediction, although there are operational automated highway alert systems. Additional information on atmospheric chemistry and structure is also available through several remote sensing platforms, and they provide a real-time observation profile of the physical-chemistry of the atmosphere as related to fog, which also involves a diagnosis of the pre-fog environment.

7.6 SUMMARY

In this chapter, frost and heatwaves, along with fog, have been presented, which are considered as hazards. At first, basic concepts and several characteristics of frost and heatwaves have been described, developing thus an understanding of these temperature extremes. Then risk identification of frost and heatwaves has been examined, including frost quantification, namely modelling and forecasting methods, as well as FEWS. In addition, heatwaves quantification has been also addressed through indices and HEWS. Similarly, fog modelling and assessment has also been presented. The following section has covered risk assessment, including mainly frequency analysis of both frost and heatwaves. Finally, risk management of frost, heatwaves and fog has been considered involving protection and mitigation methods.

REFERENCES

Bagdonas A., Georg J. C. and Gerber J. F. (1978). Techniques of Frost Prediction and Methods of Frost and Cold Protection. Tech Note 157, 487, WMO.

Bampzelis D., Dalezios N. R. and Pikoulas E. (2006). Severity – duration – frequency relationships of heatwaves in Greece. In: International Conference on: Information Systems in Sustainable Agriculture, Agroenvironment and Food Technology, University of Thessaly (UTH), 20–23 September 2006, Volos, Greece, sponsored by HAICTA and EC, Dalezios and Tzortzios (eds), pp. 1055–1061.

Beniston M. (2009) Trends in joint quantiles of temperature and precipitation in Europe since 1901 and projected for 2100. *Geophysical Research Letters*, **36**, L07707.

Blanc M. L., Geslin H., Holzberg I. A. and Mason B. (1963). Protection Against Frost Damage. WMO, Technical Note 51.

Changnon S. A., Kunkel K. E. and Reinke B. C. (1996). Impacts and responses to the 1995 heat wave: a call to action. *American Meteorological Society Bulletin*, **77**(7), 1497–1506.

Croft P. J. (2013). Fog hazards. In: Encyclopedia of Natural Hazards, P. T. Bobrowsky (ed.), Springer, Dordrecht, pp. 342–346.

Dalezios N. R. (1987). Development of a watershed system using estimation theory. In: Proceedings, 3rd Greek Hydrotechnical Conf., Greek Hydrotechnical Union, 7–9 October, Thessaloniki, pp. 621–630.

Dalezios N. R. (2015). Agrometeorology: Analysis and Simulation (in Greek). Kallipos: Libraries of Hellenic Universities (also e-book), ISBN: 978-960-603-134-2, p. 481.

Dalezios N. R. and Eslamian S. (2016). Drought assessment and management for heat waves monitoring. In: Book Chapter 10 in Vol. 3 of 3-Volume Handbook of Drought and Water Scarcity (HDWS), S. Eslamian (ed.), Taylor and Francis, Oxfordshire, UK. (accepted, in press).

Dalezios N. R. and Lavrediadou E. E. (1994). Frost severity-duration-frequency relationships. In: Proceedings, 2nd Greek Scient. Conf on Meteorology, Climatology and Atmospheric Physics, 29–30 September, Thessaloniki, pp. 27–34.

Dalezios N. R. and Lavrediadou E. E. (1995). Features of frost-affected areas from digital meteosat IR images. *Advances in Space Research*, **15**(11), 123–126.

Dalezios N. R., Loukas A., Vasiliades L. and Liakopoulos H. (2000). Severity-duration-frequency analysis of droughts and wet periods in Greece. *Hydrological Sciences Journal*, **45**(5), 751–770.

Domenikiotis C, Dalezios N. R., Loukas A. and Karteris M. (2002). Agreement assessment of NOAA/AVHRR NDVI with Landsat TM NDVI for mapping burned forested areas. *International Journal of Remote Sensing*, **23**, 4235–4246.

Domenikiotis C., Spiliotopoulos M., Kanellou E. and Dalezios N. R. (2004). Mapping of temperature – related areas in Greece for the study of radiation frost. In: 7th Panhellenic Geographical Conference of the Hellenic Geographical Society, 14–17 October 2004, Mytilene, pp. 74–81.

Domenikiotis C., Spiliotopoulos M., Kanellou E. and Dalezios N. R. (2006). Classification of NOAA/AVHRR images for mapping of frost affected areas in Thessaly, Central Greece. In: International Symposium 'GIS and Remote Sensing: Environmental Applications', EC, Dalezios and Dobesch (eds), University of Thessaly (UTH), 7–9 November 2003, Volos, pp. 25–32.

EM-DAT (2012). The OFDA/CRED International Data Base, Université catholique de Louvain, Brussels, Belgium. http://www.emdat.be.

Epstein Y. and Moran D. S. (2006). Thermal comfort and the heat stress indices. *Industrial Health*, **44**(3), 388–398.

Fiala D., Lomas K. J. and Stohrer M. (2001). Computer prediction of human thermoregulatory and temperature responses to a wide range of environmental conditions. *International Journal of Biometeorology*, **45**, 143–159.

Havenith G., Fiala D., Błazejczyk K., Richards M., Bröde P., Holmér I., Rintamaki H., Benshabat Y. and Jendritzky G. (2012). The UTCI-clothing model. *International Journal of Biometeorology*, **56**, 461–470.

Hogg W. H. (1971). Spring frosts. *Agriculture*, **78**(1), 28–31.

Höppe P. (1999). The Physiological equivalent temperature in a universal index for the bio-meteorological assessment of the thermal environment. *International Journal of Biometeorology*, **43**, 71–75.

Houghton F. C. and Yaglou C. P. (1923). Determining equal comfort lines. *Journal of the American Society of Heating and Ventilating Engineers*, **29**, 165–76.

IPCC (2007). Climate change 2007: impacts, adaptation and vulnerability. In: The Intergovernmental Panel on Climate Change, M. L. Parry, O. F. Canziani, J. P. Palutikof, P. J. Van Der Linden and C. E. Hanson (eds), Cambridge University Press, Cambridge, UK, p. 976.

Jendritzky G. and Birger T. (2009). The thermal environment of the human being on the global scale. *Global Health Action* **2**. Online: globalhealthaction.net. doi:10.3402/gha.v2/0.2005.

Kalkstein L. S., Jamason P. F., Greene J. S., Libby J. and Robinson L. (1996). The Philadelphia hot weather – health watch / warning system: development and application, summer 1995. *Bulletin AMS*, **77**(7), 1519–1528.

Kalma J. D., Laughlin G. P., Caprio J. M. and Hamer P. J. C. (1992). Advances in Bioclimatology. The Bio Climatology of Frost, **Vol. 2**, Springer, Berlin.

Kunkel B. A. (1980). Controlling fog. *Weatherwise*, **33**(3), 117–123.

Mayer H. and Höppe P (1987). Thermal comfort of man in different urban environments. *Theoretical and Applied Climatology*, **38**, 3–49.

Palecki M. A., Changnon S. A. and Kunkel K. E. (2001). The nature and impacts of the July 1999 heat wave in the Midwestern United States: learning from the lessons of 1995. *Bulletin AMS*, **82**(7), 1353–1367.

Smith K. (2013). Environmental Hazards: Assessing Risk and Reducing Disaster. 6th edn. Springer, Berlin, Heidelberg, New York. ISBN0-415-22463-2, p. 478.

Snyder R. L. and Paulo de Melo-Abreu J. (2005). Frost Protection: Fundamentals, Practice, and Economics, Vol. 1. FAO Environment and Natural Resources Series 10. Rome. http://www.fao.org/docrep/008/y7223e/y7223e00.htm.

Stolwijk J. A. J. and Hardy J. D. (1977). Control of body temperature. In: Handbook of Physiology, Section 9, Reactions to Environmental Agents, H. K. Douglas (ed.), American Physiological Society, Bethesda, MD, pp. 45–69.

Tetzlaff G. (2013). Heat waves. In: Encyclopedia of Natural Hazards, P. T. Bobrowsky (ed.), Springer, Dordrecht, pp. 447–451.

Tsiros E., Bampzelis D., Domenikiotis C. and Dalezios N. R. (2008). The chronicle of the heatwave and forest fire of Thessaly in June 2007. In: 9th Hellenic Conference of Meteorology, Climatology and Atmospheric Physics, Arist. Univ. of Thessaloniki (AUTH), sponsored by AUTH and Association, 28–31 May, Thessaloniki, pp. 953–960.

Ventskevich G. Z. (1958). Agrometeorology. Translated from the Russian by the Israel Programme for Scientific Translation, Jerusalem, 1961.

Webb L. and Snyder R. L. (2013). Frost hazards. In: Encyclopedia of Natural Hazards, P. T. Bobrowsky (ed.), Springer, Dordrecht, pp. 363–366.

Weihs P., Staiger H., Tinz B., Batchvarova E., Rieder H., Vuilleumier L., Maturilli G. and Jendritzky G. (2012). The uncertainty of UTCI due to uncertainties in the determination of radiation fluxes derived from measured and observed meteorological data. *International Journal of Biometeorology* **56**, 537–555.

WMO (2010). Guide to Agricultural Meteorological Practices. WMO – No134, p. 799.

Yaglou C. P. and Minard D. (1957). Control of heat casualties at military training centers. *American Medical Association Archives of Industrial Health*. **16**, 302–316.

Chapter 8

Climatic Hazards and Health

Nicolas R. Dalezios, Panagiotis T. Nastos and Antonia N. Daleziou

8.1 CLIMATE AND CUMULATIVE HAZARDS

There is a large number of climate hazards with several and important impacts, which justifies the current concerns, as well as the future challenges for the global climate (Hobbs, 2005). Indeed, there are already very large differences in the magnitude and nature of hazard losses in several parts of the world, which are attributable not only to atmospheric and environmental, but mainly to socioeconomic factors (Degg, 1992). It is recognized that there is an increasing emphasis, which has shifted to the nature and alleviation of impacts and involves frequencies, magnitudes, causes, effects and behavior of climatic hazards, as well as perception, planning, preparedness, mitigation and control (UNISDR, 2015). Nevertheless, identification and assessment of hazardous events is not an easy task, although certain criteria are normally used, which may include disruption of services or communications and transportation, as well as property damage and economic loss, among others.

In general, it is not easy to distinguish between atmospheric and non-atmospheric factors, which cause and produce climatic hazards (Maarouf & Munn, 2005). Indeed, increasing diachronic losses emphasize on the significance of socioeconomic factors resulting in vulnerability of communities to hazard events. It is mentioned that a broad distinction can be made between tropical cyclones affecting large areas and severe local storms, along with their associated weather extremes, which involve a sudden impact of very large energy amounts discharged over relatively short periods. On the other hand, specific hazard features result into cumulative hazards when they reach or exceed threshold magnitudes (Gentilli, 1979). Typical examples are heatwaves, cold spells, flood-producing rains, frosts, fogs, droughts, high winds, snow and ice associated with extratropical low-pressure systems, as well as climate change impacts. Furthermore, certain climatic hazards originate from human activity. Indeed, these hazards include biological hazards, impacts to human health, the possible risk of accidental modification of climatic patterns, as well as acid rain impacts on natural ecosystems.

8.1.1 Climate hazards

Potential atmospheric hazards are considered thunderstorms, tornadoes, tropical and extra-tropical cyclones, lightning, hail, snow, drought, fog, temperature extremes, strong winds, air pollution, as well as climatic change and its impacts. Hazards may arise either from single-element extremes, such as excessively low temperatures causing physiological cold stress or from a combination of elements, such as tropical cyclones with strong wind, storm surge and torrential rain, which cause threats to properties and the population. Annual global economic losses attributed to meteorological disasters have indicated an increase from the 1960s to the early 1990s to almost $90 billion, whereas insured losses increased to over $50 billion (Bruce, 1994).

Sudden-impact hazards. **Tropical cyclones** can be the most deadly and harmful storms on Earth. They are usually characterized by low predictability and fast movement. The major threats to properties and population are caused by three distinct hazards, namely torrential rain, storm waves and surges and gust winds. **Severe local storm** hazards are usually widespread, not easily predictable, very difficult to avoid and prevent, occasionally resulting in loss of lives and property damages. There may be about 2000 active thunderstorms around the world at any time, which can cause severe local damages. For example, a thunderstorm squall can develop gust winds up to 185 km/h, and are usually associated by heavy and intense rainfall, large hail, or lightning. Isolated storms usually affect small areas, however, there may be several such storms at any time in a particular region. It is recognized that accurate forecasting is significant due to their potential association with flash floods, tornadoes, lightning, downbursts and strong winds.

8.1.2 Cumulative hazards

Several atmospheric hazards and disasters are caused by cumulative events, where each of them is not usually considered hazardous. For example, several successive very hot days can prove disastrous, especially in areas with not frequent heatwaves in association with high atmospheric humidity. On the other hand, one dry day or even one dry year is not necessarily considered a drought, but successive and abnormally dry years can cause disastrous environmental and societal impacts.

Heat and cold. Continuous and excessive heat may result into more illness and mortality than any other weather-related cause (Nastos & Matzarakis, 2012; Kalkstein and Greene, 1997). Indeed, heat-related death rates are usually higher in urban areas than in rural areas. It is stated that this is essentially the result of climatic modification and heat retention in urban areas, the heat island effect, as well as the pollution trapping and concentrating effects of stagnant atmospheric conditions of heatwaves, along with heat stress. Needless to say, death rates are usually higher among the aged. Moreover, effective surveillance systems are required, to alert the public that potentially dangerous excessive hot or cold weather is imminent. Indeed, an acceptable surveillance system could be based on the identification of high risk air masses historically associated with increased mortality (Kalkstein & Greene, 1997). Specifically, synoptic weather classification schemes have been used in heat stress and cold warning systems worldwide. Nevertheless, accurate minimum and/or maximum temperature forecasts can also help the grower for decision making about protection measures against frost or heatwaves (Dalezios & Eslamian, 2016).

Snow and ice. The snow hazard is usually affected by a number of environmental and weather conditions, such as wind, topography or even road surface materials. Indeed, on heavy traffic conditions, even a shallow snow cover, say 20 mm, may be sufficient to affect traffic flow, resulting into potentially higher costs due to work losses, late delivery of goods, losses of consumer sales and losses in public transportation, in addition to increasing accident and casualty rates. Similarly, ice storms can result into major social and economic disruption. Specifically, aircraft icing has been identified as the main cause of several aircraft accidents during winter storms.

Drought. There is not universally accepted definition of drought, which can be defined only in terms of a particular need. Indeed, among the extreme meteorological events, drought is possibly the most slowly developing and long lasting event. The driving factor is the accumulated lack or deficiency of precipitation, which may happen quickly or it may take months before the impacts become apparent. Climate is changing and droughts are becoming more frequent and/or more severe. However, there have always been droughts and existing records show that such events are part of the variability of nature (Landsberg, 1982). Moreover, society must be prepared to cope with the impacts of drought at any time. Impacts in the past have been exacerbated by the absence of coping mechanisms, which justifies the need for planning drought preparedness and mitigation measures.

Fog. Fog is a hazard in transportation over land, sea, and air. Indeed, many multiple-vehicle accidents occur when visibility is severely restricted on highways. Moreover, the aviation industry loses millions of dollars each year when fog causes aircraft diversions and delays, inconveniencing thousands of passengers and incurring increased operating costs. It is known that visibility is one of the most difficult meteorological phenomena to forecast. Needless to say, several objective fog forecasting methods have been developed, however, they are not yet operational.

Air pollution. Many studies have pointed to a relationship between air pollution and ill-health, specifically for diseases, such as bronchitis and lung cancer, and less so for cardiovascular ailments and non-respiratory tract cancers (Lave & Seskin,

1970; Nastos *et al.* 2010). The first noticeable effect of photochemical pollution is a stinging of the eyes due to peroxyacetylnitrate gas. Moreover, carbon monoxide may cause drowsiness in low concentrations, extending to severe headaches, nausea, and collapse in high concentrations. Similarly, sulfur dioxide can lead to infections of the lower respiratory tract, especially among the elderly, the very young, and those already weakened by illness. In addition, ozone forms in photochemical reactions, damages lung tissue, increases death rates as a result of swellings in lung passages, and reduces athletic performance. Furthermore, acid deposition from the atmosphere is a major worldwide environmental concern, particularly in parts of northern and western Europe, North America and China. It is known that originates from the release of sulfur and nitrogen oxides by industry and transport. Indeed, oxidation and hydrolysis of the oxides produce sulfuric and nitric acids or related sulfates and nitrates, which are transported in and eventually removed from the atmosphere in rain, dew, frost, snow, gases, particles, or fog. Possible indirect effects of the brown haze, largely composed of particulates and sulfate aerosols, can be considered the cooling of land surface, thermal inversion trapping more pollutants and an overall reduction in solar radiation, evaporation and precipitation.

8.2 CLIMATE AND HEALTH

The impact of climate on human health has been considered a significant issue due to a high association with global warming and climate change (Kalkstein, 2005). Indeed, intervention in the indoor environment in terms of heating and air-conditioning, may significantly affect the atmospheric and microclimatic conditions, where humans are exposed. Moreover, climate influences human health indirectly through its effects on several environmental components, such as hydrological cycle, ecosystems, food species, as well as disease agents and vectors.

Human biology can deal with climate variability in terms of degree or rate of change. However, large short-term fluctuations in weather can cause acute adverse effects, usually resulting into increased death rates (WHO/WMO/UNEP, 1996). Indeed, it is quite common the notion of a "temperature threshold", which represents the temperature beyond which human health declines significantly (Auliciems *et al.* 1998). It is stated that these threshold temperatures are relative rather than absolute with varying frequency, implying that the notion of a "heatwave" is relative on an interregional scale and depends on the exceedance frequency of the local threshold. Specifically, case studies on summer mortality indicate that several cities in temperate climates, where hot and severe, but infrequent, weather conditions, show a sharp rise in total mortality during unusually hot weather (Kalkstein & Greene, 1997). As a result, the physiological and behavioral "shock value" of very high temperature is significant. In tropical cities, however, the hottest periods are less unique, as there is little variation from the mean, which may result into reduced impact of very hot episodes on human health. Moreover, cold-related mortality and morbidity shows similar findings, however, infectious agents and their quick spread prevail for winter health. Nevertheless, weather variability is the most significant factor in developing winter weather health problems (Keatinge *et al.* 1997).

A synoptic climatological classification methodology is usually followed to identify meteorological conditions that exceed certain threshold values. The synoptic approach identifies "air masses", which affect individuals. As a result, the interaction of several meteorological factors can be assessed simultaneously, as well as their impact on human health can also be considered. In addition, there is evidence that the elderly and very young people or individuals living in poverty, as well as urban populations in developing countries, such as China and northern India, are particularly vulnerable to heat stress. Moreover, poor housing conditions, the exacerbation of stress because of the "urban heat island", and the lack of access to air-conditioning are usually considered primary causes (Mestel, 1995). Despite the fact that the most direct impact of heat stress on the human body is the onset of heat exhaustion or heat stroke, there is evidence that morbidity and mortality increases in connection with hot weather seem to be associated with a variety of causes. Typical examples are deaths from cardiovascular, respiratory, and immune system disorders, as well as accidents, which appear to increase during stressful weather conditions. On the other hand, heat stroke and heat exhaustion represent only a small proportion of the mortality increase (IPCC, 1996).

Needless to say, direct exposure to hostile weather conditions, such as lightning strikes and tropical cyclones, can cause death to large numbers of people very rapidly. Nevertheless, the largest direct weather-related killer is considered extreme heat and cold. Indeed, the combination of temperature, wind and humidity produce an "apparent temperature", which is the perceived temperature to the human body (Steadman, 1984). Specifically, the body, if exposed to heat, can increase radiant, convective, and evaporative heat loss through methods, such as perspiration and vasodilation, i.e. enlargement of blood vessels (Diamond, 1991). Moreover, several days with continuing exposure can cause acclimatization to oppressive conditions (Kilbourne, 1992).

It is stated that the relationship between weather conditions and cold-related mortality is not clear. In particular, the existence of a threshold temperature is less evident in winter than in summer. Indeed, although deaths from cardiovascular disease occur in both seasons, influenza, pneumonia and increased mortality from a variety of accidents appear to be of greater significance in winter than in summer (Langford & Bentham, 1995). In fact, the population in temperate climates may be less accustomed to extreme cold, whereas in colder climates, behavioral responses, such as cold avoidance, may be a dominant thermoregulatory process. In summary, the relationship between climate and human health is multifaceted and complex. Nevertheless, several aspects of interest include the potential synergism between weather, pollution and human health, or stratospheric ozone depletion and sea level rise.

8.2.1 Climate change and health

Climate change is a sign of unsustainable conditions, which justifies an ecological approach of public health. Indeed, population health is a long-term success index for effective management of natural and social environments. A global climate change is expected to affect the structure and functioning of many ecosystems, as well as the biological health of many plants and living organisms. Typical examples are the retreat of glaciers and sea ice, the earlier occurrence of bird-nesting and insect migrations (Michael & Woodruff, 2005). Nevertheless, health impacts and vulnerability in human populations are expected to vary as a function of available resources, local conditions, technological advances and implementation of effective administrative measures. Needless to say, it is anticipated that several health impacts could be beneficial. For example, the seasonal wintertime peak in deaths that currently occurs in temperate regions is expected to be reduced due to milder winters. Moreover, a further temperature increase in hot regions could reduce the viability of disease-transmitting mosquito populations, which currently occurs. There are several health impacts of climate change, which may cause climate-induced changes in the frequency or severity of health risks. For example, the mortality levels during heatwaves, the impacts of economic dislocation and possible population displacement, the productivity reduction of local agro-ecosystems, the range and seasonality of infectious diseases, or the health consequences of altered freshwater supplies. Nevertheless, at the present time, there is little empirical evidence that climate change has already started to affect human health. Table 8.1 presents the potential health impacts of climate change. Moreover, Figure 8.1 shows a schematic diagram of pathways by which climate change affects health.

Table 8.1 Potential Health Consequences of Climate Change.

Threat	Consequence
Increased frequency and intensity of heat waves	Increased mortality from heat waves
Altered distributions of aeroallergens	increased frequency and severity of allergic diseases and symptoms
Altered distributions of vectors of infectious	Spread and increased frequency of diseases infectious diseases
Increased air pollution	Increased morbidity and premature mortality
Changing agricultural yields	More undernourished people in low-income countries
Social and economic disruptions, displacement	Various, with acute and chronic consequences

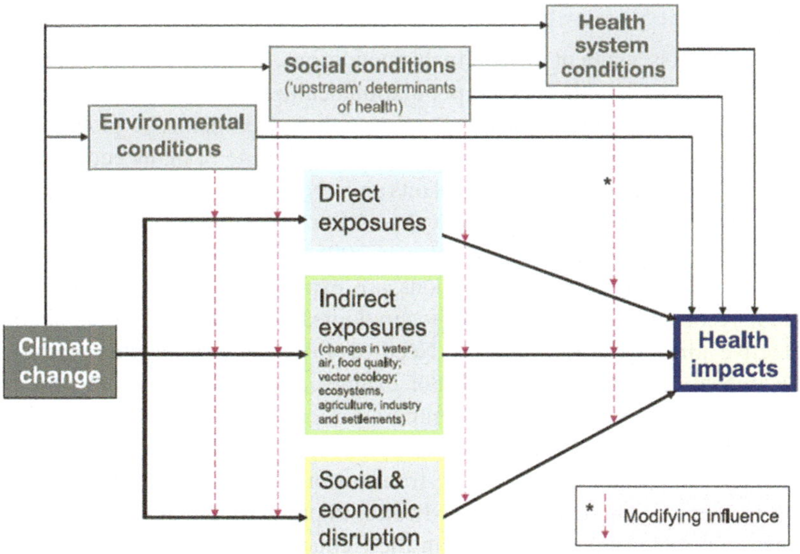

Figure 8.1 Schematic Diagram of Pathways by which Climate Change Affects Health, and Concurrent Direct-Acting and Modifying (Conditioning) Influences of Environmental, Social, and Health System Factors (from Samet, 2010).

Direct impacts. Direct impacts include those, which are related to increases in extreme weather events, such as floods, tropical cyclones, storm-surges and droughts, and those, which are related to changes in exposure to very hot and very cold weather extremes, respectively. Moreover, climate change is expected to directly increase the production of certain air pollutants, such as tropospheric ozone, which is affected by both temperature and level of sunlight, as well as various aeroallergens, which contribute to the development of asthma, hay fever and other allergic disorders. Indeed, daily temperature directly affects the pattern of daily deaths and hospitalization of the population. It is recognized that death rates increase both with greater heat and greater cold. Specifically, in warm temperate and tropical regions, the overall number of temperature-related deaths is expected to increase. On the other hand, in cool temperate regions, climate change is expected to result in a decrease in winter mortality due to less severe winters, which may compensate the increases in summer heat-related mortality (Langford & Bentham, 1995). Indeed, the net balance of future changes in hot and cold effects varies substantially between different geographic regions, such as between urban and rural populations, and the states of economic development. Nevertheless, although there is potentially great impact of such events on deaths, injuries and consequent diseases, such as infections, malnutrition and mental health disorders, an estimate of health impacts, is still considered indicative.

Indirect impacts. The indirect health impacts are specifically related to changes in regional food-producing systems and changes in the transmission patterns of infectious diseases. It is anticipated that, in the longer term, it is probable that these indirect impacts on human health would have greater magnitude than the more direct impacts. Indeed, these impacts are caused by ecological mechanisms. Moreover, for vector-borne infectious diseases the geographical range and abundance of vector organisms are affected by several meteorological parameters, such as temperature, precipitation, humidity, surface water and wind, biotic factors, such as vegetation, host species, predators, competitors, and parasites, as well as human interventions. It is true that the rate of maturation of the pathogen within the vector is sensitive to temperature. This means, for example, that an increase of 1°C across different parts of the temperature range would yield very different increases in transmissibility.

There is considerable uncertainty about the nature of climatic trends, which depends on the rate of greenhouse gas concentrations increase and on the particular models and scenarios used to predict future climate. Assessments of future climate scenarios based on global circulation models (GCMs) indicate that an increase in global average temperature and related changes in precipitation patterns would cause a net increase in the geographical range of malaria-transmitting mosquito species. Moreover, temperature-related changes in the life-cycle dynamics of both the vector species and the pathogens, such as protozoa, bacteria and viruses, is expected to increase the potential transmission of many vector-borne infectious diseases, such as malaria (mosquito), dengue fever and leishmaniasis (sand-fly). Models indicate that dengue fever would extend its range and seasonality. For example, a combination of heavy rainfall and high temperatures have been used in Australia to predict epidemics of Ross River virus disease, which is transmitted by mosquitoes, with an expected increase in future transmission in several regions.

Climate change is also expected to affect the transmission of waterborne infectious diseases through several mechanisms, including rainfall events, which result into flooding that causes contamination of drinking water supplies. Bacterial water-borne illnesses, such as gastroenteritis due to coliform bacteria, giardiasis, and cholera may also be affected. It is assessed that warmer temperatures would tend to increase the summer seasonal peaks of food-borne bacterial enteric infections, such as those due to Salmonella and Campylobacter. Moreover, the sensitivity of child diarrhea to variations in climatic conditions has been clearly explained. Finally, climate change is likely to affect crop yields, livestock health and resultant food products, as well as fisheries. Such downturns are expected to increase the number of malnourished people in the world, which is currently an estimated 830 million, by at least several percent overall, with higher percentage decreases in Sub-Saharan Africa and South Asia.

Effects of economic, social and demographic disruption. Several small island states in the Pacific region face situations that are likely to be experienced, at some level, in other parts of the world as well. Specifically, in these islands, there is a growing concern about sea-level rise and disturbance of natural and managed food systems, due to the depletion of freshwater stocks, arable land, and damage to coastal industries. The anticipated health impacts of economic and social disruptions due to climate change may not become evident for several decades.

8.2.2 Climate change and infectious diseases

Many of the biological organisms linked to the spread of infectious diseases are especially affected by fluctuations in climate, namely thermal and moisture parameters. Most of these fluctuations are part of regular climatic variability, since diseases show a seasonal pattern. Indeed, climate affects the vector biology in many ways. For example, warmer temperatures accelerate vectors' metabolic processes, thus affecting their nutritional needs, such as the increasing frequency of blood-feeding vectors. Moreover, high humidity prolongs the survival of most arthropods, whereas low humidity might cause some vectors to feed more frequently to offset dehydration. In addition, temperature changes may affect the distribution of many arthropod vectors, since there is a geographical limitation in the ranges of tolerance of the organisms. Furthermore, the presence or absence of breeding sites can be determined by precipitation. In summary, climate is a major factor in vector development and subsequent disease transmission. Nevertheless, early impacts of climate change on health include several infectious diseases, as well as health impacts of temperature extremes and impacts of extreme climatic and weather events.

Vector-borne diseases constitute a major concern due to potential adverse effects of climate change. Important determinants of vector-borne disease transmission include: (1) vector survival and reproduction; (2) the vector's biting rate; and (3) the pathogen's incubation rate within the vector organism. Vectors, pathogens and hosts, each survive and reproduce within a range of optimal climatic conditions, starting with the most significant, namely temperature and precipitation, whereas sea level, wind, and daylight duration are also important. Moreover, human exposure to waterborne infections occurs by contact with contaminated drinking water, recreational water, or food. Rainfall can influence the transport and dissemination of infectious agents, whereas temperature affects their growth and survival. Indeed, there are three categories of approaches for the linkages between climatic conditions and infectious disease transmission. The first examines evidence from the recent past of relationships between climate variability and infectious disease occurrence. The second looks at early indicators of already-emerging infectious disease impacts of long-term climate change. The third uses the above evidence to create predictive models to estimate the future burden of infectious disease under projected climate change scenarios. The main types of predictive models used to assess future climatic impacts on infectious diseases include statistical, process-based, and landscape-based models. It is clear that the above three types of models address different approaches.

Environmental conditions, which promote a vector or contribute to its spatial expansion, increase the potential for infection by the agent, such as warming, which expands the geographical area, where an infectious agent is present. There is empirical evidence, which links such epidemics, as cholera and Rift Valley fever, to climate change, whereas there is still uncertainty about the impacts of climate change on malaria. Nevertheless, there is significant historical evidence of the relationships between climatic conditions and infectious diseases. Specifically, malaria constitutes a major public health concern and seems likely to be the vector-borne disease most sensitive to long-term climate change, since malaria varies seasonally in highly endemic areas. It is stated that the link between malaria and extreme climatic events has long been studied in India. Specifically, excessive monsoon rainfall and high humidity has been identified early on as a major influence, enhancing mosquito breeding and survival. In addition, recent analyses have shown that the malaria epidemic risk increases significantly in the year after an El Niño event. Moreover, there are two principal diseases associated with aeroallergens, namely allergic rhinitis, also referred to as hay fever, and asthma, where climate change might increase the risk of exacerbation through altered local and regional pollen production. Indeed, warming already has caused an earlier occurrence of the spring pollen season. It may also increase the duration of the pollen season, change the spatial distribution of vegetation and possibly alter pollen production (Metz *et al.* 2007). Prolonged and intense exposure to aeroallergens could result in more severe disease, greater morbidity, and even mortality from asthma.

In summary, of a number of infectious diseases, which are closely tied to climate, there are at least three that may become particularly troublesome if global warming continues. At first, **malaria**, which is transmitted by mosquitoes, causes the death of millions annually, at the present time. The malarial parasite requires a temperature of at least 15°C to complete its development within the mosquito. Moreover, increased use of irrigation has the potential for creating numerous new breeding grounds for mosquitoes. Second, **tripanosomiasis** (sleeping sickness) is a major disease of humans and their domestic animals in Africa. Research has indicated that mortality rates of the tsetse fly show high correlation with humidity, and to a lesser extent with temperature. Studies also indicate that a very small change in temperatures can alter the range of the tsetse fly significantly. Third, **dengue fever** is widespread in Asia, Oceania, Australia, tropical America, and the Caribbean. It is caused by one of four distinct viruses, which are transmitted by a mosquito. Dengue is a particularly insidious disease, which is characterized by the abrupt onset of fever, severe headache, muscle and bone pain, and sometimes hemorrhaging of blood vessels. At present, a number of models are available to evaluate and estimate mosquito populations that transmit dengue, and are being used to control pesticide application.

8.2.3 Climate change mitigation and adaptation to health issues

Since the time scale involved is of the order of centuries or longer, it is difficult to demonstrate stress–response relationships. However, there are several cases of climatic adaptation to health issues. For example, small insects often have spherical or cylindrical bodies to reduce heat loss on cold nights. Similarly, desert plants and animals have minimized their water consumption and water losses through physiological mechanisms. Moreover, bird, insect, and animal migrations depend on the regularity of annual climatic cycles. On the global scale, the *Gaia* theory of Lovelock (1979) states that the biosphere has played an essential role in the evolution of the Earth over a geological time-scale. Once established, the primitive vegetation cover was able to control the oxygen and CO_2 concentrations of the atmosphere, the temperature of atmosphere and oceans, as well as the salinity and pH of oceans.

Adaptation. For addressing the public health impacts of climate change, there is the need for "primary prevention". It is stated that the health sector needs to become more proactive in promoting solutions and advancing strategies for adaptation. The evidence on the health threat of global warming is expected to strengthen public health infrastructure and capacity. At the national level, the projected risks of climate change should motivate countries to enhance data systems and improve preparedness for addressing possibly more frequent and more severe disastrous weather events. However, it may be more difficult to motivate action at local levels. In general, the methods for addressing the health impacts of climate change are those of public health and disease control. Some consequences, such as allergic diseases, are expected to be managed through routine medical care, whereas others, such as morbidity and premature mortality from increased emissions of air pollution, could be addressed through regulatory mechanisms.

Adaptive measures. Public health refers to the approaches taken to protect and improve the health of communities, contrasting to clinical medicine, which addresses health and disease of individuals. Prevention is fundamental to public health, and the principles of prevention are directly linked to climate change. Several public health measures, which are required to track and mitigate the effects of climate change on health, can be in place. Several measures reflect the routine functions of public health and environmental management, whereas others have been refined to be more sensitive to particular issues of climate change and health, particularly related to heatwaves and infectious disease. Nevertheless, there are also immediate measures that can be undertaken. For example, places at risk for heat events should have warning systems in place, along with programs to reduce the consequences of thermal stress. Risk assessment methods, including disease estimation, can remain the major tools for estimating the need for implementation of adaptive strategies and quantifying their benefits, given the complexity and great variety of factors affecting the health outcomes of concern. At the national level, the health impacts of climate change could be monitored and the extent to which adaptation strategies are in place, as well as their effectiveness, could be assessed.

Temperature warning systems. The tools necessary for protecting people from heat stress are available and include extensive temperature measures and forecasting of weather conditions associated with dangerous heat stress. The numerous epidemics of heat-caused deaths have identified those who need to be protected during heatwaves. Since just heat has to be avoided, heat watch systems can be implemented and can prove to be effective. The approach involves the characterization of conditions in a particular location that are likely to produce dangerous heat stress and trigger a protective response from public health and local authorities.

Air quality monitoring. Air quality regulations or guidelines, along with extensive air quality management programs, can be adopted to control air pollution levels. Such programs have already resulted into an improvement of air quality throughout several decades in a number of developed countries. However, air pollution still poses a continued threat to public health, since a substantial proportion of the world's population is exposed to outdoor air pollutants at high concentrations.

8.3 BIOLOGICAL AND HEALTH HAZARDS

Biological hazards are defined as processes of organic origin or those conveyed by biological vectors, including exposure to pathogenic micro-organisms, toxins and bioactive substances, which may cause the loss of life or injury, property damage, social and economic disruption or environmental degradation (UNISDR, 2015). Indeed, biological hazards, also known as biohazards, refer to biological substances that pose a threat to the health of living organisms, mainly of humans. The biological element can occur naturally in the environment most of the time, however, they frequently impact human populations when favorable conditions exist. Such conditions include parasites, viruses, bacteria, fungi and protein. In general, there are three major routes of entry for these micro-organisms into a human body, namely through the respiratory system, transmission through contact with body fluids of the infected or contact with contaminated objects (UNISDR, 2015). Nevertheless, the harmful effects posed to human health by these biological hazards are mainly of three types, namely infections, allergy and poisoning.

Health hazards. Health hazard is any organism, chemical, condition, or circumstance that may cause injury or illness (HaSPA, 2012). Indeed, health hazard is considered a danger to health resulting from exposure to environmental pollutants, such as asbestos or ionizing radiation, or to a life-style choice, such as chemical abuse or cigarette smoking. Moreover, health hazard can be any substance that causes measurable changes in the human body. Specifically, the public and people exposed at work place must be informed of the possible change in body function and its symptoms. Regarding chemicals, a substance is considered a health hazard if it can be documented that acute or chronic effects may occur in connection with use of or exposure to that chemical. In fact, development and, in particular, industrialization have made significant positive contributions to health, including greater personal and social wealth, as well as vastly improved health and education services, transportation and communication (UNISDR, 2015). In general, on the global scale, people are living longer and are healthier than they were centuries or even decades ago. However, industrialization has also had adverse health impacts for both workforces and the general population. Nevertheless, these have been caused either directly by exposure to safety hazards and harmful agents, or indirectly through environmental degradation locally and globally.

Environmental and occupational health. Environmental health hazards, such as occupational health hazards, could be biological, chemical, physical, biomechanical or psychosocial in nature (UNISDR, 2015). Indeed, environmental health hazards include traditional hazards of poor sanitation and shelter, as well as agricultural and industrial contamination of air, land, water and food. These hazards cause health impacts, ranging from catastrophic direct effects, to chronic effects, to subtle, as well indirect effects. There are several examples of environmental diseases, some of which are not easily detectable diachronically. Meanwhile, over a billion people in the world lack access to safe drinking water and over 600 million are exposed to ambient levels of sulphur dioxide that well exceed recommended levels. Moreover, the pressure on agriculture and food production, since population and per capita demand increase, is expected to result into a greater burden on the environment. Environmental health impacts, thus, include the indirect effects of industrial disruption of adequate food and housing, as well as the degradation of the global systems on which the health depends.

The main link between the workplace and the general environment is that the hazard source is usually the same, whether it is an agricultural or an industrial activity. In order to control the health hazard, a common approach may be used and considered effective. Indeed, if an acceptable result or product can be produced with a less toxic chemical, the choice of such a chemical can reduce or even eliminate the health risk, such as the choice of non-chemical pest-control methods. At the present time, it is well recognized that the scientific knowledge and training required to assess and control environmental health hazards use basically the same skills and knowledge required to address health hazards within the workplace (HaSPA, 2012). Toxicology, epidemiology, occupational hygiene, ergonomics, safety engineering are the basic environmental tools, along with risk assessment and management. Thus, occupational and environmental health are strongly linked by common and well established methodologies.

In summary, occupational and environmental health are associated and linked by common features as follows: (1) the source of the health threat is usually the same; (2) common methodologies exist particularly in health assessment and exposure control; (3) occupational epidemiology contributes to the knowledge of environmental exposure impacts; (4) occupational disease has effects on well-being at home and the community, and also environmental pathology has effects on worker productivity; (5) there is a scientific need to consider total exposures in order to determine dose-response relationships; (6) there is efficiency in human resource development and utilization gained by such a linkage; (7) there are improvements in exposure control decisions; (8) there is greater consistency in standard setting facilitated by the link; and (9) the link between environmental and occupational health enhances the incentive for rectification of hazards to both the workforce and the general public.

8.3.1 Classification of biohazards

Biological substances are classified as hazardous due to toxicity. Indeed, toxicity is one of four factors used to classify hazardous wastes and refers to the degree of damage the substance can do to a living organism, i.e. the effect on the whole organism. Common examples of biological hazards include: malaria, dengue fever, meningitis, influenza, pest infestations, zoonoses, such as HIV, H5N1 virus (Bird flu), H1N1 (Swine Flu), the plague, anthrax, cholera, leptospirosis and medical wastes. The United States Centers for Disease Control and Prevention (USCDC, 2009) categorizes various diseases in levels of biohazard, ranging from Level 1 being minimum risk, to Level 4 being extreme risk. These levels are briefly described as follows.

Biohazard Level 1: Bacteria and viruses, as well as some cell cultures and non-infectious bacteria. At this level precautions against the biohazardous materials in question are minimal, most likely involving gloves and some sort of facial protection.

Biohazard Level 2: Bacteria and viruses that cause only mild disease to humans, or are difficult to contract via aerosol in a lab setting, such as hepatitis A, B, and C, some influenza A strains, Lyme disease, salmonella, mumps, measles, scrapie, dengue fever, HIV.

Biohazard Level 3: Bacteria and viruses that can cause severe to fatal disease in humans, but for which vaccines or other treatments exist, such as anthrax, West Nile virus, tuberculosis, typhus, yellow fever, or malaria.

Biohazard Level 4: Viruses that cause severe to fatal disease in humans, and for which vaccines or other treatments are not available, such as Ebola virus and other hemorrhagic diseases.

8.3.2 Pandemics

A pandemic is an epidemic occurring on a scale, which crosses international boundaries. Indeed, a pandemic is an epidemic of infectious disease that has spread through human populations across large regions, such as continents or even worldwide. A widespread endemic disease that is stable in terms of how many people are getting sick from it is not a pandemic. Furthermore, flu pandemics generally exclude recurrences of seasonal flu.

The World Health Organization (WHO, 2005) has developed a six-stage classification that describes the process by which a novel influenza virus moves from the first few infections in humans through to a pandemic. This starts with the virus mostly infecting animals, with a few cases where animals infect people, then moves through the stage where the virus begins to spread directly between people, and ends with a pandemic when infections from the new virus have spread worldwide and it will be out of control until it is stopped. A disease or condition is not a pandemic simply because it is widespread or causes death to many people, but it must also be infectious. For example, cancer is responsible for many deaths, however, is not considered a pandemic, since the disease is not infectious or contagious. Throughout history, there have always been pandemics. The most recent pandemics include the HIV pandemic, and the 1918 and 2009 H1N1 pandemics. Current pandemics are HIV and AIDS. Specifically, in 2006, the HIV prevalence rate among pregnant women in South Africa was 29.1%. It is stated that the AIDS death toll in Africa may reach 90–100 million by 2025.

8.4 INSECT HAZARDS

Insects have been an integral part of human culture throughout history. It is true that insects cause a hazard to people. Typical example is in harvesting honey from bees, which is the biggest insect killer of humans in the developed world (Weinstein, 2013). As a result, the relationship between humans and insects is considered complex. Indeed, the emphasis is on insects that are "judged by man to cause harm to himself, his crops, animals, or his property" (Dent, 1993). A brier presentation follows of categories, such as stings and allergies, bites and disease, as well as physical and imagined hazards.

Stings and allergies. The Hymenoptera (bees, ants, and wasps) constitute the greatest direct hazard to humans of any insects, due to the fact that their stings are both venomous and allergenic. Indeed, people are at risk due to the acute pain of the venom injected with a bee sting, but also due to the anaphylactic shock, which potentially leads to death, if an allergy is developed to the same venom. It is worth mentioning that no insect actively pursues humans, with the exception of hematophagous, which are blood-feeding insects. As a result, bites and stings are largely administered in self-defense, although never proactively. In fact, swarms of bees protecting their hive can rarely cause death in humans, by the absolute number of stings, even in the absence of allergy. Specifically, worrisome are aggressive varieties commonly called "killer bees" (Weinstein, 2013). Moreover, a number of insects cause or exacerbate respiratory allergies in humans. Nevertheless, occupational exposure to particular species of insects can lead to several unexpected allergic reactions, from skin irritation to anaphylaxis. By far the majority of morbidity and mortality cases from insect stings is a result of allergic reactions, since only insect relatives, such as spiders or scorpions, possess venom potent enough for a single animal to cause death by envenomation.

Bites and disease. Such as with stings and allergies, it is not the bite of an insect that poses a hazard to humans, but the diseases that can be transmitted by biting insects. Indeed, disease transmission by biting insects constitutes a public health problem, since more people die annually of mosquito-borne disease than of any other disease. Typical examples are agricultural communities often situated near water bodies, where mosquitoes breed, as well as hungry swarms that can potentially constitute insect hazards. It is recorded that close to 300 million people catch mosquito transmitted malaria annually and about a million people, mostly children, die (WHO, 2015). Indeed, malaria parasites are multiplied in the host, causing the symptoms of infection and sometimes death. Moreover, a dozen other diseases transmitted by mosquitoes, including dengue fever. Thus, it is clear that mosquitoes are considered by far the most hazardous insects to humans globally. In addition, some other insects, such as tsetse flies, sand flies, or fleas, and insect relatives, such as ticks, have a similar ecology. As a result, hematophagous insects can be feared in areas that harbor such diseases, where precautions should be taken against being bitten. In general, the risk can be drastically reduced by simple measures, such as avoiding peak biting times, namely dawn and dusk, or wearing light colored, loose, long-sleeved clothing.

Physical impacts. Insects can pose direct hazards, which are caused by their pure body mass, and indirect hazards that result from losses in human productivity. In general, based on economic, medical, or aesthetic criteria, an outbreak can be considered to be any situation, where insect numbers reach unusually high levels, and outbreaks are driven by both abiotic and biotic triggers. Indeed, a typical example of an abiotically driven insect hazard is the interaction between climate and locusts. In particular, locusts, like all insects, are cold blooded, and, as a result, the rate of development is temperature-dependent. Moreover, locust is also humidity or rainfall dependent and female locusts require soft wet soil in order to successfully lay their eggs underground. Similarly, swarms can contain millions of individuals, being tens of kilometers long and traveling at over 100 km per day, stripping the ground of all vegetation when landing to feed. Apparently, food security is threatened by such large masses of insects, which have caused historical famines (Weinstein, 2013).

The biotic triggers of outbreaks are anthropogenic. Natural predators of insects may be accidentally removed by changing land use, such as the case of destroying bat habitats. The elimination of such useful creatures removes their nightly mosquito control service, thus producing an increased risk of disease transmission. Other common examples include the spraying of agricultural insecticides. The indirect hazard created is then one of threatening food security or other industries, such as timber production. Moreover, importation of potential pest species into new environments can also pose a hazard. In general, in anthropogenical ecosystems, there is a lack of biodiversity. As a result, insect pests are more likely to escape from the biological control, which is usually provided not only by predation, but also by interspecific competition, such as for the same food resources. Indeed, abiotic or climatic variables often enhance the impact of such biological mechanisms, since pest species can be easily adapted to exploit environments with rapidly changing conditions. Consequently, it is likely that the number and extent of insect hazards is expected to increase with global climate change.

Imagined hazards. As human society has become diachronically more urbanized, insects have gradually become more hostile. In general, at the present time, there is a dislike to insects associated with fear, which results from inadequate information and lack of more regular contact. Such reluctances cover the full spectrum of appropriate apprehension when faced with the possibility of a bee sting, ranging from clinical insect phobias to psychotic delusions of insect attacks (Weinstein & Slaney, 2004).

In summary, there is a paradox of attributing a hazardous nature to insects, both real and imagined. At first, mosquito-borne disease causes the death to over a million people every year. Specifically, in countries with these diseases, people do not seem to fear insects, but accept them as part of everyday life. On the other hand, in most developed countries, there are few or no diseases transmitted by insects, and the risk of dying from a sting is less than one in a million. However, insects are feared to the point of psychiatric disorders. Finally, it seems likely that insects will continue to be over-rated as a hazard to human health and wealth.

8.5 EPIDEMIOLOGY OF DISASTERS

Epidemiology is defined as the study of disease distribution, as well as their risk factors, within populations. The epidemiology of disease in natural disasters is affected to a large extent by the capacity of health services (Burnham, 2013). Moreover, communicable diseases are the diseases that can be transmitted either through direct contact with an infected individual or indirectly through a vector, whereas non-communicable diseases are non-transmissible and include chronic illnesses. For example, communicable diseases that are observed after the occurrence of natural disasters are respiratory infections, hepatitis and sexually transmitted diseases. Similarly, examples of non-communicable diseases include injuries and mental health conditions. Indeed, high density populations have a higher risk of injury and disability, following a natural disaster. On the other hand, endemic diseases are commonly present within a population. These endemic diseases can become epidemic after a natural disaster, either directly or indirectly in association with the event, as a result of the collapse of health and social services (Burnham, 2013). Indeed, during natural disasters, structural and nonstructural components of health services are at risk. For example, in Indonesia, 122 hospital and health services were lost during the 2004 tsunami, although even when facilities are left intact after a disaster, the health workforce may be incapacitated due to their own injuries or personal losses (Burnham, 2013).

Epidemics of communicable diseases are more a characteristic of Complex Humanitarian Emergencies (CHE) than of natural disasters. In CHE situations, populations are usually displaced, there are poor hygienic conditions, there is food insecurity and the access to health services is limited, thus the occurrence of communicable disease conditions is more possible (Watson *et al.* 2007). Indeed, when there is prolonged food insecurity, illness and death occur more frequently, particularly among children. Specifically, in natural disasters, the more common infections are outbreaks of conjunctivitis among people, who are temporarily housed in schools and public buildings, as well as head lice among children. In fact, populations who are forced to be displaced from their usual residence are also more likely to be at increased risk of illness and death (Wilder-Smith, 2005). Moreover, conditions, such as cholera, meningitis and dysentery occur rarely after natural disasters, but if they do occur, they can overcome completely health services and create widespread fear or even panic to the public. Furthermore, in conditions of over-crowding in temporary housing, there is an increased risk of communicable diseases, such as measles, meningitis and tuberculosis (Burnham, 2013).

It is understood that the nature of the disaster affects the type of disease transmission after a natural disaster, but the most important risk determinant is usually the duration of population displacement following the disaster. Indeed, if following a natural disaster a population gets displaced and a new epidemic condition appears, it is usually associated with population displacement (Jawaid, 2008). In addition, if conditions, such as cholera and meningitis are endemic in a location, these same conditions might also appear after a natural disaster, if the population is displaced and there are no supporting health services. Moreover, in countries, where infectious agents are the primary causes of illness and death, communicable disease outbreaks are more probable to occur after a natural disaster. Specifically, in developed countries, where populations have a higher life expectancy and chronic illnesses are the primary cause of death, complications of non-communicable diseases, such as diabetes, may increase after disasters, if patients get cut off from their medications (Burnham, 2013). Indeed, populations, who are physically exhausted, may become more susceptible to disease outbreaks following a disaster. Communities with low immunization coverage are also at higher risk of outbreaks after a natural disaster. Typical example is Indonesia, where there was a tetanus outbreak following the tsunami of 2004, due to poor immunization coverage (Jeremijenko *et al.* 2007).

8.5.1 Diseases associated with each type of disaster

Earthquakes may create a demand for injury care and occasionally complex orthopedic procedures, although usually they do not lead to direct increase of disease transmission. On the other hand, flooding, as well as other disasters that affect water and sanitation conditions, are more likely to result into disease transmission. Specifically, wells and water sources can be directly contaminated with human or animal fecal material, which can lead to an increase of transmission of fecal-oral infections. Moreover, reductions in water availability for personal hygiene can also lead to transmission of fecal-oral infections, as well as an increase prevalence of skin conditions. Indeed, contamination with human fecal material is usually more risky for disease transmission than contamination with animal fecal material (MacKenzie *et al.* 1994). In addition, flooding can indirectly lead to outbreaks of malaria and dengue fever, which is possible due to an increase in the breeding sites of vectors of disease and also since populations and vectors are coming into more close proximity. In summary, natural disasters do not usually lead to large outbreaks of epidemic diseases, although there is the possible exception of flooding (Burnham, 2013).

8.5.2 Mitigation and prevention

The need for health services increases after a natural disaster, although the capacity of health care facilities to provide services usually decreases dramatically. It is stated that most countries have national disaster plans. However, it is recognized the significance for districts, provinces and even individual health care facilities to develop their own disaster response plans. Nevertheless, it has been documented that facilities, which have their own emergency response plans, are more able to provide services compared to facilities without plans (Burnham, 2013). Indeed, prevention of disease in public health focuses on three stages, namely primary, secondary and tertiary prevention. The overall objective of prevention strategies is to prevent the disease from occurring and in case it does occur, to limit its spread, if possible. The same basic prevention concepts are used when managing diseases after a natural disaster.

Primary prevention focuses on preventing the exposure of a population to potential pathogens, in order to avoid the occurrence of the disease. Typical examples of primary prevention strategies in controlling possible disease outbreaks after a disaster are the maintenance of sanitation conditions, the water safety ensuring, immunization coverage of the population, as well as information availability for those at risk about exposure or infection prevention. Moreover, primary prevention strategies include prevention of complications, which might occur, in persons that require regular medical treatment for their chronic diseases and prevent any exposure to hazardous substance that might be released during natural disasters (Burnham, 2013). **Secondary** prevention includes the treatment of any disease that occurs following a disaster in order to control any outbreak and limit the potential spread of infection. Indeed, in order to prevent a disease from having serious impacts, secondary prevention strategies focus on establishing standard protocols and procedures for outbreak control, as well as adopting an effective consideration and treatment of clinical disease. Specifically, the latter includes rescuing and transferring persons with surgical injuries, treating communicable diseases, such as malaria, meningitis or tetanus. Moreover, secondary prevention includes management of the health care needs of patients with chronic illnesses, such as hypertension, diabetes and heart disease, in order to prevent deterioration of existing conditions, especially in populations with large numbers of older persons or where many people are being treated for HIV or tuberculosis (Burnham, 2013).

Tertiary prevention involves the long-term treatment of conditions as a result of natural disasters. A typical example is the physical rehabilitation of persons injured during earthquakes, which may take several years.

Surveillance systems. Surveillance systems are significant in order to understand the escalating epidemiology of disease within a population as a result of a natural disaster. It is known that surveillance consist of a systematic collection of basic data, which are analyzed and interpreted in order to assist in the management of public health interventions and also to evaluate the effectiveness of the operational efforts (WHO, 2005). Indeed, such surveillance systems can schedule and organize reporting from the operation and functioning of health care facilities, or they can develop as the result of the initial quick assessment surveys. Specifically, the establishment of surveillance systems is one of the first emergency priorities in public health and it is of great importance that the collected information is directly transferred to public health action. Needless to say, by establishing efficient and well-functioning disease surveillance systems after a natural disaster, a prevention or limitation of the public health impact might be achieved, as well as an assessment of the public health response to the disaster can be conducted (Burnham, 2013).

8.6 BIOCLIMATOLOGICAL CONCEPTS AND METHODS

Bioclimatology is the study of the relationships between climate and living organisms. The field is vast and brings together scientists from many disciplines. Bioclimatology is frequently divided into human, plant (agricultural and forest), and animal bioclimatology (Maarouf & Munn, 2005; Dalezios, 2015). Other subdivisions include aerobiology, which is the behavior of airborne living material, phenology, urban bioclimatology, air pollution bioclimatology, tourism and recreation bioclimatology, mountain bioclimatology, electromagnetic and ionization bioclimatology, and bioclimatological rhythms. Bioclimatology has many important practical applications for human comfort, agricultural yields, regional land-use planning, forest management, building research, and so forth.

8.6.1 Human bioclimatology

Atmospheric variables that may affect humans include heat, cold, wind, humidity, solar radiation (especially UV-B radiation), air pollution, pressure, negative ions, electromagnetism, and biorhythms. In the case of the first six factors, the existence of stress–response relationships has been clearly demonstrated (Nastos & Matzarakis, 2012). However, with respect to the last four factors the results obtained are still controversial, even though studies began more than a hundred years ago, particularly at the great European health spas, where people went to seek relief from arthritis, respiratory ailments, and allergies. Heat stress and cold stress continue to demonstrate the most obvious effects of weather and climate on people, and will be discussed in the following paragraphs as good examples of the state of the art in bioclimatology. Warm-blooded mammals must keep their inner body temperature within a very narrow range (around 37°C) or irreparable harm ensues. The body gains heat by its own metabolism; gains (or loses) heat from (or to) its surroundings by radiation, conduction, and convection; and loses heat by evaporation (breathing and sweating). There are two types of metabolic heat stress: *basal metabolic heat* (released when the body is at rest) and *muscular metabolic heat* (released, in addition to basal metabolic heat, during periods of work or body exercise). Basic heat transfer models can predict the heat balances of simple volumes, such as spheres or cylinders. However, the human body has a complex shape and some parts of the body, such as nose, cheeks and earlobes, are fully exposed whereas other parts are covered with clothing. Therefore, the study of heat and cold stress is not a straightforward problem in thermodynamics. Recently, more advanced heat transfer models have been developed and are being tested for use in a wide range of environmental conditions (www.biometeorology.org/; Fiala *et al.* 2012).

Heat stress is associated with various combinations of the following conditions: (1) high muscular activity; (2) high solar radiation, particularly in the tropics at high elevations; (3) high infrared radiation, e.g. near a blast furnace in a steel mill; (4) air temperatures greater than body temperature (37°C), producing a net gain in body heat from convection; (5) high humidity, reducing the rate of evaporational cooling from the body; and (6) strong winds in combination with (4) and (5), increasing the convective heat gains of the body. The physiological(involuntary) mechanisms used to cope with heat stress are: (1) dilating of blood vessels near the surface of the skin, increasing the flow of blood near the skin and increasing the heat exchange from the body to its surroundings; (2) increased sweating, resulting in evaporational cooling; and (3) increased respiration (equivalent to panting of dogs). The voluntary mechanisms are: (1) avoiding strenuous activity; (2) avoiding direct sunlight and strong infrared sources; (3) switching to lightweight clothing; (4) changing diet to reduce basal metabolic heat production; and (5) remaining in air-conditioned buildings.

A large number of theoretical and empirical indices (Table 8.2) have been developed in order to correlate the environmental conditions to human heat sensation (Fanger, 1972), taking into consideration the heat balance of the human body (Epstein & Moran, 2006). The proper choice of heat stress indices for assessing human thermal comfort at outdoor public spaces is of particular importance for the best approach of the actual thermal comfort of visitors to these places. From the spatial distribution of the selected indices, it is possible to identify locations with favorable or unfavorable thermal comfort conditions and in association to the formation of the area, useful conclusions could be considered for potential interventions, thus improving the biometeorological conditions (Matzarakis & Nastos, 2011). During the last years, many attempts have been carried out to formulate a reliable and user-friendly index for the assessment of the physiological thermal response of the human body to the climatic conditions, but only the Physiological Equivalent Temperature (PET) and the Universal Thermal Climate Index (UTCI) seem to meet these requirements.

Table 8.2 Biometeorological indices (from Epstein & Moran, 2006).

Year	Index	Authors
1905	Wet bulb temperature (Tw)	Haldane
1916	Katathermometer	Hill *et al.*
1923	Effective temperature (ET)	Houghton & Yaglou
1929	Equivalent temperature (Teq)	Dufton
1932	Corrected effective temperature (CET)	Vernon & Warner
1937	Operative temperature (OpT)	Winslow *et al.*
1945	Thermal acceptance ratio (TAR)	Ionides *et al.*
1945	Index of physiological effect (Ep)	Robinson *et al.*
1946	Corrected effective temperature (CET)	Bedford
1947	Predicted 4-h sweat rate (P4SR)	Mc Ardel *et al.*
1948	Resultant temperature (RT)	Missenard *et al.*
1950	Craig index (I)	Craig
1955	Heat stress index (HIS)	Belding & Hatch
1957	Wet-bulb globe temperature (WBGT)	Yaglou & Minard
1957	Oxford index (WD)	Lind & Hellon
1957	Discomfort index (DI)	Thom
1958	Thermal strain index (TSI)	Lee & Henschel
1959	Discomfort index (DI)	Tennenbaum *et al.*
1960	Cummulative discomfort index (CumDI)	Tennenbaum *et al.*
1960	Index of physiological strain (Is)	Hall & Polte
1962	Index of thermal stress (ITS)	Givoni
1966	Heat strain index (corrected) (HIS)	Mc Karns & Brief
1966	Prediction of heart rate (HR)	Fuller & Brouha
1967	Effective radiant field (ERF)	Gagge *et al.*
1970	Predicted mean vote (PMV)	Fänger
1970	Prescriptive zone	Lind
1971	New effective temperature (ET*)	Gagge *et al.*
1971	Wet globe temperature (WGT)	Botsford
1971	Humid operative temperature	Nishi & Gagge
1972	Predicted body core temperature	Givoni & Goldman
1972	Skin wettedness	Kerslake
1973	Standard effective temperature (SET)	Gagge *et al.*
1973	Predicted heart rate	Givoni & Goldman
1978	Skin wettedness	Gonzales *et al.*
1979	Fighter index of thermal stress (FITS)	Nunneley & Stribley
1981	Effective heart strain index (EHSI)	Kamon & Ryan
1982	Predicted sweat loss (msw)	Shapiro *et al.*
1985	Requires sweating (SWreq)	ISO 7933

(*Continued*)

Table 8.2 Biometeorological indices (from Epstein & Moran, 2006) (*Continued*).

Year	Index	Authors
1986	Predicted mean vote (modified) (PMV*)	Gagge *et al.*
1996	Cummulative heart strain index (CHSI)	Frank *et al.*
1998	Physiological strain index (PSI)	Moran *et al.*
1999	Modified discomfort index (MDI)	Moran *et al.*
1999	Physiological equivalent temperature (PET)	Höppe
2000	Standard effective temperature (modified) (SET*)	Pickup and De Dear
2001	Environmental stress index (ESI)	Moran *et al.*
2005	Wet-bulb dry temperature (WBDT)	Wallace *et al.*
2005	Relative humidity dry temperature (RHDT)	Wallace *et al.*
2009	Universal Thermal Climate Index (UTCI)	Bröde *et al.*

Physiologically equivalent temperature (PET) is described in VDI (1998) guideline 3787 part I. The VDI guideline is edited by the German Association of Engineers (VDI). PET is based on the Munich Energy-balance Model for Individuals (MEMI), which describes the thermal conditions of the human body in a physiological relevant way (Höppe, 1999). This index requires meteorological information of air temperature, humidity, wind speed and all short- and long-wave radiant fluxes related to the human body by Tmrt (mean radiant temperature) and non-meteorological components of activity level, clothing type and physiological adaptation to a particular environment (usually unknown) (VDI, 1998). The PET assessment scale (Table 8.3) is derived by calculating Fanger's (1972) PMV for varying air temperatures in the reference environment using the settings for the PET reference person (height: 1.75m, weight: 75kg, age: 35 yrs and sex: male; work metabolic rate 80 W of light activity) (Matzarakis *et al.* 1999). The range of PET studies cover global to local and micro scale (Nastos & Matzarakis, 2013) (Figure 8.2).

Table 8.3 Physiological Equivalent Temperature (PET) for different grades of thermal sensation and physiological stress on human beings (Matzarakis *et al.* 1999).

PET (°C)	Thermal sensation	Physiological stress level
<4	very cold	extreme cold stress
4–8	cold	strong cold stress
8–13	cool	moderate cold stress
13–18	slightly cool	slight cold stress
18–23	comfortable	no thermal stress
23–29	slightly warm	slight heat stress
29–35	warm	moderate heat stress
35–41	hot	strong heat stress

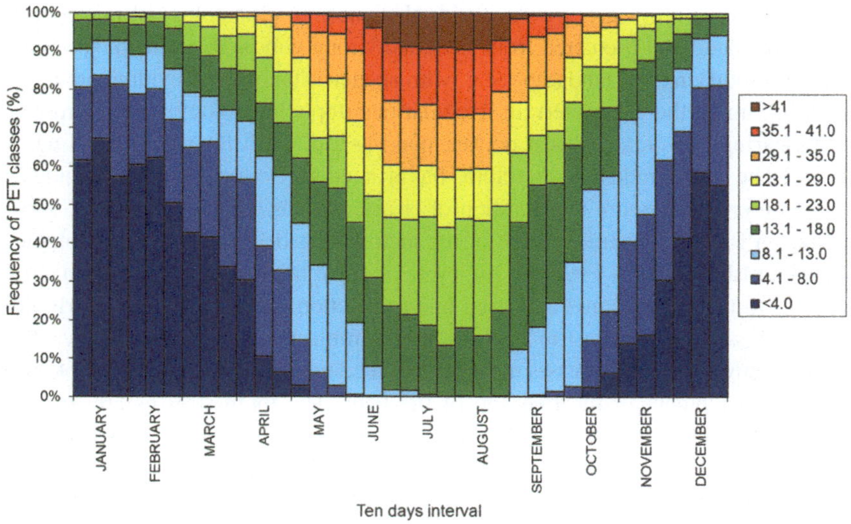

Figure 8.2 Bioclimatic diagram for Athens University Campus for the period 1999–2007; Frequencies (%) of the nine PET classes for 10-days period of the mean year (Nastos and Matzarakis, 2013).

A recent thermal index is the Universal Thermal Climate Index (UTCI) (Jendritzky *et al.* 2012). UTCI is used in daily forecasts and warnings of extreme weather, bioclimatic mapping, urban and regional planning, environmental epidemiology and climate impacts research. The UTCI represents an equivalent temperature that equals the ambient air temperature that results in the same level of 7 crucial physiological parameters, which an average human would experience in a reference environment (Fiala *et al.* 2012). The meteorological input is the same as for PET, except clothing type, which is air temperature dependent and, thus, UTCI differs from former heat budget models, such as PET, by the application of a state-of-the-art clothing model. Table 8.4 presents the UTCI assessment scale, which classifies UTCI in terms of thermal stress.

8.6.2 Plant bioclimatology

Humans have been managing (or in some cases mismanaging) renewable resources for many centuries. The goal is to increase crop and timber yields without causing long-term degradation of soils and forests. Climate has a direct effect on plants; it provides the primary source of energy; it governs the temperature of plant organs; and it controls water loss by transpiration. The main climatological stresses that adversely affect vegetation are drought, flood, hail, strong winds, and frost. However, atmospheric processes are indirectly important in seed dispersal, pest outbreaks, air pollution concentration (including acidic rains), soil erosion, and soil moisture.

Since many cash crops are grown outside their normal climatological ranges, they must be irrigated and protected from frost and strong winds. Bioclimatology assists either by modifying the landscape (e.g. with shelterbelts) or by providing advice on how to optimize the use of terrain irregularities (topoclimatology). For example, a forest clearing can create a frost pocket, making it difficult for young seedlings to survive. The forest industry, therefore, has a need to evaluate topoclimatological features in order to assure regeneration of clear-cut areas. A single example, apple-growing in Britain, illustrates some of the factors considered in a bioclimatological study (Landsberg, 1980). First, the life cycle of apple trees must be described: from seedlings to mature orchards and from bud production (late summer and early fall) through blossoming (spring) to fruiting (late summer). Next, the sensitivity of an orchard to environmental stresses must be evaluated with respect to each phenological stage of development of the tree and of the fruit. The main stresses are drought, unseasonably mild winters, spring frost, wet and cold summers, and weather sensitive pest outbreaks. These several stresses have an effect on both quantity and quality of the harvest. Agricultural and forest bioclimatology has become increasingly significant in the recent climate change debate. Agricultural lands and forests request large amounts of carbon; therefore improved management practices are needed to partially offset the rising levels of atmospheric CO_2. Climate change may also alter the pattern of food and timber production, either for the better in some regions, or for the worse in others.

Table 8.4 UTCI Assessment Scale: UTCI categorized in terms of thermal stress (Bröde *et al.* 2012).

UTCI (°C)	Stress Category
> +46	extreme heat stress
+38 to +46	very strong heat stress
+32 to +38	strong heat stress
+26 to +32	moderate heat stress
+9 to +26	no thermal stress
+9 to 0	slight cold stress
0 to −13	moderate cold stress
−13 to −27	strong cold stress
−27 to −40	very strong cold stress
< −40	extreme cold stress

8.6.3 Animal bioclimatology

Bioclimatology seek to quantify the direct and indirect impacts of climate on animals, particularly domestic ones. In the case of farm animals, the objective is to improve the quality and quantity of meats and dairy products, as well as to increase the work output of "beasts of burden". Direct effects are linked to air temperature, humidity, wind speed, and thermal radiation. These linkages affect animal health, growth, milk and wool production, reproduction, and performance, in general. For example, heat and cold stress can have marked effects on milk yield of lactating cows, depending on breed, feed intake and degree of acclimatization. Conception rates of dairy cows are also sensitive to seasonal fluctuations. Climatic extremes, such as droughts, floods, violent winds, heatwaves, and severe winter storms can result in injury and death of vulnerable animals. Indirect effects include climatic influences on quantity and quality of feed stuffs, such as pastures, forages, and grains, as well as the severity and distribution of livestock diseases and parasites. Bioclimatology also undertakes studies on wild animals to determine the effects of atmospheric stresses, such as acidic deposition, long-range transport of toxic chemicals, weather disasters, and more recently UV-B radiation and climate change. Ecologists need this information to help understand population changes, species diversity and ecosystem health.

8.6.4 Phenology

Blooming wildflowers, falling leaves, migrating birds and insects, spawning fish, hibernating animals, freezing ponds and rivers, and the like, are all influenced primarily by climatic conditions. Bioclimatology undertakes phenological studies to understand the role of climate variables in the dynamics of plant and animal natural cycles. The relationships between climate variables and the timing of these phenophases provide useful information to farmers, gardeners, horticulturists, and beekeepers, e.g. first growth and flower dates, last frost of spring or first frost of fall, planting and harvesting dates, and appearance of insect or weed pest species. Wildlife managers also need information on bird and animal migration dates, growth stage dates of various plant and animal species, dates of critical lake and soil temperatures, and breeding activities and nesting/denning dates. Phenological studies have become increasingly valuable in recent years, because trends in phenology may serve as natural indicators of global climate change (Lechowicz, 2001).

8.6.5 Aerobiology

Airborne living material is transported by the wind, sometimes for thousands of kilometers. Field naturalists are interested in bird and insect migrations, and in large-scale movements associated with the life cycles of pollen, rusts, and spores, which are of economic and health significance. For example, pollen and other allergy-causing materials may be transported from countryside to city, or from state to state, causing health problems at considerable distances from the source regions. Computerized models are often used to predict the dispersion, movement and deposition of viruses and other biological agents. Aerobiology contributes to understanding the atmospheric part of the life cycles of the pests mentioned above, and to an activity called "integrated pest management".

In summary, although bioclimatology considers stress–response relations for more than a century, nature unlocks its secrets slowly. Nevertheless, future prospects for meaningful research are promising. The main problem to be overcome is the difficulty that specialists encounter when working with specialists in other fields. For example, medical doctors and climatologists have traditionally not communicated with one another except at the non-specialist level.

8.7 SUMMARY

Chapter 8 has covered climatic hazards associated with health and health issues. At first, a conceptual description of climate and cumulative hazards has been presented with references to health. Then climate and health have been presented including climate change and health, infectious diseases and mitigation measures. The following sections have covered comprehensive presentations of biological and health hazards, insect hazards and epidemiology of disasters. Finally, bioclimatological concepts and methods are considered, involving classification and bioclimatological indices used in monitoring health-related issues.

REFERENCES

Auliciems A., deDear R., Fagence M. *et al.* (Eds.) (1998). *Advances in Bioclimatology, 5: Human Bioclimatology*. Springer, New York.

Bruce J. P. (1994). Natural disaster reduction and global change. *American Meteorological Society Bulletin*, **75**(10): 1831–1835.

Burnham G. M. (2013). Epidemiology of disease in natural disasters, In: Encyclopedia of Natural Hazards, Peter T. Bobrowsky (ed.), Springer, Dordrecht, 285–288.

Changnon S. A., Kunkel K. E. and Reinke B. C. (1996). Impacts and responses to the 1995 heat wave: a call to action. *AMS Bulletin*, **77**(7), 1497–1506.

Dalezios N. R. (2015). AGROMETEOROLOGY: Analysis and Simulation (in Greek). KALLIPOS: Libraries of Hellenic Universities (also e-book), Athens, Greece, ISBN: 978-960-603-134-2, 481 p., Nov 2015.

Dalezios N. R. and Eslamian S. (2016). Drought assessment and management for heat waves monitoring. In: Book Chapter 10 in Vol. 3 of 3-Volume Handbook of Drought and Water Scarcity (HDWS). S. Eslamian (ed.). Taylor and Francis, Oxfordshire, UK (accepted, in press).

Degg M. (1992). Natural disasters: recent trends and future prospects. *Geography*, **77**: 198–209.

Dent D. (1993). Insect Pest Management. CAB International, Wallingford, UK.

Landsberg H. E. (1982). Climatic aspects of droughts. *American Meteorological Society Bulletin*, **63**(6): 593–596.

Diamond J. (1991). *The Rise and Fall of the Third Chimpanzee*. Radius, London.

Epstein Y. and Moran D. S. (2006). Thermal comfort and heat stress indices. *Industrial Health*, **44**(3), 388–398.

Fanger P. O. (1972) Thermal comfort. Analysis and applications in environmental engineering. McGraw-Hill, New York.

Fiala D., Havenith G., Brode P., Kampmann B. and Jendritzky G. (2012) UTCI-Fiala multi-node model of human heat transfer and temperature regulation. *International Journal of Biometeorology*, **56**(3), 429–441.

Gentilli J. (1979). Atmospheric factors in disaster: an appraisal of their role. In Heathcote R.L. and Thom B.G. (eds), *Natural Hazards in Australia*. Australian Academy of Science, Canberra, pp. 34–50.

Health and Safety Professionals Alliance (HaSPA), (2012). Biological hazards: The Core body of knowledge for Generalist OHS Professionals. Safety Institute of Australia (SIA) Ltd, Tullamarine, Victoria, Australia, p. 30.

Hobbs J. (2005). Climate hazards, In Encyclopedia of World Climatology, John E. Oliver (ed.), Springer, UK, 233–243.

Höppe P. (1999). The physiological equivalent temperature – a universal index for the biometeorological assessment of the thermal environment. *International Journal of Biometeorology,* **43**(2), 71–75.

IPCC (Intergovernmental Panel on Climate Change), (1996). *Climate Change 1995: Chapter 3: impacts, adaptations and mitigation of climate change: scientific-technical analysis*. Watson R.T., Zinyowera M. C. and Moss R. H. (eds.), Cambridge University Press, New York.

Jawaid A. and Zafar, A. M. (2001). Disease and dislocation, the impact of refugee movements on the geography of malaria in NWFP, Pakistan. *Social Science & Medicine*, **52**, 1042–1055.

Jendritzky G., De Dear R. and Havenith G. (2012). UTCI - Why another thermal index? *Int International Journal of Biometeorology,* **56**, 421–428.

Jeremijenko A., McLaws M. L. and Kosasih H., (2007). A tsunami related tetanus epidemic in Aceh, Indonesia. *Asia PacificJournal of Public Health*, **19**, Spec no. 40–44.

Kalkstein L. S. and Greene, J. S. (1997). An evaluation of climate/mortality relationships in large U.S. cities and the possible impacts of a climate change. *Environmental Health Perspectives*, **105**(1), 84–93.

Kalkstein L. S. (2005). Human health and climate, in Encyclopedia of World Climatology, John E. Oliver (ed.), Springer, UK, 407–411.

Keatinge W. R., Donaldson G. C., Bucher K. *et al.* (1997). Cold exposure and winter mortality from ischaemic heart disease, cerebrovascular disease, respiratory disease, and all causes in warm and cold regions of Europe. *Lancet*, **349**, 1341–1346.

Kilbourne E. M. (1992). Illness due to thermal extremes. In Last J. M. and Wallace R. B., (eds.), *Public Health and Preventive Medicine*, 13th edn. Appleton, Norwalk, Lange, pp. 491–501.

Landsberg J. J. (1980). From bud to bursting blossom: weather and the apple crop. *Weather*, **34**, 394–407.

Langford I. H. and Bentham G. (1995). The potential effects of climate change on winter mortality in England and Wales. *InternationalJournal of Biometeorology*, **38**: 141–147.

Lave L. B. and Seskin E. P. (1970). Air pollution and human health. *Science*, **169**, 723–733.

Lechowicz M. J. (2001). Phenology. *Encyclopedia of Global Environmental Change*, vol. 2, John Wiley, Chichester, pp. 461–465.

Lovelock J. (1979). *GAIA: a new look at life on earth*. Oxford University Press, Oxford.

MacKenzie W. R., Hoxie N. J., Proctor M. E., Gradus S., Blair K. A., Peterson D. E., Kazmierczak J. J., Addiss D. G., Kim R., Fox K. R., Rose J. B. and Davis, J. P. (1994). A massive outbreak in Milwaukee of cryptosporidium infection transmitted through the public water supply. *The New England Journal of Medicine*, **331**, 161–167.

Maarouf A. R. and Munn R. E. (2005). Bioclimatology, In Encyclopedia of World Climatology, John E. Oliver (ed.), Springer, UK, 158–164.

Matzarakis A. and Nastos P. T. (2011). Human-biometeorological assessment of heat waves in Athens. *Theoretical and Applied Climatology,* **105**(1–2), 99–106.

Matzarakis A., Mayer H. and Iziomon M. G. (1999) Applications of a universal thermal index: physiological equivalent temperature. *International Journal of Biometeorology*, **43**(2), 76–84.

Mestel R. (1995). White paint. *New Scientist*, 25 March, pp. 34–37. Mills D. M. (1995). A climatic water budget approach to blackfly population dynamics. *Publications in Climatology*, **XLVIII**(2).

Metz B., Davidson O. R., Bosch P. R., Dave R. and Meyer L. A. (eds.), (2007). Climate Change 2007: Mitigation of Climate Change. Contribution of Working Group III to the Fourth Assessment Report of the Intergovernmental Panel on Climate Change. Cambridge University Press, Cambridge, UK, New York.

Michael A. J. and Woodruff R. E. (2005). Climate change and human health, in Encyclopedia of World Climatology, John E. Oliver (ed.), Springer, UK, 209–213.

Nastos P. T., Paliatsos A. G., Anthracopoulos M. B., Roma E. S. and Priftis K. N. (2010). Outdoor particulate matter and childhood asthma admissions in Athens, Greece: a time-series study. *Environmental Health*, **9**(1), article 45, 1–9.

Nastos P. T., Matzarakis A. (2012). The effect of air temperature and human thermal indices on mortality in Athens, Greece. *Theoretical and Applied Climatology* **108**(3–4), 591–599.

Nastos P. T. and Matzarakis A. (2013). Human Bioclimatic Conditions, Trends, and Variability in the Athens University Campus, Greece. AdvMeteorol 2013(Article ID 976510):8 pages, http://dx.doi.org/10.1155/2013/976510.

Samet J. (2010). Public Health: Adapting to Climate Change. Issue Brief 10-06. Publisher: Resources for the Future, Washington, D.C., p. 15.

Steadman R. C. (1984). A universal scale of apparent temperature. *Journal of Climatology and Applied Meteorology*, **23**, 1674–1687.

UNISDR (2015). Reading the Sendai framework for disaster risk reduction 2015–2030. UNISDR, Geneva, Switzerland, p. 34.

US Centers for Disease Control and Prevention (USCDC), (2009). Biosafety in microbiological and biomedical laboratories (BMBL), 5th edition. Retrieved from: http://www.cdc.gov/biosafety/publications/bmbl5 (accessed date: 10 May 2016).

VDI 3787, Part I (1998). Environmental meteorology, Methods for the human biometeorological evaluation of climate and air quality for the urban and regional planning at regional level. Part I: Climate. Beuth, Berlin 39.

Watson J. T., Gayer M. and Connolly M. A. (2007). Epidemics after natural disasters. *Emerging Infectious Diseases*, 13, 1–5.

Weistern P. (2013). Insect hazards, In Encyclopedia of Natural Hazards, Peter T. Bobrowsky (ed.), Springer, Dordrecht, 540–542.

Weinstein P. and Slaney D. (2004). Psychiatry and insects: phobiasand delusions of insect infestations in humans. In Capinera, J. L. (ed.), Encyclopeadia of Entomology. Kluwer, Dordrecht, pp. 1845–1849.

Wilder-Smith A. 2005. Tsunami in South Asia: what is the risk of post-disaster infectious disease outbreaks? *Annals of the Academy of Medicine, Singapore*, 34, 625–631.

WHO/WMO/UNEP (1996). *Climate change and human health*. McMichael A. J., Haines A., Slooff R. and Kovats S. eds. WHO, Geneva.

WHO 2005. Epidemic-prone disease surveillance and response after the tsunami in Aceh province, Indonesia. *Weekly Epidemiological Record*, 80(18), 160–164.

WHO (2015). Fact Sheet: World Malaria Report, December. http://www.who.int/malaria/media/world-malaria-report-2015/en/ . Accessed 25 Oct 2016.

Chapter 9

Wildland fires

Alan Ager, Mark Finney, Kostas Kalabokidis and Peter Moore

Forest and rangeland fires are responsible for the loss of invaluable human life, infrastructure, ecological and landscape values, and property every year (Bowman *et al.* 2013; Stephens *et al.* 2014). Wildland fire is also an important ecological process in many ecosystems, and thus socioecological conflicts arise in many settings on a global basis (Fischer *et al.* 2016). The problem of negative impacts is especially acute in regions where human development is proximal to fire-prone landscapes (Paveglio *et al.* 2009; Bailey, 2013; Darques, 2015). In recent decades, global fire activity has dramatically increased and fire risks have been amplified by a number of factors including climate change (Pausas & Fernandez-Munoz, 2012; Jolly *et al.* 2015; Westerling, 2016), exodus from rural areas (Fernandes *et al.* 2014), human development into fire-prone wildland areas (Bailey, 2013), and increases in anthropogenic ignitions (Fernandes *et al.* 2014; Salis *et al.* 2014). A good example is the fire situation in Mediterranean-type ecosystems where fire-prone natural environments in terms of vegetation, topography, weather and human geography have created a high frequency and high intensity fire regime with catastrophic impacts (Moreira *et al.* 2011; Pausas & Fernandez-Munoz, 2012; Regos *et al.* 2014).

Mitigating losses from natural disasters including wildfire is a complex problem that requires in-depth analyses of their biophysical causes, geographic distribution, and damage inflicted. Such analysis guides the development of policy settings, planning, implementation and institutional capacities to manage landscapes, and to plan and build infrastructure to reduce damage and loss. In the case of wildland fires, risk governance systems (e.g., regional prevention organizations) must incorporate actions that achieve the harmonization of land planning, technology, natural resources management, economic development, and interdisciplinary research that reduces socioeconomic and ecological impacts from natural disasters.

At the core of natural disaster management is the use of risk assessment and management, to measure and communicate uncertain future events in terms of damage and loss the consequences of the event (Corotis & Hammel, 2010). Typically, risk assessment methods are applied to the problem of low probability, high consequence, and highly stochastic events. The term risk is generally used to measure the chance of damage and loss, or more specifically the expected value of the conditional probability of the event occurring and the consequences of the event[1]. Risk assessments are most useful when predicted outcomes are uncertain, but possible outcomes can be described and their likelihoods can be estimated (Haynes & Cleaves, 1999).

9.1 WILDFIRE RISK CONCEPTS

The application of risk to wildfire issues is relatively new in comparison to natural disturbances such as floods, earthquakes, and tsunamis. The demand for risk-based tools and assessments has expanded substantially among public agencies responsible for managing impacts from wildfires on human and ecological resources, and new approaches and methods are being continually developed to support a wide range of wildfire risk management issues (Miller & Ager, 2013). The growth in risk science for wildland fire management has resulted from the growing and global incidence of uncharacteristic large wildfires that overwhelm suppression resources and inflict enormous damage including loss of human life (Cruz *et al.* 2012; Adams, 2013; Attiwill & Binkley, 2013). Wildfire risk assessments are now widely used to analyze the timing, location, and potential impacts of wildfires at a range of scales (Calkin *et al.* 2010; Miller & Ager, 2013), and to guide the development of a wide spectrum of risk management strategies, including: (1) landscape fuel reduction activities (Ager *et al.* 2010b), (2) fire prevention programs (Kalabokidis *et al.* 2016), (3) community wildfire protection planning (Jakes *et al.* 2011), and (4) managing fires to reduce future fires (Regos *et al.* 2014). Fire is a global disturbance process and thus land managers have no choice but to manage fires to manage landscapes at risk from damaging wildfires. In fact, it could be argued that fire management is a form of risk management.

There are many approaches to risk assessment in fire-prone regions around the world (Gonzalez *et al.* 2005; Roloff *et al.* 2005; Scott, 2006; Loboda & Csiszar, 2007; Catry *et al.* 2009; Martínez *et al.* 2009; Atkinson *et al.* 2010; Calkin *et al.* 2010; Verde & Zezere, 2010; Miller & Ager, 2013; Kline *et al.* 2015; Oliveira *et al.* 2016). In the United States, risk-based tools and planning frameworks support a wide range of fire management activities ranging from individual wildfire incidents (Noonan-Wright *et al.* 2011) to national policy implementation (Andrews *et al.* 2007, Calkin *et al.* 2011; Finney *et al.* 2011b). These latter advances have been enabled by high performance fire simulation systems for estimating burn

[1]From the Society for Risk Analysis (SRA) http://www.sra.org/sites/default/files/pdf/SRA-glossary-approved22june2015-x.pdf; accessed 20th July 2016.

probabilities from large fire events (Finney *et al.* 2011b) using conterminous geospatial fuels data (Rollins, 2009), automated networks of weather stations (Zachariassen *et al.* 2003), and spatial data describing social and ecological values affected by wildfires (Calkin *et al.* 2010).

Despite the many technical advances in risk assessment, it is important to recognize that evaluating risk associated with natural hazards needs to consider the interaction between biophysical and social factors (Corotis & Hammel, 2010; Fuchs *et al.* 2011), since risk results from the interaction between wildfires and human values (e.g., see Blaikie *et al.* 1994). Thus, wildfire risk management needs to consider quantitative evaluations of biophysical risk (i.e., based on prediction), as well as subjective evaluations of risk (i.e., based on perception) since both influence human behavior in preparing for and in response to predicted adverse events (Kasperson & Kasperson, 1996; Slovic, 1999; Corotis & Hammel, 2010; Lindell & Perry, 2012). Biophysical approaches to risk assessment (i.e., risk analysis) focus on statistical estimation of the probability and magnitude of adverse events, defining magnitude on the basis of economic, ecological, or social impacts (Brillinger *et al.* 2006; Corotis & Hammel, 2010; Jones & Corotis, 2012). Social aspects of risk assessment generally concern defining human risk perceptions based on qualitative social processes, including identifying community values exposed to wildfire (van Aalst *et al.* 2008; Everett & Fuller, 2011; Moser & Ekstrom, 2011; Williams *et al.* 2012; Olsen *et al.* 2013).

In this chapter, we review recent advances in quantifying fire risk factors and applying risk science to wildland fire management problems. We include a discussion of new technologies that have fostered wider application of risk assessment methods. Although we recognize the importance of both social and biophysical aspects of risk management, we focus on the latter and refer the reader to recent papers that discuss emerging social science contributions to risk assessment (Fischer & Charnley, 2012; Fischer *et al.* 2014; Charnley *et al.* 2015; Fischer *et al.* 2016).

9.2 DEFINITIONS AND STANDARDS FOR WILDFIRE RISK

Risk is a word commonly used to express the chance of bad things happening. In this section, we will define concepts and terminology either adopted from risk science or developed by the science and technical community to describe social, ecological, and economic impacts from wildfires. The wildfire community has widely applied the term risk to describe fire in terms of its timing, location, frequency, and potential impacts. We start with Society for Risk Analysis (SRA)[1] definitions: (1) risk is the potential for realization of unwanted, adverse consequences to human life, health, property, or the environment; and (2) the estimation of risk is based on the expected value of the conditional probability of an event occurring times the consequence of the event given that it has occurred. With these definitions, risk is the expectation of loss, and includes some assessment of (1) likelihood of the event, (2) expected intensity, and (3) one or more impacts depending on the susceptibility of the values of interest. The terms risk, hazard, exposure, threats, vulnerability, and fire danger are widely used in wildfire literature. The term hazard is frequently interchanged with risk, and should not be. The hazards of weather, fuel and ignition combine for a wildfire to occur. If any one of them is absent, or "low" in the case of fire weather danger, the wildfire cannot occur; the "hazard" does not create a "risk".

Wildfire exposure concerns the general description of potential wildfire activity in relation to social, environmental and economic values of concern, and is a precursor to more detailed risk analyses where losses are predicted with associated probabilities (Finney, 2005). Exposure analyses are part of risk assessments and can reveal much of the same spatial patterns without the complexity of predicting fire effects on specific human and ecological values. A threat is a type of risk (e.g., wildfire, flood, earthquake, etc.). Vulnerability describes the potential impact of threats and considers the adaptive capacity of the system. For instance, a vulnerability assessment for a wildland-urban interface (WUI) would consider adaptive capacity of a community to face wildfire threats via adaptations it has adopted such as fire proofing structures and other mitigation efforts. Fire danger describes the short-term outlook for fire occurrence (days, weeks), as derived from short-term weather forecasts (Bradshaw *et al.* 1983; Forestry Canada Fire Danger Group, 1992; Vasilakos *et al.* 2007). Fire danger ratings may also include an assessment of fire behavior (e.g., Haines Index[2]), but fire intensity and effects are not considered.

Risk assessment concerns practical application (Figure 9.1) of mapping patterns of risk in space and time. Risk mitigation is the process of ingesting risk assessment to develop strategies to manage risk. For instance, risk assessment includes comparative studies to examine how natural and anthropogenic fire hazards occur by an identification of all possible vulnerabilities through the analysis of past trends, events and developments, combined with future probability and frequency of emergencies and disasters. Risk mitigation then outlines alternative scenarios for managing a wide gamut of impacts and long-term consequences through reducing hazards and reducing the exposure of values.

Risk management is a process-based framework within which risk assessment and mitigation activities occur. In 2009 the International Organization for Standardization (ISO) released "ISO 31000:2009 Risk management – Principles and guidelines"[3]. This standard identifies that risks can have consequences in terms of economic performance and professional reputation, as well as environmental, safety and societal outcomes. It notes that managing risk effectively helps organizations to perform well in an environment full of uncertainty. The standard provides principles, a framework and a process for managing risk, and can be used by any organization regardless of its size, activity or sector to increase the likelihood of achieving objectives, improve the identification of opportunities and threats, and effectively allocate and use resources for risk treatment.

[2]http://www.wfas.net/index.php/haines-index-fire-potential—danger-34; accessed 23rd July 2016.

[3]From http://www.iso.org/iso/home.htm; accessed 24th June 2016.

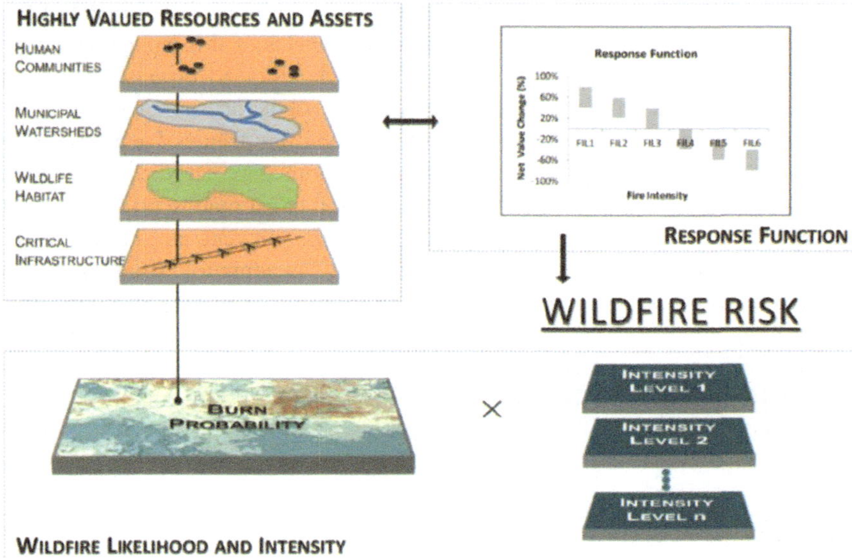

Figure 9.1 Wildfire risk assessment framework from Scott *et al.* (2013), showing the combination of probability and value change to yield risk.

9.3 QUANTIFICATION OF WILDFIRE RISK

Quantification of risk (as defined above) for stochastic wildfire events must consider the probability, intensity, and size of wildfires across a large area or at a specific location. Many factors determine the likelihood and intensity of the fire including weather, fuel type, fuel spatial patterns, ignition timing, and location. Moreover, a myriad of factors determine fire impacts to structures, people, and natural systems; this is a challenging problem and the primary reason why wildfire risk remained largely qualitative for many years. There is a formal definition of risk used by insurance companies to capture these factors and estimate the product of probability and consequence. The actuarial risk definition yields an *expectation* for net value change E[nvc] from wildfires (Finney, 2005) – that is, the weighted average of a distribution of value changes or effects caused by wildfires (Figure 9.1) which is expressed as an equation:

$$E[nvc] = \sum_i \sum_j P(f_i) RF_{ij} \tag{9.1}$$

Equation (9.1) identifies that risk is the product of fire probability at the ith intensity $P(f_i)$ and consequence of that intensity expressed through the Response Function RF_{ij} for each intensity and j_{th} value of concern. For each intensity there will be a response for a particular value (trees, houses, etc.), and across intensities there will be a distribution of those responses that can be summed to find the expected value – or expectation. If either the probability or the response (the impact of the wildfire) is zero, then there is no risk. A few key aspects of this equation need to be noted. First, the summation of products (of probability X response) yields an estimate of static risk – or extant risk – that assumes all factors remain constant for some period of time. Clearly this is not true for longtime frames (years) because vegetation changes as can other factors. The second point concerns the implied time frame to these calculations introduced by the method of estimating probabilities. It is possible to estimate probabilities for a day, a month, or a year, etc. Thirdly, there is an explicit spatial extent for the probabilities that can apply to a small area such as a grid-cell, or a larger area such as defined by a fire or watershed boundaries or other land unit. These and other issues are discussed in later sections, but aside from probabilities, the consequences part of the equation depends upon the values for which risk is of interest. Not all values are affected by wildfire equally – homes, crops, and the variety of forest types – and all respond differently across the range of fire intensities, hence the response function for the range of the *i*th intensities.

9.3.1 Wildfire likelihood

The actuarial definition of risk using the expected net value change E[nvc] is common for all kinds of perils such as house fires or earthquakes. The challenge for wildfires has been in estimating the terms in the equation, especially wildfire probabilities. As noted above, there are implied time frames, spatial extents, and other conditional properties with estimated wildfire probabilities that affect their meaning and interpretation. Obviously, wildfire damage and loss derives from fires that burn land area, and this occurs because fires grow or spread away from their ignition location. The size of fires that can burn in a particular region can be highly variable depending on where they start and how far they travel and their duration (how many hours/days they burn) (Figure 9.2). The probability in equation (1) is produced from land area burned, rather than solely from ignition likelihood. Obviously a larger fraction of land area is burned by large fires, so the probabilities of burning any place on a landscape increase with fire size. An example of this is shown in Figure 9.3 as the cumulative fraction of land area burned by historical fires of different sizes; this graph demonstrates that 95% of the land area is burned by 5% of fires. Thus, large fires strongly affect burn probability and risk, often more so than ignition frequencies, although spatial variation in ignition density can have considerable effect on burn probabilities. The risk from large fire events results from environmental factors driving fire growth, which are very difficult to account for compared to probabilities for other insurable perils which largely depend upon the properties or uses of the insured entity, home or automobile for example. The growth of fires is highly variable because of the large number of permutations of time- and space-varying factors that determine how fires grow across heterogeneous terrain, vegetation patterns, and sequences of weather (i.e., wind direction, wind speed, and moisture). In particular, the sequence of weather conditions is critical to the growth of fires in spatially varying landscapes not just the combination. For example, if weather types A, B, C each represent a fuel moisture and wind combination that produces distinctively different fire behaviors, then the sequences A -> B -> C and C -> A -> B will produce different growth and intensity patterns for a fire starting in the same location. The number of permutations from *n* weather types is *n-factorial* (*n!*) which produces, for example, 326,880 permutations from 10 weather types. Realistically, there are many more weather types, such as might be defined by 3 moisture levels, 8 wind directions, and 5 wind speeds, which would produce 120 combinations. Modeling all of these variations in weather and ignition locations to generate a complete set of fire growth realizations would be an enormous task. Thus, methods that short-cut the complete set of combinations must be devised to approximate or estimate the variability resulting from these weather sequences and ignition locations.

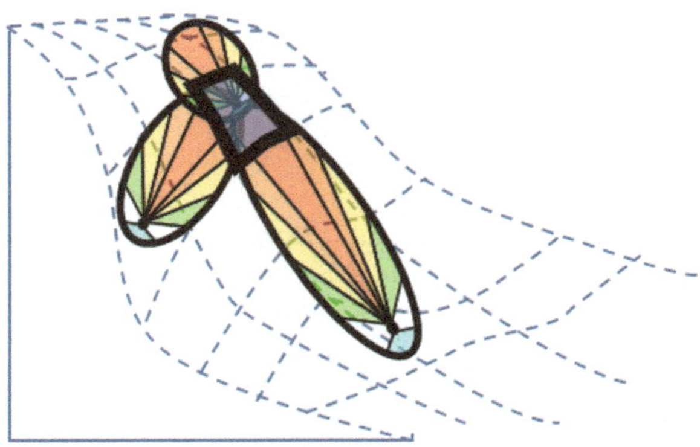

Figure 9.2 A given point on the landscape may be affected by fires starting from many surrounding locations and impacted by various intensities. Elliptical fires demonstrate a radial pattern of intensities and depend on the relative direction the fire encounters a given point on the landscape.

Two main methods have been attempted to efficiently represent the variation in weather for the purpose of using wildfire simulation to represent the 2-Dimensional (2-D) growth of wildfires and the intensity patterns within the burned areas. The simplest short-cut method is to identify the fewest weather conditions associated with the largest fires and simulate only those. This method generates *conditional* probabilities of burning, meaning that they require first the condition of having a large fire somewhere in the analysis area, and then determine the relative probability of burning one place or another rather than absolute probabilities which would imply some time frame (year, month, etc.). This is more easily justified in places where wildfire suppression is largely successful because most large wildfires will occur only under the driest and windiest conditions and burn until weather moderates (fires under moderate or mild conditions are usually suppressed). So, using historical fires as a guide, we can identify the associated burn duration and direction and wind/moisture conditions for the largest fires (Ager *et al.* 2007a; Ager *et al.* 2010a; Salis *et al.* 2013). For example, with a single very dry moisture scenario and two wind conditions, and fires burning for one day each, we may have only two scenarios. Wildfire growth can be simulated from ignition points (e.g., random, or selected from an ignition density grid) for just two weather conditions – selected according to their historical frequency – for some large number of runs. The number of fires that must be simulated depends upon the burn probabilities. Very low probabilities imply a small fraction of area burned per fire and thus many fires must be simulated (Miller & Ager, 2013).

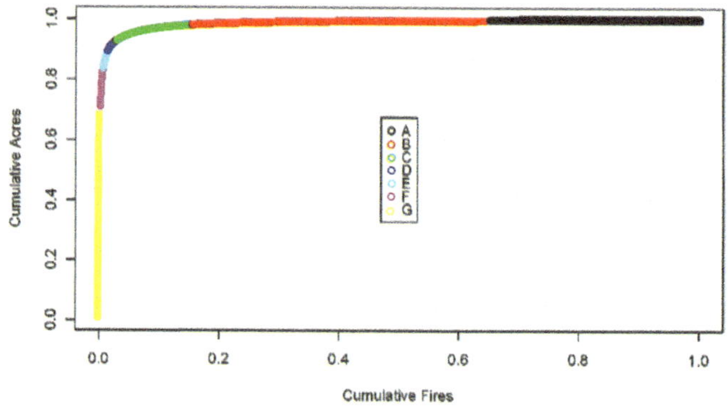

Figure 9.3 Cumulative size distribution showing that 5% of the fires are responsible for 95% of the burned area.

A second method is to stochastically sample from the permutations of weather sequences and ignition locations to generate a sufficiently large number of realizations to estimate the distribution of fire sizes. This method is considerably more complicated than the one described above, but generates absolute probabilities for a particular time period (e.g., annual). Examples of this approach are described by Finney *et al.* (2011a, 2011b) for the entire landscape. In both cases, artificial weather sequences are generated using a time-series model of a fuel moisture index. Details are found in the original literature sources, but the general strength of such an approach is that it combines ignition probability and fire growth potential to estimate absolute probability distributions that extend beyond the historical observations of fire sizes and fire frequencies.

It is difficult, if not impossible, to understand wildfire risk without having records of historical wildfires. It is also important to understand fire causes. People light fires for many reasons including practical and beneficial reasons in support of ecological processes or livelihoods; some fires are accidental, and others are deliberately set to cause damage (Moore *et al.* 2003). In some regions, wildfire ignitions are dominated by anthropogenic activities, while lightning generates most fires in others. In either case, wildfire records are the basis of almost all risk analyses by any method, either as input data for analysis and certainly for comparison of estimates. Many statistical modeling approaches have been developed to predict the probability of ignition or of large fires based on cause, weather, location, etc. (see review by Miller & Ager, 2013). Fire records are not without their own sources of uncertainty and error which are well described by Short (2014) for the United States; and errors and omissions affect all further uses of the data. Nevertheless, from historical fires we can begin to estimate the frequency distribution of fire size, including the maximum size, the seasonality of fire occurrence and burned area, the cause and location, as well as the weather conditions associated with the start and spread of each fire. Spatial information on ignition location and the extent of fire area (fire footprint) may provide information on fire spread directions and shapes, and the relative growth rate of the fires in different fuel types and terrain. Spatial information is critical to fire simulation methods described below.

There are few countries that have a full set of fire history and spatial information on ignition location. With improvements in the sophistication of the sensors being used on satellites, the numbers of satellites, and the increasing detail of the remote sensing data being collected, analyses of fire history are feasible. There are several data sets available for analysis and use including for "active fire data" (also referred to as "hotspots") that include satellite data sets, which extend back to 1981 in some cases, and repeat coverage for other remote sensing platforms. For example, the Moderate Resolution Imaging Spectroradiometer (MODIS) instrument was launched in 1999 and 2000 and for 15 years has been a source of hotspot data. The MODIS satellites and the Landsat satellite series provide area burned. Landsat represents an extensive continuously acquired collection of space-based moderate-resolution land remote sensing data, having been in operation since the early 1970s. These data are being used to generate fire records and history; for example, Indonesia has prepared burned area for the country using Landsat for the period 2000–2012 and is extending the analysis to 2013–2015.

However, having only historical wildfire records, it is possible to get a sense of the average annual burn probability by summing the area burned for all years and dividing by the total area of the land and number of years in the record:

$$P_{\text{burning}} = BurnedArea/TotalLandArea/TotalYears \tag{9.2}$$

Estimates of burn probability with this calculation are made by Littell *et al.* (2009) for vegetation zones of the western United States. Similarly, the annual ignition probability can be estimated:

$$P_{\text{ignition}} = TotalIgnitions/TotalLandArea/TotalYears \tag{9.3}$$

These calculations can also be performed for each year or decade or month for example, to estimate the variability in burn or ignition probability. There are several examples of statistical modeling of burn probability and ignition probability where historical weather indices were used to predict monthly large fire activity in the western United States (Preisler & Westerling, 2007). In other work, Preisler *et al.* (2004) analyzed ignitions and large fires using historical data on fire activity and associated weather variables. Parisien *et al.* (2012) developed a statistical model for the probability of burning from historical ignition data using climate and vegetation. Data on causes and proximity to spatial attributes (roads, etc.) can also be used in statistical modeling. If P_{burning} for a location is greater than P_{ignition} then the chances of that location burning is more from fires starting remotely than from fires starting locally. This is the most common situation and highlights the contribution of fire growth to quantitative estimates of risk.

Fire records are also necessary for estimating fire size distributions. These distributions provide fascinating characterization of the overall fire occurrence situation for an area. It is well known that fire size distributions tend to form a negatively sloped straight line when plotted on logarithmic axes (Figure 9.4). The exponent in the equation for this distribution is the slope, which is a *power function* of the fire size, and tells us about the ratio of the number of small fires to large fires. The slope of the line is flatter for areas with large fires (fewer small ones for each large one) and steeper for areas with many small fires. Many explanations have been offered for the seemingly universal power-law behavior of fire size distributions (Malamud *et al.* 1998). Some of these explanations assume that large fires are limited in frequency by the burn patterns of many small fires. However, over large landscapes (relative to fire sizes) over long-time frames (compared with the rate of fuel recovery), such interaction is not possible unless the burning rate is very high (burn probabilities > 0.2 for example). Another explanation is that opportunities for fires to grow are made by the combined patterns of the landscape and sequences of weather which are themselves distributed in power-law fashion. This was suggested by simulations for a single landscape condition (Finney *et al.* 2011a) where no fire-pattern interaction could occur, yet fire size distributions were remarkably close to historical observations.

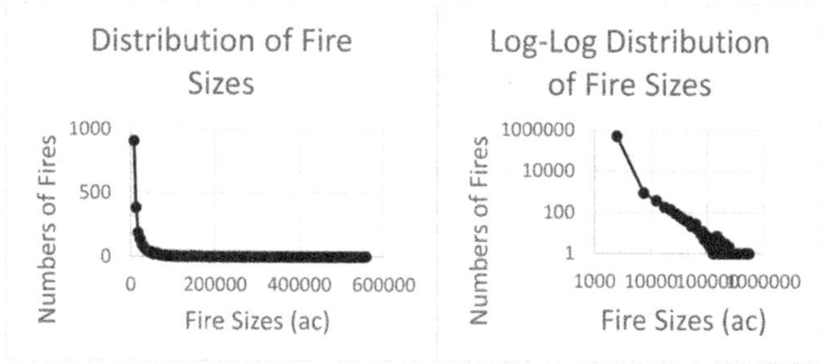

Figure 9.4 Fire size distributions on linear and logarithmic axes (data from Short, 2014).

Estimates of burn probability can come from the historical data as described above, but these reflect averages for an entire land area and contain no information about intensity. To attempt to resolve finer-scale spatial patterns of burning and intensity, computational methods are needed. There are a number of possible methods that depend on the risk being assessed in terms of the spatial and temporal dimensions and absolute or conditional probabilities. Table 9.1 provides some example approaches.

Table 9.1 Example approaches to estimating wildfire probabilities for risk assessments.

Risk Objective	Spatial Scale	Temporal Scale	Examples
1. Pixel- or Grid-Based Risk	Entire Landscape	Annual	Finney *et al.* 2011b, Ager *et al.* 2012, Ager *et al.* 2014
	Entire Landscape	Conditional (upon having a fire)	Finney 2005, Ager *et al.* 2010a, Salis *et al.* 2014, Ager *et al.* 2015
	Per Fire	Fixed Time Period (day, week, etc.)	Finney *et al.* 2011a
2. Polygon- or Fire-Based Risk	Entire Landscape	Annual	Scott *et al.* 2013, Ager *et al.* 2014, Scott and Thompson 2015
	Per Fire	Fixed Time Period (season, day, week, etc.)	Scott *et al.* 2012

The main difference between pixel- or polygon-based risk analyses is that pixel calculations are performed independent of the fire extent that burns them. Probabilities are calculated as the number of times each pixel burns in a given time period. Polygon-based analysis depends upon the contiguous areas burned by each fire, because some effects depend greatly upon the size and intensity of the burned area and pattern in a given event (hydrology, fire suppression).

9.3.2 Fire intensity

Equation (9.1) set out earlier describes risk as the product of fire probability at the *i*th intensity $P(f_i)$ and consequence of that intensity expressed through the Response Function RF_{ij} for each intensity and j_{th} value of concern. Depending upon the assets or values at a location, the intensity would be an important variable to consider. Low intensity fires are easier to suppress and may cause little damage to large trees, and buildings may be more defendable by firefighters if the intensity is low. Fire intensity is represented as a distribution since wildfires can burn a particular location at a variety of intensities depending upon the fuel available to the fire in that specific location and for a distance surrounding it, the weather at the time the fire arrives (wind speed and moisture content for example, or arriving at night vs. late afternoon) or the direction the fire approaches from (heading, backing, etc.).

Fire intensity results from the interactions of biophysical factors (physical geography) such as vegetative fuels, topography/terrain and weather (Pyne *et al.* 1996). Fire intensity is strongly influenced by the amount and condition of fuel available to burn (leaf litter, bark, leaves and branches). Many factors including the quantity, size, density, quality, continuity and moisture content of vegetation determine the availability of fuel for combustion (Heinsch & Andrews, 2010). Topography modifies the general climate over the landscape and thereby affects fuel availability. Weather is a critical element of the fire environment, as it can at times overshadow the influence of vegetation and topography. Fire weather includes wind, humidity, rainfall and temperature conditions as they influence fire ignition and behavior (Schroeder & Buck, 1970). The strong impact of fuel loadings on fire behavior, and our ability to manipulate them for risk reduction purposes has stimulated a large body of research on human impacts on fuel loadings. For example, for many years United States forest managers allowed the accumulation of large amounts of fuel in Western forests by attempting to totally exclude fire; eventually, this created conditions for very destructive wildfires that are proving impossible to contain. In some tropical forests conventional logging practices have encouraged harmful dry season fires through accumulation of large amounts of logging waste and forest drying caused by increased canopy openings. In other cases, damaging fires are a symptom of the same underlying causes that drive land use/land cover (LULC) change, forest loss, and degradation: i.e., perverse economic incentives; ill-defined or inequitable land tenure; failure to enforce laws and regulations; failure to recognize and respect customary law; lack of economic opportunities for rural dwellers living in and around protected areas; and weak or under-resourced government institutions. These factors influence how land and forests are exploited and managed, and therefore influence fuels or changes to them and affect both the likelihood of harmful wildfires to occur and their destructive potential.

9.3.3 Fire susceptibility

The Response Function RF_{ij} for each intensity and j_{th} value of concern quantifies the net change in value for a particular resource or value across the range of fire intensities. Thus, it concerns the degree to which population, built environment, and socioeconomic activities are susceptible to damage from wildfire events given a fire at a particular intensity (Chen *et al.* 2003; Miller & Ager, 2013). Within this context, fire susceptibility refers to the human and ecological geography of potential wildfire impacts. The values which may be damaged by wildfire and suffer loss will be impacted by the wildfire through flames, radiant heat, ember attack, and smoke. They include a wide range of potential damage and loss including human health, fatalities, infrastructure, agriculture, forest resources, ecological values, water, social disruption, and trauma for people living in wildfire affected areas. Additional adverse impacts from wildfires may include financial cost of suppression activities, post-fire rehabilitation and restoration, and rebuilding efforts and compensation. Each of these values and costs can have different characteristic responses to different fire intensities (Figure 9.5).

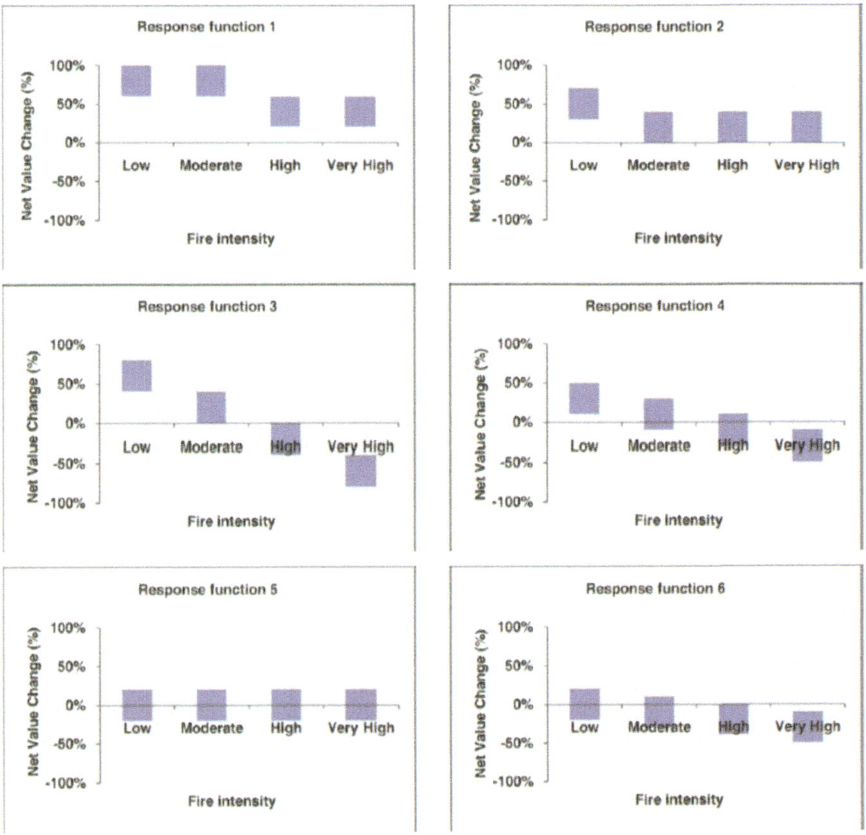

Figure 9.5 Response functions for various values (taken from Calkin *et al.* 2011).

The response functions vary considerably for the different resources but the *net change* is the fraction of the existing value that changes because of fire of a given intensity from –1.0 to 1.0 (Calkin *et al.* 2011). The use of response functions for wildfire risk analysis was first demonstrated in Ager *et al.* (2007a) to examine fuel treatment effects on risk to northern spotted owl habitat (*Strix occidentalis* caurina) in central Oregon, United States. Response functions were developed that quantified threshold flame lengths at which fire effects eliminated owl habitat and were combined with burn probabilities to estimate the risk of habitat loss under different fuel management scenarios. Follow up studies demonstrated similar development of response functions for carbon (Ager *et al.* 2010a) and old growth forests (Ager *et al.* 2010b) (Figure 9.6). A more generalized approach to response functions was developed in Lee *et al.* (2011) and expanded in Scott *et al.* (2013) (Table 9.2) where value changes were defined using percentages and expert opinion for a range of specific highly valued resources and assets (HVRA). The percentages estimate change to the value of each asset caused by a given wildfire intensity. The positive numbers in the table indicate beneficial effects of fire at a particular intensity. Each percentage is multiplied by the probability of intensity at that level and summed to estimate expected net value change.

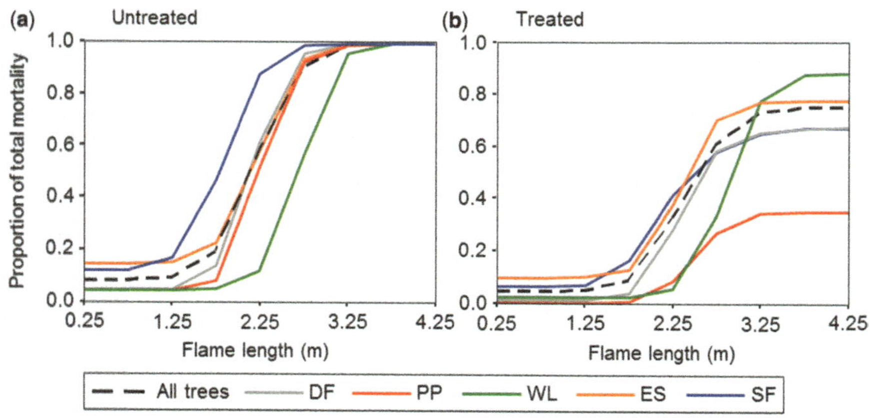

Figure 9.6 Example wildfire response functions for western United States conifers developed with the forest vegetation simulator (Dixon, 2002), as described in Ager *et al.* (2010b). Images show the comparison of (**a**) untreated and (**b**) treated modeled mortality of large trees by flame length for all trees and select species for treated stands within the maximum treatment area (66% of forested lands). DF: Douglas-fir, PP: ponderosa pine, WL: western larch, ES: Engelmann spruce, SF: subalpine fir.

Table 9.2 List of highly valued resources and assets (HVRA) to fire by intensity level (FIL); from Scott *et al.* (2013).

HVRA Name	Sub-HVRA Name	FIL 1	FIL 2	FIL 3	FIL 4	FIL 5	FIL 6
Investments	Game and Fish feedgrounds	–50	–70	–90	–100	–100	–100
	Special use permit areas	–50	–70	–90	–100	–100	–100
	Trail heads/boating sites	0	–10	–20	–30	–40	–50
	Campgrounds/picnic areas	0	–10	–20	–55	–75	–75
	Cabins/guard stations	–50	–70	–90	–100	–100	–100
	Oil and gas development	–10	–20	–40	–80	–100	–100
	Communication sites	0	–30	–60	–80	–100	–100
	Power lines	–10	–20	–40	–80	–100	–100
	Whitebark pine plus trees	–10	–70	–100	–100	–100	–100
Wildland–Urban Interface	WUI defense zone	0	–50	–75	–100	–100	–100
	Protection FMU	10	0	–25	–50	–50	–50
Watershed	Municipal Watershed (DFC 4)	20	0	–20	–50	–75	–100
Timber base	Desired future condition 1B	20	–20	–50	–80	–100	–100
	Desired future condition 10	50	25	10	0	–25	–50

9.4 WILDFIRE RISK MANAGEMENT

Quantitative risk analysis is very useful for displaying existing risk but is perhaps most powerful as a tool for identifying the scale, location and degrees of management options for mitigating undesired wildfire consequences (or enhancing desired benefits). Conceptually, mitigation of risk can be achieved by changing any of the terms of the risk equation (9.1) – specifically the probabilities of the hazards creating a wildfire, the intensities of the wildfires that result, and the response or susceptibility of given values or resources to wildfire at a given intensity (Figure 9.7a). Intensity distributions can be changed by managing fuel and vegetation structure, and new simulations would change the proportion of different intensities burning at each treated area as shown in Figure 9.7b. Simply changing the intensity distribution affects the overall risk. Alternatively, or in addition, the susceptibility of values or assets can be enhanced such that fires of a given intensity result in less reduction in net value (Figure 9.7c). Changing fuel conditions and enhancing resistance of values to wildfire is the most effective combination of methods.

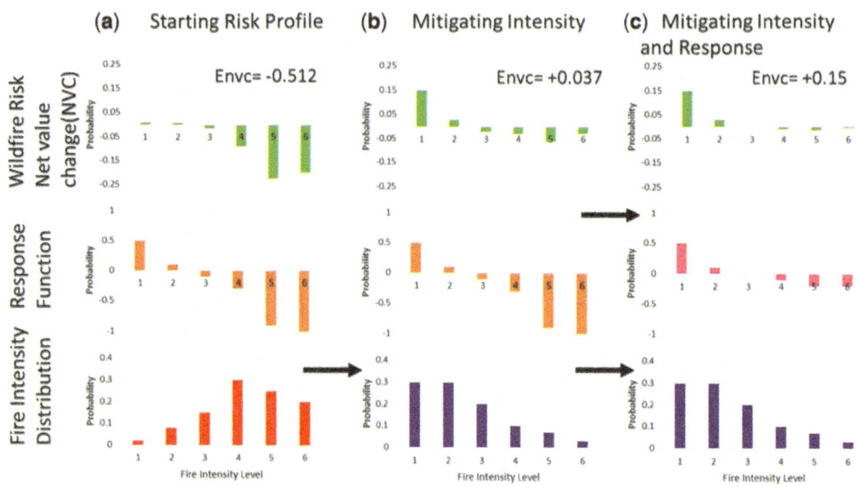

Figure 9.7 Illustration of risk mitigation from (**a**) the starting profile by changing (**b**) probability distribution of fire intensity through fuel treatment, and (**c**) by also changing the response function to reflect reduced susceptibility of assets to wildfire intensities. Note that the expected net value change E[nvc] is the sum of all net value change and increases with mitigation.

Systematic risk management includes risk assessment, development of mitigation strategies, and risk communication (Thompson *et al.* 2016). The article by Thompson *et al.* (2016) reviews these components with respect to wildland fire and offers guidelines for managing risk by an organization. Risk management in Australia and New Zealand adopted ISO standard 31000, and land and fire management agencies are now undertaking the process of setting out the requirements, means, and methods. The standard was very similar to the previous Australian/New Zealand standard that had been developed and used. The Strategic Bushfire Management Plan of the Australian Capital Territory (ACT Government, 2005) and subsequently the Canadian Wildfire Strategy identified aspects of fires that are important in prevention efforts, and the focal points for risk reduction; specifically, fires ignite, THEN spread through fuels, THEN impact assets (human built or environmental) (Moore, 2005). This suggests that we have opportunities to (1) prevent/reduce ignitions, (2) prevent/reduce the chance for fires to spread, and (3) prevent/reduce the negative impacts on assets. Consequently, the assessment of risks in terms of hazards and values needs to target these three aspects, which interact and can be interdependent. Their attributes include:

- Weather – severity (wind speed, relative humidity), duration, and dynamics (wind direction changes, increases/decreases in weather variables).
- Ignition – location of the ignition, time of day (when fires start), number (how many fires start), and distribution (where the fire(s) start).
- Fuel – hazard level (how much fuel, what type of fuel, fuel availability), extent/continuity (across the landscape), flammability (readiness to burn), and landscape features (terrain and geography, ecosystem variability).

Thus, risk management needs to consider the interactions of these components in terms of their influence on how a wildfire can adversely impact assets and values. If any one of the exposure factors (probability and intensity) is insufficient to support a wildfire, such as benign weather, no ignitions or low fuel, there is little or no wildfire risk. Conversely, the higher the values for exposure factors (probability and intensity) the higher the susceptibility of the values at risk, the greater the risk. Risk management concerns manipulating exposure factors and susceptibility to reduce the impact of a wildfire.

In the case of wildfire risk management in and around developed areas, risk management is focused on: (1) civil protection and emergency planning, including adequate water and road systems for fire protection; (2) building codes to reduce susceptibility; (3) defensible space around and within structures; and (4) land-use planning and zoning. A civil protection and emergency planning framework was described by Kalabokidis *et al.* (2012) that used individual components for *Analysis – Prevention – Preparedness – Response – Recovery* indicators, to be usable in fire risk and crisis management activities including:

- Research and analysis that generates assessments of the existence and changes in both hazards and the susceptibility of values, in time and space.
- Preventive measures that aim at reducing fire ignitions, e.g., personnel and volunteers' training, effective legislation regarding property land development, and law enforcement.

- Preventive measures that reduce fuel hazard through fuel and land management.
- Preparedness measures that promote the existence of an agency able to initiate promptly and effectively direct suppression of any forest fire at its ignition, with sufficient firefighting force for direct suppression, including dispatch systems for fire suppression, operation of lookout towers, and patrols.
- Raising public awareness and communication of oncoming fire risk, and access administrative response arrangements.
- Emergency situation assessment and evacuation of threatened areas, if appropriate.
- Response operations – rescue, relief, and recovery.

9.5 FIRE RISK GEO-INFORMATICS

Holistic fire risk assessment of hazards and values should compose quantitative indices of wildfire behavior and effects with spatial layers of meteorological, vegetative, topographic and socioeconomic information that will eventually develop geographical fire potential indices (Kalabokidis *et al.* 2002). The use of Geographic Information Systems (GIS) and remote sensing technology supports the input, management, processing, spatial analysis, cartographic modeling, and visualization of complex environmental data, referenced in space and time, for wildfire risk assessment. Such an information and computing infrastructure, developed a priori, can provide for timely and realistic assistance in fire prevention planning and real-time fire suppression operations that will contribute to public safety, maintain natural resources physically and aesthetically intact, and allow people to live more safely in "natural" environments.

New technologies of geo-informatics (GIS, decision support systems, digital technology, remote sensing, etc.) and electronics (remote automated weather stations, detection cameras, etc.) contribute to more effective organization for environmental protection with prompt detection and risk assessment, methodical observation of biophysical and socioeconomic parameters and decision support management (Kalabokidis *et al.* 2013; Kalabokidis *et al.* 2014). GIS technology has evolved into a powerful tool to statistically analyze complex and multi-faceted physical phenomena and mechanisms (sometimes very much chaotic in behavior), including anthropogenic parameters. Use of modern geo-informatic procedures and decision support systems contribute to better understanding and explanation of landscape wildfire dynamics (Chuvieco & Congalton, 1989; Kalabokidis *et al.* 2016). The advantages that prompt the choice of GIS as a tool for analyzing wildfire, human and environmental factors at the landscape level include (Kalabokidis *et al.* 2012):

- Input and analysis of descriptive and spatial data in digital format.
- Correlation and analysis of peripheral statistical data to respective spatial data.

- Development of models for managing and processing data.
- Processing large amounts of varying types of data that cannot be achieved through traditional methods.
- Presentation of results at any scale and map type.

As hazards and vulnerabilities are spatially and temporally distributed, risk is inherently a dynamic phenomenon, and risk assessment should address both the degree of risk and its spatiotemporal distribution. There is a particular need for land management planning systems that can simulate the combined impacts of wildfire, fuel management, and public policies over time to understand what levels of investments are needed to alter trajectories in wildfire risk (Spies *et al.* 2014; Conlisk *et al.* 2015). Fire risk management requires the use of large volumes of data that change continuously over time and space, creating both the need and the opportunity to automate the tasks (Yuan, 1997). For example, integrated satellite and ground technologies can be applied using advanced geo-informatic tools and models for the inventory, mapping and monitoring of geomorphology, LULC, atmospheric and physical processes, and anthropogenic influences. Data from various Earth Observation Satellites are used to extract geographic information useful to study and monitor wildfires. In addition to satellite image processing, this involves extraction of LULC maps from existing data and mapping using geo-database management, advanced remote sensing, and automated cartography techniques.

Considerable effort has been devoted to integrating fire models and GIS to create risk assessment and planning tools to support fire management. Planners and fuel specialists use risk assessment tools to (1) map fire risk to important social and ecological values; (2) prioritize investments in hazardous fuel reduction among forests, and (3) design and test fuel management projects aimed at reducing surface and canopy fuels and potential wildfire impacts (Miller & Ager, 2013; Ager *et al.* 2014). Risk planning systems (e.g., ArcFuels; Vaillant & Ager, 2014) couple desktop GIS with wildfire and vegetation simulation models to streamline the planning process. Fire models are used to study fire behavior over a range of scales, from a localized fuel type (e.g., "forest stand", 5–50 ha) where fuel treatment activities including thinning, fuels mastication, and prescribed burning are simulated, to large landscapes (1,000–50,000 ha) where large numbers of fires are simulated to estimate burn probabilities. In the latter case, simulation of multiple fires at local and landscape scales are performed to analyze uncertainty associated with wildfire events in terms of timing, location, intensity, and duration.

9.6 FIRE MODELS TO SUPPORT WILDFIRE RISK MANAGEMENT

The inherent complexity of risk assessment and fire management planning has led to a rapid increase in the application of fire behavior modeling software in both research and operational contexts (Miller & Ager, 2013). Simulation models are widely used to characterize fire behavior and examine the potential effectiveness of proposed fire management programs (Table 9.3). Landscape fire spread models are used to examine how fuel treatment patterns change spatial patterns in burn probability and large fire spread (Finney *et al.* 2007). Integrated landscape analysis packages have been developed to stitch together fire behavior models with vegetation databases to streamline risk assessments (Ager *et al.* 2011). Details on fire behavior simulation models are found in the original literature sources and summary descriptions elsewhere (e.g., Finney, 2006; Stratton, 2006; Andrews *et al.* 2007; Ager *et al.* 2011; Noonan-Wright *et al.* 2011; Kalabokidis *et al.* 2016).

Table 9.3 Fire simulation models used as part of risk and fire behavior assessments developed in the United States.

Model and Citation	Description	Case Studies
1. Forest Vegetation Simulator (FVS) (Dixon, 2002; Crookston & Dixon, 2005)	Individual-tree, distance-independent growth and yield model	Ager *et al.* 2007a, Ager *et al.* 2007b, Finney *et al.* 2007, Ager *et al.* 2010a
2. Fire and Fuels Extension to FVS (FVS-FFE) (Rebain, 2010)	Stand-level simulations of fuel dynamics and potential fire behavior over time	DeRose and Long 2009, Hurteau and North 2009
3. NEXUS (Scott, 1999)	Stand-level spreadsheet that links surface and crown fire prediction models	Perry *et al.* 2004, Agee and Lolley 2006, Roccaforte *et al.* 2008
4. BehavePlus (Heinsch & Andrews, 2010)	Stand-level fire behavior, fire effects, and fire environment modeling system	Glitzenstein *et al.* 2006, Page and Jenkins 2007, Schmidt *et al.* 2008, Vaillant *et al.* 2009
5. BehavePlus (SURFACE module) (Andrews, 2005)	Stand-level fire behavior, fire effects, and fire environment modeling system	n/a
6. FlamMap (Finney, 2006)	Landscape-level fire behavior mapping and analysis program	Moghaddas *et al.* 2010
7. FARSITE (Finney, 1998)	Fire spread simulator	Roccaforte *et al.* 2008, Schmidt *et al.* 2008, Moghaddas *et al.* 2010
8. FOFEM (Reinhardt & Dickinson, 2010)	Stand-level first order fire effects modeling system	Keane *et al.* 2010
9. FSim (Finney *et al.* 2011b)	Large fire simulator	Ager *et al.* 2014

Models developed in the United States to support wildfire risk management activities include NEXUS (Scott, 1999), Fire and Fuels Extension to the Forest Vegetation Simulator (FVS-FFE) (Rebain, 2010), FARSITE (Finney, 1998), FlamMap (Finney, 2006), BehavePlus (Heinsch & Andrews, 2010), and FSim (Finney *et al.* 2011b). In addition, the model FSPro is used for active fire risk management as part of the Wildland Fire Decision Support System (Noonan-Wright *et al.* 2011). A number of supporting models and software are used to estimate appropriate weather, fuel moisture, and other input variables required to run the fire behavior models (USDA, 1996; Zachariassen *et al.* 2003; Butler *et al.* 2006; Stratton, 2006). The fire behavior models simulate one-dimensional fire behavior as part of a spreading line fire (Rothermel, 1972; Albini, 1976; Van Wagner, 1977; Albini, 1979; Anderson, 1983; Rothermel, 1991). All the aforementioned models integrate Rothermel's models (1972, 1991) for predicting surface and crown fire rates of spread, with Van Wagner's (1977, 1993) or Scott and Reinhardt's (2001) crown fire transition and propagation models. These models output a battery of fire behavior metrics (e.g., rate of fire spread, fireline intensity, flame length, crown fire activity). Limitations of the fire models are discussed in a number of papers (McHugh, 2006; Varner & Keyes, 2009; Cruz & Alexander, 2010; Mell *et al.* 2010). Fire effects (versus fire spread) models are also a key part of risk assessments and provide a system for quantitating susceptibility of values to wildfires. Fire effects models have been built to estimate tree mortality, carbon loss, soil impacts, and other ecosystem services (Butler & Dickinson, 2010; Kavanagh *et al.* 2010; Massman *et al.* 2010; Reinhardt & Dickinson, 2010; Stephan *et al.* 2010).

One particularly useful model is FlamMap[4], which provides a fire modelling platform that provides the analytical capacity required to address fuel management and risk reduction problems at a range of scales. FlamMap uses eight grid themes that quantify fuel canopy characteristics, surface fuel models, and topography. Canopy fuel is represented by crown bulk density, canopy closure, height to live crown, and average height. Surface fuel is described by fuel models (Anderson, 1982; Scott & Burgan, 2005) that characterize: (1) loadings for live and dead fuels (by size class), (2) surface-area-to-volume ratio for live and dead fuels, (3) fuel bed depth, (4) moisture of extinction, and (5) heat content. Fuel models are classified into groupings according to the dominant carrier of fire (grass, grass and brush, brush, timber with vegetative understory, timber litter and slash). Fuel models are either selected in the field using guides, or obtained from data sources such as LANDFIRE (Rollins, 2009).

There are three types of fire simulation procedures within FlamMap: (1) basic fire behavior at the pixel scale; (2) minimum travel time (MTT) fire spread (Finney, 2002) for individual fires; and (3) MTT for multiple fires to generate burn probabilities. Basic fire behavior is modeled by independently burning each pixel to generate rate of spread, flame length, fireline intensity, and crown fire activity (Table 9.4). Surface fire rate of spread (R, m min^{-1}) is calculated using the Rothermel

[4]http://www.firelab.org/project/flammap; accessed 23rd July 2016.

(1972) spread equation. Crown fire activity has three categories: surface fire, passive crown fire, and active crown fire. Crown fire behavior metrics can be calculated with Finney (1998) or Scott and Reinhardt (2001) models. In the second type of simulation procedure, discrete fire events are simulated using the two-dimensional MTT fire growth algorithm. MTT calculations assume independence of fire behavior among neighbor cells but are dependent on the ignition locations, resolution of calculations, and simulation time. The MTT algorithm replicates fire growth by Huygens' principle where the growth and behavior of the fire edge is modeled as a vector or wave front (Knight & Coleman, 1993). This method results in less distortion of fire shape and response to temporally varying conditions than techniques that model fire growth from cell-to-cell on a gridded landscape (Finney, 2002). Extensive application has shown that Huygens' principle as implemented in the MTT algorithm can be used to replicate fire perimeters on heterogeneous landscapes (Finney *et al.* 2011a). In the third type of simulation procedure, multiple fires are simulated with the MTT fire growth model to generate burn probabilities. The program simulates random ignitions for a fixed burn period and a conditional burn probability is then calculated as the ratio of the number of times a pixel burns to the total number of fires simulated. Burn probability in FlamMap measures the probability that a pixel will burn given one random ignition within the study area under the modeled weather conditions and burn period. Thus, the burn probability and resulting risk obtained from FlamMap simulations represent conditional estimates.

Table 9.4 Example fire behavior output variables from FlamMap.

Fire Behavior Value	Simulation Type	Output Type	Units
1. Spread Vectors	Static	Vector	m min^{-1}
2. Major Paths	MTT	Vector	n/a
3. Flow Paths	MTT	Vector	n/a
4. Flame Length	Static	ASCII	m or ft
5. Rate of Spread	Static/MTT	ASCII	m min^{-1} or ft min^{-1} or ch hr^{-1}
6. Fireline Intensity	Static/MTT	ASCII	kW m^{-1} or BTU ft^{-1} sec^{-1}
7. Heat per Unit Area	Static	ASCII	kW m^{-2} or BTU ft^{-2} sec^{-1}
8. Crown Fire Activity	Static	ASCII	Index (0, 1, 2, or 3)
9. Midflame Windspeed	Static	ASCII	km hr^{-1} or mi hr^{-1}
10. Horizontal Movement Rate	Static	ASCII	m min^{-1} or ft min^{-1} or ch hr^{-1}
11. Max Spread Direction	Static	ASCII	Radians or degrees
12. Elliptical Dimension	Static/MTT	ASCII	m min^{-1} or ft min^{-1} or ch hr^{-1}
13. Arrival Time	MTT	ASCII	min
14. Node Influence	MTT	ASCII	Frequency
15. Burn Probability	MTT	ASCII	Interval (0–1)

The FSim program (Finney *et al.* 2011b) was later developed and enhanced risk assessments by providing: (1) annualized estimates of burn probability; (2) seasonal trends in risk; (3) suppression effects on fire growth and containment; and (4) a mechanistic linkage between daily fire weather and fire growth. In brief, FSim generates wildfire ignitions and perimeters for a large number (e.g., 50,000) of hypothetical wildfire seasons using statistical relationships between historical Energy Release Component (ERC) (i.e., a fire danger rating index) and ignition probability (Finney *et al.* 2011b). Weather data (ERC, wind speed, wind direction) are derived from historical records obtained from remote automated weather stations (Zachariassen *et al.* 2003). FSim predicts and simulates wildfires on a daily time step using synthetic ERC streams generated from time series analysis (Finney *et al.* 2011b). The synthetic ERC streams incorporate seasonal trends, ERC autocorrelation (dependency of a day's ERC value on previous days), and daily standard deviations. Daily wind speed by direction data are randomly sampled from a monthly joint probability distribution also derived from the weather stations. Fire containment is based on probabilistic relationships between containment success and serial correlations in ERC (Finney *et al.* 2009). FSim uses the same fuels and topographic data as FlamMap.

9.7 EPILOGUE

Better understanding and integrated analysis of holistic institutional approaches and frameworks could provide possibilities for proactively reducing fire risks and protecting life, property, and the environment. Wildfires have both regional and local detrimental effects on the environment (i.e., flora, fauna, soils, water resources, hygiene, air quality, and aesthetics) and human life and property (e.g., civilians, firefighters, agriculture, livestock and forage production, houses and other buildings) in rural and urban areas, so improved knowledge may generate positive economic and social impacts. More specifically, the improvement of scientific knowledge about technological and physicochemical fire processes involved, may reduce environmental degradation, optimize ecosystem management strategies, contribute to the design of emergency response procedures, and help to assess wildfire's role in ecosystem and landscape damage, loss and air pollution at various scales. Such ecological values and the quest for sustainable development require both information and management options to mitigate wildland fire hazards and risks associated with fast changing environments.

REFERENCES

ACT Government (2005). Strategic Bushfire Management Plan, Version 1. Act Government, p. 72.

Adams M. A. (2013). Mega-fires, tipping points and ecosystem services: managing forests and woodlands in an uncertain future. *Forest Ecology and Management*, **294**, 250–261.

Agee J. K. and Lolley M. R. (2006). Thinning and prescribed fire effects on fuels and potential fire behavior in an eastern Cascades forest, Washington, USA. *Fire Ecology*, **2**, 3–19.

Ager A. A., Finney M. A., Kerns B. K. and Maffei H. (2007a). Modeling wildfire risk to northern spotted owl (Strix occidentalis caurina) habitat in Central Oregon, USA. *Forest Ecology and Management*, **246**, 45–56.

Ager A. A., McMahan A. J., Barrett J. J. and McHugh C. W. (2007b). A simulation study of thinning and fuel treatments on a wildland-urban interface in eastern Oregon, USA. *Landscape and Urban Planning*, **80**, 292–300.

Ager A. A., Finney M. A., McMahan A. J. and Cathcart J. (2010a). Measuring the effect of fuel treatments on forest carbon using landscape risk analysis. *Natural Hazards and Earth System Sciences*, **10**, 2515–2526.

Ager A. A., Vaillant N. M. and Finney M. A. (2010b). A comparison of landscape fuel treatment strategies to mitigate wildland fire risk in the urban interface and preserve old forest structure. *Forest Ecology and Management*, **260**, 166–167.

Ager A. A., Vaillant N. M. and Finney M. A. (2011). Integrating fire behavior models and geospatial analysis for wildland fire risk assessment and fuel management planning. *Journal of Combustion*, p. 19. doi: 10.1155/2011/572452.

Ager A. A., Buonopane M., Reger A. and Finney M. A. (2012). Wildfire exposure analysis on the national forests in the Pacific Northwest, USA. *Risk Analysis*, **33**, 1000–1020.

Ager A. A., Day M. A., McHugh C. W., Short K., Gilbertson-Day J., Finney M. A. and Calkin D. E. (2014). Wildfire exposure and fuel management on western US national forests. *Journal of Environmental Management*, **145**, 54–70.

Ager A. A., Barros A. and Day M. A. (2015). Understanding the transmission of wildfire risk on a fire prone landscape: a case study from Central Oregon. *Geophysical Research Abstracts*, **17**, 15128.

Albini F. A. (1976). Estimating Wildfire Behavior and Effects. INT-30, Gen. Tech. Rep. INT-GTR-30. U.S. Department of Agriculture, Forest Service, Intermountain Forest and Range Experiment Station, Ogden, UT, p. 92.

Albini F. A. (1979). Spot Fire Distance From Burning Trees- a Predictive Model. General Technical Report INT-56. USDA Forest Service, Intermountain Forest and Range Experiment Station, Ogden, UT, p. 73.

Anderson H. E. (1982). Aids to Determining Fuel Models for Estimating Fire Behavior. INT-122, Gen. Tech. Rep. INT-122. U.S. Department of Agriculture, Forest Service, Intermountain Forest and Range Experiment Station, Ogden, Utah, p. 22.

Anderson H. E. (1983). Predicting Wind-driven Wildland Fire Size and Shape. Research Paper INT-305, U.S. Department of Agriculture, Forest Service, Intermountain Forest and Range Experiment Station, Ogden, UT, p. 32.

Andrews P. L. (2005). Fire danger rating and fire behavior prediction in the United States. In: Proceedings of the 5th NRIFD Symposium on Forest Fire Protection, November 30-December 2, 2005, Mitaka. Tokyo, Japan. pp. 106–117.

Andrews P. L., Finney M. A. and Fischetti M. (2007). Predicting wildfires. *Scientific American*, **297**, 47–55.

Atkinson D., Chladil M., Janssen V. and Lucieer A. (2010). Implementation of quantitative bushfire risk analysis in a GIS environment. *International Journal of Wildland Fire*, **19**, 649–658.

Attiwill P. and Binkley D. (2013). Exploring the mega-fire reality: a "Forest Ecology and Management" conference. *Forest Ecology and Management*, **294**, 1–3.

Bailey D. (2013). National dialogue needed about WUI fires. *Wildfire*, **22**, 6–7.

Blaikie P., Cannon T., Davis I. and Wisner B. (1994). At Risk: Natural Hazards, People's Vulnerability and Disasters. Routledge, London.

Bowman D. M. J. S., O'Brien J. A. and Goldammer J. G. (2013). Pyrogeography and the global quest for sustainable fire management. *Annual Review of Environment and Resources*, **38**, 57–80.

Bradshaw L. S., Deeming J. E., Burgan R. E. and Cohen J. D. (1983). The 1978 National Fire-Danger-Rating System: Technical Documentation. General Technical Report INT-169. U.S. Department of Agriculture, Forest Service, Intermountain Forest and Range Experiment Station, Ogden, UT, p. 44.

Brillinger D. R., Preisler H. K. and Benoit J. (2006). Probabilistic risk assessment for wildfires. *Environmetrics*, **17**, 623–633.

Butler B. W. and Dickinson M. B. (2010). Tree injury and mortality in fires: developing process-based models. *Fire Ecology*, **6**, 55–79.

Butler B., Finney M., Bradshaw L., Forthofer J., McHugh C., Stratton R. and Jimenez D. (2006). WindWizard: A new tool for fire management decision support. In: Fuels Management - How to Measure Success: Conference Proceedings, P. L. Andrews and B. W. Butler (Comps), March 28–30, Portland, OR. USDA Forest Service, Rocky Mountain Research Station Proceedings RMRS-P-41, pp. 787–796.

Calkin D. E., Ager A. A. and Gilbertson-Day J. W. Eds. (2010). Wildfire Risk and Hazard: Procedures for the First Approximation. Gen. Tech. Rep. GTR-RMRS-235. U.S. Department of Agriculture, Forest Service, Rocky Mountain Research Station., Fort Collins, CO, p. 62.

Calkin D. E., Finney M. A., Ager A. A., Thompson M. P. and Gebert K. G. (2011). Progress towards and barriers to implementation of a risk framework for federal wildland fire policy and decision making in the United States. *Forest Policy and Economics*, **13**, 378–389.

Catry F. X., Rego F. C., Bação F. L. and Moreira F. (2009). Modeling and mapping wildfire ignition risk in Portugal. *International Journal of Wildland Fire*, **18**, 921–931.

Charnley S., Poe M. R., Ager A. A., Spies T. A., Platt E. K. and Olsen K. A. (2015). A burning problem: social dynamics of disaster risk reduction through wildfire management. *Human Organization*, **74**, 329–340.

Chen K., Blond R. and Jacobson C. (2003). Towards an integrated approach to natural hazards risk assessment using GIS: with reference to bushfires. *Environmental Management*, **31**, 546–560.

Chuvieco E. and Congalton R. G. (1989). Application of remote sensing and geographic information systems to forest fire hazard mapping. *Remote Sensing of Environment*, **29**, 147–159.

Conlisk E., Syphard A. D., Franklin J. and Regan H. M. (2015). Predicting the impact of fire on a vulnerable multi-species community using a dynamic vegetation model. *Ecological Modelling*, **301**, 27–39.

Corotis R. B. and Hammel E. M. (2010). Multi-attribute aspects for risk assessment of natural hazards. *International Journal of Risk Assessment and Management*, **14**, 437–458.

Crookston N. L. and Dixon G. E. (2005). The forest vegetation simulator: a review of its structure, content, and applications. *Computers and Electronics in Agriculture*, **49**, 60–80.

Cruz M. G. and Alexander M. E. (2010). Assessing crown fire potential in coniferous forests of western North America: a critique of current approaches and recent simulation studies. *International Journal of Wildland Fire*, **19**, 377–398.

Cruz M. G., Sullivan A. L., Gould J. S., Sims N. C., Bannister A. J., Hollis J. J. and Hurley R. J. (2012). Anatomy of a catastrophic wildfire: the Black Saturday Kilmore East fire in Victoria, Australia. *Forest Ecology and Management*, **284**, 269–285.

Darques R. (2015). Mediterranean cities under fire. A critical approach to the wildland-urban interface. *Applied Geography*, **59**, 10–21.

DeRose R. J. and Long J. N. (2009). Wildfire and spruce beetle outbreak: simulation of interacting disturbances in the central rocky mountains. *Ecoscience*, **16**, 28–38.

Dixon G. E. Comp. (2002). Essential FVS: A User's Guide to the Forest Vegetation Simulator. Internal Report, U. S. Department of Agriculture, Forest Service, Forest Management Service Center, Fort Collins, CO, p. 226. (Revised: 2 November 2015).

Everett Y. and Fuller M. (2011). Fire safe councils in the interface. *Society and Natural Resources*, **24**, 319–333.

Fernandes P. M., Loureiro C., Guiomar N., Pezzatti G. B., Manso F. T. and Lopes L. (2014). The dynamics and drivers of fuel and fire in the Portuguese public forest. *Journal of Environmental Management*, **146**, 373–382.

Finney M. A. (1998). FARSITE: Fire Area Simulator - Model Development and Evaluation. Res. Pap. RMRSRP-4. U.S. Department of Agriculture, Forest Service, Rocky Mountain Research Station, Fort Collins, CO, p. 47.

Finney M. A. (2002). Fire growth using minimum travel time methods. *Canadian Journal of Forest Research*, **32**, 1420–1424.

Finney M. A. (2005). The challenge of quantitative risk analysis for wildland fire. *Forest Ecology and Management*, **211**, 97–108.

Finney M. A. (2006). An overview of FlamMap fire modeling capabilities. In: Fuels Management - How to Measure Success: Conference Proceedings, P. L. Andrews and B. W. Butler (Comps), March 28–30, Portland, OR. USDA Forest Service, Rocky Mountain Research Station Proceedings RMRS-P-41, pp. 213–220.

Finney M. A., Seli R. C., McHugh C. W., Ager A. A., Bahro B. and Agee J. K. (2007). Simulation of long-term landscape-level fuel treatment effects on large wildfires. *International Journal of Wildland Fire*, **16**, 712–727.

Finney M. A., Grenfell I. C., McHugh C. W. (2009). Modeling large fire containment using generalized linear mixed model analysis. *Forest Science*, **55**, 249–255.

Finney M. A., Grenfell I. C., McHugh C. W., Seli R. C., Trethewey D., Stratton R. D. and Brittain S. (2011a). A method for ensemble wildland fire simulation. *Environmental Modeling and Assessment*, **16**, 153–167.

Finney M. A., McHugh C. W., Grenfell I. C., Riley K. L. and Short K. C. (2011b). A simulation of probabilistic wildfire risk components for the continental United States. *Stochastic Environmental Research and Risk Assessment*, **25**, 973–1000.

Fischer A. P. and Charnley S. (2012). Risk and cooperation: managing hazardous fuel in mixed ownership landscapes. *Environmental Management*, **49**, 1192–1207.

Fischer A. P., Kline J. D., Ager A. A., Charnley S. and Olsen K. A. (2014). Objective and perceived wildfire risk and its influence on private forest landowners' fuel reduction activities in Oregon's (USA) ponderosa pine ecoregion. *International Journal of Wildland Fire*, **23**, 143–153.

Fischer A. P., Spies T. A., Steelman T. A., Moseley C., Johnson B. R., Bailey J. D., Ager A. A., Bourgeron P., Charnley S., Collins B. M., Kline J. D., Leahy J. E., Littell J. S., Millington J. D. A., Nielsen-Pincus M., Olsen C. S., Paveglio T. B., Roos C. I., Steen-Adams M. M., Stevens F. R., Vukomanovic J., White E. M. and Bowman D. M. J. S. (2016). Wildfire risk as a socioecological pathology. *Frontiers in Ecology and the Environment*, **14**, 276–284.

Forestry Canada Fire Danger Group (1992). Development and Structure of the Canadian Forest Fire Behavior Prediction System. Inf. Rep. ST-X-3, Forestry Canada, Ottawa, Ontario, p. 64.

Fuchs S., Kuhlicke C. and Meyer V. (2011). Editorial for the special issue: vulnerability to natural hazards-the challenge of integration. *Natural Hazards*, **58**, 609–619.

Glitzenstein J. S., Streng D. R., Achtemeier G. L., Naeher L. P. and Wade D. D. (2006). Fuels and fire behavior in chipped and unchipped plots: implications for land management near the wildland urban interface. *Forest Ecology and Management*, **236**, 18–29.

Gonzalez J. R., Palahi M. and Pukkala T. (2005). Integrating fire risk considerations in forest management planning in Spain – a landscape level perspective. *Landscape Ecology*, **20**, 957–970.

Haynes R. and Cleaves D. (1999). Uncertainty, risk, and ecosystem management. In: Ecological Stewardship, Volume III, W. T. Sexton, A. J. Malk, R. C. Szaro and N. C. Johnson (eds), Elsevier Science Ltd, Oxford, UK, pp. 413–429.

Heinsch F. A. and Andrews P. L. (2010). BehavePlus Fire Modeling System, Version 5.0: Design and Features. Gen. Tech. Rep. RMRS-GTR-249. U.S. Department of Agriculture, Forest Service, Rocky Mountain Research Station, Fort Collins, CO, p. 111.

Hurteau M. and North M. (2009). Fuel treatment effects on tree-based forest carbon storage and emissions under modeled wildfire scenarios. *Frontiers in Ecology and the Environment*, **7**, 409–414.

Jakes P. J., Nelson K. C., Enzler S. A., Burns S., Cheng A. S., Sturtevant V., Williams D. R., Bujak A., Brummel R. F., S. Grayzeck-Souter, and Staychock E. (2011). Community wildfire protection planning: is the Healthy Forests Restoration Act's vagueness genius? *International Journal of Wildland Fire*, **20**, 350–363.

Jolly M., Cochrane M. D., Freeborn P. H., Holden Z. A., Brown T. J., Williamson G. J. and Bowman D. M. J. S. (2015). Climate-induced variations in global fire danger from 1979 to 2013. *Nature Communications*, **6**, 7537.

Jones J. M. and Corotis R. B. (2012). The regional consequences of individual natural hazard events. *International Journal of Risk Assessment and Management*, **16**, 78–111.

Kalabokidis K. D., Gatzojannis S. and Galatsidas S. (2002). Introducing wildfire into forest management planning: towards a conceptual approach. *Forest Ecology and Management*, **158**, 41–50.

Kalabokidis K., Xanthopoulos G., Moore P., Caballero D., Kallos G., Llorens J., Roussou O. and Vasilakos C. (2012). Decision support system for forest fire protection in the Euro-Mediterranean region. *European Journal of Forest Research*, **131**, 597–608.

Kalabokidis K., Athanasis N., Gagliardi F., Karayiannis F., Palaiologou P., Parastatidis S. and Vasilakos C. (2013). Virtual Fire: a web-based GIS platform for forest fire control. *Ecological Informatics*, **16**, 62–69.

Kalabokidis K., Athanasis N., Vasilakos C., Palaiologou P. (2014). Porting of a wildfire risk and fire spread application into a cloud computing environment. *International Journal of Geographical Information Science*, **28**, 541–552.

Kalabokidis K., Ager A., Finney M., Athanasis N., Palaiologou P. and Vasilakos C. (2016). AEGIS: a wildfire prevention and management information system. *Natural Hazards and Earth System Sciences*, **16**, 643–661.

Kasperson R. E. and Kasperson J. X. (1996). The social amplification and attenuation of risk. *Annals of the American Academy of Political and Social Science, Challenges in Risk Assessment and Risk Management*, **545**, 95–105.

Kavanagh K. L., Dickinson M. B. and Bova A. S. (2010). A way forward for fire-caused tree mortality prediction: modeling a physiological consequence of fire. *Fire Ecology*, **6**, 80–94.

Keane R. E., Drury S. A., Karau E. C., Hessburg P. F. and Reynolds K. M. (2010). A method for mapping fire hazard and risk across multiple scales and its application in fire management. *Ecological Modelling*, **221**, 2–18.

Kline J. D., Ager A. A. and Fischer A. P. (2015). A conceptual framework for coupling the biophysical and social dimensions of wildfire to improve fireshed planning and risk mitigation. In: 13th International Wildland Fire Safety Summit & 4th Human Dimensions of Wildland Fire Conference, Managing Fire, Understanding Ourselves: Human Dimensions in Safety and Wildland Fire, Boise, ID.

Knight I. and Coleman J. (1993). A fire perimeter expansion algorithm based on Huygens' wavelet propagation. *International Journal of Wildland Fire*, **3**, 73–84.

Lee D. C., Ager A. A., Calkin D. E., Finney M. A., Thompson M. P., Quigley T. M. and McHugh C. W. (2011). Appendix A: Comparative Risk Assessment. A National Cohesive Wildland Fire Management Strategy, Report to Congress. Wildland Fire Leadership Council, Washington, DC, pp. 13–32.

Lindell M. K. and Perry R. W. (2012). The protective action decision model: theoretical modifications and additional evidence. *Risk Analysis*, **32**, 616–632.

Littell J. S., McKenzie D., Peterson D. L. and Westerling A. L. (2009). Climate and wildfire area burned in western U.S. ecoprovinces, 1916–2003. *Ecological Applications*, **19**, 1003–1021.

Loboda T. and Csiszar I. A. (2007). Assessing the risk of ignition in the Russian Far East within a modeling framework of fire threat. *Ecological Applications*, **17**, 791–805.

Malamud B. D., Morein G. and Turcotte D. L. (1998). Forest fires: an example of self-organized critical behavior. *Science*, **281**, 1840–1842.

Martínez J., Vega-García C. , and Chuvieco E. (2009). Human-caused wildfire risk rating for prevention planning in Spain. *Journal of Environmental Management*, **90**, 1241–1252.

Massman W. J., Frank J. M. and Mooney S. J. (2010). Advancing investigation and physical modeling of first-order fire effects on soils. *Fire Ecology*, **6**, 36–54.

McHugh C. W. (2006). Considerations in the use of models available for fuel treatment analysis. In: Fuels Management-How to Measure Success: Conference Proceedings, 28–30 March 2006; Portland, OR. Proceedings RMRS-P-41, U.S. Department of Agriculture, Forest Service, Rocky Mountain Research Station, Fort Collins, CO, pp. 81–105.

Mell W. E., Manzello S. L., Maranghides A., Butry D. and Rehm D. B. (2010). The wildland-urban interface fire problem: current approaches and research needs. *International Journal of Wildland Fire*, **19**, 238–251.

Miller C. and Ager A. A. (2013). A review of recent advances in risk analysis for wildfire management. *International Journal of Wildland Fire*, **22**, 1–14.

Moghaddas J. J., Collins B. M., Menning K., Moghaddas, E. E. Y. and Stephens S. L. (2010). Fuel treatment effects on modeled landscape-level fire behavior in the Northern Sierra Nevada. *Canadian Journal of Forest Research*, **40**, 1751–1765.

Moore P. (2005). Fires – are they really unexpected? *Parks*, **15**, 37–44.

Moore P., Hardesty J., Kelleher S., Maginnis S. and Myers R. (2003). Forests and Wildfires: Fixing the Future by Avoiding the Past. XII World Forestry Congress, Quebec, Canada.

Moreira F., Viedma O., Arianoutsou M., Curt T., Koutsias N., Rigolot E., Barbati A., Corona P., Vaz P., Xanthopoulous G., Mouillot F. and Bilgili E. (2011). Landscape - wildfire interactions in southern Europe: implications for landscape management. *Journal of Environmental Management*, **92**, 2389–2402.

Moser S. C. and Ekstrom J. A. (2011). Taking ownership of climate change: participatory adaptation planning in two local case studies from California. *Journal ofEnvironmental Studies and Sciences*, **1**, 63–74.

Noonan-Wright E. K., Opperman T. S., Finney M. A., Zimmerman G. T., Seli R. C., Elenz L. M., Calkin D. E. and Fiedler J. R. (2011). Developing the US wildland fire decision support system. *Journal of Combustion* **2011**, p. 14. Article ID 168473.

Oliveira T. M., Barros A. M. G., Ager A. A. and Fernandes P. M. (2016). Assessing the effect of a fuel break network to reduce burnt area and wildfire risk transmission. *International Journal of Wildland Fire*, **25**, 619–632.

Olsen C., Kline J. and Ager A. (2013). Objective and perceived fire experience and fire risk: what influences WUI homeowners' willingness to conduct Firewise activities? Presented at the 19th International Symposium on Society and Resource Management. International Association for Society and Natural Resources, Estes Park, Colorado, USA.

Page W. and Jenkins M. J. (2007). Predicted fire behavior in selected mountain pine beetle-infested lodgepole pine. *Forest Science*, **53**, 662–674.

Parisien M. A., Snetsinger S., Greenberg J. A., Nelson C. R., Schoennagel T., Dobrowski S. Z. and Moritz M. A. (2012). Spatial variability in wildfire probability across the western United States. *International Journal of Wildland Fire*, **21**, 313–327.

Pausas J. G. and Fernandez-Munoz S. (2012). Fire regime changes in the Western Mediterranean Basin: from fuel-limited to drought-driven fire regime. *Climatic Change*, **110**, 215–226.

Paveglio T., Jakes P., Carroll M. and Williams D. (2009). Understanding social complexity within the wildland-urban interface: a new species of human habitation? *Environmental Management*, **43**, 1085–1095.

Perry D. A., Jing H., Youngblood A. and Oetter D. R. (2004). Forest structure and fire susceptibility in volcanic landscapes of the eastern High Cascades, Oregon. *Conservation Biology*, **18**, 913–926.

Preisler H. K. and Westerling A. L. (2007). Statistical model for forecasting monthly large wildfire events in western United States. *Journal of Applied Meteorology and Climatology*, **46**, 1020–1030.

Preisler H. K., Brillinger D. R., Burgan R. E. and Benoit J. W. (2004). Probability based models for estimation of wildfire risk. *International Journal of Wildland Fire*, **13**, 133–142.

Pyne S. J., Andrews P. L. and Laven R. D. (1996). Introduction to Wildland Fire. 2nd edn. Wiley & Sons, New York.

Rebain S. A. (2010). The Fire and Fuels Extension to the Forest Vegetation Simulator: Updated Model Documentation. Internal Report, U. S. Department of Agriculture, Forest Service, Forest Management Service Center, Fort Collins, CO, p. 361.

Regos A., Aquilue N., Retana J., De Caceres M. and Brotons L. (2014). Using unplanned fires to help suppressing future large fires in Mediterranean forests. *PLoS ONE*, **9**, e94906.

Reinhardt E. D. and Dickinson M. B. (2010). First-order fire effects models for land management: overview and issues. *Fire Ecology*, **6**, 131–142.

Roccaforte J. P., Fulé P. Z., and Covington W. W. (2008). Landscape-scale changes in canopy fuels and potential fire behaviour following ponderosa pine restoration treatments. *International Journal of Wildland Fire*, **17**, 293–303.

Rollins M. G. (2009). LANDFIRE: a nationally consistent vegetation, wildland fire, and fuel assessment. *International Journal of Wildland Fire*, **18**, 235–249.

Roloff G. J., Mealey S. P., Clay C., Barry J., Yanish C. and Neuenschwander L. (2005). A process for modeling short- and long-term risk in the southern Oregon Cascades. *Forest Ecology and Management*, **211**, 166–190.

Rothermel R. C. (1972). A Mathematical Model for Predicting Fire Spread in Wildland Fuels. Intermountain Forest and Range Experiment Station Research Paper. INT-1 15, U.S. Department of Agriculture, Forest Service, p. 40.

Rothermel R. C. (1991). Predicting Behavior and Size of Crown Fires in the Northern Rocky Mountains. INT-438, Research Paper INT-438, U.S. Department of Agriculture, Forest Service, Intermountain Research Station, Ogden, UT. p. 46.

Salis M., Ager A., Arca B., Finney M. A., Bacciu V., Duce P. and Spano D. (2013). Assessing exposure of human and ecological values to wildfire in Sardinia, Italy. *International Journal of Wildland Fire*, **22**, 549–565.

Salis M., Ager A. A., Arca B., Finney M. A., Munoz-Lozano O., Alcasena F. J., Bacciu V. and Spano D. (2014). Spatiotemporal changes in wildfire patterns in Sardinia. In: 2nd Annual Conference on Climate Change: Scenarios, Impacts and Policy. Societa Italiana per le Scienze del Clima, Venice, Italy.

Schmidt D., Taylor A. H. and Skinner C. N. (2008). The influence of fuels treatment and landscape arrangement on simulated fire behavior, southern Cascade Range, California. *Forest Ecology and Management*, **255**, 3170–3184.

Schroeder M. J. and Buck C. C. (1970). Fire Weather. USDA Forest Service, Washington, DC.

Scott J. H. (1999). NEXUS: a system for assessing crown fire hazard. *Fire Management Notes*, **59**, 21–24.

Scott J. H. (2006). An analytical framework for quantifying wildland fire risk and fuel treatment benefit. In: Fuels Management – How to Measure Success: Conference Proceedings, P. L. Andrews and B. W. Butler (Comps), March 28–30, Portland, OR. USDA Forest Service, Rocky Mountain Research Station Proceedings RMRS-P-41, pp. 169–184.

Scott J. H. and Burgan R. E. (2005). Standard Fire Behavior Fuel Models: A Comprehensive Set for use with Rothermel's Surface Fire Spread Model. USDA Forest Service General Technical Report, Rocky Mountain Research Station RMRS-GTR-153, p. 72.

Scott J. H. and Reinhardt E. D. (2001). Assessing Crown Fire Potential by Linking Models of Surface and Crown Fire Behavior. Research Paper RMRS-RP-29. U.S. Department of Agriculture, Forest Service, Rocky Mountain Research Station, Fort Collins, CO, p. 59.

Scott J. H. and Thompson M. P. (2015). Emerging concepts in wildfire risk assessment and management, pp. 196–206. In: Proceedings of the large wildland fires conference; R. E. Keane, M. Jolly, R. Parsons and K. Riley (eds), 2015, May 19–23, 2014; Missoula, MT. Proc. RMRS-P-73. U.S. Department of Agriculture, Forest Service, Rocky Mountain Research Station, Fort Collins, CO, p. 345.

Scott J. H., Helmbrecht D. J., Parks S. A. and Miller C. (2012). Quantifying the threat of unsuppressed wildfires reaching the adjacent wildland-urban interface on the Bridger-Teton National Forest, Wyoming, USA. *Fire Ecology*, **8**, 125–142.

Scott J. H., Thompson M. P. and Calkin D. E. (2013). A Wildfire Risk Assessment Framework for Land and Resource Management. Gen. Tech. Rep. RMRS-GTR-315. U.S. Department of Agriculture, Forest Service, Rocky Mountain Research Station. p. 83.

Short K. C. (2014). A spatial database of wildfires in the United States, 1992–2011. *Earth System Science Data*, **6**, 1–27.

Slovic P. (1999). Trust, emotion, sex, politics, and science: surveying the risk-assessment battlefield. *Risk Analysis*, **19**, 689–701.

Spies T. A., White E. M., Kline J. D., Fischer A. P., Ager A., Bailey J., Bolte J., Koch J., Platt E., Olsen C. S., Jacobs D., Shindler B., Steen-Adams M. M. and Hammer R. (2014). Examining fire-prone forest landscapes as coupled human and natural systems. *Ecology and Society*, **19**, 9.

Stephan K., Miller M. and Dickinson M. B. (2010). First-order fire effects on herbs and shrubs: present knowledge and process modeling needs. *Fire Ecology*, **6**, 95–114.

Stephens S. L., Burrows N., Buyantuyev A., Gray R. W., Keane R. E., Kubian R., Liu S., Seijo F., Shu L. and Tolhurst K. G. (2014). Temperate and boreal forest mega-fires: characteristics and challenges. *Frontiers in Ecology and the Environment*, **12**, 115–122.

Stratton R. D. (2006). Guidance on Spatial Wildland Fire Analysis: Models, Tools, and Techniques. RMRS-GTR-183, Gen. Tech. Rep. RMRS-GTR-183. U.S. Department of Agriculture, Forest Service, Rocky Mountain Research Station, Fort Collins, CO, p. 15.

Thompson M. P., MacGregor D. G., Calkin D. E. (2016). Risk Management: Core Principles and Practices, and Their Relevance to Wildland Fire. Gen. Tech. Rep. RMRS-GTR-350. U.S. Department of Agriculture, Forest Service, Rocky Mountain Research Station, Fort Collins, CO, p. 29.

USDA (1996). (United States Department of Agriculture) KCFAST: Kansas City Fire Access Software User's Guide. U.S. Department of Agriculture, Forest Service, Fire and Aviation Management, Washington, DC.

Vaillant N. and Ager A. (2014). ArcFuels: an ArcMap toolbar for fuel treatment planning and wildfire risk assessment. *Fire Management Today*, **74**, 21–23.

Vaillant N., Fites-Kaufman J., Reiner A. L., Noonan-Wright E. K., Dailey S. N. and F. USDA (2009). Effect of fuel treatments on fuels and potential fire behavior in California, USA, National Forests. *Fire Ecology*, **5**, 14–29.

van Aalst M. K., Cannon T. and Burton I. (2008). Community level adaptation to climate change: the potential role of participatory community risk assessment. *Global Environmental Change*, **18**, 165–179.

Van Wagner C. E. (1977). Conditions for the start and spread of crown fire. *Canadian Journal of Forest Research*, **7**, 23–34.

Van Wagner C. E. (1993). Prediction of crown fire behavior in two stands of jack pine. *Canadian Journal of Forest Research*, **23**, 442–449.

Varner J. M. and Keyes C. R. (2009). Fuels treatments and fire models: errors and corrections. *Fire Management Today*, **69**, 47–50.

Vasilakos C., Kalabokidis K., Hatzopoulos J., Kallos G. and Matsinos Y. (2007). Integrating new methods and tools in fire danger rating. *International Journal of Wildland Fire*, **16**, 306–316.

Verde J. C. and Zezere J. L. (2010). Assessment and validation of wildfire susceptibility and hazard in Portugal. *Natural Hazards and Earth System Sciences*, **10**, 485–497.

Westerling A. L. (2016). Increasing western US forest wildfire activity: sensitivity to changes in the timing of spring. *Philosophical Transactions of the Royal Society B-Biological Sciences*, **371**, 20150178.

Williams D. R., Jakes P. J., Burns S., Cheng A. S., Nelson K. C., Sturtevant V., Brummel R. F., Staychock E. and Souter S. G. (2012). Community wildfire protection planning: the importance of framing, scale, and building sustainable capacity. *Journal of Forestry*, **110**, 415–420.

Yuan M. (1997). Use of knowledge acquisition to build wildfire representation in Geographical Information Systems. *International Journal of Geographical Information Science*, **11**, 723–745.

Zachariassen J., Zeller K. F., Nikolov N. and McClelland T. (2003). A Review of the Forest Service Remote Automated Weather Station (RAWS) Network. Gen. Tech. Rep. RMRS-GTR-119. U.S. Department of Agriculture, Forest Service, Rocky Mountain Research Station, Fort Collins, CO, p. 153.

Part 4

Geophysical Hazards Methodologies

Chapter 10

Geological hazards

Pavel Liščák, Marek Biskupič, Josef Richnayvsky and Martin Bednarik

10.1 MASS MOVEMENT HAZARDS

Two main counter-forces are shaping the morphology of the Earth's surface – Earth's internal heat vs gravity. While the internal energy of the Earth drives orogenic processes of mountains formation, the gravitational force is acting in the sense of surface levelling. These gravity-induced processes are termed as mass wasting. They include erosion, abrasion, sliding, etc. and the transporting agents are water, air and ice.

10.1.1 Slope deformations

Slope deformation (synonyms: slope failure, gravitational slope deformation, gravitational re slope failure) is a resulting form of slope movement generated by gravitational force, which led to a formation of a body differing from the surrounding rock environ due to change in shape, location or volume, or internal structure.

Landslide is a type of slope failure, which evolved as a result of gravitation movement of rock or soil masses along one or several shear planes. The main parts of landslide are depicted in Figure 10.1.

Slope movement is a geodynamic process, during which transport of rock masses occurs as a result of the effect of gravitational force of the Earth or the Moon. The result of the process is a slope failure. In civil engineering, sensu stricto, the slope movement doesn't comprise the cases, when the rock masses are transported by transportation media (water, ice, snow or wind).

Sliding a relatively short-term, glide down-slope movement of rock mass along one or a series of shear planes, in which a part of sliding masses is relocated above intact rocks, thus creating a landslide accumulation.

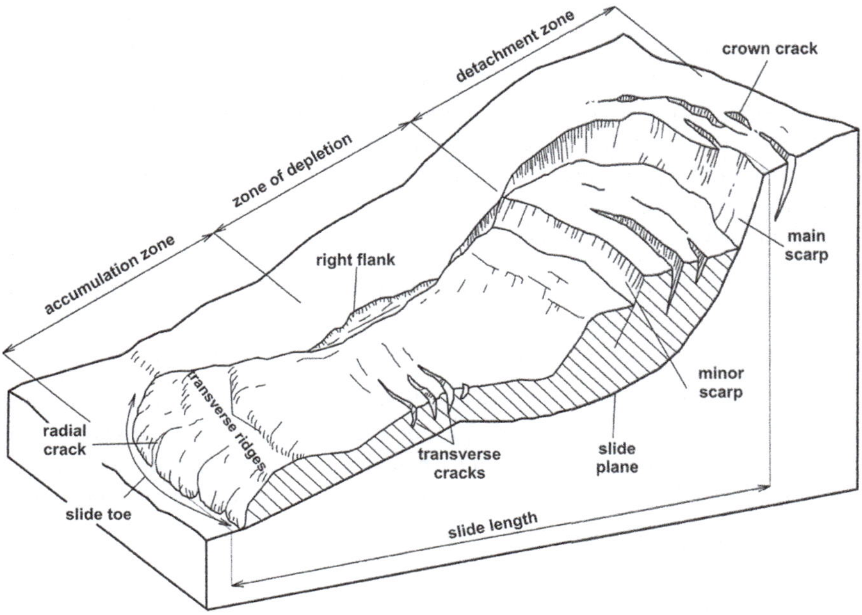

Figure 10.1 Landslide and its major morphologic signs (modified according to Petro *et al.* 2008)

Slope stability is a capability of natural or artificial slope (cut, embankment, levee, etc.) to sustain certain angle. Slope stability is affected by geometric (slope angle, bedding etc.), hydrogeological and climatic (groundwater table level, precipitation totals, effective precipitation, etc.) conditions and engineering geological conditions (bulk gravity, shear strength). Slope stability is assessed in the form of stability degree.

Slope stability degree is a numerical expression of a ratio between passive forces – friction strength, which act against rock disturbance, and the active forces-shearing stress, tending down-slope pull. Slope stability degree is marked by a symbol F and is calculated by the following formula:

$$F = \frac{\sum P}{\sum A} \tag{10.1}$$

where $\sum P$ is a sum of passive forces acting in the slope, $\sum A$ is a sum of active forces acting in the slope.

10.1.2 Snow avalanches

Snow avalanches are typical mass downslope movements and are a part of bigger group called natural hazards. Mass of snow moving downslope driven by gravitational force is usually referred to as snow avalanche. Snow avalanches can range from only a few meters long, almost harmless sluffs, to several kilometres long, disastrous avalanches capable of destroying a whole village.

Topography of the slope and meteorological condition are the most important factors and, thus, determine when and where the naturally triggered avalanches will be released. On the other hand, human trigged avalanches include a third factor, which is us, the human being. There are two basic types of snow avalanche classification based on the following:

(1) type of triggering mechanism and motion (loose snow, slab and glide avalanche);
(2) type of snow in avalanche (wet, powder and mixed avalanches).

Despite the fact that snow avalanches do not have the impact of large natural hazards as earthquakes, floods and volcanic eruptions, they still can be a threat to humans and infrastructure in mountainous areas. Worldwide annual statistics count approximately 250 fatalities caused by snow avalanches. An example of one of the biggest avalanches is shown in Figure 10.2.

Figure 10.2 One of the biggest avalanches in Carpathian mountain range was released from the slopes of Prislop in Žiarska valley (Western Tatras) in March 2009.

10.1.3 Ice avalanches

Ice avalanches are common glacial hazard occurring in glaciated areas. They belong to typical gravity driven mass wasting process. Ice avalanche occurs when a large mass of ice breaks off from a glacier, drops downslope and bursts into pieces of ice (Alean, 1984; Margharet & Funk, 1999). Many high alpine hanging glaciers produce ice avalanches as normal ablation process on steep slopes. The main preconditions for ice avalanching are: steep slope (critical > 25°), sufficient ice mass and certain degree of instability within the ice mass (Salzman *et al.* 2004). Most of the hanging glaciers fulfil the required conditions. Hanging glaciers are small type of glaciers originating from the steep mountain faces or wall of the glacier valley. Large ice masses do not detach from steep hanging glaciers very often. Such cases are extremely rare, occurring usually during the melt season, when the stability reaches its critical state. Ice avalanches are comparable to snow avalanches, however, they occur less frequently.

Combined ice – snow avalanches together with rock falls can gain high destructive potential with very long run-outs (Margharet & Funk, 1999). Major ice avalanche catastrophes usually draw public attention. In 1970, earthquake in Peru induced rock/ice avalanche from the slopes of Nevado de Huascaran. The consequent mass wasting process killed approximately 20,000 people (Plafker, 1978). Another large scale combined rock/ice was recorded in Northern Caucasus – Kolka region in September 2002. This event took the life of 140 people and caused large destruction (Kääb *et al.* 2003).

Despite the fact that occurrence of ice avalanche is low compared to other geological hazards in combination with other mass wasting processes, they can have disastrous potential. It can be assumed that with ongoing climatic change and glaciers retreat, it is expected to hear more about ice avalanches in the near future.

10.2 LANDSLIDES

Mass movements can be divided according to major (mechanism of movement, rate of movement) and minor features (age, activity degree, genesis, slope structure) into groups and types. In Slovakia the most adopted classification is according to Nemčok *et al.* (1974), which distinguishes slope movements based on the rate of the movement: creeping, sliding, flowing and falling.

10.2.1 Landslides classification

Worldwide, the most adopted is the classification by Cruden and Varnes (1996), which distinguishes six groups (Table 10.1).

Falls are abrupt movements of masses of geological materials, such as rocks and boulders, that become detached from steep slopes or cliffs, undercut by eroding agents like stream, glacier, ocean, lake or wind loaded with aeolian sediments. The fall is the fastest of slope movement types with velocities reaching

100–200 km hour^{-1} (terminal velocity). Separation occurs along discontinuities, such as fractures, joints, and bedding planes, and movement occurs by free-fall, bouncing, and rolling. Falls are strongly influenced by gravity, mechanical weathering, and the presence of interstitial water. Repeated falls in one place over certain time period result in an accumulation of fragments and blocks termed as talus.

Table 10.1 Schematic landslide classification adopting the classification by Cruden and Varnes (1996).

Movement type		Type of Sliding Material		
		Rocks	Soils (Engineering Soils)	
			Non-Cohesive in Prevail	Cohesive in Prevail
Fall		Rock fall	Debris fall	Soil fall
Topples		Toppling of stony blocks	Debris topple	Soil topple
Sliding	Along rotational shear plane Along planar shear plane	Rock slide	Rock slump Earth block slide Earth slide	Earth slide
Lateral movement (block displacement)		Rock spread	Rock spread	Earth spread
Flow		Rock flow, rock avalanche	Debris flow, debris avalanche	Earth-flow, (soil creep)
Complex – combination of two or more prevailing types of movement				

Toppling failures are distinguished by the forward rotation of a block about some pivotal point, below or low in the block, apart from the rock massif, under the actions of gravity and forces exerted by adjacent blocks, by fluids in cracks (freezing and thawing), by temperature changes or by wedge effect of roots.

There are five basic categories of **flows** that differ from one another in fundamental ways.

Debris flow. Sudden channelled flows of large masses of weathered material (sand, clay, fragments of rock), vegetation and water in a form of a slurry down the slope due to extremely intense precipitation (long-term rainfall, heavy rains) or by a sudden snow melt. The debris flows are often accompanied by landslides and rock avalanches. The debris flows are formed mainly in high mountains with steep slopes over 30°.

Debris avalanche. This is a variety of very rapid to extremely rapid debris flow, often related to the catastrophic collapses from an unstable side of a volcano.

Earthflow represents a movement of colluvial fine-grained material, substantially wetted and saturated with water, forming a bowl or depression at the head.

Mudflow is a much faster earthflow consisting of material that is wet enough to flow rapidly and that contains at least 50 percent sand-, silt-, and clay-sized particles. They are triggered by extremely heavy rains or snow thawing.

Creep or **flowage** is the imperceptibly slow, steady, downward movement of slope-forming soil or rock. Its ultimate cause is gravity. Movement is caused by shear stress sufficient to produce permanent deformation, but too small to produce shear failure. Typical manifestations of creep are curved tree trunks ("drunken forest"), bent retaining walls, fractured engineering structures, tilted poles or fences, and small soil ripples or ridges.

Lateral spreads occur typically on very gentle slopes or flat terrain. The dominant mode of movement is lateral extension accompanied by shear or tensile fractures. The failure is caused by liquefaction of sensitive clays or sandy clays, usually triggered by rapid ground motion, for instance due to an earthquake.

10.2.2 Landslide causes

Essentially the mass movement occurs whenever the downward acting gravity and resulting shearing stress overcomes the forces resisting the sliding or flow – shear and friction strength. When shearing stress exceeds friction or shear strength, sliding occurs. This scenario may occur both due to naturally generated change of stability conditions and improper anthropogenic interventions into the slope.

10.2.2.1 Geological causes

The following list may be used: (a) weak or sensitive materials, (b) weathered materials, (c) sheared, jointed, or fissured materials, (d) adversely oriented discontinuity (bedding, schistosity, fault, unconformity, contact, and so forth), (e) contrast in permeability and/or stiffness of materials, and (f) earthquake.

10.2.2.2 Morphological causes

This class may include the following: (a) tectonic or volcanic uplift, (b) glacial rebound, (c) fluvial, wave, or glacial erosion of slope toe or lateral margins, (d) subterranean erosion (solution, piping), (e) deposition loading slope or its crest, (f) vegetation removal (by fire, drought), (g) thawing, (h) freeze-and-thaw weathering, and (i) shrink-and-swell weathering.

10.2.2.3 Human causes

This class may include: (a) excavation of slope or its toe, (b) loading of slope or its crest, (c) drawdown (of reservoirs), (d) deforestation, (e) irrigation, (f) mining, (g) artificial vibration, (h) water leakage from utilities.

10.3 SNOW AVALANCHES

Avalanches are usually released on the slope, where the gravitational force on the snowpack exceeds the internal cohesive force between the individual snow layers and the mass of snow starts to move downslope. Basically, there are two types of avalanches: (1) avalanches starting from a single point – **loose snow avalanches** and (2) avalanches in a form of cohesive snow layer – **slab avalanches**. The slab avalanches are the ones, which are the most dangerous to people and property. As a consequence of various precipitation events, the snowpack consists of several layers. The main prerequisite for triggering the slab avalanche is the presence of weak layer within the snowpack. The weak layer consists of snow grains, which have very low internal cohesion within the layer. On contrary, with normal snow layers, weak layers are usually not as hard and they tend to collapse easily. Release of snow slab starts with fracture, which propagates within the weak layer across the snowpack and the so-called snow slab is triggered. Consequently, the snow mass of a slab is triggered and driven downslope by gravitational force. The causes of initial fracture can be various ranging from precipitation, temperature change, cornice fall or additional load by winter traveller or explosive. The snow is the main substance in snow avalanches, but during their motion the avalanche entrains other substances, such as air, soil, rock debris, trees, etc. Multiple avalanche experiments have confirmed that dry slab avalanches can gain speeds of 50–100 km/h depending on terrain topography. However, powder avalanches are the ones that can reach maximum velocities of 200–300 km/h (Figure 10.3).

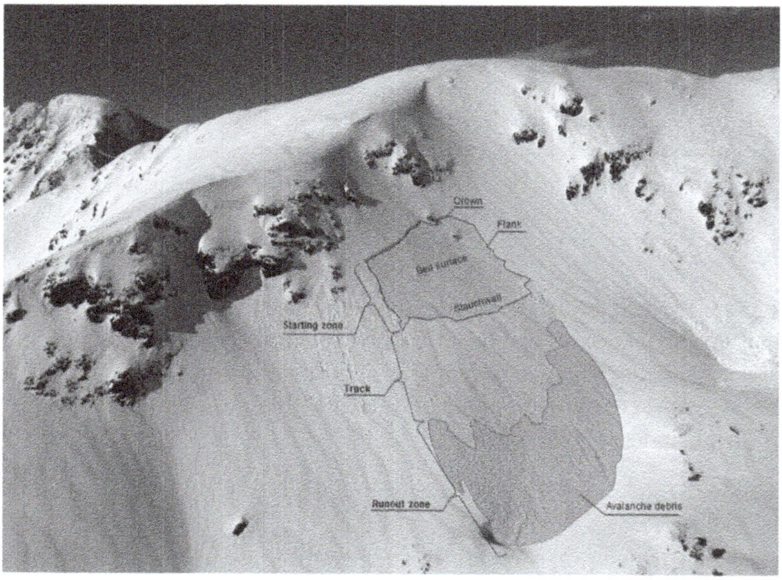

Figure 10.3 Parts of an avalanche slide path.

Besides the snow avalanches in the mountains, roof avalanches occur equally, however, these types are far less of a concern. No snow avalanche would ever be triggered without a steep avalanche-prone slope. While the snow is the sliding medium, mountain slopes provide sliding surface for every avalanche. The vast majority of avalanches are released from slopes with an angle between 30°–45°. Usually, slopes less steep than 30° are just not steep enough to produce an avalanche, but they might be avalanche-prone only during extremely unstable conditions or for special avalanche types (e. g. slush flows). Slopes steeper than 45° are subject of frequent sluffs and small avalanches and they do not collect enough snow to trigger large snow avalanches. According to worldwide statistics (Jamieson, 2000) the most avalanche-prone slopes are 30°–45° with 37° as prime avalanche angle. Other factors, which affect the avalanche predisposition of certain slope are: aspect, land cover, slope shape curvature and elevation.

Avalanche starts in starting zone also sometimes referred to as a release or trigger zone. After the fracture propagates in the snow, slab starts to move and gain speed. Crown face height and crown length have great importance, thus, they determine the volume, which gives initialization to the slab. Avalanches gain the highest velocities in track and here they entrain more snow, soil or rocks. In a very steep terrain, large slab avalanches can turn into very destructive powder avalanches. Mixture or air and snow at high velocities create an air blast with very destructive force. When the slope becomes less steep or surface friction exceeds certain limits the avalanche starts to decelerate. This happens in the runout zone, where the avalanche stops and thus creates avalanche debris. Depending on the size of avalanche (Table 10.2), avalanche debris can range from tens of centimetres to tens of meters.

10.3.1 Types of snowpack

Snow is the crucial building material of every avalanche and thus it influences the triggering and avalanche motion. The type of snowpack greatly depends on climate. In general, there are three different snowpack types based on climate: maritime, intermountain, and continental. Snow packs in maritime climate are usually thick, strong and warm. On the other hand, continental snowpack's are thin, cold and weak. Intermountain snowpack are somewhere in between of maritime and continental. Of course, there are no strict boundaries and in certain period of winter season there might be continental thin and cold snowpack in coastal mountains and vice versa. This phenomenon is often visible on small scale, when it is possible to find two different snowpack types just a few meters from each other. For example, wind scoured areas can show presence of continental snowpack and just below the ridge on the lee side there is a maritime thick snowpack.

According to many previous studies and observations, snowpack in coastal mountains tends to be over 3 meters deep with relatively warm temperatures near to freezing point (–5°C to +5°C). The new snow falls at high rates with high densities while avalanches occur immediately during or just after the storms. The weak layers often form at the transition between new and old snow and include low cohesion layers or graupel with a short life span. The intermountain snowpack is, on the average, 1.5 to 3 meters deep with temperatures ranging from –15°C to –3°C. Weak layers, such as faceted snow or buried surface hoar, can persist for several days. Also the avalanche activity and snowpack instabilities can remain for longer periods (weeks). The thinnest (<1.5 metres) and the coldest (–10°C to –30°C) snowpack is the one in continental climates. The new snow falls at low rates and low densities. The most common weak layers consist of faceted snow, depth and surface hoar, which can last for prolonged periods of time. Even many days after the storms, either naturally or human triggered avalanche activity can be observed.

Table 10.2 Classification of avalanches by size.

Classification/ Size		Destructive Potential	Run-Out	Typical Length and Volume
Size 1	Sluff	Typically does not bury a person, could carry a person away (danger of falling)	Stops typically before the end of a slope	<50 m 100 m³
Size 2	Small avalanche	Could bury, injure or kill a person	Stops typically at the end of a slope	50–200 m 1.000 m³
Size 3	Medium avalanche	Could bury and destroy a car, damage truck, destroy a small building or break a few trees	Could traverse a flat terrain (considerably below 30°) over a distance of typically less than 50 m	200–600 m 10.000 m³
Size 4	Large avalanche	Could destroy a railway, large truck, several buildings or a small forest area of approx. 4 hectares	Could traverse a flat terrain (considerably below 30°) over a distance of typically more than 50 m	1 km 100.000 m³
Size 5	Very large avalanche	Could gouge the landscape, catastrophic destructive potential	Reaches the valley floor, the largest run out distances known	2–3 km >100.000 m³

10.3.2 Avalanche formation and motion

Generally, avalanches can be triggered either as loose snow or a slab. In the second case, the slab is released after fracture propagates across certain area and creates initial volume that starts to slide downslope. As a result of bed roughness, slabs break into smaller blocks and particles. As the avalanche motion continues, the particles become smaller due to their collision among themselves and surface. As the velocity increases, the particles become smaller and smaller. After the particles are broken into fine elements or the avalanche runs for longer distance, the motion develops into a flow. High density material of snow particles and air flow at the bottom of the avalanche creates the so-called core. In the case of dry snow entrained in avalanche, the low density dust cloud is being created around the core. It has been estimated that one third to one half of the volume of the core represents the snow particles and the rest is the air. On the contrary, in the dust cloud, only about 1% of the space is filled with snow particles and 99% is air. Estimating the speed of avalanche and its impact, pressures is subject of exhaustive research and studies. Basically, the speed depends on the terrain roughness, resistive forces between the core and surface and resistance between the dust and air. Of course, most of the frictional resistive forces are generated at the bottom of the avalanche between the core and sliding surface. All the rubbing and collision between particles generate heat that results in the production of small amount of water. Avalanche decelerates and stops in the runout zone. Water causes snow particles to freeze together. This results in debris hard to penetrate with shovels and probes making avalanche rescue very demanding. It is well known that avalanche motion plays a crucial role in avalanche modelling and estimating runout distances. Figure 10.4 shows the sequence of an avalanche motion in different time steps calculated by numerical model RAMMS.

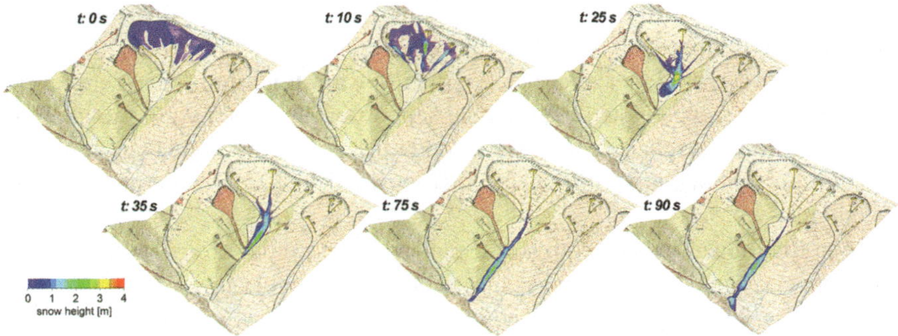

Figure 10.4 Sequence of an avalanche motion in different time steps calculated by numerical model RAMMS.

Snow avalanches usually entrain a large amount of snow from the slope during their motion (Sovilla *et al.* 2001). The entrainment affects the motion and it should be taken into account by the model. The mechanisms of entrainment can be different depending on the avalanche type (whether it is a heavy dense avalanche or a light dry one, or a powder one). In the same avalanche there can be various mechanisms of entrainment (Gauer & Issler, 2004). The entrainment mechanism for heavy dense avalanches differs from that for light dry ones. Numerous studies have shown that avalanches can entrain 10 times more of snow than the initial slab volume.

10.4 SLOPE MOVEMENTS MITIGATION

Depending upon the infrastructure under threat of slope failure and the magnitude of slope failure, the measures are implemented to safeguard the slope stability, or to protect a slope. In the case of sudden slope failures, **emergency remedial measures** are realized immediately. The aim is to stop or at least slow down the slope movement and to alleviate the damage to a minimum scale. The principal strategy is to reduce the water content, or pore pressure, in the sliding masses and thus to increase the frictional resistance to sliding. The prevention against infiltration of surface water in the sliding masses can be reached by covering the surface with impervious material, by clogging the cracks in the landslide body with impervious material and diverting surface runoff above the slope by system of drainage ditches. The subsurface water content can be reduced by intense pumping of water from the existing wells. In the case of relatively pervious rocks a system of sub-horizontal drainage wells can also help to carry the water out of the slide area.

In the case of small-scale landslides the stability can be increased by additional surcharge material at the toe of the slope. In the case of large investments, for instance construction of highways, tunnels, railways or development of urban areas, **remedial measures** are implemented. They are usually preceded by mapping and survey, which shall clarify the engineering geological conditions and identify areas at risk and shall suggest effective stabilizing measures.

Among the essential measures belongs the **slope modification**. This includes (1) reduction of the slope angle, (2) additional support of the slope foot (accumulation zone) by material surcharge, and (3) reduction of load on the slope by removal of a certain portion of the sliding material from the detachment zone and zone of depletion.

Slope drainage includes both surface and subsurface dewatering of the slope. The surface drainage shall prevent the slope against infiltration of precipitation water into the slope and divert the water from the parts above the sliding slope. The subsurface drainage is realized by sub-horizontal boreholes, sometimes combined with gravel drainage piles or walls.

Technical stabilization measures involve anchoring combined with (micro) pile retaining walls. In the case of smaller slides in silty- and clayey rock masses the stabilizing-drainage ribs are often applied combined with gabion walls. Rock bolts are used to stabilize rockslides and rocky slopes. As a passive measure on steep rocky slopes, mainly above the communications, protective barrier or catch fences are used. They are made of flexible mesh so that the energy of debris flow is dissipated. The protective galleries are built as overshoots above roads to prevent them from debris flows and snow avalanches. Material simply slides over the roof of the protective structure and continues down the hillside.

Protection of near surface soil against weathering and erosion includes ground covers or other vegetation planted. The most preferred are fast growing plants with sturdy and extensive root systems. The alternative protection offer thin sheets of concrete, gravel or geotextile.

10.5 AVALANCHE MITIGATION

It is very demanding and so far impossible to forecast avalanches with very precise timing. As the avalanches occur very suddenly there is no way to expect them in advance of days or weeks. Therefore, the mitigation and protection measures have to be adapted to the very specific avalanche behaviour. Avalanches failing in remote uninhabited locations do not pose serious threat to human settlements and infrastructure. The mountain settlements, roads and mountain infrastructure are the most vulnerable to avalanches. Damage from interaction of avalanches and human activities can be prevented by controlling the avalanches either by explosives and permanent structure placing into avalanche path or evacuations (Figure 10.5).

Figure 10.5 Snow fences placed in the release zone are used prevent avalanche to trigger.

The aim to avalanche control by explosives is to artificially and periodically trigger smaller avalanches by exposing the snowpack to the air blast from explosion.

Explosion, which is most effective over the snowpack, creates forces, which put additional loading and stress on the snowpack, which tends to slide and create avalanches (Figure 10.6). Regular control by explosives prevents the creation of large and disastrous avalanches, which could threaten human infrastructure. Artificial avalanche triggering can be done either by using permanent installations (towers mast or lifts) with reserve of explosives, which are usually remotely triggered. These installations are placed in the release zones of avalanche paths. The advantage is that they can be used in almost any kind of weather unless the lower placed infrastructure (road, lifts etc.) are not in use or operation. Another and more flexible approach is to carry explosives by helicopter, but this way is limited by weather and flight conditions. Sometimes, hand charges are being used mostly around ski slopes and lifts.

Figure 10.6 Hand charges assembled in to the area of avalanche starting zone.

Permanent structures, which are placed either in release or run-out zones are often used to protect human settlements and roads. Installations in run-out zones (snow fences, supporting structures) tend to prevent avalanche initiation, while the installations in track and run-out zones (deviation, retarding dams, snow sheds, galleries, and splitting wedges) tend to deflect and decelerate avalanche in motion or they directly protect from the influence of avalanches. These installations are the ones most expensive and they do have large impact on the nature. On the other hand, they are the most effective. By far the most nature friendly method is to use forest and its natural avalanche protection feature. However, this requires reforestation in the avalanche path, which can be done only until the upper border of treeline and it is rather cost and time-consuming.

10.6 MASS MOVEMENT HAZARD AND RISK ASSESSMENT

The assessment of landslide hazard and risk is predominantly important in urbanised zones included into land use plans, which assume certain socioeconomic and technological progress of a region. Especially, during the last decade, a number of scientific works have been devoted to statistical approaches to landslide hazard using map algebra tools implemented within the geographic information systems. This methodology includes the assessment of landslide hazard, identification of the elements at risk, assessment of vulnerability and the assessment of landslide risk (Figure 10.7). The application of Geographic Information Systems (GIS) represents an innovative, modern and prospective approach in the assessment of hazards and risk for larger territorial units. This section is focused on the theoretical description of the most common quantitative methods used for landslide hazard or susceptibility assessment.

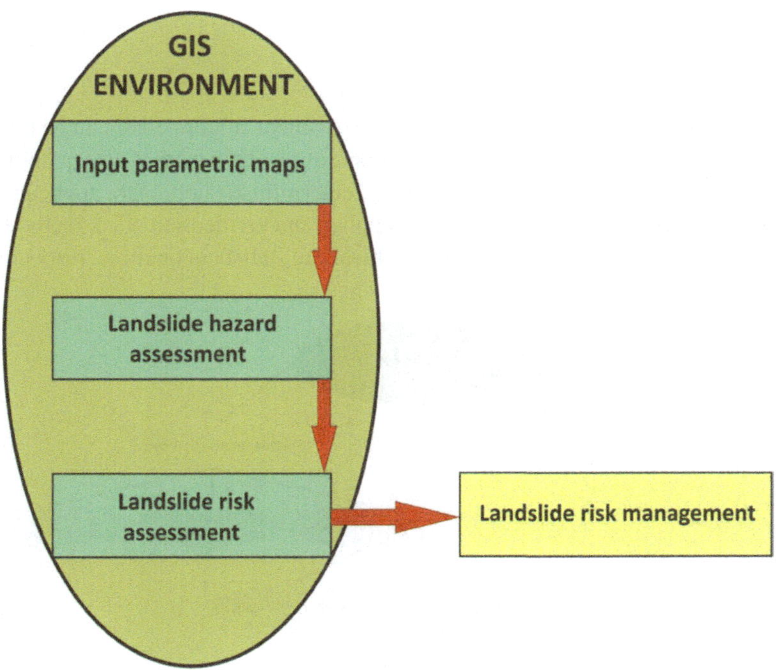

Figure 10.7 Flowchart of processing the landslide risk issue.

10.6.1 Risk terminology

The terminology used here is based on the conventional terminology according to Varnes (1978, 1984), van Westen (1993), and Aleotti and Chowdhury (1999).

Landslide hazard is defined as the probability of occurrence of landslide event within specific area and time period. **Vulnerability** reflects the capacity of the system and its elements to react to the occurrence of a harmful phenomenon in the form of a loss or damage. **Element at risk** is an element of the assessed system, where loss and damage may appear during the occurrence of a potentially harmful phenomenon. **Risk** represents probable losses and damage on the endangered elements of the environment caused by the occurrence of a potentially damaging phenomenon. Leroi (1996) defines the risk as an expected loss (toll of casualties, loss of property and interrupted economic activities) due to a specific hazard in a given area and time period. As a mathematical expression, the risk is a product of hazard and vulnerability. Four basic axioms are needed before we start to assess landslides hazard:

(1) landslide will occur under identical geological, geomorphological, hydrogeological and climatic conditions as in the past;
(2) the triggering factors of landslides may be defined and analysed;
(3) the degree of landslide hazard must be quantifiable;
(4) slope deformations must be classifiable.

10.6.2 Methods of landslides hazards assessment

Generally, the methods of landslide hazard assessment (Figure 10.8) can be divided into qualitative and quantitative. This section covers the group of the quantitative methods, particularly the statistical analyses, deterministic approach, application of fuzzy logic and neural networks. The most applied and verified statistical methods are the multivariate statistical analysis and the bivariate statistical analysis, respectively.

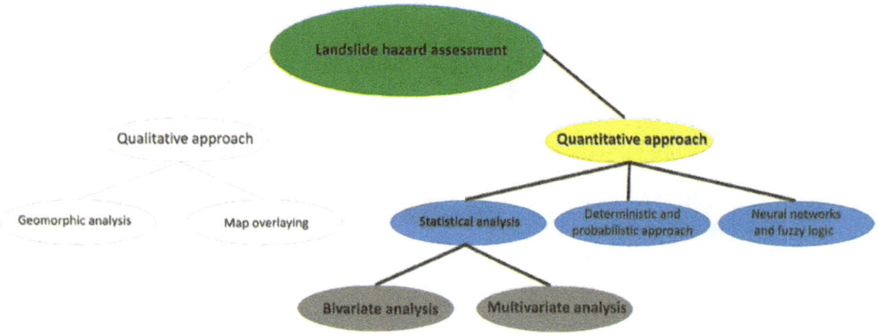

Figure 10.8 Methods of landslide hazard assessment.

The task of predicting the landslide hazard is to delimitate a certain degree of probability, the spatial delimitation of the occurrence of a landslide event as well as its intensity and time frequency. The spatial delimitation of landslides may be predicted with high probability, based on the analyses of relevant factors, using mentioned quantitative methods.

10.6.2.1 Bivariate statistical analysis

This method presents a statistical combination of each input parametric map with the landslide inventory map. The bivariate statistical analysis works with one dependent variable – landslide inventory map and one independent variable – the individual input parametric maps. The result of the combination is the delimitation of the total number of grid cells with or without landslides within the individual classes of input parametric maps, calculated as areal units or as percentage. The double combinations are stored in a table form, where one of the numbers represents the parametric map class and the second shows the presence or absence of landslides (0–false, 1–true).

Based on the combinations, each parametric map must be reclassified secondarily. During the secondary reclassification the existing classes in each parametric map are attributed with new numerical values representing statistically defined landslide probability. The top numerical value is attributed to the most landslide susceptible class and the lowest numerical value means that the class is the least susceptible to landslides. The consistency of the technical preparation of the parametric maps, especially the differences in the position accuracy, superposition and grid geometry, has a great influence on the accuracy of the results. The result of the bivariate statistical analysis is the landslide hazard map which is a weighted sum of the secondarily reclassified parametric maps. The equation for the final sum is as follows:

$$y = \sum_{i=1}^{n} C * V_i \tag{10.2}$$

where: y – the value in the final landslide hazard map, i – the individual parametric maps, C – class value, V_i – weight of the relevant parameter.

There are different approaches to the determination of input parameter weights. In principle, they may be divided into two major groups. The first group is the subjective approach based on expert opinion – analytical hierarchy process (AHP), Fuller triangle; and the second group is based on the mathematical approach – entropy model, frequency ratio (FR).

The result of the weighted sum is a continuous interval of values representing the landslide hazard value in the study area. The continuous interval must be classified as a level of landslide hazard. The most convenient are five classes corresponding to very low, low, medium, high and very high degree of landslide hazard (Bednarik *et al.* 2010; Constantin *et al.* 2010). The literature contains the following classifications:

(a) Expert-based classification – it is based on expert experience, is very subjective and very misleading when applied within large regions.
(b) Binary classification – it uses the 0.5 threshold in distinguishing two extreme cases, the presence or absence of landslides. However, this approach does not define more prone or less prone zones to landslides.

(c) Natural breaks classification – a conventional classification method, mainly in the case of data with evident breaks and discrete data.
(d) The standard deviation – it works with the mean determining the class median and uses the standard deviation (+ and –) to determine the extent of the individual landslide susceptibility classes.
(e) The equal interval method – it equally distributes the final sum into three or five used classes.

Figure 10.9 below provides the principle of the bivariate analysis determining the parameter weight values.

10.6.2.2 Multivariate statistical analysis

The most used multivariate analyses are the conditional analysis and logistic regression analysis, both provide a comparable and satisfactory prediction.

10.6.2.2.1 Conditional analysis

Multivariate statistical analysis is based on the mutual combination of all input parametric maps (lithology, slope, altitude, slope deformations) and a subsequent application of the information within areas, where no landslides have been registered (Figure 10.10). During this analysis the geohazard assessment does not include the weighting process.

In the case of the conditional analysis (Clerici, 2002) the outcome is a table containing the mutual combination of all input parametric maps, including all category combinations arising from the superposition of all the input maps. These combinations create new spatial units, the so-called unique conditional units (UCU), in the final map. For example, if there is the category 6 (10–15°) in the slope angle map, category 4 (slope sediments) in the lithology map and class 3 (spruce forest) in the land use map, the final unique conditional unit is the combination 6–4–3.

In the case of the multivariate analysis, no secondary reclassification and parameter weighting is needed. The final combinations containing landslides (value 1, true in the landslide map) are ordered based on the calculated occurrence density – the ratio of the landslide UCU cell number to the total area (the descending result gives the least favourable combinations for landsliding). The shell script in GIS GRASS may be used to classify all the combinations into the relevant class number (3 or 5) that expresses the degree of landslide susceptibility or hazard (Clerici, 2002; Bednarik *et al.* 2005).

10.6.2.2.2 Logistic regression analysis

Logistic regression models are used to predict a dependent variable Y that has a discrete distribution (it gains values from 0 to 1) and has a non-linear relationship to the predictor X. It estimates the probability of a specific event, while using the

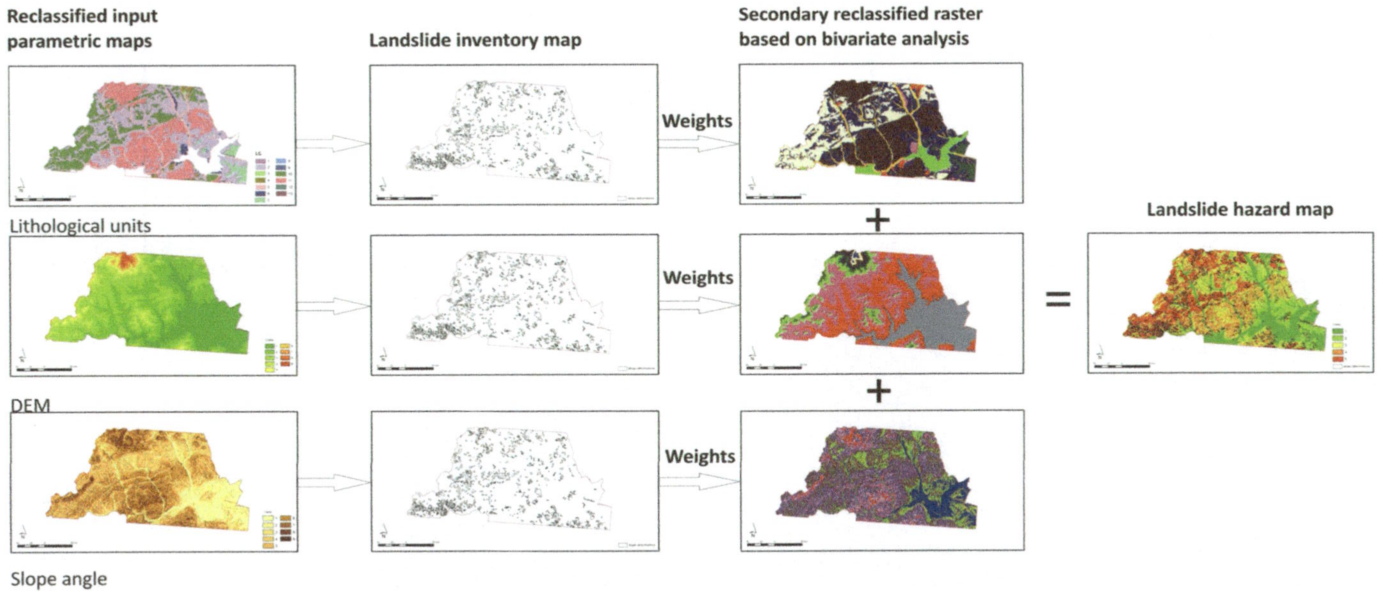

Figure 10.9 Principle of the bivariate statistical analysis determining the parameter's weight.

input factors related to the presence or absence of the event. The input factors are also processed in the form of parametric maps in GIS and as such they enter the process of statistical evaluation using the map algebra. Making use of the logistic regression for modelling the landslide hazard or susceptibility there are mutual combinations of all input parameters influencing the slope stability, the predictor X in this case, and their subsequent comparison with the landslide registration map, the dependent variable Y in this case. The logistic function equation is as below:

$$f(y) = \frac{e^y}{e^y + 1} = \frac{1}{1 + e^{-y}} \tag{10.3}$$

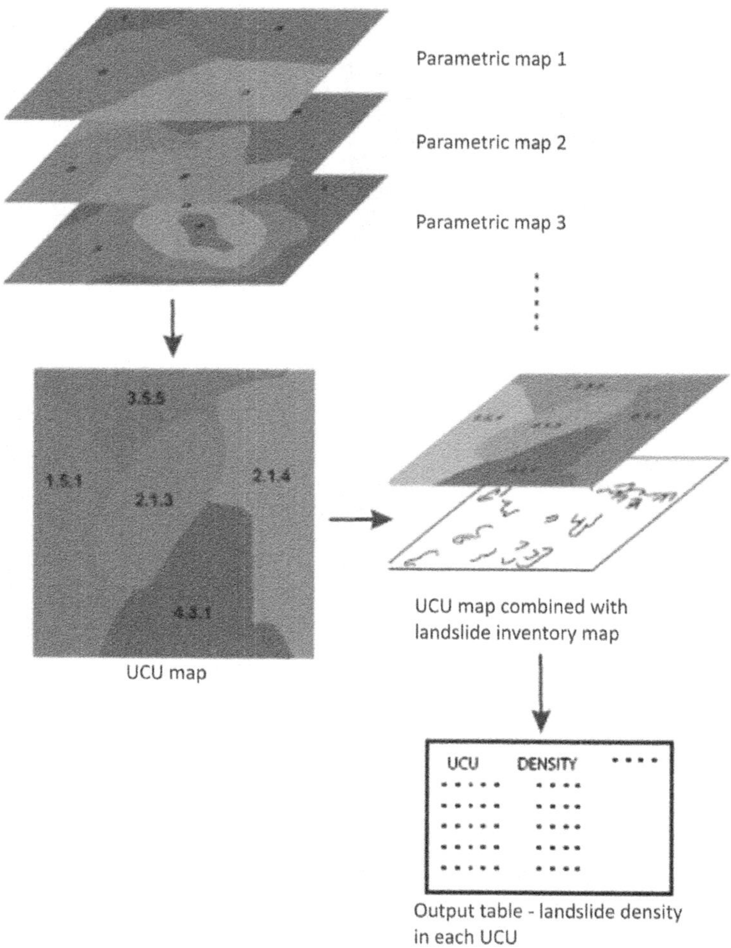

Figure 10.10 Principle of the multivariate statistical analysis (according to Pauditš, 2006).

where y is the input and $f(y)$ is the output. The variable y is affected by a certain set of independent variables, while $f(y)$ represents the probability of the given result arising from the set of variables. The variable y expresses the rate of the overall contribution of all independent variables used in the model. With regard to the amount of the independent variables $p(x_1, x_2, x_3 \ldots., x_p)$, the vector $X = (x_1, x_2, x_3 \ldots., x_p)$ is defined. The conditional probability that a landsliding event occurs is represented by the equation $P(y = 1/X)$. The course of the logistic function of the multiple models in logistic regression is defined as follows:

$$y = \beta_0 + \beta_1 x_1 + \beta_2 x_2 + \beta_3 x_3 + \ldots\ldots. + \beta_k x_k, \tag{10.4}$$

where β_0 is the constant in the equation and $\beta_1, \beta_2, \ldots., \beta_k$ are the coefficients of the variables $x_1, x_2, \ldots., x_k$, respectively. The probability $P(y = 1/X)$ may be expressed as:

$$P(y = 1/X) = \frac{1}{1 + e^{-(\beta_0 + \beta_1 x_1 + \beta_2 x_2 + \ldots.. + \beta_k x_k)}} \tag{10.5}$$

10.6.2.2.3 Deterministic approach

Landslide hazard assessment by deterministic analyses requires a relatively high precision and variability of input parameters. For that reason, the application of the method in most cases has been limited to the small areas on a large scale with homogeneous geomorphologic and geological conditions. This method is not appropriate for small and medium scale given the lack of detailed input data, mainly geological, geotechnical and hydrogeological data (groundwater table level data). Required data have been determined in the field or the laboratory. However, the processing of input data, for example the spatial variability of data, sampling error or material properties, could induce the many limitations of the method.

The use of GIS tools for the deterministic approach allows the simulation of multiple scenarios, based on the hypothesis of the variability trigger factors and compilation of relatively reliable landslide hazard maps.

Deterministic models are based on physical laws of conservation of mass, energy or momentum (Terlien *et al.* 1995). The only one model that calculates the stability of the slope of each unit cell in GIS environment is separately infinite slope stability model. Using this geotechnical model, it is necessary to make some simplifications (Jelínek & Wagner, 2007):

(1) availability of detailed and sufficiency geotechnical, hydrogeological and morphological input data must be available;
(2) the method should be applied on small areas in a large scale, in consequence the absence of detailed physical and mechanical soil properties of large areas;
(3) the method is suitable mostly for assessment of shallow landslides;

(4) considering that the landslide material has been sliding along the planar slip-plane area parallel to the surface and soil layers merge into a single layer;
(5) the simple infinite slope stability model is used;
(6) assumption of quasi-homogeneous geological and geomorphologic conditions.

Assessing the landslide hazard of the selected area by deterministic method, is expressed as a factor of safety (FS), calculating for all the slopes and final classes of landslide hazard; divided by the degree of Factor of safety. Degree of FS is given by the ratio of the sum of the passive forces and active forces. When the final value of FS is less 1.0 then the stability conditions are not stable and mass movement may occur. The following formula may be used to calculate value of FS (according to Bruden & Prior, 1979, in van Westen, 1993):

$$FS = \frac{c_{ef} + (\gamma - m\gamma_w)z\cos^2\beta\tan\varphi_{ef}}{\gamma z\sin\beta\cos\beta} \qquad (10.6)$$

where:

c_{ef} – the effective cohesion [kPa];
φ_{ef} – the effective angle of shearing resistance [°];
γ – the unit weight of soil [$kN \cdot m^{-3}$];
γ_w – the unit weight of water [$kN \cdot m^{-3}$];
β – the slope angle [°];
z – the depth of slip surface below the terrain [m];
m – hw/z – the ratio of the height of groundwater table level above the slip surface *hw* [m] and the slip surface depth *z* [m].

An example of landslide hazard map based on deterministic approach is showed on Figure 10.11 (Kralovičová *et al.* 2014). As a study area landslide prone territory in Slovakia – the western boundary of Nitrianska pahorkatina Upland, part between the towns of Hlohovec and Sered' is assessed.

10.6.2.2.4 Fuzzy logic

Fuzzy logic is a branch of mathematics derived from fuzzy set theory. It is a quite young discipline, less than fifty years of history. Basic operation and mathematical principles of fuzzy logic are described in many articles; basics are in manuscript of Zadeh published in 1965 "Fuzzy sets". Logical statements are expressed by function of membership, values are in the range from 0 to 1. In classical Boolean logic and predictive logic statements are valued as binary (dichotomic) 1 or 2; true or false; belongs or doesn't belong. Function of membership (FP) in fuzzy logic allows partial membership and associate membership to the set, so that it is extended to the whole interval <0; 1>, including both border values. Fuzzy logic allows mathematically expressed concepts like "slightly," "quite" or "a lot". Therefore, it is preferable for a number of real decision-making tasks.

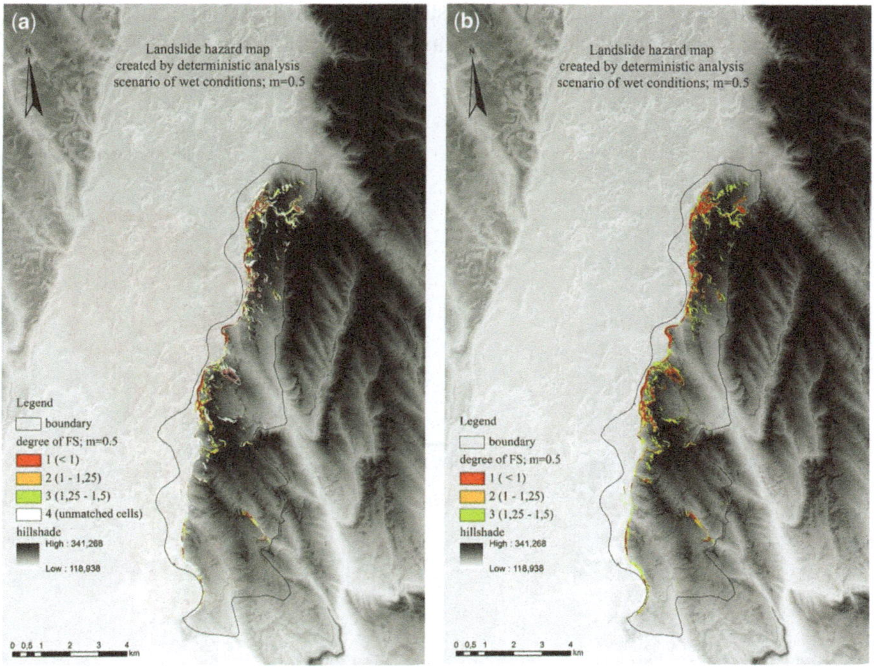

Figure 10.11 Landslide hazard maps for scenario of wet condition (m = 0.5) created by overlaying of partial hazard maps **(a)**, map only with the same degree of FS calculated in each partial hazard map, **(b)**, cells with FS value calculating less than 1.0 in each partial hazard map.

In a process of landslide hazard assessment, fuzzy logic defines the instability factors as members of a set reaching from 1, expressing the highest level of hazard or susceptibility, to 0 defining the lowest level of landslide hazard or susceptibility. The classes of individual parametric maps can be associated with fuzzy membership values based on expert opinion. Fuzzy represents membership widely defined, not strictly as probability.

Fuzzy logic has various operators – AND, OR, SUM, GAMMA etc., which allow to combine input parametric maps in several phases. Figure 10.12 shows an example of using fuzzy logic in landslide susceptibility assessment; study area here is a part of Flysch zone in northeast part of Slovakia.

10.6.2.2.5 Neural networks (NN)

Application of NN for interpretation data from remote sensing (RS) has been motivated by the ability to efficiently handle very large amounts of data from various sources. The rapid increase of NN applications in RS, mainly due to their skills: working more accurately as, e.g. statistical classifiers, especially when the feature space and complex data sources have different statistical distribution works faster than other techniques.

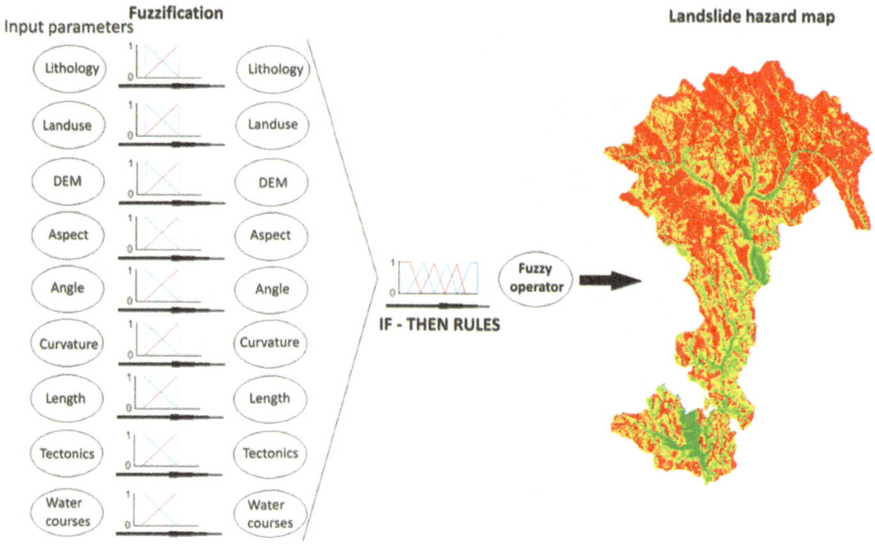

Figure 10.12 Landslide susceptibility map using fuzzy operator OR; an example from northeast part of Slovakia.

For landslide hazard assessment, multilayer forward neural network, the so-called Multilayer Perceptron (MLP) and the learning algorithm of back propagation error are used. MLP, as the name implies, consists of a series of layers, each consisting of a set of nodes (neurons). In the forward neural network, there are only forward connections between neurons, each neuron of one layer sends the signals to each neuron of the next layer, the connections to the previous layer or in one layer do not exist.

A neural network consists of a number of interconnected nodes. Each node is a simple processing element that responds to the weighted inputs it receives from other nodes. The arrangement of the nodes is called network architecture (Figure 10.13). The receiving node sums the weighted signals from all the nodes that it is connected to in the preceding layer (Pradhan & Lee, 2009). Neural network gets its ability to transform input data into output during the learning process. Learning is a fundamental and essential characteristic of neural networks. As already mentioned, the most used learning algorithm for multi-layered NN is a method of back propagation error (back propagation training algorithm).

Figure 10.14 shows an example of application NN approach in landslide hazard assessment; study area here is a central north part of Slovakia – Žilina region (Tornyai *et al.* 2016).

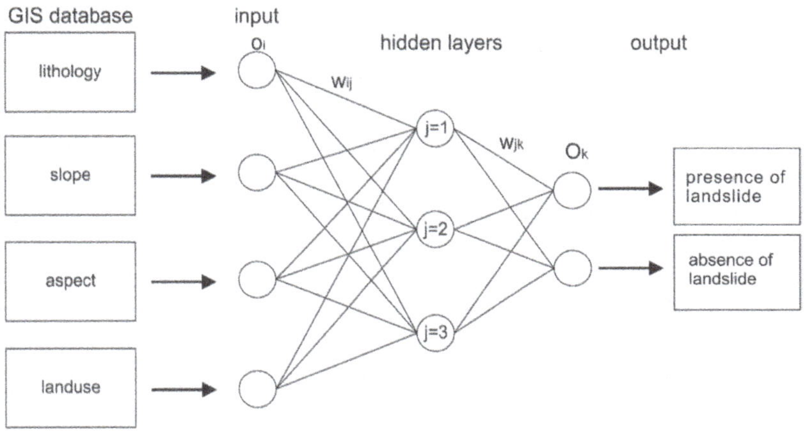

Figure 10.13 Basic architecture of neural network.

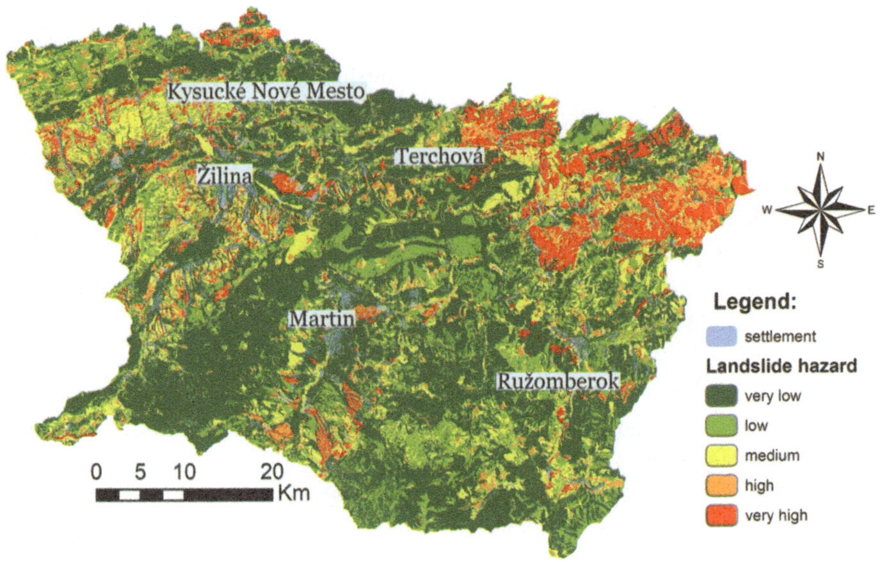

Figure 10.14 Landslide hazard map created using NN approach.

10.6.3 Landslide risk assessment

It is important to assess the risk especially in urbanised zones or areas included in land use plans for socio-economic, technological and other purposes. Fundamental element in landslide risk assessment is evaluation of vulnerability with respect to the elements at risk. Among the elements at risk, there are primarily the population, buildings, economic activities, public services, infrastructure, etc.

10.6.3.1 Vulnerability assessment

The vulnerability assessment of the individual elements at risk is an indispensable part of landslide risk assessment and it helps to understand the interaction between the given landslide event and the affected elements at risk.

The elements at risk are most frequently defined from the current land use parametric map. The elements at risk are usually assessed based on the material vulnerability, while only potential direct losses caused by landslide hazard are taken into consideration. Indirect losses that may manifest in a certain time horizon after the given hazard activation, such as interrupted infrastructure and thus the interruption of the economic activities, e.g. due to a fallen road, generally are not included in the regional studies.

Having identified the elements of risk, the further vital and logical step is the financial evaluation that serves as the background for the calculation of vulnerability. The financial evaluation is usually processed in official prices as the market price for the individual elements at risk changes in dependence on the specific market demand.

Vulnerability is a simple product of the spatial distribution of the element at risk and the relevant price. It, thus, expresses the extent of the potential damage that occurs in a case of activation of a predicted landslide hazard.

The procedure of the material vulnerability calculation may be as follows:

(1) classification of the evaluated elements at risk, e.g. in accordance with municipality cadastre maps, represented by a code;
(2) delimitation of the risk element spatial distribution according to the level of landslide hazard map;
(3) calculation of vulnerability as a product of the given risk element price and its spatial distribution within selected landslide hazard levels.

10.6.3.2 Landslide risk map

The landslide risk map shows the expected financial losses caused by slope deformations and makes use the results of prediction and analyses based on the landslide hazard assessment. Landslide risk presents simple multiplication of hazard and vulnerability assessment. The map of landslide risk is made using the following equation:

$$r = h * v \tag{10.7}$$

where: r – is the landslide risk value (e.g. expressed in Euros per pixel – 10 × 10 m), h – is the landslide hazard level, v – is the vulnerability of the elements at risk (per pixels).

The example used herein is the landslide risk map for the wider surroundings of the town of Hlohovec in Slovakia. The result is the "Euro – pixel" map mirroring the landslide risk in the model area (Figure 10.15). The landslide risk values are divided into 5 categories:

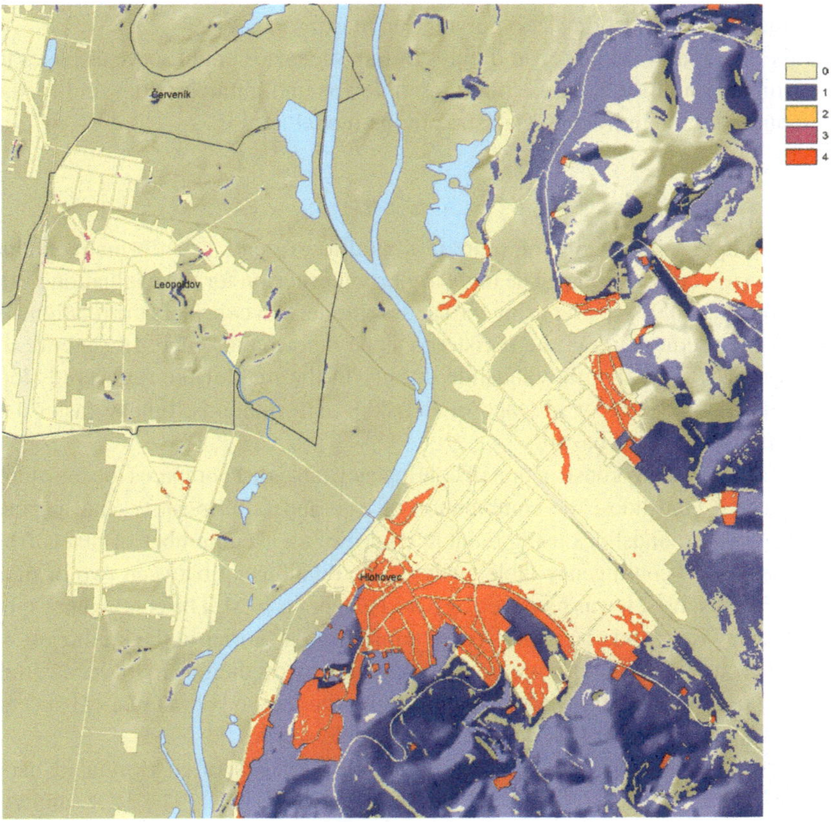

Figure 10.15 Landslide risk map – an example from Hlohovec city (southwest Slovakia).

– category 0 – represents the zero value of the landslide risk per pixel,
– category 1 – Eur 30 to 100/pixel,
– category 2 – Eur 1 300/pixel,
– category 3 – Eur 2 000/pixel,
– category 4 – Eur 3 000/pixel.

The category 0 concerns areas that are not exposed to landslide hazard. The elements of vulnerability have been defined for such areas, but have zero landslide risk values. The category 1 covers the landslide risk value for arable land and forests in the model area. The elements of risk in the category 1 are in the 4th and 5th landslide hazard degree based on the hazard prediction map. The categories 2, 3 and 4 represent the landslide risk within the built-up areas (thus the high values arising from the material vulnerability values for the built-up areas). Similarly, they are exposed to the 4th and 5th landslide hazard degree, which arise from the above mentioned landslide hazard prediction map.

In the first place, such prepared maps of risk scenarios could serve as bases for the optimisation of the land use planning in towns and municipalities and further development of the individual cadastres in the model area.

10.6.3.3 Verification of the prediction maps

Having compiled the prediction map, it is vital to evaluate its informative value, this means to verify the map. The most important criterion for the evaluation of the prediction map quality is preparing the success model that considers the relationship between the prediction and the registered slope deformation map. In general, the success model compares the landslide density in the map of the registered slope deformations (presence or absence of the slope deformations, binary raster 1/0) with levels of landslide hazard.

The base element is construction of contingency table that compares the registered and predicted landslides, the combination of true and false, positive and negative classifications of landslides using random choice of an even number of pixels from both rasters. The contingency table analyses the relationship between two or more categorical variables. It was first used by Karl Pearson in 1904. The contingency table has that many rows and columns as is the number of categories in the prediction.

The literature discusses numerous techniques of prediction map verification. The most used are mostly the procedures using the methods of statistical success rates and ROC (Receiver Operating Characteristic) curves.

Correctness, accuracy or preciseness of the model is often evaluated using the ROC curves (ROC – Receiver Operating Characteristic). The ROC analysis was developed during the Second World War to evaluate the success rate of radar receivers to detect targets. The size of the area under curve (AUC) defines the overall quality of the prediction model; the larger the area, the more successful model. The maximum area of the chart is 1 (ideal model); the area for a model with 50% success rate has AUC = 0.5 (trivial model). It means that the closer the area's size to 1, the more precise the model is. The ROC curve is constructed according to the contingency tables and their number corresponds to the number of the cut-off – threshold values. Subsequently, the true and false positive values (TP and FP) are calculated for each contingency value. Those values then define the shape of the ROC curve. The closer the ROC curve to the upper left corner, the higher the quality of the evaluated model is, and, thus, the higher AUC. Figure 10.16 shows the ROC curves to verify the model area in already mentioned Žilina region calculated using the bivariate and multivariate statistical analyses and neural networks. The size of the AUC varies from 0.852 to 0.924; together with the steepness of the curve itself, points to the success of the prediction models.

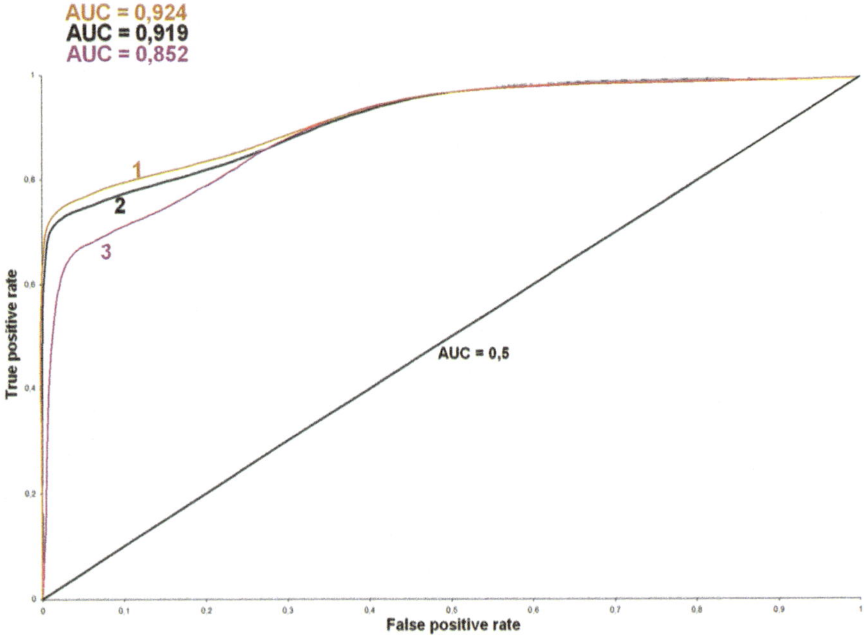

Figure 10.16 ROC curves: (1) artificial neural network, (2) multivariate statistical analysis, (3) bivariate statistical analysis.

10.7 SNOW AVALANCHE MODELLING

10.7.1 Geoinformation technologies integration into the snow and avalanche research

Avalanches are regular effects of the mountain environment. This spontaneous gravitational shift of snow down a mountain slope is a natural mountain phenomenon, which often endangers natural sources, settlements, infrastructure and unfortunately very often also human lives. Therefore, the effort to understand all the processes connected with the instability of a snow cover and the related effort to minimize the negative consequences of avalanches is natural. It seems to be very important to predict an avalanche danger and to recognize the conditions leading to its increase. However, snow is one of the most variable natural materials for mathematical description. For that reason, an avalanche dynamics modelling requires the use of complex, physically based numerical modelling tools and modern efficient computer technology. It is the development of computer and geoinformation technologies (GIT) that stand behind the great progress in the field of mass movements modelling.

The confirmation of this is the development of various modelling tools specialized exactly on the simulation and prediction of mass movement (snow) in a complex terrain. RAMMS and AVAL 1D, (developed by WSL Institute for Snow and Avalanche Research SLF), SAMOS-AT (developed by Austrian Service for Torrent and Avalanche Control), ELBA+ (developed at the University of Natural Resources and Applied Life Sciences in Vienna) belong among the most complex modelling tools nowadays. A considerable research in the avalanche dynamics in Europe was realized also in SATSIE project (Avalanche Studies and Model Validation in Europe), which is supported by the European Commission (MN2L and D2FRAM models). An accurate prediction of runout distances, flow velocities and impact pressures (Figure 10.17) in general three-dimensional terrain is the driving motivation for the development of dynamical mass movement models (Christen, 2013).

Avalanche dynamics modelling is in close connection with hydrological modelling, which deals with the formation and attributes of a snow cover (Figure 10.18).

It is possible to connect these two fields of modelling on the level of inputs and outputs with the help of GIS. Snow precipitation defines the distribution and attributes of a snow cover. These are the important factors, which define the regional distribution of potential avalanche release areas. However, snow avalanches can significantly affect back the hydrological terms of a specific area. The accumulation of avalanche deposit at the bottom of valleys can affect the process of snow melting. Depending on the conditions, snow melting can be accelerated or decelerated. There is a time shift of maximum water flow in the waterways fed from these valleys. At avalanches, a substantial mass of snow descends from higher to lower altitudes. Therefore, snow comes into warmer environment (in the consequence of the temperature gradient). Furthermore, an avalanche deposit can comprise different foreign materials and dirts that are gradually accumulated on the surface of a drift. The darker surface can absorb a bigger amount of sun radiation, what accelerates the melting process. The effect of this is that the maximum water flow comes earlier. When the snow in an avalanche drift is too dense, its melting can be significantly decelerated. The effective area of an avalanche drift is much smaller than the surface of the snow, which would not be released by the avalanche. These are the factors, which slow down the melting process and delay the maximum water flow.

10.7.2 Physically based numerical tools for avalanche dynamics modelling

Nowadays, the physically based numerical tools are the most widespread group of tools and they represent the most complex tool for quantitative analysis of a studied system. They are much less encumbered with the simplifying assumptions used in

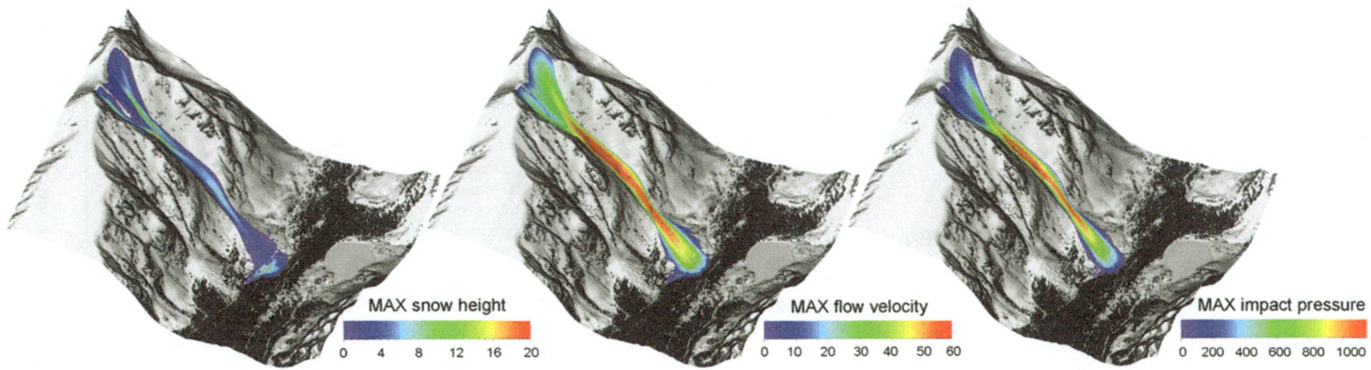

Figure 10.17 Avalanche dynamics modelling allows to predict maximal snow height, flow velocity and impact pressure of the avalanches.

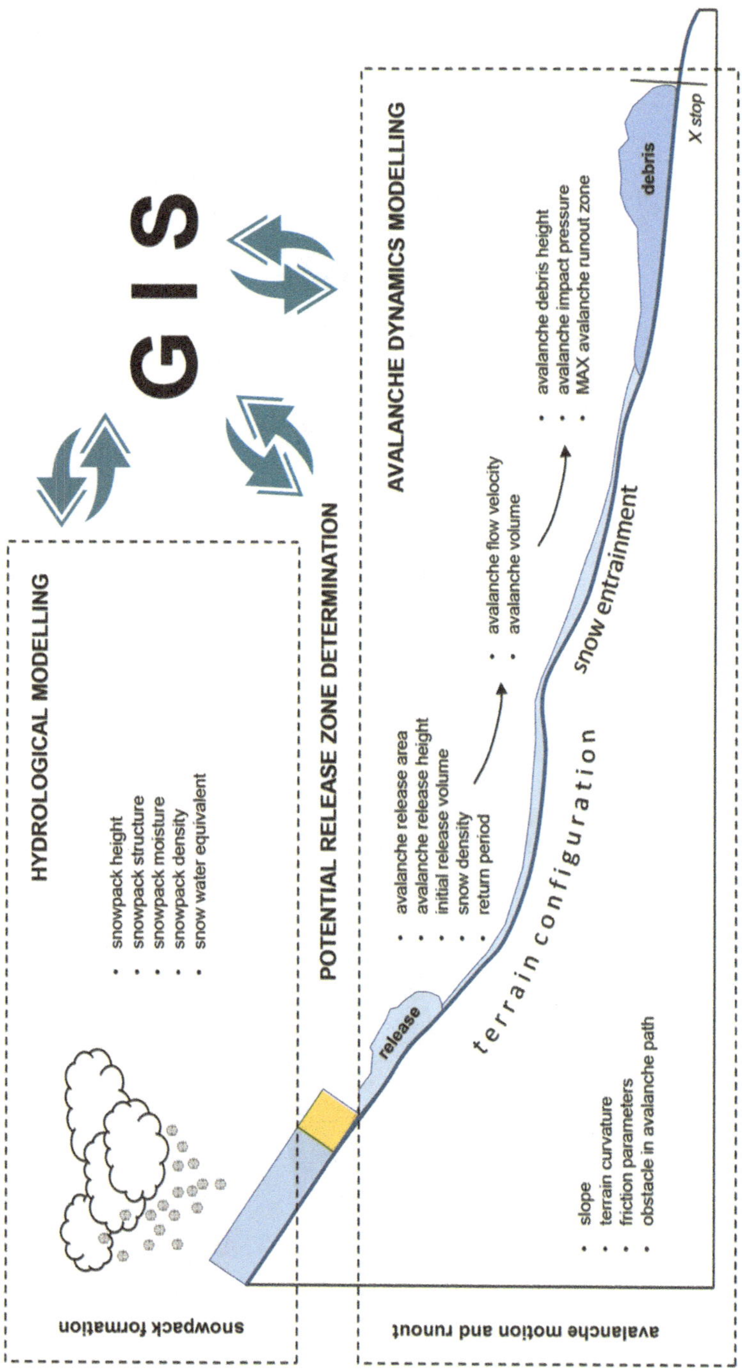

Figure 10.18 Avalanche dynamics modelling is in a close connection with hydrological modelling.

analytical tools. Therefore, they are more appropriate for solving more complicated problems in more difficult conditions (Unucka, 2001). The main aim of avalanche dynamics studies is the answer to the question: in what way, how fast and how far does an avalanche move and what destructive potential is this movement connected with. The potential of dynamic numerical models is significant especially for the identification of potential avalanche runout distances. This identification is crucial for the evaluation of an avalanche danger. Furthermore, the avalanche impact pressure can be estimated by avalanche dynamics modelling (Figure 10.19) that brings a completely new dimension to the evaluation of an avalanche danger.

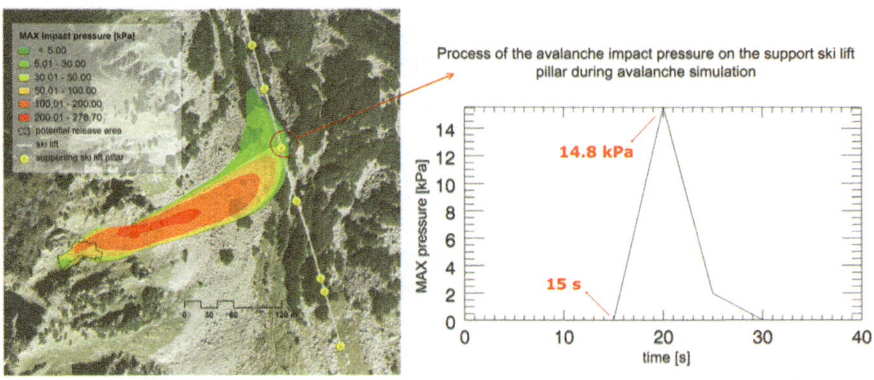

Figure 10.19 Avalanche impact pressure estimation brings a new dimension to the evaluation of an avalanche danger.

The definition of a possible avalanche path, the demarcation of potential avalanche runout distances and the estimation of avalanche destructive force represent an important basis for spatial planning and for projecting and dimensioning of avalanche protecting structures (Figure 10.19). Avalanche dynamics calculations are used to predict an extreme avalanche event and to delineate the different hazard zones (Gruber *et al.* 1998).

The determination of avalanche runout distances by numerical models is the process, which requires a lot of input data of high quality. The most important is the quality and accuracy of digital elevation model (DEM). The best results are achieved with DEM derived from LIDAR scanning. Another important group of input data are the parameters of avalanche release area (delimitation of release area, release height, snow density). These parameters underly the size, the volume and the type of an avalanche flow, which is consequently influenced by a terrain surface roughness, snow entrainment, the shape of an avalanche path, the presence of forests and other obstacles in an avalanche path. Also the selected spatial resolution of individual input layers is considerably important for the outcome of modelling. It is obvious, that the accuracy of DEM is significantly decreasing according to the lowering of spatial resolution (increasing pixel size). It is necessary to define the spatial resolution with regard to a terrain character, so that all terrain features, which can significantly influence an avalanche flow, are included (Figure 10.20). This is also the principal problem, which restricts the usage of these tools at the modelling of small avalanches.

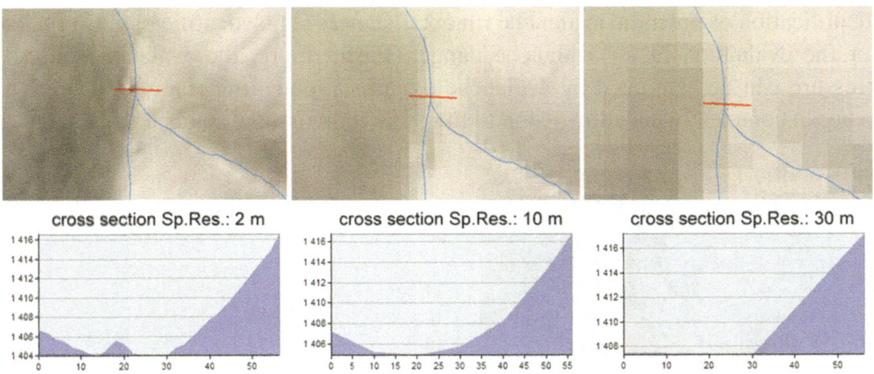

Figure 10.20 Important terrain element with avalanche flow influence was not included into DEM with spatial resolution more than 2.

10.7.3 Model calibration and verification

The biggest danger of modelling is the possibility to easily generate the outputs that have little in common with reality. A model is always similar to the hypothesis, on which it is based, and to the data, which are input into it. However, every model works on a certain level of reliability and it is necessary to verify and test the acquired prognosis. A model calibration and the verification of results must be an inseparable part of each modelling. In the case of avalanche dynamics modelling, a calibration refers especially to the adjustment of terrain friction coefficients. According to these coefficients, an avalanche flow accelerates or decelerates. However, the avalanches in individual mountain areas have a very specific progress, due to a different combination of local, meteorological, geomorphological and climatic conditions. Therefore, it is not always possible to use the calibration coefficients relevant for one area also for runout distances modelling in another area. In the process of calibration, the selected calibration coefficients are set according to user's experience and detail knowledge of the local conditions in an examined area. The only possibility to set these parameters is to study the behaviour of a real system in the past with the intention to gather the information about initial conditions and about the reaction of the system to these conditions. An effort to set calibration coefficients in such a way that the range of a simulated avalanche agrees with the maximum range of a past avalanche is very frequent at avalanche runout distances. Such a method is known as a back-calculation (Figure 10.21).

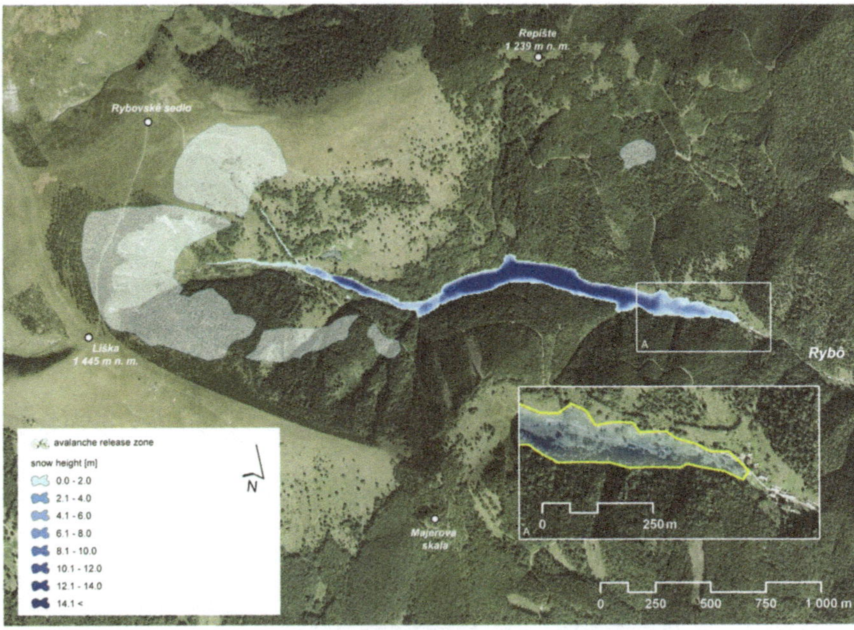

Figure 10.21 Back calculation and reconstruction of the avalanche in settlement Rybô, February 6, 1924.

The verification is used in order to evaluate the fact, whether a model simulates the behaviour of a real phenomenon in an acceptable rate. Every result of modelling should be complemented with the information about achieved simulation accuracy. There are several statistical methods that can be used for verification. The verification of the avalanche front line localization, but also the verification of the whole shape of an avalanche path is relevant for the modelling of avalanche runout distances. Also the verification of the snow height in a drift is important for a back-calculation. The places, in which the check figures are measured in a terrain, should be chosen with regard to local terrain features and a drift character. Only when a model is properly calibrated and its results are verified accurately enough, it can be used for the simulation of various cases.

10.7.4 Avalanche danger zoning

It is required to carefully examine and evaluate an avalanche danger when planning any human activities in avalanche prone areas (construction of buildings, roads, railways, ski-lifts, ski slopes, power lines and others). This includes a precise estimation of avalanche runout distances, impact pressure and return periods of the highest possible number of potential avalanches. The mistakes in these estimations

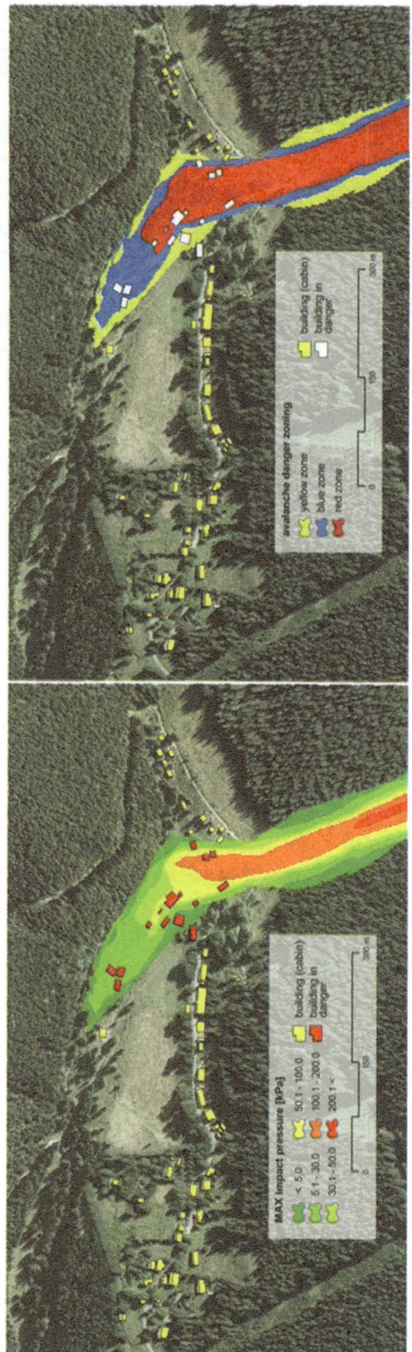

Figure 10.22 Avalanche danger zones according to maximal impact pressure reached by avalanche, settlement Magurka in Low Tatras, Slovakia.

can lead to great material damages and to losses of human lives. Also a too conservative evaluation of an avalanche danger can lead to unpractical restrictions, which inhibit the development of the region. This evaluation should be periodically examined, minimally once in 10 years (McLung & Schaerer, 2006).

The division of an area into different zones, according to their potential avalanche hazard, was established for the first time in Switzerland in 1961. The definition of individual avalanche danger zones was legislatively set and this Swiss model was established also in Austria, Italy, France, the USA and Canada. In accordance to this definition there are four zones (Figure 10.22) of potential danger (McLung & Schaerer, 2006):

- *Red Zone = High Hazard*: No new structures and buildings are permitted in this zone. Existing buildings must be protected either by control structures or reinforcement, and evacuation plans must be in place.
- *Blue Zone = Moderate Hazard*: New residences may be permitted in this zone, but they must be protected. No lift terminals, lodges, schools, or buildings that attract large crowds of people are allowed.
- *Yellow Zone = Low Hazard*: In this zone, structural measures against powder avalanches may be recommended.
- *White Zone = No Hazard*: No restrictions apply to development. No avalanches are reasonably expected to reach this zone.

10.8 SUMMARY AND CONCLUSIONS

The mass movements occur when the shearing stress acting on rock masses exceeds the shear strength of the material. The susceptibility of slope to failure is determined by a combined effect of pre-conditions (geology) and factors like slope angle and moisture content. There are various triggering factors, which generate sudden failures, such as intense heavy precipitation, natural (seismicity) or artificial (explosives, traffic) vibration, steepening of slopes (by undercutting), deforestation. The mass movements are distinguished based on various criteria, where the most commonly used is their rate. The slowest mass movement is creep. More rapid movements are termed landslides and they can be further subdivided based on the involvement of rock or soil in sliding, on the shape of shear plane, etc. Flows are even more rapid movements and they are again subdivided upon the character of the load. Fall is the quickest mass movement occurring on the very steep or overhang slopes. GIS assessment is effective tool in the hazard assessment processing. Prior to undertaking any area surveys, it is essential that the system is established for identifying and describing hazards and the infrastructure at risk. Once the assessment has been completed, hazard ranking can be carried out and the inevitable remedial measures can be designed.

The use of innovative geoinformation technologies opens a new perspectives for mapping and appraisal of natural hazards also in the field of avalanche prevention.

Their main aim is to evaluate an avalanche hazard with regard to the planning of human activities in mountains. The number, the importance and also the accuracy of avalanche dynamics modelling tools are developing in accordance to the development of computer technology. Such models, connected with a precise terrain research and study, represent a great contribution to the appraisal of an avalanche hazard in mountains.

REFERENCES

Alean J. C. (1985). Ice avalanches: some empirical information about their formation and reach. *Journal of Glaciology*, **31**(109), 324–333.

Aleotti P. and Chowdhury R. (1999). Landslide hazard assessment: summary review and new perspectives. *Bulletin of Engineering Geology and Environment*, **58**(1), 21–44.

Bednarik M., Clerici A., Tellini C. and Vescovi P. (2005). Using GIS GRASS in evaluation of landslide susceptibility in Termina valley in the Northern Apennines (Italy). In: Proceedings of the Conference on Engineering Geology: Forum for Young Engineering Geologists, M. Moser (ed.), April 6th–9th 2005, DGGT Erlangen-Nürnberg, Fridrich – Alexander – University of Erlangen- Nürnberg, pp. 19–24.

Bednarik M., Magulová B., Matys M. and Marschalko M. (2010). Landslide susceptibility assessment of the Kraľovany–Liptovský Mikuláš railway case study. *Physics and Chemistry of the Earth*, **35**(3–5), 162–171.

Christen M., Bartelt P. and Kowalski J. (2008). Calculation of dense snow avalanches in three-dimensional terrain with the numerical simulation programme RAMMS. In: International Snow Science Workshop 2008, Whistler, BC, Canada.

Clerici A. (2002) A GRASS GIS based shell script for landslide susceptibility zonation by the conditional analysis method. In: Proceedings of the Open Source GIS GRASS Users Conference, M. Ciolli and P. Zatelli (eds), Trento, Italy.

Constantin, M., Bednarik, M., Jurchescu, M. C. and Vlaicu, M. The landslide susceptibility assessment using the bivariate statistical analysis and the index of entropy in the Sibiciu Basin (Romania). *Environmental Earth Sciences*, DOI 10.1007/s12665-010-0724-y, Published online: 05 September 2010.

Cruden D. M. and Varnes D. J. (1996). Landslide types and processes. In: Landslides: Investigation and Mitigation, Special Report 247, A. K. Turner and L. R. Schuster (eds), National Academy Press, Washington, DC, pp. 26–75. ISBN 0-309-06151-2.

Gruber U., Bartelt P. and Haefner H. (1998). Avalanche hazard mapping using numerical Voellmy-fluid models. In: 25 Years of Snow and Avalanche Research (NGI), International Symposium Held in Voss May 1998, Norges Geotekniske Institutt, pp. 117–121.

Jelínek R. and Wagner P. (2007). Landslide hazard zonation by deterministic analysis (Veľká Čausa landslide area, Slovakia). *Landslides*, **4**(4), 339–350, ISSN 1612-510X.

Kääb A., Paul F. and Huggel C. (2003). Glacier monitoring from ASTER imagery: accuracy and applications. In: Proceedings EARSeL Workshop, Observing our Cryosphere from Space, March 11–13, Bern, pp. 43–53.

Kern M. A., Bartelt P. and Sovilla B. (2010). Velocity profile inversion in dense avalanche flow. *Annals of Glaciology*, **51**, 27–31.

Kralovičová L., Bednarik M., Trangoš I. and Jelínek R. (2014). Landslide hazard assessment using deterministic analysis – a case study from the Chmiňany landslide. *Slovak Geological Magazine*, **14**(1), 41–63, ISSN 1335-096X.

Leroi E. (1996). Risk maps at different scales: objectives, tools and developments. In: Landslides, K. Senneset (ed.), Balkema, Rotterdam, pp. 35–51, 2014.

Margreth S. and Funk M. (1999). Hazard mapping for ice and combined snow/ice avalanches: two case studies from the Swiss and Italian Alps. *Cold Regions Science and Technology*, **30**(1–3), 159–173.

McLung D. and Schaerer P. (2006) The Avalanche Handbook, 3rd edn. The Mountaineers Books, Seattle, USA.

Pauditš P. (2006). Landslide Susceptibility Assessment Using Statistical Methods within GIS. Geol. Práce, Spr. 112, ŠGÚDŠ, Bratislava, pp. 41–58: (In Slovak).

Plafker G. and Ericksen F. E. (1978). Nevados Huascaran avalanches, Peru. In: Rockslides and Avalanches, Natural Phenomena, B. Voight (ed.), Elsevier, Amsterdam, pp. 277–314.

Pradhan B. and Lee S. (2009). Landslide risk analysis using artificial neural network model focussing on different training sites. *International Journal of Physical Sciences*, **4**(1), 001–015, ISSN 1992 - 1950 © 2009 Academic Journals.

Salzmann N., Kääb A., Huggel C., Allgöwer B. and Haeberli W. (2004). Assessment of the hazard potential of ice avalanches using remote sensing and GIS-modelling. *Norwegian Journal of Geography*, **58**, 74–84.

Schweizer J., Jamieson B. and Schneebeli M. (2003). Snow avalanche formation. *Reviews of Geophysics*, **41**, 2.1–2.25.

Terlien M. T. J., Van Ash T. W. J. and Van Westen C. J. (1995). Deterministic modeling in GIS – based landslide hazard assessment. In: Geographical Information Systems in Assessing Natural Hazard, Kluwer Academic, Dordrecht, pp. 55–77, ISBN 0-7923-3502-3.

Tornyai R., Bednarik M. and Havlín A. (2016) Application of neural network to assess landslide hazard and comparison with bivariate and multivariate statistical analyses. *Acta Geologica Slovaca*, **8**(1), 109–118.

Tremper B. (2001). Staying Alive in Avalanche Terrain, 1st edn. The Mountaineers Books, Seattle, WA.

Unucka J. (2010). *Možnosti propojení GIS a environmentálních modelů pro potřeby krizového řízení a ochrany přírody* (GIS and Environmental Models Connectivity for the Needs of Crisis Management and Nature Protection). Habilitation thesis, VŠB – Technical University of Ostrava, Czech Republic.

Van Westen C. J. (1993). GISSIZ – Training Package for Geographic Information Systems in Slope Instability Zonation. Part 1: Theory. Project on Geo-Information for Environmentally Sound Management of Natural Resources (ITC Publication Nb. 15). UNESCO – International Institute for Aerospace Survey and Earth Sciences (ITC). ISBN 90-6164-078-4.

Varnes D.J. (1978). Slope movement types and processes. In: Landslides: Analysis and Control. Transportation Research Board. Special Report 29, National Research Council, Washington, D C, pp. 170–197.

Varnes D. J. (1984). Landslide Hazard Zonation – a Review of Principles and Practise. UNESCO, Paris, p. 63.

Chapter 11

Tectonic hazards: Earthquakes

Ioannis Kassaras and Danai Kazantzidou-Firtinidou

11.1 PRIMARY AND SECONDARY EARTHQUAKE HAZARDS

Earthquake hazards comprise any natural phenomenon associated with earthquakes. Earthquakes are known as the shaking of the Earth's surface, producing significant impacts on both physical and urban environments with severe socioeconomic aspects. Most of the earthquakes are generated at the boundaries of the lithospheric plates, which float over the mantle's asthenosphere, converging or diverging. Friction caused by the plates interaction builds up stresses that, when released, produce ground faulting that radiates through the lithosphere producing complex seismic waves, which, in turn, can affect the near built environment.

Although earthquakes mainly control the morphology of the Earth's surface, they would unlikely be considered of major significance in the absence of their effects on the anthropogenic environment, primary and secondary. Primary earthquake hazards at a site regard the ground shaking due to the passage of the seismic waves (dynamic deformation) and/or the ground displacements (static deformation) in the vicinity of the causative fault. Secondary earthquake hazards are the after-earthquake effects caused by the primary ones and may often be more catastrophic. Such hazards are ground failures, fire, landslides, rock and snow avalanches, liquefaction, flooding, tsunamis and seiche that have been frequently reported to follow the occurrence of strong earthquakes. The measure of earthquake hazards at a site mainly depends on the size and type of the seismic rupture, its distance from the site and the geological structure between the source and the site's surface that may impact the seismic energy.

Given that earthquake prediction is still infeasible, the major task in seismological research is the understanding of earthquake phenomena and their consequences on the natural and anthropogenic environment, with the purpose of mitigating them by providing valid and timely information, to be used for effective earthquake planning and decision-making processes. Primary and secondary effects are related with vulnerability, which is defined as a set of prevailing or consequential physical and sociopolitical conditions that affect a community's ability to mitigate, prepare for, or respond to an earthquake hazard (ADCP, 2003). Earthquake hazards, structural vulnerabilities and exposed values, when combined, yield a region's exposure to seismic risk.

11.2 EARTHQUAKE RISK

Earthquake or Seismic Risk is defined as the potential physical, economic, social and environmental losses due to earthquake hazards that may occur in a specified period of time. It is the function of Vulnerability (V), Hazard (H), and Loss (L) (Figure 11.1):

(1) Seismic Hazard (H) is defined as the probability that an earthquake will occur at a given area, within a given window of time, and with ground motion intensity exceeding a given threshold.
(2) Vulnerability (V) characterizes the expected endurance of a structure exposed to the area's seismic hazard.
(3) Loss (L) is two-folded, physical and socioeconomic. Physical loss represents the Seismic Damage, defined as the most probable structural damage state under a certain expected earthquake ground excitation. Socioeconomic loss illustrates the expected consequences on the exposed socioeconomic values and the population due to the probability of an imminent earthquake.

Earthquake hazards caused the most deadly natural disaster in the last two decades, resulting in over 800,000 deaths and 1.7 million injuries (EM-DAT, 2014). The matter of earthquake hazards is particularly crucial nowadays, as manifested by extremely high losses during several recent seismic crises (i.e. Sumatra, 2004; Haiti, 2010; Christchurch, 2011; Japan, 2011; Nepal, 2015). Reason for this is in principal (a) the rapid urbanization of large parts of the population (USDA, 1994), resulting in growth of cities towards areas susceptible to earthquake hazards, (b) the economically driven reduction of the suburban construction behavior, as largely viewed in less-developed countries, (c) the increasing exposure into technological hazards in the developed countries, i.e. nuclear power plants (as it happened in the Fukushima nuclear plant during the 2011 Japan earthquake). Urban earthquake loss is increasingly a problem of developing countries, whereas future earthquakes will have even more disastrous consequences, unless seismic safety is improved.

Earthquake early warning systems aim to prevent loss of life and reduce the economic impact of disasters, by enabling protective actions within a short time span after the generation of hazardous seismic waves at a remote site. Despite the increasing interest for developing and installing early warning systems

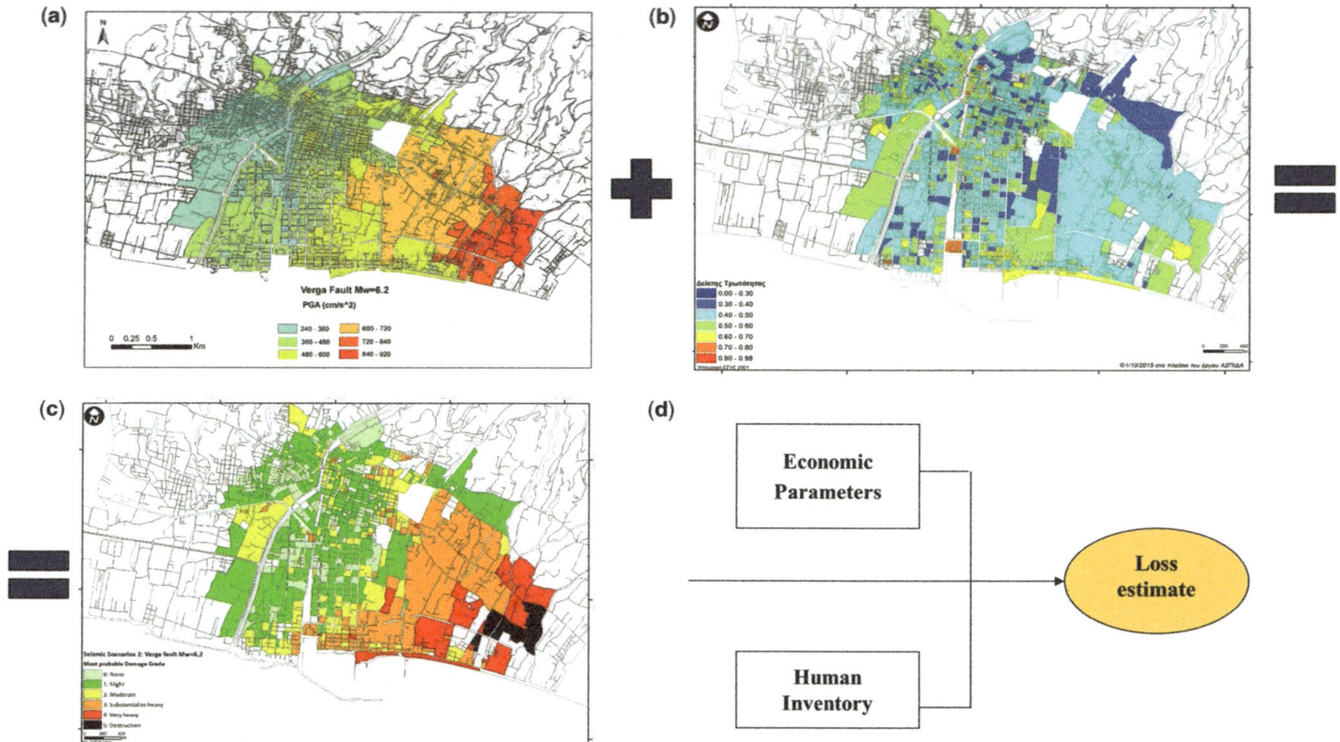

Figure 11.1 (**a**) Seismic hazard map in terms of PGA for an earthquake scenario for the city of Kalamata (southern Greece, Kassaras & Kazantzidou, 2015), (**b**) structural vulnerability distribution, (**c**) expected damage distribution and (**d**) conceptual generic loss estimation approach for the current damage scenario.

worldwide, such systems to be effective need to actively involve the population at risk, facilitate public education and awareness and ensure a constant state of preparedness. The above are hardly feasible especially in the developing countries and, therefore, effective protection should in priority derive from seismic design regulations contained in seismic codes, which intend to ensure that structures can resist seismic forces.

Recent catastrophic earthquakes have proven that the generic provisions of the seismic codes, in general, underestimate the seismic hazard potential in earthquake prone areas, mainly due to soil conditions and near field effects causing non-linear soil-structure interaction; as a consequence, seismic codes should at times be reconsidered. In this respect, thoroughly investigated soil-structure interaction effects and vulnerability estimates are perquisite. On the other hand, separate investigations of only the primary earthquake processes might lead to a misjudgment of the general natural risk for an area and hence secondary earthquake effects should be taken up in the frame of a multi-hazard approach.

In the next sections, the state of the art in the seismic risk analysis and management is reviewed, in terms of methodologies regarding the assessment of the seismic hazard, the seismic risk and the financial and other losses from earthquake damage.

11.3 METHODS OF SEISMIC HAZARD ASSESSMENT

Probabilistic Seismic Hazard Analysis (PSHA) and Deterministic Seismic Hazard Analysis (DSHA) together with its modern enhanced variant, Neo-Deterministic Seismic Hazard Analysis (NDSHA), are the commonly used approaches in seismic hazard assessment (Figure 11.2). The factors on which the choice between the two methods, PSHA or DSHA is the optimal one, is still a matter of debate that mainly concerns the objectives of the hazard/risk assessment, the seismic potential of the target area and the available input data (McGuire, 2001).

The expected intensity of ground shaking at a given site, in terms of peak ground acceleration (PGA), velocity (PGV) and displacement (PGD) or any other parameter relevant to seismic engineering, is the main aspect of any SHA. These can be defined as functions in time and/or frequency domain resulting from the convolution of four sub-functions, (a) the source, (b) the ray-path, (c) the site effects, and (d) instrumental effects. Instrumental effects are typically negligible and hence omitted due to the high dynamic range of the modern equipment. Therefore, the ground motion at frequency (ω) and time (t), $A(\omega,t)$, is given by:

$$A(\omega,t) = B(\omega,t)C(\omega,t)D(\omega,t) \tag{11.1}$$

where $B(\omega,t)$, $C(\omega,t)$ and $D(\omega,t)$ are the contributions due to the source, the ray-path, and the local site conditions, respectively, each one being treated differently in the SHA methods.

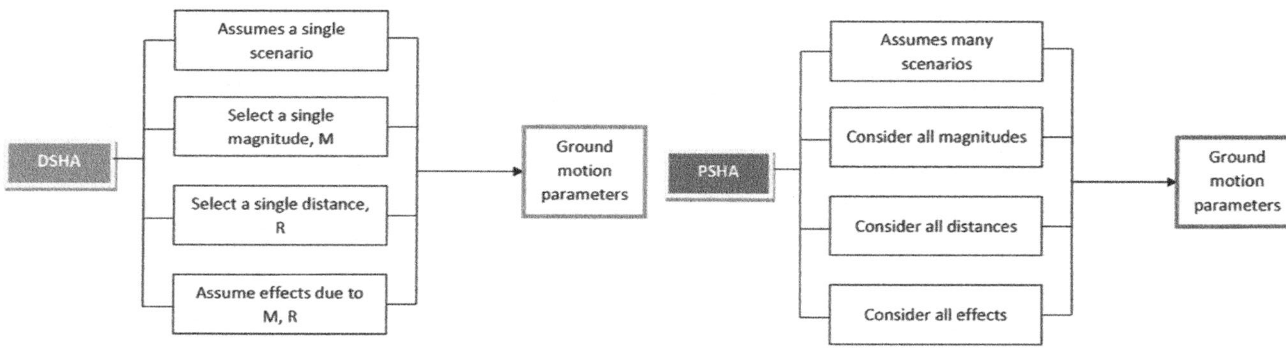

Figure 11.2 Chart showing the generic modules involved in the deterministic (DSHA) and the probabilistic (PSHA) approaches.

The seismic source: From a geological point of view, a seismic source is adequately constrained by three independent source parameters (any two from: dimension, L, average slip, D, and static moment, M_0; and either the mean stress, σ, or strain energy change, E). However, to describe the dynamic process of earthquakes, more source parameters are required, i.e. radiated energy, seismic efficiency, rupture duration, source time function, stress drop and rupture propagation, directivity and hanging wall effect. Certainly, the knowledge of all these constraints is impossible and hence various assumptions on the physics of the source are unavoidable for any SHA.

The seismic ray-path: In isotropic media, normal to the wave-fronts are seismic ray-paths, carrying the energy released at the source to the Earth's surface. Their properties highly depend on the characteristics of the source-to-surface elastic medium, which can largely affect their amplitude and frequency. As the source-site distance increases travel-path effects become more important. In SHA, ray-path effects are approached by GMPEs (Ground Motion Prediction Equations) defined by simple attenuation relations derived using recorded strong-motion data. More sophisticated techniques deal with the simulation of seismic waves propagating in 2D and 3D media taking into account geometrical spreading, scattering, anelastic attenuation, critical reflections, topographic and basin effects.

The site-effects: It is described by the local geological conditions in an area, regarding the soil, structure properties and the surface morphology configuration. Site conditions, which may significantly vary from one building block to the other, can drastically impact the seismic motion by generating amplification, de-amplification, liquefaction and soil-structure resonance, which in turn may be proven catastrophic. The effect of soft soil deposits on seismic waves is frequency dependent and strongly influenced by their relative thicknesses, seismic velocities, water content and pore pressure, impedance contrast, resonant amplification, energy focusing, near-site scattering and directional site resonance, anelastic attenuation and topography. Rock sites, although usually thought not to experience large amplification, have been found to exhibit high variability and at times a strong response. In the near field, the strains generated by earthquakes can be large, producing non-linear behavior of the ground that may explain the difference between expected high amplifications due to site effects and the lower amplifications recorded by large earthquakes and vice-versa.

11.3.1 Probabilistic Seismic Hazard Analysis (PSHA)

PSHA was at first introduced by Cornell (1968) in terms of ground motion's annual probability of excess at a site. In PSHA, sources denote faults or areal sources, within which earthquakes are assumed equally likely to occur at any location. Their parameterization consists of four primary steps, involving identification and characterization of all hazardous sources for the target site, description of seismicity of each source, determination of the motion intensity at each source and probabilistic calculations towards the recurrence period of the resulted maximum magnitude earthquake and ground motion excess.

A recurrence relationship, or equivalently an earthquake probability distribution, indicates the chance of an earthquake of a given size to occur anywhere inside the source during a specified period of time. Given the latter, probabilistic methods can be viewed as inclusive of all deterministic events with a finite probability of occurrence. A typical recurrence relationship is the Gutenberg-Richter equation derived for the whole Earth (Gutenberg & Richter, 1944):

$$\log N = a - bM \tag{11.2}$$

where N is the cumulative number of earthquakes of a given magnitude or larger that is expected to occur during a specified period.

Next in PSHA is the estimation of the ground motion parameters (i.e. PGA, PGV, PGD) produced by earthquakes of different size occurring in each seismic source. For this, GMPEs, or attenuation relationships are used (e.g. Makropoulos, 1978; Makropoulos & Burton, 1985). Finally, hazard at a site is produced assuming the effects of earthquakes of varying size and location within earthquake sources at different occurrence probabilities by their integration into a curve that shows the probability of exceeding different levels of ground motion during specified time periods.

By definition, PSHA lacks consideration about the physical process behind earthquakes, which leads only to some generic implications, not adequate enough to predict the most hazardous near field earthquake motions within a reference time period. Recent studies showed that PSHA includes several cons, i.e. not based on a valid earthquake source model, includes uncertainty in source-site distance, the ground motion attenuation is not treated correctly (e.g. Klügel, 2007), does not adequately consider site effects, the few hundred years earthquake catalogues are extrapolated for much longer return periods, events below the completeness magnitude are neglected. To overcome some of the above shortcomings, new ideas about PSHA have been proposed by Graves *et al.* (2010), who implemented full 3D wave propagation simulations to forecast ground motions by various ruptures, in substitution of attenuation relations.

11.3.2 Deterministic Seismic Hazard Analysis (DSHA)

DSHA develops one or more earthquake motions linked to the distinct location and characteristics of all significant potential earthquake sources that might affect the site, hence providing a straightforward framework for the worst-case scenarios, e.g. the largest magnitude at the closest distance, defined as the Maximum Credible Earthquake (MCE). The method prerequisites the accurate study of geological and site effects on seismic ground motion by making use of observational data and implementation of modern theories for the forward and the inverse problems.

DSHA provides more accurate near field constraints with respect to PSHA and hence it is strongly recommended where the consequences of failure are intolerable. The method is sensitive to the selection of the MCE, often based on PSHA and in turn subjected to the aforementioned biases. The matter is treated in the enhanced NDSHA (Neo-Deterministic SHA) by configuration of only the largest events reported in the earthquake catalogues into spatial cells, where the maximum magnitude event is assigned. A smoothing procedure follows to account for location uncertainty and source dimensions and thereby, uncertainties related with the magnitude completeness of the earthquake catalogues are successfully treated (e.g. Panza *et al.* 2010).

The next step in DSHA is the use of the selected sources for the generation of synthetic seismograms. The stochastic simulation of strong ground motion (Boore, 1983) is a procedure frequently applied in DSHA for synthesizing ground motions as an alternative for regions lacking recordings of strong earthquakes. In general, a stochastic simulation traces the evolution of variables that can change stochastically (randomly) with certain probabilities. In the method, the spectrum of strong ground motion $R(f)$ at a site is given by the product (e.g. Boore, 1996):

$$R(f) = CS(f)A(f)D(f)I(f) \tag{11.3}$$

where C is a scaling factor, $S(f)$ the source spectrum, $A(f)$ the site amplification function, $D(f)$ the seismic energy diminution and $I(f)$ is a factor comprising the instrument response and filters used in the procedure. The stochastic model applies for both point source and finite-fault. The latter requires expansion of the point source theoretical assumptions, to capture the finite dimensions of discrete sub-sources, which sum into a large earthquake; the procedure accounts for faulting geometry, distributed rupture and rupture inhomogeneity that can strongly influence the duration, frequency content and amplitude of simulated ground motions.

Within the context of NDSHA, the scientific trend is to compute realistic synthetic seismograms taking into account source complexity, 2D and 3D travel-path effects and detailed site effects, utilizing seismological, geological, geophysical, and geotechnical data (e.g. Panza *et al.* 2010; Peresan *et al.* 2010). The approach of hybrid broadband ground motion simulation combines a deterministic approach at low frequencies ($f < 1$ Hz) with a stochastic approach at high frequencies ($f > 1$ Hz) (Graves & Pitarka, 2010). At frequencies below 1 Hz, the methodology is deterministic and combines theoretical source rupture kinematics with wave propagation effects utilizing theoretical full waveform Green's functions, towards producing ground motion time series, (e.g. Zeng *et al.* 1994). Above 1 Hz, the method is semi-stochastic using a stochastic representation of the source radiation combined with a simplified theoretical representation of wave propagation and scattering effects.

The major advantage of NDSHA with respect to DSHA and PSHA is that the simultaneous treatment of the seismic source and the seismic wave propagation through deterministic relations does not require the use of any attenuation relation (Figure 11.3). One disadvantage of this method is the limited availability of full kinematic rupture models and the demand for computation of theoretical Green's functions.

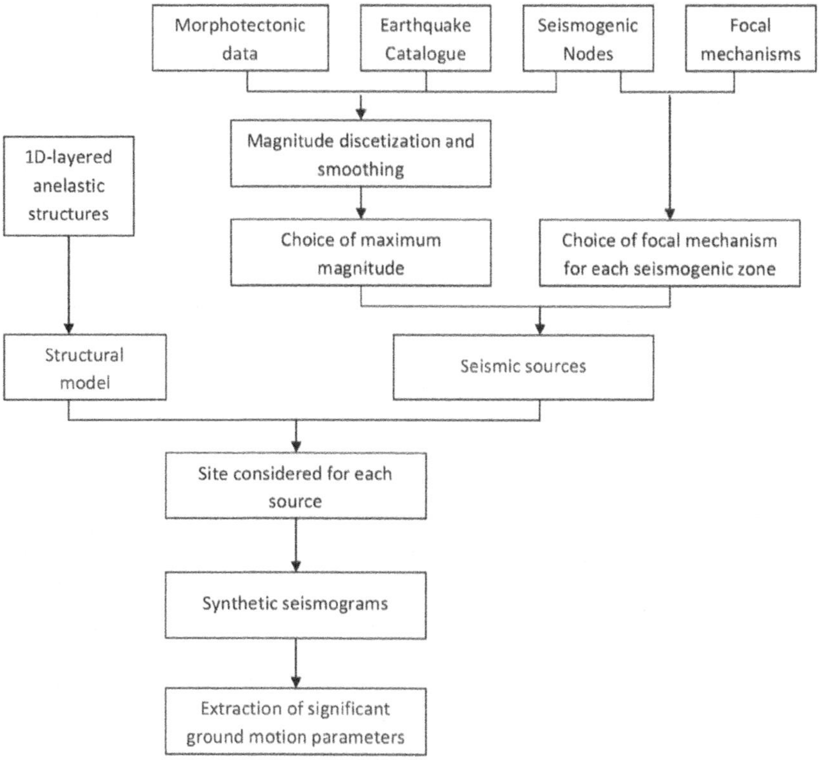

Figure 11.3 Modules involved in NDSHA (modified after Panza *et al.* 2001).

The results of a DSHA are sensitive to the choice of the threshold magnitude. Moreover, it provides no information about the likelihood or frequency of occurrence of the controlling earthquake, i.e. the Maximum Credible Earthquake (MCE) and, if such information is required, a PSHA must be undertaken to better quantify the seismic hazard.

Definitely, the question about which method, DSHA or PSHA to be selected, is not straightforward as it depends on the number and the type of unknown factors involved in the problem under formulation. DSHA and PSHA analyses should be complementary and the strength of one over the other depends on the earthquake mitigation decisions to be made, on the seismic environment, and on the relevant scope (McGuire, 2001).

11.4 METHODS OF SEISMIC VULNERABILITY ASSESSMENT

Structural vulnerability is defined as the tendency of a structure to experience damage under a certain level of ground shaking. It is generally an intrinsic parameter of the studied structural system; it mainly depends on its structural, mechanical and geometrical characteristics, although the soil conditions are often responsible for modifying its structural behavior. Methods of vulnerability analysis are divided into two main categories, *Empirical* and *Analytical.*

Empirical or *Macroseismic* methods are based on statistical observations of recorded damage data by strong past earthquakes. *Analytical* methods are based on the mechanical calculation of the building structural response. Both can be used in *Hybrid* methods, which combine statistical data with appropriate nonlinear analyses (Kappos *et al.* 2006). Each approach is characterized by limitations, subsequently the choice of the most feasible one depends predominantly on the level of information available and on the extent of the area under examination (Dolce *et al.* 1994).

The impact assessment of seismic vulnerability was initiated in San Francisco (NOAA, 1972; ATC-13, 1985) and has been widely extended by the end of the '90s. An analytical estimation of the structural damage in urban scale has been standardized by HAZUS (1997), which is applied worldwide for the establishment of catastrophe emergency plans. In Europe, the RISK-UE model (Milutinovic & Trendafiloski, 2003), launched after the disastrous Izmit and Athens earthquakes in 1999. The method adapted the HAZUS methodologies into the European building typologies included into the European Building Typology Matrix (BTM). According to RISK-UE, two approaches have been proposed:

(a) A macroseismic (empirical) vulnerability method based on post-seismic field observations. The method, following the Damage Probability Matrix (DPM) method principles, identifies six damage grades (Table 11.1)
(b) A simplified "mechanical" method, relating limit-state mechanical properties of the buildings, such as the inter-story drift capacity, with peak ground motion parameters (Calvi *et al.* 2006). The method is based on the HAZUS methodology, where 4 damage states are recognized (Table 11.1).

11.4.1 The macroseismic (empirical) method

In the empirical method, a Building Typology Matrix (BTM) is given by the most probable vulnerability index V_I* for each building typology, together with the bounds of the plausible $[V_I^-, V_I^+]$ and the possible $[V_I^{min}, V_I^{max}]$ index range. Following the idea that the seismic behavior of the building does not only depend on its structural system (type, material, age), but also on its position in aggregates, its plan and vertical irregularity, its height, its state of preservation and possible retrofitting interventions, behavior modifier factors (ΔV_m) are introduced to the generic index V_I*.

Table 11.1 Damage Grades and Characterization.

Damage State LM1	Label LM1	Damage State LM2	Label LM2	Description	Post-EQ Assessment
D0	None	DS1	Slight	No damage	Green (practically undamaged)
D1	Slight			Negligible to Slight damage	Green (practically undamaged)
D2	Moderate	DS2	Moderate	Slight structural, moderate non-structural	Yellow (repairable, non-dangerous)
D3	Substantial to heavy	DS3	Severe Extensive	Moderate structural, heavy non-structural	Yellow (repairable, non-dangerous)
D4	Very heavy	DS4	Complete Collapse	Heavy structural, very heavy nonstructural	Red (non-repairable, dangerous)
D5	Destruction			Very heavy structural, total or near total collapse	

Moreover, local construction techniques and site effects can lead to additional vulnerability modifiers (ΔV_R and $\Delta V_{S,}$ respectively) (Giovinazzi & Lagomarsino, 2004). After the appropriate modifications the final Vulnerability index becomes:

$$\overline{V_I} = V_I^* + \Delta V_m + \Delta V_R + \Delta V_S \tag{11.4}$$

Regarding the site modifiers, reference is made to the Elastic Response Spectrum of EN1998 (CEN, 2004) for different soils, in order to take into account information about the dynamic behavior of a structure. Kassaras and Kazantzidou (2015) propose the use of the output response spectra from selected seismic hazard scenarios combined with the elastic period of the buildings as following (Figure 11.4):

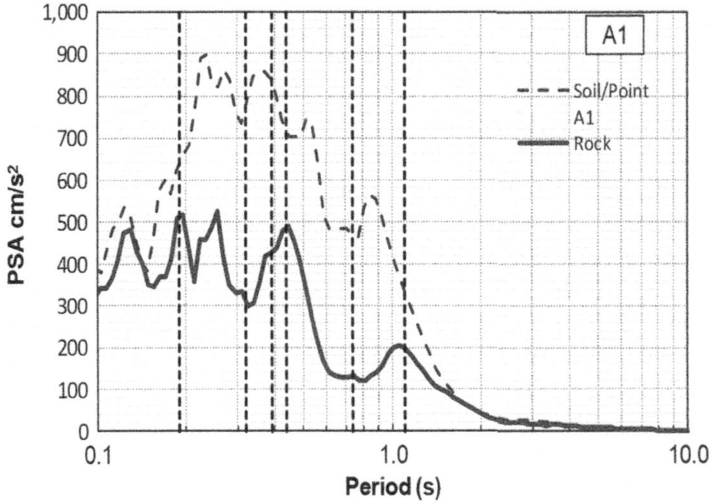

Figure 11.4 Response spectrum of the soil and the rock for a seismic scenario of the city of Kalamata, Greece (Kassaras & Kazantzidou, 2015). The dashed vertical lines indicate the eigenvibration period of the main structural categories in Kalamata.

Giovinazzi and Lagomarsino (2004) suggest the soil-structure interaction effect, as well as the soil's response amplification to be considered upon the macroseismic intensity by the ratio of the spectral acceleration (Sa) on the soil, where the building is found, with respect to the Sa on the bedrock, corresponding to the building eigenvibration period (EN1998, CEN, 2004):

$$f_{PGA} = \frac{Sa[T_1]_{\text{Soil } K}}{Sa[T_1]_{\text{rock}}} \tag{11.5}$$

The Intensity Increment (ΔI) due to the spectral amplifications is (Guagenti & Petrini, 1989):

$$\Delta I = \frac{ln(f_{\text{PGA}})}{0.602} \tag{11.6}$$

The Vulnerability Increment (ΔV_s), which corresponds to the Intensity Increment ΔI is:

$$\Delta V_S = \frac{\Delta I}{6.25} \tag{11.7}$$

The latter is implemented as a modifier to the Vulnerability Index.

11.4.2 The mechanical method

The mechanical method (LM2 method of Risk-UE) is based on the HAZUS (1999) approach with modifications regarding the European masonry typologies and the seismic design of buildings according to EN1998 (CEN, 2004). Four damage states are adopted (Table 11.1) and, for consistency reasons with the empirical method, the no-damage state has been also included. The method is defined as a capacity spectrum method, according to which a capacity curve represents the overall force-displacement efficiency of the structure in terms of estimated engineering parameters (design, yield and ultimate strength performance).

Similarly to the non-linear detailed analyses methods, the performance of a building hit by an earthquake, is identified by an intersection point, referred to as the "performance point" (PP), between the equivalent Single Degree of Freedom (SDOF) capacity curve, representative of the non-linear lateral performance of a structure, and the earthquake demand curve. The latter is the seismic input, expressed in terms of demand spectrum, based on the 5% damped building-site specific response spectrum, appropriately reduced for the structure's inelastic response.

The performance point (PP) represents the condition for which the seismic capacity of the structure is equal to the seismic demand imposed on the structure by the specified ground motion. Its location on the capacity curve between the limit is then used to estimate the expected damage state of each building.

Similarly to the macroseismic method, the buildings are classified into separate typologies for which representative capacity curves exist. Numerous detailed analyses, taking into account the prevalent collapse mode, the mechanical material characteristics, geometrical features and experts' judgment have led to idealized capacity curves for each building typology (Figure 11.5a).

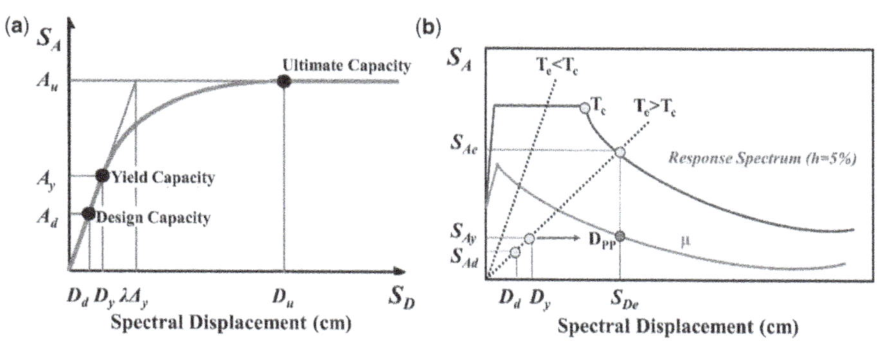

Figure 11.5 (**a**) Building capacity model according to Risk-UE (Milutinovic & Trendafiloski, 2003) and (**b**) EN1998-1 (CEN, 2003) procedure for the evaluation of the performance point with equal displacement approximation.

Two main control points are needed to characterize the curve:

(1) Yield Capacity [YC (V_y, D_y)]: It is the lateral load resistance strength of the building before the development of nonlinear response of the structural system. It depends on the design strength coefficient (C_s) (as a fraction of building's weight) and the overstrength factor relating design strength to yield strength (γ). The yielding displacement is calculated through $V = \left(\frac{2\pi}{T}\right) D$ relationship.

(2) Ultimate capacity [UC (V_u, D_u)]: It is the maximum strength of the building when the global structural system has reached fully plastic state. It depends on the overstrength factor relating the yield strength to the ultimate one (λ).

For seismically designed structures the capacity spectrum is given in terms of spectral acceleration S_a and displacement S_d, through the dynamic characteristics of the structure (elastic period, first mode shape φ_i and considered lumped floor mass m_i):

$$S_a = V / a_1 \quad \text{and} \quad S_d = D / a_1 \tag{11.8}$$

where

$$a_1 = \frac{\left[\Sigma m_i \varphi_i\right]^2}{\Sigma m_i \, \Sigma m_i \varphi_i^2} \tag{11.9}$$

The elastic response spectrum, obtained from the effective seismic Codes or through microzonation analysis for given seismic scenarios is the key tool for the method. It characterizes the ground motion, summarizing the peak response of all possible linear SDOF to a particular component of ground motion. The "equal displacement approximation" is the simplest technique for defining the target displacement (D_{PP}) into the elastic-to-perfectly plastic representation of the capacity spectrum (Figure 11.5b).

The coordinates of the Performance Point (PP) are:

$$D_{PP} = D_y \cdot \mu \quad \kappa\alpha\iota \quad S_{aPP} = S_{ay} \tag{11.10}$$

Other methods (ATC-40, 1996; FEMA-440, 2005) apply different methods for the definition of the PP, using reduced damped spectrum or corrective coefficients.

11.5 METHODS OF EARTHQUAKE PHYSICAL LOSS ESTIMATION

For the two methods of vulnerability assessment, *macroseismic* and *mechanical*, respectively, *Macroseismic* and *Analytical* Earthquake Physical Loss (or structural damage) Analysis Methods exist, respectively, and are differently calculated. Cross-validation of the two methods has revealed a good association (e.g. Lagomarsino & Giovinazzi, 2006) and a common feature is the similar analytical formulation of the damage states, as soon as they refer to the same building typologies.

11.5.1 Macroseismic (empirical) physical loss estimation

Recent models mainly follow the Giovinazzi & Lagomarsino (2004) approach, who proposed a semi-empirical function that correlates the mean expected damage grade μ_D with the seismic intensity I and the vulnerability index V_I:

$$\mu_D = 2.5\left[1 + \tan\text{h}\left(\frac{I + 6.25\overline{V_I} - 13.1}{2.3}\right)\right] \tag{11.11}$$

where μ_D is a discrete variable characterized by 6 damage grades. The probability p_k of occurrence of a certain average damage grade (k = 0–5) is evaluated according to the Cumulative Distribution Function (CDF) of the beta-β distribution:

$$p_k = P_\beta(k + 1) - P_\beta(k) \tag{11.12}$$

The mean damage grade (μ_D) calculated for each building, or each building block is defined as the one that corresponds to the highest probability of occurrence, expressed within a range of plausibility, according to the vulnerability definition.

Therefore, empirical formulae connect peak ground motion parameters derived in the framework of PSHA or DSHA with MMI intensities (I_{MSK}, I_{EMS}, etc). Wald *et al.* (1999), based on the regression of peak acceleration and velocity amplitudes with observed intensity for eight significant US earthquakes, proposed a descriptive table permitting the creation of rapid instrumental intensity maps according to PGA and PGV recordings, adopted by the US Geological Survey (USGS).

Fragility curves express the probability $P(D \geq D_k)$ that the expected damage of a structure will exceed a certain grade during the ground shaking in terms of seismic intensity or mean peak ground acceleration. The cumulative probability beta distribution CDF (P_β) is used as follows:

$$P\left(D \geq D_k\right) = 1 - P_\beta(k) \tag{11.13}$$

The probability of occurrence of different damage states is, this way, expressed for every structural typology for defined seismic intensities.

Fragility curves best express the probabilistic risk assessment principles and can be then applied for further loss estimations (Figure 11.6a). Post-seismic observations can be of major importance for validation and calibration of fragility curves of given structural typologies (Figure 11.6b). As an example, in Figure 11.7, the simulated macroseismic damage distribution (Kassaras *et al.* 2015) is plotted together with the observed damage during the 2003 strong earthquake in Lefkas (Ionian Islands, Greece). A very good correlation is observed, validating the vulnerability estimates of the method.

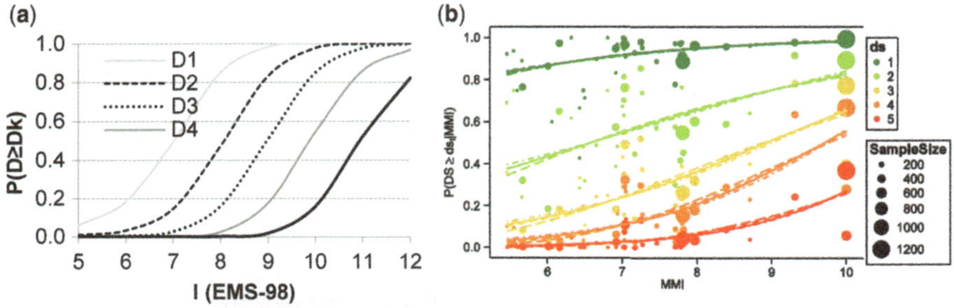

Figure 11.6 (**a**) Fragility curves for RC2 buildings in Kalamata (Kassaras & Kazantzidou, 2015), (**b**) Fragility curves from regression analysis based on empirical data for Irpinia earthquake (Rossetto *et al.* 2014).

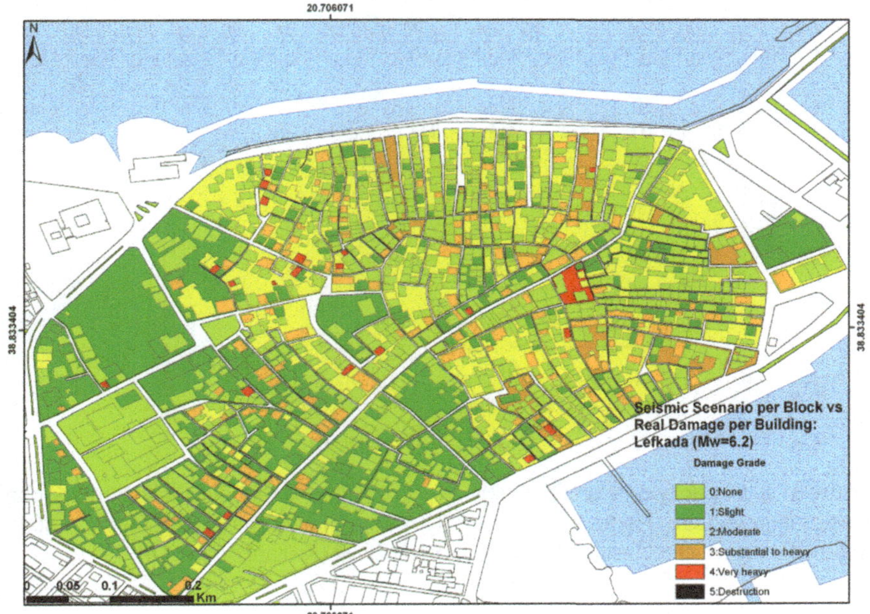

Figure 11.7 Damage distribution per block for the 2003 Lefkada earthquake simulation vs observed damage characterization per building according to the post-seismic investigation.

11.5.2 Mechanical SDE

As previously described, structural damage is identified by the position of the performance point (PP) on the capacity curve (Figure 11.5b, Figure 11.8). According to Risk-UE-LM2 the following damage thresholds are identified:

$$\begin{aligned} S_{d1} &= 0.7D_y \\ S_{d2} &= \lambda D_y \\ S_{d3} &= \lambda D_y + 0.25\left(D_u - \lambda D_y\right) \\ S_{d4} &= D_u \end{aligned} \tag{11.14}$$

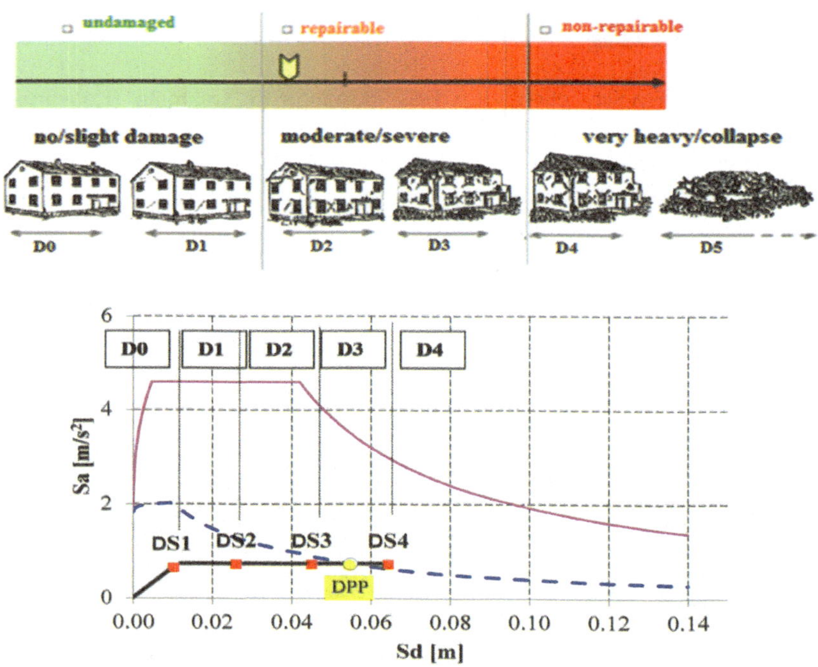

Figure 11.8 Capacity curve of RC buildings with elastic and reduced inelastic response acceleration spectra. Performance point (DPP) is defined among the damage state thresholds for which classification is illustrated above (Kazantzidou *et al.* 2015).

The damage thresholds can be alternatively defined in terms of overstrength and ductility factors provided by different design seismic Codes. In Figure 11.8, the damage states thresholds are illustrated on the capacity curve together with their characterization.

The fragility models is a main outcome of the mechanical method, used to predict conditional probabilities for a building (or a typology) of being within $P[D = ds]$ or exceeding $P[D > ds]$ specific damage states for given levels of ground excitations. In Figure 11.9a, fragility curves represent damage states: Minor (D1), Moderate (D2), Severe (D3) and Collapse (D4). The assessed target displacement (performance point – D_{PP}) is marked in terms of Spectral Displacement S_d. The probability that S_d exceeds the defined damage state thresholds S_{dk} is evaluated by the lognormal cumulative distribution:

$$P\left[ds \middle| S_d\right] = \Phi\left[\frac{1}{\beta_{ds}} \ln\left(\frac{S_d}{\bar{S}_{ds}}\right)\right] \tag{11.15}$$

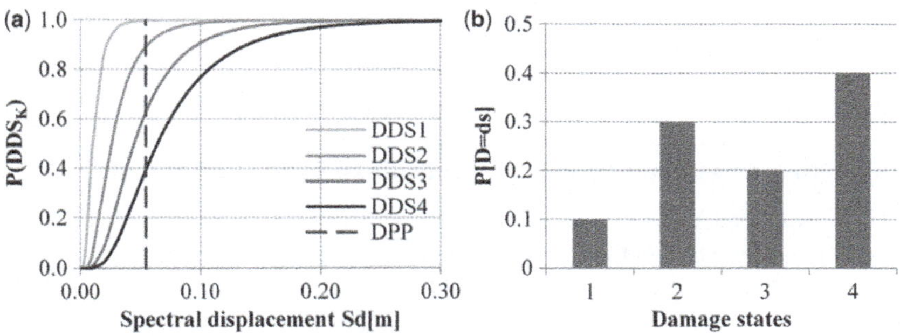

Figure 11.9 (**a**) Fragility curves for RC building and (**b**) corresponding damage probability matrix for S_d = DPP (Kazantzidou *et al.* 2015).

where $\bar{S}_{ds}$ is the median value of spectral displacement at which the building reaches a certain threshold of the damage state ds, β_{ds} combines the uncertainty in the damage state threshold ($\beta_{M(Sds)}$), the variability of the capacity curve (β_C) and the spatial variability of the seismic demand (β_C):

$$\beta_{ds} = \sqrt{\left[CONV\left(\beta_C, \beta_D\right)\right]^2 + \left(\beta_{M(Sds)}\right)^2} \tag{11.16}$$

11.6 SOCIOECONOMIC LOSS ESTIMATION (SLE)

SLE models combine seismic hazard and vulnerability models to estimate the extent of likely damage and the socioeconomic consequences from a range of seismic events. SLE inventory includes buildings and infrastructure (e.g. roads, bridges). The output comprises economic costs of the expected damage to the built environment and estimates of injuries and deaths. Most of these models are contained in commercial software packages that have been developed mainly by the insurance industry.

Models publicly available for response planning and mitigation purposes also exist. One of the few standardized procedures worldwide and mostly used is the HAZUS model, developed and applied by the US Federal Emergency Management (FEMA). HAZUS (1999) is a natural hazard loss estimation method, mainly implemented through GIS software. It deals with nearly all aspects of the built environment and a wide range of different types of losses. It is capable for computing damage to residential, commercial, industrial buildings and to essential facilities, as a function of the ground motion intensity and probability of occurrence of each damage state.

Social losses regard casualties, displaced households and shelters need. Economic losses are *direct* and *indirect*. Direct losses regard repair and replacement costs, building content loss, activity interruption. Indirect losses concern supply shortages and time of reconstruction. Output for both commercial and public models are summarized in Table 11.2.

Table 11.2 Example Outputs from Loss Estimation Models.

Commercial Loss Estimation Models	Public Loss Estimation Models
Maximum expected cost for a portfolio of insured risks	Losses associated with damage to buildings (structural, nonstructural, contents, and inventory damage)
Expected annual losses (average annual loss) of a specific insured portfolio	Distribution of building damage (damage state) by occupancy and building type
Impact of mitigation factors on expected annual losses and maximum expected claims	Losses to and post-earthquake functionality of transportation and utility lifelines and essential facilities
Probable maximum loss of a specific insured property (i.e. mortgage securitization)	Regional economic impacts (e.g. direct and indirect business interruption)
N/A	Injuries, deaths, and shelter requirements

Direct Economic loss is a common task for commercial and public models. Typically, it is evaluated by the probability P(ds| IM) of the building sustaining a damage state ds for a given intensity IM, as computed for Damage Probability Matrices based on the fragility models (Figure 11.9b):

The mean damage cost (or loss) ratio is defined by:

$$E\left(C\middle|IM\right) = \sum\nolimits_{i=1}^{4} DF_i \times P\left[d_{si}\middle|IM = S_d\right] \tag{11.17}$$

where DF_i is the cost replacement factor that reflects the ratio between the cost to repair a structure with respect to the cost of a new construction. The loss ratio can be expressed by means of vulnerability curves for all IM intensities. Construction and repair costs may vary significantly among different countries and periods and, thus, attentive local research should take place. The damage cost (or repair cost) CS, for a given IM is calculated knowing the cost of a new construction or replacement (BRC):

$$\mathrm{CS} = \mathrm{E}\left(\mathrm{C}\middle|\mathrm{IM}\right) \times \mathrm{BRC} \tag{11.18}$$

The above equations refer to individual structural typologies. For the estimation of the mean loss function of all typologies of an area, the average mean damage ratio is calculated. For the determination of the total cost of building damage the sum of the structural CS and non-structural CNS damage of all buildings, all typologies and all occupancy classes, is calculated:

$$\mathrm{CBD} = \Sigma_i\,\mathrm{CS}_i + \mathrm{CNS}_i \tag{11.19}$$

Non-structural damage, acceleration sensitive damage (damage to ceilings, mechanical and electrical equipment) and drift sensitive (walls partition, glass breaking) is similarly calculated.

It is admitted that even limited earthquake physical damage is capable of imposing a chain reaction transmitted throughout the regional economy. Effects on the commercial sector, potentially vulnerable to the interruption in its operation, constitute the *Indirect Economic Loss.* Indirect Loss Assessment is based on the economic analysis module regarding the flows of goods and services among industries, households, governments, investment and exports. This analysis module is modified in a more sophisticated manner to include earthquake impacts (Boisvert, 1992).

Casualties caused by a postulated earthquake can be modeled by developing a tree of events leading to their occurrence. HAZUS foresees four injury severity levels. An example of assessing the expected number only of occupants passed away in a building during an earthquake is given; the loss model is simulated by an event tree combining probability estimates (P_i) for each branching event including damage state and casualties (Figure 11.10):

$$\mathrm{P}_{\mathrm{killed}} = \mathrm{P}_A \times \mathrm{P}_E + \mathrm{P}_B \times \mathrm{P}_F + \mathrm{P}_C \times \mathrm{P}_G + \mathrm{P}_D \times \left(\mathrm{P}_H \times \mathrm{P}_J + \mathrm{P}_I \times \mathrm{P}_K\right) \tag{11.20}$$

with

$$\mathrm{P}_{\mathrm{killed}\,|\,\mathrm{collapse}} = \mathrm{P}_D \times \mathrm{P}_I \times \mathrm{P}_K \quad \text{and} \tag{11.21}$$

$$\mathrm{P}_{\mathrm{killed}\,|\,\mathrm{no-collapse}} = \mathrm{P}_A \times \mathrm{P}_E + \mathrm{P}_B \times \mathrm{P}_F + \mathrm{P}_C \times \mathrm{P}_G + \mathrm{P}_D \times \mathrm{P}_H \times \mathrm{P}_J \tag{11.22}$$

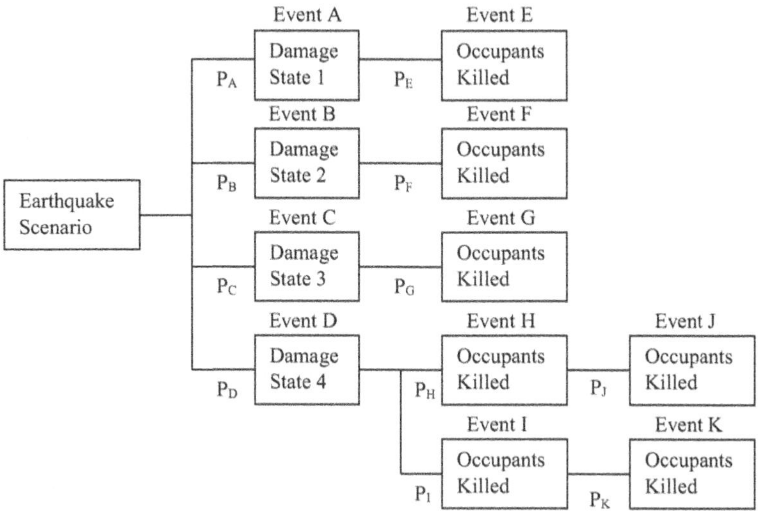

Figure 11.10 Casualty modelling event Tree (HAZUS, 1999).

Evaluation of the probability of each branch constitutes the main effort in the earthquake casualty modelling. There are three types of data used in a casualty model: (a) scenario of earthquake striking hour, (b) population distribution, inventory and damage state probabilities (P_A to P_D in Figure 11.10) and (c) population distribution within occupancy categories for three time zones. The probabilities P_E to P_K are mainly evaluated according to past earthquakes, available data statistics and expert judgment.

The expected number of occupants killed ($EN_{\text{occupants killed}}$) is a product of the number of occupants of the building at the time of earthquake ($N_{\text{occupants}}$) and the probability of an occupant to be killed (P_{killed}).

$$EN_{\text{occupants killed}} = N_{\text{occupants}} \times P_{\text{killed}} \tag{11.23}$$

Within the direct social losses, the *shelter model* is implemented, including the number of displaced households (loss of habitability) and the number of people requiring short-term shelter. The methodology is based on the consideration that all dwelling units located in buildings being in the complete damage state are uninhabitable. The probability of relocation is separately computed for single-family (SF) and multi-family (MF) units, based on the probability of occurrence of the most severe damage states (P2–P4) and different weight factors based on regional peculiarities.

$$\%SF = wS2 \times \%P2 + wS3 \times \%P3 + wS4 \times \%P4 \tag{11.24}$$

$$\%MF = wM2 \times \%P2 + wM3 \times \%P3 + wM4 \times \%P4 \tag{11.25}$$

By applying an occupancy rate (for single-family #SFU and multi-family #MFU units), the total number of displaced households (#DH) for the given number of households (HH) is calculated from census data by the following formula:

$$\#\mathrm{DH} = \left(\#\mathrm{SFU} \times \%\mathrm{SF} + \#\mathrm{MFU} \times \%\mathrm{MF}\right) \times \left(\frac{\#\mathrm{HH}}{\#\mathrm{SFU} + \#\mathrm{MFU}}\right) \tag{11.26}$$

Default values for all modules are implemented in HAZUS for US territories. When HAZUS vulnerability and loss models are used for other countries, particular attention should be given on different construction practices, design and quality that should be considered after appropriate region-specific fragility curves. Regarding loss estimates, appropriate injury classification scales must be established, incorporating different levels of the regions' earthquake preparedness and resilience capacity.

11.7 TSUNAMI RISK ESTIMATION

Tsunamis are sea waves with extremely long wavelength, periods ranging from minutes to hours and small height when travelling across the open sea. When reaching shallow water their height increases and it rushes onshore causing a rapidly rising flood, often consisting of multiple waves. Tsunamis are generated when the seafloor abruptly deforms from different reasons and the overlaying water is vertically displaced. Abrupt seafloor movement at converging plate boundaries is the most common causative mechanism of tsunamis, particularly due to the faulting mechanism, involving in principle vertical sea bottom displacement. Apart from the extended impact of the waves, the destructive power of the driven debris may also produce major damage to the coastal area.

Much like terrestrial seismic hazard and risk maps, recent technical studies of tsunami modelling are being performed in order to determine, other than deterministic worst-case scenarios, probabilistic tsunami return periods that would lead to tsunami hazard maps in terms of inundation distribution. Tsunami modelling takes into account several parameters, i.e. off-shore bathymetry, source location, direction of wave propagation, coastal configuration, etc., in order to predict wave amplitudes and their inundation area in realistic return periods. In parallel, effort is being made to interpret the effect of the tsunami to the urban environment leading to vulnerability estimations.

The field of quantitatively assessing tsunami risk is young. It is only during the last decade or so, that impact research has been accomplished in this field, although there are still unknown factors about how the built environment reacts to the tsunami loading. A validated method to assess tsunami vulnerability is the Papathoma Tsunami Vulnerability Assessment model (PTVA) (Dominey-Howes & Papathoma, 2007; Dall'Osso *et al.* 2009), which applies for any urban and building scale. PTVA model takes into account factors as inundation amplitude, construction material, number of floors, building orientation, state of preservation, position of buildings, building surroundings and land cover, which lead to vulnerability scores applied on a vulnerability scale ranging from 'low' to 'very high' rates. More refined models (Wiebe, 2014) include fragility curves, developed by relating near-coast water depth to the probability of urban failure.

Based on this increasing knowledge, global effort is being carried out to predict and mitigate tsunami effects. In USA, FEMA has recently embedded tsunami inundation return intervals into Flood Insurance Rate Maps (CGS, 2015). In addition, building design standards against tsunamis is envisaged to be incorporated into the International Building Code by 2018. FEMA is currently working towards developing a tsunami modelling insertion to the HAZUS program for damage and loss prediction. These models in combination with Tsunami Warning Systems (TWS), operating globally, furnish coastal districts against tsunami impacts.

11.8 EARTHQUAKE RISK MANAGEMENT AND PREPAREDNESS

In the frame of earthquake risk management, early warning systems are in general major elements of environmental risk reduction, preventing loss of life and reducing the economic and material impact of disasters. An Early Warning System (EWS) is defined as a chain of communication, event detection and decision systems working together to forecast and signal threatening disturbances, providing time for the response system to prepare for the adverse event and to minimize its impact (Waidyanatha, 2010). Warning can enable protective actions regarding the public, business, and delicate actions such as medical services, emergency responders and power infrastructure.

Earthquake Early Warning Systems (EEWS) make use of monitoring systems to alert devices and people when an earthquake capable to affect their location occurs. Since seismic waves travel slower than internet and phone communication, sensors installed near hazardous seismogenic sources detect strong ground motion and broadcast a warning of imminent strong shaking to more distant areas before the shaking arrives. EEWS are tailor-made for the system of faults in the region of interest.

The concept relies on the different propagation speed of the seismic wave types; fast traveling P-waves trigger alert signals, allowing some time (seconds to minutes) for precautionary actions to be taken before the arrival of slower but stronger/catastrophic shear and surface waves. Advanced computers enhance the rapid determination of the earthquake location and magnitude upon receiving the alert message and the expected arrival time and intensity of shaking at a certain location.

A complete and effective early warning system supports four main functions: (a) risk analysis, (b) technical monitoring and warning service, (c) dissemination of meaningful warnings, and (d) response capability (ISDR, 2006). Failure in any part can mean failure of the whole system. Therefore, EEWS should be built upon existing national and regional capacities, and complement broader initiatives aimed at disaster preparedness and mitigation. However, considerable shortcomings and gaps remain, especially in developing countries, where basic capacities, equipment and resources are often not available, while at the human level, public awareness of risks, timely and understandable dissemination of warnings, community preparedness and resilience still fail to raise (UN, 2005).

Significant global progress has been made in the ability to assess risks and to generate and to communicate predictions and warnings as a result of growing scientific and technological developments. EEWS are currently operating in several countries, while others are developing them with increasing rates. Japan currently has the most sophisticated early warning system in the world that was initially developed for slowing and stopping high-speed trains prior to strong ground shaking. The system was successfully operated during the devastating 1995 Kobe earthquake and since then it has been extended throughout the country with the deployment of a dense network of on- and off-shore seismic and geodetic instruments to rapidly detect earthquakes. USGS has been working for several years to develop EEWS for the US, with the help of several cooperating organizations. Other EEWS exist in Mexico, Turkey, Romania, China, Italy and Taiwan.

11.9 DISCUSSION

Earthquake risk is a public safety issue that requires appropriate mitigation measures to protect citizens, property, infrastructure and the built cultural heritage. Current state-of-the-art involves the implementation of multi-disciplinary modules covering the physical environment, urban areas and infrastructures. Each of these modules employs several types of cross-cutting fields, i.e. instrumentation and monitoring, methods to assess and reduce vulnerability, disaster scenarios and loss modelling. The integration of these modules leads to holistic models suitable for disaster mitigation.

Computer tools have been developed allowing for the simulation of real time damage scenarios, with the specific aim to support local Civil Protection departments and government officials for the emergency management, during the first hours after an earthquake. The evaluation of deterministic damage scenarios, by convolving the results from deterministic hazard analysis, exposure and vulnerability assessment, aids towards the estimation of the expected consequences to buildings and people.

Seismic Risk Models have been implemented during the last decades at various spatial scales that exhibit a significant potential to improve emergency response planning and risk mitigation (Table 11.3). Such initiatives are increasingly popular globally, whereas numerous techniques are applied, making use of modern data acquisition technologies combined with innovative analyses, promoting outcomes with weighty societal benefits. Risk assessments and risk maps help to motivate people, prioritize early warning system needs and guide preparations for response and disaster prevention activities.

Table 11.3 Summary description of current state-of-the-art Projects related to ERMs.

Project	Summary
LESSLOSS www.LESSLOSS.org	Addresses natural disasters, risk and impact assessment, natural hazard monitoring, mapping and management strategies, improved disaster preparedness and mitigation, development of advanced methods for risk assessment, methods
EFEHR www.efehr.org	A sustainable community resource for European Earthquake Hazard and Risk. The EFEHR web platform provides access to data, models, tools and expertise relevant for assessment of seismic hazard and risk in Europe.
GEM www.globalquakemodel.org	Develops and implements high-quality earthquake risk assessment tools, collect and generate earthquake risk datasets, methods and guidelines, in an attempt to combine seismic hazard, elements at risk, physical and socio-economical vulnerability, leading to integrated ERMs.
CAPRA www.ecapra.org	It provides modular software for probabilistic risk analysis related to natural hazards and the design of risk-financing strategies.
RASOR www.rasor-project.eu	A platform under development to perform multi-hazard analysis for the full cycle of disaster management. A scenario-driven query system simulates future scenarios based on existing or assumed conditions and compares them with historical scenarios.
Copernicus http://emergency.copernicus.eu	EU program aimed at developing information services based on satellite and in situ data. Its objective is to monitor and forecast the state of the environment on land, sea and in the atmosphere. The information provided by Copernicus, regard mainly secondary effects of earthquakes, such as fires or tsunamis.

State-of-the-art assessment of seismic risk requires systematic collection and analysis of the dynamics and variability of seismic hazard and vulnerabilities that, apart from the buildings structural behavior, arise from socioeconomic processes (i.e. urbanization, rural land-use change, technological risks). A common field of the current state-of-the-art in the earthquake risk problem is the multidisciplinary approach employing massive datasets and their combination using sophisticated tools for producing holistic models to be used by the public domain and the individuals.

However, due to the uncertainties inherent in any risk estimation methodology, such models have been often poorly confirmed, especially in cases of urban areas affected by near strong earthquakes. Uncertainty arises to some extent from incomplete scientific knowledge concerning earthquakes and their effects upon buildings and facilities and quite often there are not enough data available at sufficient levels to be comparable for all specific tasks. Inaccurate inventories of the built environment, demographics and economic parameters add to the uncertainty. Complications largely arise from the sparsity of the available data and the difficulty in obtaining datasets of uniform quality due to the differences in monitoring methods. Uncertainties also result from the approximations and simplifications that are necessary for comprehensive analyses, summing into significant biases in loss estimates (NRC, 2006).

Identifying the seismic potential in terms of crustal deformation and active faulting, sites where a local amplification of seismic shaking will occur, and the buildings that will be the weakest under the seismic shaking, is the best strategy that allows effective defense against earthquake damage at an affordable cost, by applying selective strengthening only to the structures that need it. In this regard, macroseismic data from historical earthquakes are of great importance, especially for countries with a long seismological history, since it facilitates correlations with engineering indices and hence the appropriate parameterization of seismic sources.

So much of the above are compliant to demonstrations, generic provisions of existing seismic codes have often been proven grossly misleading, with ground motion parameters and macroseismic effects found far higher than predicted (Wyss & Rosset, 2013) (Figure 11.11). Accordingly, each of the global and regional models, produced by generic and/or incomplete assumptions, are useful but only of being a reference for further detailed studies of the possible areas identified to be at risk. For these earthquake prone regions, a thorough evaluation of the seismic risk is mandatory, as well as the appropriate definition of the design input, especially as it has been observed that spectral acceleration of strong ground motion often exceeds the elastic design spectra thresholds (Figure 11.12). Microzonation studies are effective tools towards the accurate estimation of seismic excitations affected by local conditions.

Precise seismic risk models are usually not available, especially in less developed countries, several of which adopt risk models constrained for other countries, as the best case scenario. Indeed, losses are most effectively reduced through informed decisions guided by thoroughly-researched understanding of the hazards, risks, and cost of mitigation. Effective reduction of seismic threats can only be achieved by setting off new and existing scientific research toward addressing gaps in vulnerability, producing innovative products, and connecting experts with users of their science.

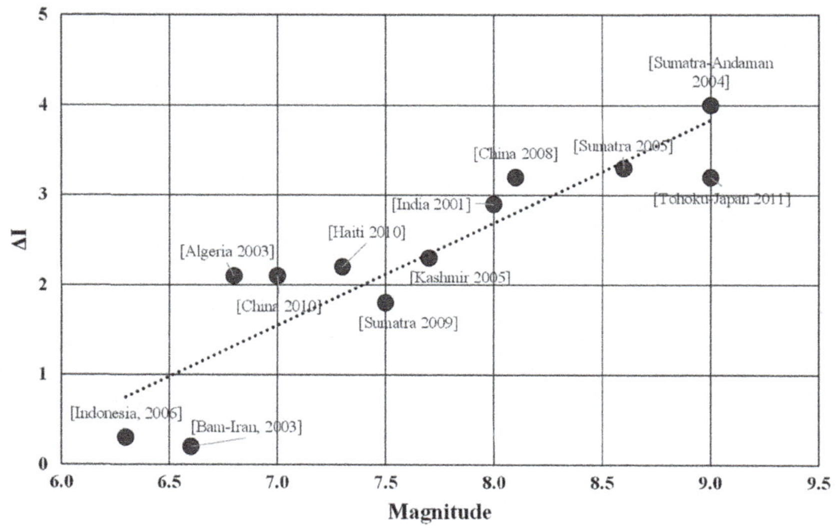

Figure 11.11 Diagram presenting the magnitudes of the deadliest earthquakes occurred during the period 2000–2011, and the corresponding seismic intensity differences ΔI, among the observed values and predicted by GSHAP. Drawn after Kossobokov & Nekrasova (2011).

11.10 SUMMARY

In summary, the following actions-scheme is proposed to be applied for earthquake prone urban regions (Figure 11.13):

(1) Elaboration of microzonation studies for estimating the seismic response of soil layers together with their liquefaction susceptibility for strong earthquake excitations.
(2) Combined deterministic and probabilistic ground motion valuation in terms of PGA, PGV, PGD, Seismic Intensity, Fourier and Response Spectra.
(3) Construction of shakemaps containing the spatial distribution of the above parameters for buildings specific vibration periods.
(4) Vulnerability analysis of the buildings using empirical and mechanical methods.
(5) Mapping of the expected damage grade and damage probability on building or building block level.
(6) Construction of site-specific fragility curves and of appropriately adapted building codes for urban planning purposes.
(7) Involvement of multi-hazard and performance exceedance analyses.
(8) Assessment of social vulnerability and resilience.
(9) Loss estimation in terms of direct and indirect socioeconomic impacts. Integration of all modules into seismic risk models.

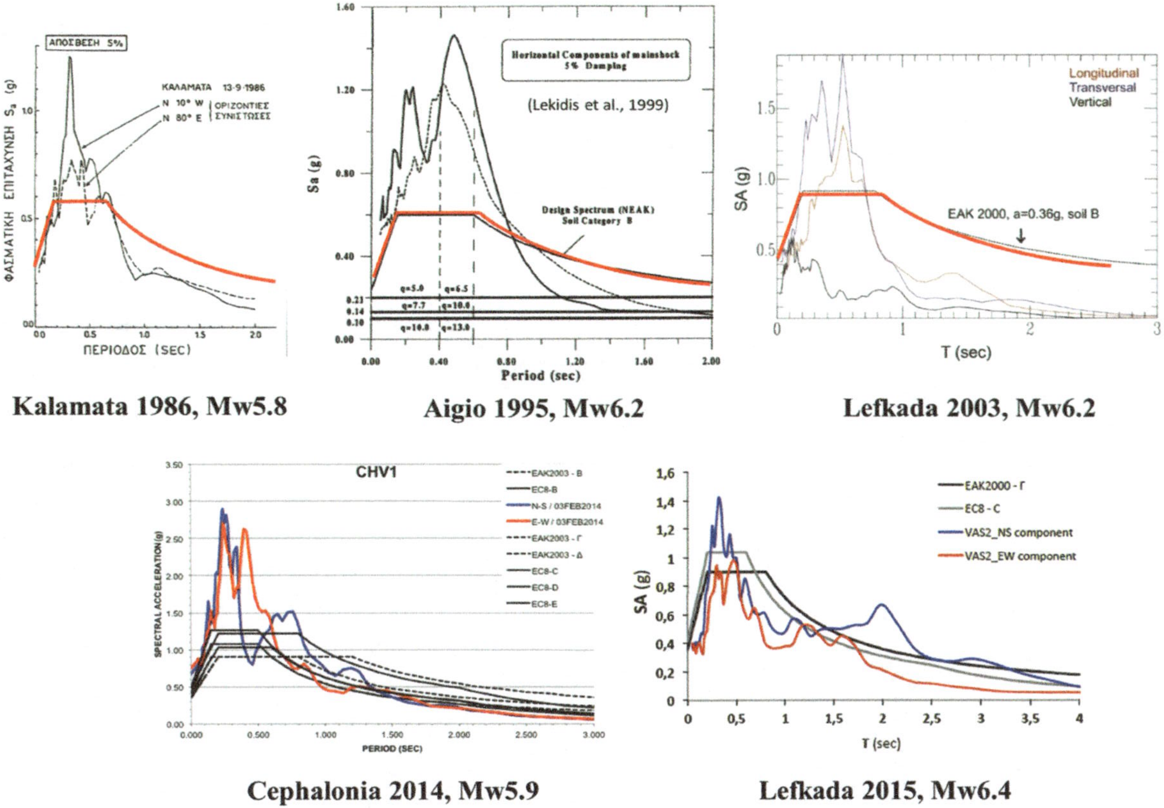

Figure 11.12 Response spectra for earthquakes occurred in Greece, found to exceed the elastic design spectra predicted by the effective seismic codes (after OASP, 1996; Lekidis, 1999; Pitilakis & Roumelioti, 2013; ITSAK, 2014, 2015).

(10) Validation of the models by historical observations, when plausible. Application of the outcomes by the central public authorities and emergency planning accordingly.
(11) Dissemination of outcomes and consecutive practical education of the public.

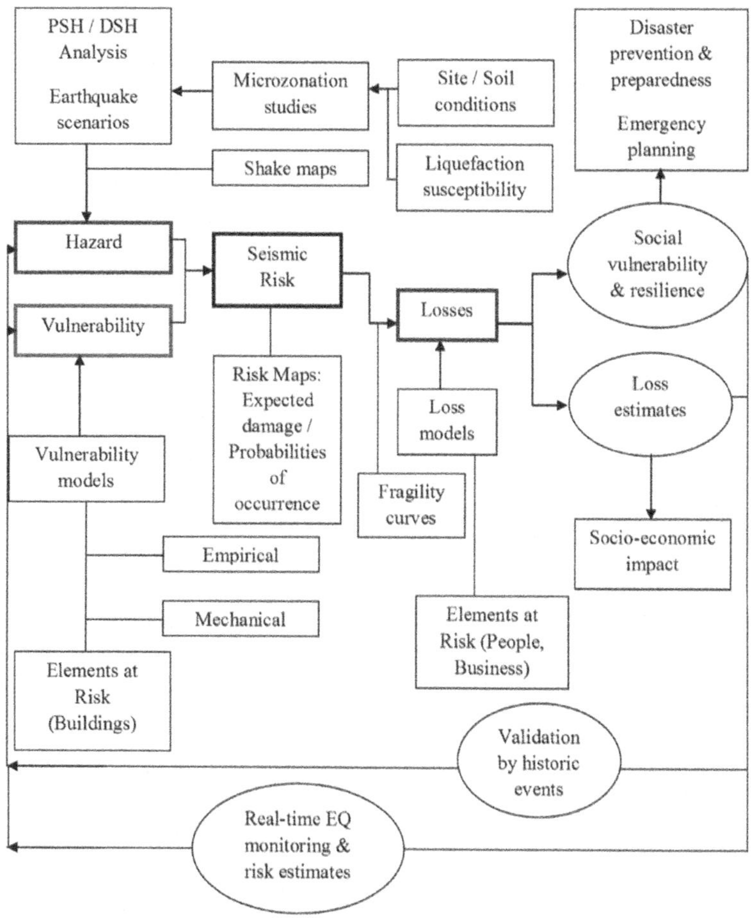

Figure 11.13 Actions-methods scheme, proposed to be applied for earthquake prone urban regions towards determining effective earthquake risk models.

Technological and scientific "tour de force" could beyond doubt guarantee a realistic approximation of the seismic risk produced by the integral of discretized sophisticated approaches. Thereafter, for each country, it is a matter of societal efficiency and national priority/capability whether existing or new earthquake risk mitigation strategies are beneficial, or may remain tabletop exercises.

REFERENCES

Asian Disaster Preparedness Center (ADPC) (2003). Third Regional Training Course on Earthquake Vulnerability Reduction for Cities. Dhaka, Bangladesh.

ATC (1985). Earthquake damage evaluation data for California (ATC-13). Appl. Technology Council, Redwood City, California.

ATC (1996). Guidelines for the Seismic Rehabilitation of Buildings, Volume I, Guidelines and Volume II, ATC-40. Prepared by the Applied Technology Council for the Building Seismic Safety Council; published by the FEMA.

Boisvert R. (1992). Indirect Economic Consequences of a Catastrophic Earthquake, Chapter 7, 'Direct and Indirect Economic Losses from Lifeline Damage', Development Technologies Inc. under FEMA Contract EMW-90-3598.

Boore D. M. (1983). Stochastic simulation of high-frequency ground motion based on seismological models of the radiated spectra. *Bulletin of the Seismological Society of America*, **73**, 1865–1894.

Boore D. M. (1996). SMSIM-Fortran programs for simulating ground motions from earthquakes: version 1.0, U.S. Geol. Surv. Open-File Rept. 96-80-A and 96-80-B, 73 p.

Calvi G., Pinho R., Magenes G., Bommer J.J., Restrepo-Vélez L.F. and Crowley H. (2006). Development of seismic vulnerability assessment methodologies over the past 30 years. *ISET Journal of Earthquake Technology*, **43**(3), 75–104.

CGS (2015). Evaluation and Application of Probabilistic Tsunami Hazard Analysis in California: Phase 1: Work Group Review of Methods, Source Characterization, and Applications of the Crescent City Demonstration Project Results, p 237.

Comité Européeen de Normalisation (2004). Eurocode 8: Design of Structures for earthquake resistance–Part 1: General rules, seismic actions and rules for buildings. prEN 1998-1: 2004, CEN, Brussels.

Cornell C. A. (1968). Engineering seismic risk analysis. *Bulletin of the Seismological Society of America*, **58**(5), 1583–1606.

Dall'Osso, F., Gonella M., Gabbianelli G., Withycombe G. and Dominey-Howes D. (2009). A revised (PTVA) model for assessing the vulnerability of buildings to tsunami damage. *Natural Hazards and Earth System Science*, **9**(5), 1557–1565.

Dolce M., Zuccaro G., Kappos A. and Coburn A. (1994). Report of the EAEE Working Group 3: Vulnerability and risk analysis. In Proc. of 10th European Conference on Earthquake Engineering, Vienna, **4**, pp. 3049–3077.

Dominey-Howes D. and Papathoma M. (2007). Validating a tsunami vulnerability assessment model (the PTVA Model) using field data from the 2004 Indian Ocean Tsunami. *Natural Hazards*, **40**, 113–136.

EM-DAT (2014). The international disaster database. http://www.emdat.be/database.

FEMA-440 (2005). *Improvement of Nonlinear Static Seismic Analysis Procedures*. NEHRP Guidelines, Washington, DC.

Giovinazzi S. and Lagomarsino S. (2004). A macroseismic method for the vulnerability assessment of buildings. In Proceedings of the 13th WCEE, Vancouver, Paper No 896.

Graves R. W. and Pitarka A. (2010). Broadband ground-motion simulation using a hybrid approach. *Bulletin of the Seismological Society of America*, **100**(5A), 2095–2123.

Graves R., Jordan T., Callaghan S., Deelman E., Field E., Juve G., Kesselman C., Maechling P., Mehta G., Milner K., Okaya D., Small P. and Vahi K. (2010). CyberShake: a physics-based seismic hazard model for Southern California. *Pure and Applied Geophysics*. DOI: 10.1007/s00024-010-0161-6.

Grünthal G. (ed.) (1998). Cahiers du Centre Européen de Géodynamique et de Séismologie: Vol. 15, European Macroseismic Scale 1998, Europ. Center for Geodyn. and Seism., Luxembourg.

Guagenti E. and Petrini V. (1989). Il caso delle vecchie costruzioni: verso una legge danni intensità. In Proc. of the 4th Italian National Conference on Earthquake Engineering, Milan.

Gutenberg R. and Richter C. F. (1944). Frequency of earthquakes in California. *Bulletin of the Seismological Society of America*, **34**, 158–188.

HAZUS (1999). *Earthquake Loss Estimation Methodology–Technical and User Manuals*. Federal Emergency Management Agency, Washington, DC.

International Strategy for Disaster Reduction (ISDR) (2006). Platform for the Promotion of Early Warning, http://www.unisdr.org/2006/ppew/whats-ew/basics-ew.htm.

ITSAK (2014). Strong Ground motion of the Feb.3, 2014 (M6.0) Cephalonia Earthquake: Effects on Soil and Built environment in combination with the Jan.26, 2014 (M6.1) event, Thessaloniki, Greece.

ITSAK (2015). Preliminary Earthquake Report–Lefkas (M6.0) on 17-11-2015. Thessaloniki, Greece.

Kappos A. Panagopoulos G., Panagiotopoulos C. and Penelis G. (2006). A hybrid method for the vulnerability assessment of R/C and URM buildings. *Bulletin of Earthquake Engineering*, **4**(4), 391–413.

Kassaras I., *et al.* (2015). Seismic damage scenarios in Lefkas old town (W. Greece). *Bulletin Earthquake Engineering*. DOI: 10.1007/s10518-015-9789-z.

Kassaras I. and Kazantzidou-Firtinidou D. (2015). Damage scenarios at the city of Kalamata. project MIS-448326, implemented under the Action, Development Proposals for Research Bodies-ASPIS-KRIPIS (in Greek).

Kassaras I., Kalantoni D., Kouskouna V., Michalaki K., Stoumpos P., Mourloukos S., Birmpilopoulos S., Makropoulos K. (2014). Correlation between damage distribution and soil characteristics deduced from ambient vibrations in the old town of Lefkada (W. Greece). Proc. of 2ECEES), Istanbul, Paper No 251.

Kazantzidou D., Bozzano C., Lestuzzi P. and Podestà S. (2015). Etude de la vulnérabilité sismique des bâtiments existants des villes de Sion et de Martigny. Centre de Recherche sur l'Environnement Alpin (CREALP), Sion, Switzerland (in French).

Klügel J. U. (2007). Comment on 'Why do modern probabilistic seismic-hazard analyses often lead to increased hazard estimates' by J. Bommer and N. Abrahamson. *Bulletin of the Seismologicla Society of America*, **97**, 2198–2207.

Kossobokov V. G. and Nekrasova A. K. (2011). Global seismic hazard assessment program (GSHAP) maps are misleading. *Problems of Engineering Seismology*, **38**(1), 65–76 (in Russian).

Lagomarsino S. and Giovinazzi S. (2006). Macroseismic and mechanical models for the vulnerability and damage assessment of current buildings. *Bulletin of Earthquake Engineering*, **4**, 415–443.

Lagomarsino S., Cattari S., Pagnini L. and Parodi S. (2010). Method(s) for large scale damage assessment, including independent verification of their effectiveness and uncertainty estimation, Deliverable D5.1 of Project S2: Development of a dynamical model for seismic hazard assessment at national scale, INGV-DPC 2007–2009.

Lekidis, V., Karakostas C., Dimitriu P., Margaris B., Kalogeras I. and Theodulidis N. (1999). The Aigio (Greece) seismic sequence of June 1995: seismological, strong motion data and effects of the earthquakes on structures. *Journal of Earthquake Engineering*, **3**, 349–390.

Makropoulos K. C. (1978). The sstatistics of llarge eearthquake mmagnitude and aan eevaluation of Greek sseismicity. PhD tthesis, Univ. of Edinburgh, Edinburgh, UK, 198pp.

Makropoulos K. C. and Burton, P. W. (1985a). Seismic hazard in Greece: I magnitude recurrence. *Tectonophysics*, **117**, 205–257.

Makropoulos K., Kaviris G. and Kouskouna V. (2012). An updated and extended earthquake catalogue for Greece and adjacent areas since 1900. *Natural Hazards and Earth System Science*, **12**, 1425–1430.

McGuire R. K. (2001). Deterministic vs. probabilistic earthquake hazards and risks. *Soil Dynamics and Earthquake Engineering*, **21**, 377–384.

Milutinovic Z. and Trendafiloski G. (2003). An advanced approach to earthquake risk scenarios with applications to different European towns. ReportWP4: Vulnerability of Current Buildings, Risk-UE. European Commission, Brussels, DOI: 10.1007/978-1-4020-3608-8_23.

NOAA (1972). A study of earthquake losses in the San Francisco Bay Area–Data and Analysis. A report prepared for the Office of Emergency Preparedness: U.S. Department of Commerce, p. 220.

OASP (1996). Systematic Presentation of Results and Conclusions of Microzonation Study Research Programs of Kalamata, Earthquake Planning and Protection (in Greek).

Panza G. F., Irikura K., Kouteva M., Peresan A., Wang Z. and Saragoni R. (Eds). (2010). Topical volume on 'advanced seismic hazard assessments'. *Pure and Applied Geophysics*, **168**, 1–2 and (3–4).

Peresan A., Zuccolo E., Vaccari F., Gorshkov A. and Panza G. F. (2010). Neo-deterministic seismic hazard and pattern recognition techniques: time dependent scenarios for North-Eastern Italy. *Pure and Applied Geophysics*, DOI: 10.1007/s00024-010-0166-1.

Pitilakis K. and Roumelioti Z. (2013). The Lefkas M6.2 2003 Earthquake. VEESD 2013, Paper No. 383.

Rossetto T., Ioannou I., Grant D. N. and Maqsood T. (2014). Guidelines for Empirical Vulnerability Assessment. GEM Technical report 2014-11 V1.0.0.

Stucchi M., Rovida A., Gomez Capera A.A., Alexandre P., Camelbeeck T., Demircioglu M. B., Gasperini P., Kouskouna V., Musson R. M. W., Radulian M., Sesetyan K., Vilanova S., Baumont, Bungum H., Fäh D., Lenhardt W, Makropoulos K., Martinez Solares J. M., Scotti O., Živi M., Albini P.,B atllo J., Papaioannou C., Tatevossian R., Locati M., Meletti C., Viganò D., Giardini D. (2013). The SHARE European earthquake catalogue (SHEEC) 1000–1899. *Journal of Seismology*, **17**, 523–544.

United Nations (UN) (2005). In larger freedom: towards development, security and human rights for all, Report of the Secretary-General, www.un.org/largerfreedom.

United States Department of Agriculture (USDA) (1994). World population density 1994–Global Population Density Map, https://commons.wikimedia.org/wiki.

Waidyanatha N. (2010). Towards a typology of integrated functional early warning systems. *International Journal of Critical Infrastructures*, **6**(1), 31–51.

Wald D., Quitoriano V., Heaton T. and Kanamori H. (1999). Relationships between peak ground accelerations, peak ground velocity and Modified Merkalli Intensity in California. *Earthquake Spectra*, **15**(3), 557–567.

Wessel P., Smith W. H. F., Scharroo R., Luis J. F. and Wobbe F. (2013). Generic mapping tools: improved version released. *EOS Trans. AGU*, **94**, 409–410.

Wiebe D. and Cox D. (2014). Application of fragility curves to estimate building damage and economic loss at a community scale: a case study of seaside, Oregon. *Natural Hazards*, **71**(3), 2043–2061.

Wyss M. and Rosset P. (2013). Mapping seismic risk: the current crisis, Nat. Hazards, DOI: 10.1007/s11069-012-0256-8.

Zeng Y., Anderson J. G. and Yu G. (1994). A composite source model for computing realistic synthetic strong ground motions. *Geophysical Research Letters*, **21**, 725–728.

Chapter 12

Tectonic hazards: volcanoes

George E. Vougioukalakis and Augusto Neri

12.1 VOLCANIC HAZARDS

Volcanic activity is one of the most exciting natural manifestations of our living planet. Volcanoes generate continuously new lithosphere in expanding plate boundaries and have a key role in recycling it in the subduction plate margins. Volcanoes created the Earth atmosphere (except oxygen component) and gave birth to life in their submarine white smokers. Volcanoes made available the most important elements for human race to survive and evolve: fire, obsidian, sulfur, fertile soils, all precious and base metals and much more.

On the other hand, explosive volcanic eruptions are one of the most impressive, violent and destructive natural events and represent a potential threat for hundreds of millions of people. Landscape beauty and soil fertility attracts populations, which settle on volcano flanks, creating, by the conjunction of hazards and population, high risk areas. The population directly at risk from volcanoes nowadays is at least 500 million, representing as much as 7% of mankind. At a larger scale, volcanic emissions (gas and ash) can affect the natural environment on the whole Earth modifying the climate for some years, inducing the so-called "volcanic winters".

During the 1990s, more than 2000 human lives were lost because of volcanic activity and 2 cities were completely devastated. During the same decade, volcanic eruptions and their effects directly affected around 3 million people and the economic losses reached several billions of Euros (€), causing disruption in entire regions and countries. The economic losses from the recent 2010 Eyjafjallajökull volcano eruption in Iceland, due to major air-traffic disruption were estimated at 1 billion € per day. At the end of the same year, the eruption of the Merapi volcano in Indonesia, killed more than 380 persons and led to the displacement of more than 400,000 people (Bignami *et al.* 2012).

Volcanic hazard consists of the probability that a specific volcanic event occurs with a specific intensity of its hazardous actions (i.e. destructive power) in a given area, within a specific time period (Blong, 1984). ***Elements or values at risk*** (sometime also called ***exposure***) correspond to all kind of assets exposed to these natural phenomena: human lives and biosphere, infrastructures and buildings, social and economic activity, etc. The level of potential damage (loss of value) of these elements at risk is expressed by their ***vulnerability***, considering it as a degree of resistance to the volcanic actions. ***Volcanic risk*** is then defined by the probability of financial, environmental and human losses caused by the volcanic eruption and it is still associated with a specific area and time period. Considering also ***coping capacities***, i.e. the community ability to recover after the specific volcanic event, volcanic risk in a given area can be expressed by the following simplified expression:

$$\text{Volcanic Risk} = (\text{Volcanic hazard}) \times (\text{Vulnerability}) \times (\text{Value at risk})/(\text{Coping capacity})$$

12.1.1 Volcano basics

Volcanoes are natural edifices, usually a mountain on land or below sea level, built up by ***lava dykes, lava flows or domes*** and ***tephra*** layers. *Lava dykes* are vertical or sub-vertical lava sheets, emplaced when magma solidify in the fractures or faults that are used as paths to the surface. *Lava flows and domes* are created during ***effusive volcanic activity***, when poor in gasses magma outflows non-explosively from volcanic vents, as a continuous medium. *Tephra layers* are deposited from ***explosive volcanic eruptions***, when rich in gasses magma is ejected into the atmosphere as a fragmented medium.

Magma is a multicomponent (in chemical species) and multiphase (liquid as melt, solid as crystals and gas) fluid, generated by *partial melting* of the *upper mantle*, in specific tectonic environments constrained by *plate tectonics theory*. Magma rises from upper mantle to the crust, mostly due to buoyancy forces. It is usually stored in *magma chambers*, at depth between 5 and 20 km. The position of the magma outflow in surface and the creation of a ***volcanic center*** is then mostly controlled by the large tectonic lineaments of the upper crust. Volcanic centers are grouped in clusters, which can form a *volcanic field*, covering areas of many square kilometers. A single active volcanic center has a typical life time of some thousand years before its extinction, whereas a volcanic field time life can last for a few million years.

Volcanic rocks are classified into three main *chemical types* according to the SiO_2 content: felsic or silicic (>62 wt%), intermediate (52–62 wt%), and mafic or basaltic (<52 wt%).

The two main *physical variables* controlling the eruptive behavior (i.e. explosive versus effusive style) are ***temperature*** (ranging from 650°C to 1200°C) and ***viscosity*** (resistance to flow), with the former having the main control on the latter.

Moreover, both are strongly influenced by the chemical composition of the magma, as well as by its crystal and gas contents. Large catastrophic explosive eruptions are associated with felsic high-viscosity/low temperature magma, whereas mafic low-viscosity/high temperature magma usually generates effusive or at most weak explosive eruptions.

Volcanic gasses (H_2O, CO_2, SO_2, H_2S, HCl, HF etc. typically ranging in 1–5 wt%), are released from magma, forming a separate vapor phase (i.e. bubbles), when pressure decreases as magma rises toward the surface of the Earth. Gas constitutes the major element driving eruptions. If bubbles can escape from the melt, as usually happens in the case of low-viscosity magma, the remaining magma rises slowly in the conduit and generates effusive eruptions (i.e. lava flows and domes). Vice versa, when the bubbles do not decouple from the melt, their expansion causes magma to accelerate upwards. The result is further decompression, magma expansion and fragmentation, feeding explosive eruptions.

Meteoric water (ground water, lake water, rain, sea water, glaciers etc.) plays also a critical role in volcanic activity. When this water is heated by magma, it flashes to steam generating an overpressure leading to explosive eruptions, called hydrothermal and hydro-magmatic. In addition, water originated from rainfall, lakes or glaciers, can remobilize loose sediment and volcanic materials generating volcanic debris flows or mudflows, called "lahars".

A volcano is defined ***active*** if it erupted in the last 10,000 years. More than 1,500 volcanoes of our planet are considered active, without taking into account the numerous submarine active volcanoes. Approximately 60 of them erupt each year (Global Volcanism Program, 2013). As there are volcanoes that can be ***dormant*** (*non-eruptive*) for time periods longer than 10,000 years and then become active again, the term ***potentially active*** volcano is also used to consider such category. Vice versa, a volcano is defined ***extinct*** when it is considered not able to erupt any more, although such possibility can never be excluded with full certainty.

12.1.2 Primary and secondary volcanic hazards and their impacts

Different types of eruptions have been defined by volcanologists (e.g. Hawaiian, Strombolian, Vulcanian, Pelean, Surtseyan, Plinian, etc.), ranging from effusive to highly explosive activity. Each type produces different volcanic phenomena, from lava flows to catastrophic explosive events, which correspond also to a different level of threat. In the following, the main phenomena, with their associated hazards and impacts, are briefly presented.

12.1.2.1 Lava flows

Lava is molten rock expelled by a volcano during an eruption. The same term indicates the resulting rock after solidification and cooling. Close to the eruptive vent, lava temperatures range from 650°C to 1200°C, being hottest and mostly mafic. Depending on its chemistry, temperature, effusion rate, viscosity and topography, lava can flow large distances (typically from less than one km to several tens of km) at greatly varying speed (from a few meters up to 60 km/h) before cooling and solidifying. Felsic lavas, being less hot and extremely viscous, typically form lava domes or thick/short length lava flows. Mafic lavas with high temperature and low viscosity tend to produce low-profile shield volcanoes, since the fluid lava flows for long distances from the vent.

Lava flows will seldom threaten human or animal life, because of their slow rate of movement. However, they cause complete destruction of buildings, infrastructure and vegetation encountered in their path. Low viscosity mafic lava flows may cause the combustion of nearby flammables, such as crops or buildings, because of their higher temperatures and low viscosity. Their relatively high speed, can occasionally threaten humans and livestock. In contrast, collapse of the steep sides of lava domes or lava flow fronts can generate gravitational landslide and in some cases produce pyroclastic density currents (see below).

12.1.2.2 Tephra fall

Tephra is a general term for fragments of volcanic rock, called also *pyroclasts*, which are ejected into the air by explosive eruptions and carried upward in volcanic columns (also called "plumes") or collapsing fountains.

Pyroclastic materials are classified according to their size in: *ash* (pyroclasts with diameter < 2 mm), *lapilli* (with diameter between 2 and 64 mm) and *blocks* (with diameter > 64 mm). According to their particle shape, composition and mode of formation, different terms are also used, such as: *bomb*, formed from fluid magma ejected into the air, that takes a fusiform shape during its trajectory path; *pumice*, light, finely vesicular volcanic rock (quenched lava foam) with typically more than 60–70 vol% gas bubbles; and *scoria*, a coarsely than pumice vesicular rock fragment. Blocks and large lapilli typically fall to the ground close to the vent. Small lapilli and ash are instead mostly carried up by buoyancy to form the volcanic plume. Fine volcanic ash can then travel hundreds to thousands of kilometers downwind forming the dispersing ash cloud (Fisher & Schmincke, 1984; Heiken & Wohletz, 1985).

12.1.2.2.1 Ballistic projectiles

Bombs, blocks and large lapilli can potentially land as projectiles more than 10 km from the vent, but typically land within 5 km. A projectile's impact energy, and thus its potential to cause damage, depends on the size, density and ejection velocity. The most damaging consequences of tephra projectiles are from direct impact: they can puncture holes in roofs, kill people or livestock and cause serious damage to crops.

Tephra not large or dense enough to penetrate roofs can, however, contribute to roof collapse through overloading or through repeated impacts that may seriously weaken a structure and leave it more vulnerable to future impacts or hazards. Hot large tephra can also ignite fires upon impact causing more serious damage than through penetration alone.

12.1.2.2.2 Ash fallout

Ash can be dispersed thousands of kilometers from source and impact human and livestock health, disrupt the aviation and critical infrastructure (e.g. transportation, telecommunications, power supplies, water and wastewater networks), damage agriculture, buildings and structures (e.g. bridges, roads). Fallout of fine volcanic ash can affect large areas and numbers of people during an eruption, with long-lasting eruptions potentially causing repeated impacts.

The level of building and infrastructure damage sustained by ash fallout depends upon structure type and ash deposit characteristics, such as chemical composition, moisture content and thickness (Baxter *et al.* 2008; Spence *et al.* 2005; Zuccaro *et al.* 2008). Thick ash deposits (typically > 200–300 mm) can be capable of causing roof collapse and weakening walls, with possible death or injury to people or livestock inside. Roof failure occurs when the ash load exceeds the resistance of the roof structure. A large non-destructive ash load may also adversely affect a building's resistance to other phenomena, such as earthquakes. Moderate to heavy falls of ash can also block out sunlight limiting emergency response. Thick ash deposits can overload and fail power lines, block drainage and irrigation systems, collapse embankments, weak bridges and render a road unusable; however, ash fallouts are unlikely to cause significant structural damage to infrastructure at medial-distal distances.

The local and regional effects of volcanic ash on the soil are also diverse. Depending on its chemical composition and thickness, ash can act either as a fertilizer or as a contaminant. Damage to crops and natural vegetation depends on the plant typology condition and morphology and can range from complete to minor damage. Nevertheless, thin tephra falls may improve soil fertility, particularly in the tropics (Wilson *et al.* 2007).

The impact of ash falls on local hydrological systems depends on the deposit thickness, grain-size distribution, geomorphology, soil permeability and climate, and, in particular, precipitation intensity. Sewerage and storm water systems can face serious damage, as ash is washed off roads, carparks and buildings into the systems. Even ash fallout layers of less than 5 mm can result in significant impacts for livestock and crops/natural vegetation, if the ash is high in chemical species, such as fluorine and acids.

Thin ash fallouts of as little as one mm can also damage building components, infrastructure and lifelines, because of their abrasive and conductive (especially when wet) properties, for example, through short-circuiting electrical systems, contaminating water supplies and causing wear on mechanical parts. Thin ash falls may also disrupt transportation systems; ash fallout of 1 to 3 mm can reduce visibility and safety on highways and can also obscure or completely cover markings on roads and can close airports. While ash is not toxic, it acts as an irritant affecting eyes and throats and exceptionally fine particles (<4 μm) and can cause major respiratory disorders (Spence *et al.* 2005; Baxter *et al.* 2008). Exposure to aerosols, i.e. particles less than 2.5 and 10 μm in diameter, can also have long-term health impacts.

12.1.2.3 Pyroclastic density currents

Pyroclast and gas mixtures flowing over the volcano topography are called Pyroclastic Density Currents (PDCs); these are denser-than-air ground-hugging clouds moving down volcano flanks at tens to hundreds of kilometers per hour, and at temperatures that can reach more than ~700°C.

PDCs are gravity-controlled, and can be generated by different mechanisms, such as:

- Gravitational collapse of the eruption column (e.g. Plinian eruptions) or tephra fountaining.
- Gravitational collapse of a felsic lava dome or of a steep-sided felsic lava flow (e.g. Merapi or Pelean eruptions), or lateral blasts.
- Hydromagmatic activity, involving explosive interaction between mafic-intermediate magma and meteoric surface or deep water (e.g. Surtseyan eruption).

PDCs can vary greatly in their density and are usually classified as diluted or concentrated PDCs. Diluted PCDs, containing only 0.1–1% of solid rock material in volume, are typically called ***pyroclastic surges*** and produce rather thin deposits (cm to dm thick). They are extremely mobile, generally travelling at tens up to hundreds of km/h. Surges can also travel even tens of km across the sea (e.g. Krakatau 1883, Indonesia), and surmount conspicuous topographic obstacles. In contrast, concentrated PDCs are referred to as ***pyroclastic flows***; they are capable of travelling at similar speeds as surges and, in some instances, up to distances of several tens of kilometers. Dense PDCs are usually topographically controlled, generally following valleys or low-lying areas.

However, PDCs are significantly complex and heterogeneous flows that can greatly vary their density in space and time also depending on several factors, such as generation mechanism, grain-size distribution, influence of topography, segregation processes within the current and ingestion of air. A dilute PDC can, therefore, become concentrated, and vice versa, and also the two types of flows can co-exist in the same current in many situations.

PDCs have caused the largest number of volcano-related fatalities worldwide among the various volcanic hazards. Warning periods for PDCs are very short. People, animals or vegetation caught in the direct path of PDCs are likely to be instantly killed as a result of trauma and burns, both external and internal (from inhalation of hot gas and ash). On the periphery, or in small dilute PDCs, people may survive and vegetation suffers heat damage, but may not be destroyed.

Damage to buildings and other structures impacted by a PDC depends upon the dynamic pressure, temperature, duration of flows and amount of solid material it carries with greater damage expected with the increase of each variable (Baxter *et al.* 2005, 2008). Failure of windows, generally the weakest component of a building, through elevated temperatures and dynamic pressures, allows inflow of hot ash, which can cause combustion of furnishings and destruction of the building through fire. For relatively low temperature PDCs (<300°C), or for areas with limited combustible material, the lateral dynamic pressure imposed on building surfaces by a PDC and the impact of entrained projectiles becomes important to determine the level of damage sustained by buildings and infrastructure. PDCs can also be destructive to road systems, to power and telecommunications networks, and to any other non-buried infrastructure elements in their area of impact (Zuccaro *et al.* 2008). The impact of PDC deposits on local hydrological systems is similar to that of ash fallout.

12.1.2.4 Gas emissions

The most abundant magmatic volatile is water (H_2O, up to about 5 wt%) followed by carbon dioxide (CO_2), sulfur species (sulfur dioxide SO_2 and hydrogen sulfide H_2S), and minor parts of carbon monoxide (CO), halogens (HF and HCl), and Radon. These volatiles are released not only during eruptions, but also from the soil and fumaroles of dormant volcanoes. Toxic volcanic gas is also released from flowing lava and through the so-called "*lava haze*", released at the contact of lava with sea water.

Exposure to gas and aerosols produced by eruptions can severely affect the health or lead to the death of humans and livestock, although direct exposure more commonly produces respiratory and eye irritation in humans and animals. Significant impacts are likely to be restricted to within about 10 km of the active vent. The interaction between volcanic gas and the atmosphere can result in "dry fog" or "vog" (i.e. volcanic fog) and/or acid rain.

Heavier than air gases, such as CO_2, an odorless gas, can be accumulated in morphological depressions or flow considerable distances down to valleys or slopes leading to the suffocation and/or poisoning of humans and animals. Such gasses, thus, have the potential to collect in building basements and expose occupants to high, potentially fatal, concentrations. The sudden release of large quantities of CO_2 during hydrothermal (phreatic) explosions, or the non-eruptive overturning of CO_2-charged lakes, has resulted in large disasters in the past.

SO_2 has a pungent odor and is highly toxic, irritating the eyes and respiratory organs. Acid rain resulting from oxidation of SO_2 with atmospheric OH can lead to decades-long damage to crops and corrosion of metallic objects. H_2S is also highly toxic and is known for its "rotten eggs" odor. Long-term exposure at relatively low concentrations (e.g. 50 ppm) can result in inflammation of the throat and, at higher concentrations (e.g. >250 ppm), in fluid buildup on the lungs and death.

If HF is released during eruptions, it can cause the disease *fluorosis*, which may affect livestock, which eat fluorine-contaminated grass even hundreds of kilometers from the source. An extreme case was the 1783 Laki fissure eruption in southern Iceland, which led to the death of more than half of Iceland's livestock. The resulting famine caused the death of one-fifth of the Icelandic population (Thordarson & Self, 2003).

Injection of volcanic aerosols into the stratosphere can affect global temperatures, triggering a "*volcanic winter*", e.g. a decrease in ground temperature of 0.2 to 0.6°C for two to five years after the event, depending directly on the SO_2 content of the plume. Injection of halogens (Cl, F, Br, I) can result in stratospheric ozone losses, triggering significant climate forcing.

12.1.2.5 Lahars

Mixing of water and loose volcanic material produces volcanic debris flows or mudflows, called "lahars", the Indonesian word for volcanic mudflows. The formation of lahars requires water, abundant loose volcanic material, and slopes, conditions that exist, in particular, in rainy tropical climates, and in the presence of glaciers and lakes. Lahars usually travel along drainage systems on volcano slopes, and can reach distances of tens to hundreds of km from their source areas.

Lahars can happen both during an eruption (i.e. primary or syn-eruptive) and after the end of eruption (i.e. secondary or post-eruptive). Syn-eruptive lahars can be generated by the explosion of a volcanic crater lake along with tephra, by lava or pyroclastic flows melting snow and/or ice and by heavy rains mobilizing freshly fallen tephra. Post-eruptive lahars can occur from a few hours to years or even millennia after the eruption. They can be generated by remobilization of tephra deposits by heavy rain, failure of hydrothermally altered volcano slopes during heavy rain, failure of volcanic material damming lakes, and similar aspects.

People or animals caught in the path of a lahar have a high risk of death from severe crush injuries, drowning or asphyxiation. In terms of human fatalities, lahars rank as the second most deadly volcanic phenomenon after PCDs and, in the 20th century, lahars have caused the most injuries of any volcanic phenomenon. Lahars have characteristics similar to riverine floods and debris flows. Their impact is potentially devastating due to the presence of large boulders being carried along in the flows. Building damage includes burial, foundation erosion, debris impact, transportation due to soil erosion and liquefaction, failure from excessive wall or roof loads, collapse, undermining and corrosion due to the acidic nature of the flow. Depending on their densities and flow velocities, lahars may destroy or bury structures and machinery. Debris accumulation around bridges, particularly those with low clearance, can cause overtopping and flooding of nearby homes, businesses and agricultural land.

12.1.2.6 Sector collapses and landslides

Volcanoes are the result of the repeated, rapid (in this context usually <10^5 years) emplacement of lava dykes, flows, domes and tephra layers in a limited area, with heights that sometimes exceed 1000 m. The discontinuities and structural weakness, associated with active tectonic lineaments, seismic and hydrothermal activity, coupled with deformation induced by intruding magma bodies, make possible the collapse of a sector of the volcano, which is a gravity driven movement of a portion of a volcano, at scales ranging from minor rockslides and rock falls to giant megaslides and sector collapses, with volumes exceeding 5000 km^3.

Sector collapses are characterized by very different velocities of the mobilized masses, ranging from slow deformation to catastrophic fast-moving landslides. Sector collapses deposits extend usually to a distance of a few km, but in some cases, mudflows and debris flows produced by sector collapses can travel for significantly greater distances.

Sector collapses can pose an infrequent, but significant, risk to populations living close by and even at great distances, because of their sudden occurrence, extent and association with tsunamis. The zone of impact of large sector collapses and landslides, suffer total destruction including humans and animals. Small rock falls may cause partial damage to infrastructure, for example, by blocking roads or collapsing a retaining wall, and humans or livestock may survive with injuries.

12.1.2.7 Volcanic earthquakes

Volcanic earthquakes are induced by and related to the movement of magma. This movement results in pressure and stress changes in the rocks around the magma. Depending on the stress field and rock properties, the rock may break (triggering earthquakes) or move aseismically (through creep movements). The greatest number of earthquakes in volcanic areas is related to magma upwelling toward the surface, and often occurs in "swarms" of hundreds to thousands of earthquakes (Aki & Ferrazini, 2000; Chouet, 2003; Roman & Cashman, 2006).

There are two general end-member categories of earthquakes that can occur at a volcano: volcano-tectonic earthquakes and long-period earthquakes. ***Volcano-tectonic earthquakes*** are produced by stress changes in brittle rock due to the injection (or withdrawal) of magma. During the last stage of an eruption, earthquakes can occur also as rock adjusts after withdrawal of magma. The magnitude of this kind of earthquakes is generally modest (<4.5 Richter scale ML), but their shallow hypocenters can produce severe damage in limited areas around the epicenters. Vice versa, ***Long-period earthquakes***, i.e. ground movement during an earthquake with a period longer than one second, are the result of pressure changes within magma, magmatic gas or groundwater, during the unsteady transport of the magma toward the Earth's surface. The magnitudes of these events are generally low (2–3 ML).

Volcanic earthquakes can cause damage to buildings and infrastructure, trigger landslides and cause loss of life. Due to the shallow hypocenters location of a volcanic earthquake, typically less than 5 km, ground movement will be particularly strong, inducing considerable damage within a few kilometers from the epicenter.

The most damaging impact of shaking is failure of masonry walls or structural frame. The failure of telecommunication and electrical structures, such as power lines, can lead to disruptions in service and high rehabilitation costs. Fire and/or flooding following an earthquake, due to ruptured gas and power lines and/or water pipes, can also cause more damage to buildings than from the original structural failure. The main reason for human casualties and injuries in earthquakes is the collapse of buildings, contents and infrastructure.

12.1.2.8 Volcano-triggered tsunamis

A tsunami is a catastrophic water wave generated by the displacement of a large volume of a body of water, such as an ocean or a large lake. Earthquake induced tsunamis are related to the seafloor movement on the fault limbs, or to large submarine landslides induced by earthquakes.

Volcanic tsunamis can be triggered by:

- Explosive submarine volcanic eruptions.
- Caldera collapses, sector collapses or landslides, either submarine or subaerial, entering the sea (or a lake).
- Pyroclastic density currents or lahars entering the sea (or a lake).

These phenomena can occur individually or combined in a series of contemporaneous events during an eruption.

The power of significant submarine explosions can generate rapid movement in the water column located above the eruptive vent and, consequently, the generation of water waves radiating from that point toward the surface of the sea. Caldera collapses are triggered in large explosive events, when a large portion of the volcanic edifice collapses into the voids left by the rapid emptying of the magma chamber beneath the volcano. In volcanic islands, sea water moves to feel the caldera depression and large tsunamis can be triggered. The same is valid for volcanic sector collapses, landslides or pyroclastic density currents and lahars entering the sea or a lake, as they rapidly displace large water volumes.

Volcanogenic tsunamis are one of the farthest-reaching volcanic phenomena, potentially impacting areas thousands of kilometers from source. In the open ocean, volcanogenic tsunami waves can travel at speeds of 600–800 km/h although wave heights may be only a few centimeters. As the tsunami approaches shallow water and low topography land, the waves travel more slowly, but their height may increase to many meters, thus, becoming very destructive. Wave heights and run-up (maximum elevation reached by the tsunami on slopes) of tens of meters can be generated by large explosive events. The run-up distribution of volcano-triggered events is characterized by high run-up along the axis of the collapse and lower run-up values on the margins.

From a damage and casualties perspective, important tsunami characteristics are the wave run-up height and velocity, inundation distance and the presence and size of debris within the tsunami wave that can cause localized damage. Drowning or trauma from collision with debris, infrastructure or trees, causes deaths and injuries. Damage to buildings and infrastructure may result from flooding, foundation erosion and failure, transport or failure by lateral dynamic pressures and missiles associated with the tsunami.

12.2 VOLCANIC HAZARD AND RISK ASSESSMENT

In the last decades, the volcanological scientific community has shown that, under some favorable circumstances, *short-term forecasting* of the onset of volcanic events is possible by monitoring some key *precursory phenomena*, i.e. some chemical and physical variable changes of the volcanic system (Scarpa & Tilling, 1996; Sparks, 2003; Tilling, 2008). However, such ***short-term volcanic hazard assessment*** is, in most of the cases, unable to give accurate information about the type, the intensity, the magnitude and the duration of the eruption to occur. It is, therefore, necessary to foresee these crucial eruptive parameters through the deep investigation of the past history of the volcano under consideration, in other words it is necessary to carry out a ***long-term volcanic hazard assessment***.

Once the hazard assessment has been completed, the next necessary step to quantify volcanic risk, is the collection and analysis of data on the ***values at risk (exposure)*** and their ***vulnerability***. Knowledge of the numbers and distribution of inhabitants, buildings, critical infrastructures and communication networks is the most essential information. Other relevant data are about land use, crop type and distribution and similar aspects. Evaluation of the physical, social, and economic vulnerability of the exposed territory is typically determined through comprehensive field surveys and engineering and economic analyses. These activities and studies consider vulnerability to the diverse specific volcanic actions thought to affect the territory, as well as their likely severity and expected impacts (e.g. Baxter *et al.* 2005, 2008; Spence *et al.* 2005; Zuccaro *et al.* 2008). The discussion of these aspects is, however, beyond the scope of this chapter.

12.2.1 Long-term volcanic hazard assessment

The assessment of volcanic hazards and associated risks requires the development and use of different methods, or approaches, as well as their integration in a coherent and holistic framework. The type and level of application of the different methods depends largely on the state of knowledge of the volcano under study, as well as on the temporal and spatial scales of the assessment carried out. In the next subsections the main methods used nowadays to assess some of the main volcanic hazards are summarized with specific reference to the assessment of long-term hazard of explosive eruptions.

12.2.1.1 Eruptive record and databases

The knowledge of the ***past activity*** of the volcano represents the fundamental starting point for carrying out any hazard assessment study. Such a past behavior has been typically taken as representative of the future activity of the volcanic system (e.g. Fisher & Schmincke, 1984; Rosi, 1996; Sparks & Aspinall, 2004). The knowledge of past activity should possibly consist of the identification of the main hazardous phenomena generated by the volcano in its history, as well as in the quantitative description of their order, timing and areas affected by the various hazardous actions. Information on the events' age is also crucial to attempt temporal forecast of future activity.

Unfortunately, several important factors prevent an accurate reconstruction of past activity including: (i) lack of preservation of eruption products due to atmospheric and environmental factors, (ii) incomplete access to the eruption products that often remain largely inaccessible (i.e. buried under subsequent products or dispersed over large areas), (iii) different interpretations of volcanic deposits that do not allow to univocally identify the type and properties of the phenomenon occurred.

All of these three limiting factors can generate remarkable, and often very large, ***uncertainties*** in the knowledge of past activity. In addition, several of these limiting factors depend on the depth of investigation carried out at the specific volcano. For instance, in the case of several well-known studies and high-risk volcanoes of the world, such as Vesuvio and Campi Flegrei (Italy) or Santorini (Greece), a quite detailed reconstruction of their eruptive records has been carried out during several decades of scientific investigations (e.g. Druitt *et al.* 1999; Cioni *et al.* 2008; Vougioukalakis *et al.* 2016). In contrast, for most part of the 1,500 volcanoes considered active worldwide, such a reconstruction is simply missing or poor due to their remote accessibility and lack of sufficient resources.

Such a knowledge of past activity is then the key ingredient for the design of ***volcanic event trees*** (ETs) able to represent, graphically and often quantitatively, the potential future evolution of the volcano (e.g. Newhall & Hoblitt, 2002; Marzocchi *et al.* 2004; Neri *et al.* 2008; Newhall & Pallister, 2014). Through the ET framework it is, in fact, possible to represent the potential evolution of a volcanic unrest, as well as the main eruptive scenarios expected along with the associated phenomena. An example of ET is shown in Section 12.3.1 with reference to Vesuvio volcano.

The quantitative mapping of volcanic hazards in space and time also requires the quantitative characterization of key ***magmatic and eruptive variables*** of past events. For instance, for explosive eruptions, the most important ones are: total erupted mass, mass flow-rate (peak value or its variation as a function of time), vent location, magma composition, physical properties of the pyroclastic mixture (e.g. grain-size distribution, particle shape, density, etc.). Other datasets, such as geophysical and geochemical data related to the location of magma storage or structural data on volcano-tectonic lineaments and fractures represent further critical information. These kinds of data, possibly with the associated accuracy, represent the key input of volcanological databases needed for hazard assessment purposes.

Volcanological databases, developed either for individual volcanoes or collecting eruption data for specific regions or on global scale, represent, therefore, one of the key outputs of the reconstruction of the eruptive record. At the same time, such databases represent the starting point for carrying out meaningful scenario simulations by using numerical models of the phenomena, as well as statistical analysis aimed to the production of probability hazard maps and forecasts (see Section 12.3). Several databases, containing at least some of these data, already exist for many years or are currently under development. Some examples are: the SMITHSONIAN database, collecting eruption data of volcanoes worldwide (Global Volcanism Program, 2013); LaMEVE, a database collecting data of volcanic eruptions with VEI ≥ 4 occurred worldwide (Crosweller *et al.* 2012); WOVOdat, a database collecting unrest data of volcanoes worldwide (Venezky & Newhall, 2007); the ESR (Eruption Source Parameters) database including key source parameters to initialize volcanic ash dispersal models (Mastin *et al.* 2009). Similar volcanological databases also exist for specific well-known volcanoes (see Newhall & Pallister, 2014, for some examples).

12.2.1.2 Modelling and simulation of volcanic processes

Volcanic processes, as many natural phenomena, involve the movement of fluids and their interaction with the surrounding environment. For volcanoes, the ascent of magma and its interactions with the surrounding rocks and the atmospheric and superficial environments determine the processes occurring during the eruption. Such processes are often extremely complex, since involve multiphase mixtures (whose properties are not always well-known) and extremely wide ranges of spatial and temporal scales and eruptive conditions. Moreover, volcanic processes are largely governed by the dynamics of the involved fluids, which are often strongly non-linear and, therefore, difficult to predict in an accurate way. Finally, such processes can be observed and characterized to a quite limited extent, due to the remote and hazardous nature of the phenomena involved.

For the above reasons, ***mathematical models*** of volcanic processes have represented a valuable tool to better understand the key controls on eruption dynamics. For instance, early works in the seventies provided first-order descriptions of magma ascent in conduit and volcanic plumes. Since then, mathematical models of volcanic phenomena have been widely developed for both explosive and effusive eruptions, as well as for their associated secondary processes. In particular, the development, mostly during the last two decades, of numerical models able to describe some of the key complexities of volcanic phenomena has represented a significant advancement and opportunity for hazard assessment studies (e.g. Sigurdsson *et al.* 2015).

As regards explosive volcanism, for instance, ***3D multiphase flow models*** of volcanic plumes and collapsing columns generating PDCs, have allowed the investigation of the transient and turbulent dynamics of these phenomena. Non-equilibrium processes associated with the multiparticle nature of the eruptive mixture, the role of large vortices and umbrella cloud in controlling the large-scale dynamics of volcanic plumes, as well as the complex influence of volcano topography on PDCs, are some of the key processes described by these models (e.g. Suzuki & Koyaguchi, 2005; Neri *et al.* 2007; Esposti Ongaro *et al.* 2012; Cerminara *et al.* 2016). Similar advances have been made in the simulation of lava flows (Harris, 2013), as well as of other hazardous processes, such as, for instance, debris avalanches (Patra *et al.* 2005) and lahars (Iverson & Denlinger, 2001).

Numerical models have allowed the representation of some of the complexities occurring during volcanic eruptions and the description of eruptive scenarios of specific interest for hazard assessment. In the following Section 12.3, a few examples of such applications are provided. A further use of such multiphase and multidimensional models is the calibration of ***simpler integral models or emulators*** used for carrying out sensitivity analyses and Monte Carlo simulations aimed at the production of probability hazard maps. For instance, a recent inter-comparison study of volcanic plume models has allowed the evaluation of the accuracy of classic plume theory models, based on the self-similarity assumption of the flow field, for the cases of weak and strong plumes (Costa *et al.* 2016). Such comparison has allowed the calibration of key unknown model parameters, such as the entrainment coefficient used by integral plume models in the description of the Eruption Source Parameters (Mastin *et al.* 2009). Similar approaches, including calibration of models through comparison of their outcomes with real events, have been followed in the development of simplified models of PDCs emplacement (Patra *et al.* 2005; Neri *et al.* 2015), lava flows (Favalli *et al.* 2005) and dispersal and deposition of volcanic ash over large distances (Beckett *et al.* 2015).

Despite the many quantitative and physics-based information provided by numerical models ***major uncertainties*** remain in their physical formulation, numerical resolution and input data. The influence of several of these uncertainties and unknown processes can be explored by sensitivity analysis of the models themselves, but many others require that further fundamental research should be carried out. For this main reason, as discussed in the next subsection, a probabilistic approach, able to account for the large variability of volcano dynamics and possibly for the main uncertainty sources affecting them, represents a necessary complement of the more classical scenario-based approach.

12.2.1.3 Probabilistic assessment and uncertainty quantification

Given the above described limitations in reconstructing the eruptive record of the volcanic system, the still limited accuracy of observations and monitoring data, as well as the, quite often, large uncertainty and errors associated with numerical modelling of phenomena, it is evident that an accurate prediction of volcanic hazards is nowadays an impossible objective. As several other natural hazards, volcanic activity is characterized by extremely complex processes and non-linear dynamics that make the evolution of the system highly sensitive to even minimum changes of the initial and boundary conditions. Moreover, large part of the volcanic system is not directly accessible, thus, requiring to hypothesize such mechanisms and controls.

The identification and quantification of such non-linear dynamics and uncertainty sources is, therefore, a key objective to quantify volcanic hazards. In general terms, the volcanic system is affected by both ***aleatoric and epistemic uncertainty*** (Woo, 1999; Aspinall, 2006). The former one, also called physical uncertainty, is associated with the natural variability of the system and governed by the physical laws controlling its dynamics. On the contrary, the latter is representative of the partial lack of knowledge of the system that prevents an accurate description of its present and past state. Despite such a profound difference between the two types of uncertainty, for volcanic systems, it is not always possible to separate them due to the large uncertainty level that prevents their accurate characterization. Nevertheless, the consideration of such uncertainty sources is needed to properly characterize the behavior of the volcano and its hazard.

Whereas the volcanic system's non-linearity can be analyzed through the development of appropriate physical and numerical models able to describe such complex behavior, the quantification of epistemic uncertainty inevitably requires some level of expert judgment. During the last decades, a number of techniques aimed to properly assess such judgment have been developed and applied to a multitude of applied sciences including volcanic hazard assessment. In particular, ***Structured Expert Judgment (SEJ)*** methods have been developed with the aim to reach a "rational consensus" in a participative and consensual manner, i.e. pooling together individual expert judgments and reconciling them in a rational way (Aspinall, 2006). This is accomplished by taking all available scientific evidence into account through a formal and objective probabilistic procedure. The approach is "democratic" in that all experts are free to express their own opinions. At the same time, different weights are ascribed to individual experts based on the experts' performance in answering known problems correctly, as verified empirically. It is worth to mention that these techniques aim to be reproducible, accountable, neutral, fair and testable in an empirical manner. In the following section, some of their applications to the assessment of volcanic hazard are briefly illustrated.

As a consequence of the above considerations, it has been recognized that the quantification and forecasting of volcanic hazards can be represented in

probabilistic terms and by accounting for some of the key ***sources of uncertainty***. Therefore, if sufficient information on past record and suited numerical models of the phenomena are available, probabilistic assessment of hazards can be carried out. Such assessment typically adopts simplified physical models of the processes in combination with appropriate ranges of variability of the initial source parameters (often also based on expert judgment). Monte Carlo sampling techniques, sometimes complemented by efficient statistical methods able to reduce the size of the computational problem, such as Latin Hypercube sampling or Polynomial Chaos Expansion techniques, are adopted to explore the effect of variation of the input variables and produce probabilistic analyses (e.g. de'Michieli Vitturi *et al.* 2015). When uncertainty information is available, double stochastic models able to describe the level of uncertainty affecting the probability estimates, are also applied (Bevilacqua, 2016). Similar probabilistic and statistical approaches can also be adopted by analyzing the temporal pattern of the eruptive record in order to produce estimates of the eruption base rate at a specific volcano or the average time, together with its confidence level, remaining to the occurrence of the next eruption (Sparks & Aspinall, 2004; Neri *et al.* 2008; Bevilacqua, 2016).

It is worth to mention, however, that such a probabilistic approach should not be seen in opposition of the most common deterministic ***scenario-based*** approach. Indeed, the latter approach remains suitable any time detailed information on the past activity of the volcano is lacking, as well as the level of knowledge of its eruptive phenomena is poor. The detailed simulation of specific scenarios can also be very informative and increase the level of our understanding of the process dynamics. In all these situations, the definition of one or more scenarios able to represent, for instance, the most likely or the maximum credible events, are very useful for a first-order hazard assessment, as well as for communication purposes (e.g. Jenkins *et al.* 2015).

12.2.2 Short-term volcanic hazard assessment

Volcanic eruptions are preceded by a variety of ***precursory phenomena,*** which accompany the rise of magma towards the surface and which are manifested, since the beginning of an ***unrest phase***. Some of these include:

- seismic activity;
- ground deformation;
- changes in the volume, temperature and chemical composition of fumarolic emanations, hot springs and soil gasses;
- changes in other geophysical parameters, such as gravity, geo-electric and geo-magnetic fields.

The continuous recording and analysis of these precursory data, registered by a ***multi-parameter monitoring network***, once compared with data registered during dormant periods of the volcano (*background level*), gives a ***volcano observatory*** the possibility to recognize and follow the onset of an unrest phase. Such data, once integrated with the knowledge of the expected eruptive behavior of the volcano, also allow civil protection authorities and volcano observatories to design appropriate alert levels and early-warning systems aimed to mitigate the potential impact of the eruption (Dean & Dehn, 2016; Dvorak & Dzurisin, 1997).

However, it must be emphasized here that, even with a very good knowledge of the past behavior of the volcano (i.e. a reliable long-term hazard assessment) and a well-operated multiparametric monitoring system, at this time it is not possible to accurately predict the evolution of the unrest phase, i.e. if the volcano will finally erupt or not. Volcanoes that come to unrest and then come back to the normal background state without an eruption to occur, have been observed in several cases in the past. Therefore, in any unrest phase, a scenario assuming that the volcano will come back to its background state after some time should be always considered. A further consequence of such major uncertainty in predicting with accuracy the onset of an eruption, is the possibility to release a ***false alarm***, i.e. an alarm that is not followed by an eruption.

In the frame of actions regarding the short-term volcanic hazard and risk assessment, the definition of ***alert levels***, associated to different evolving states of the volcano, is a crucial step. For instance, the US Geological Survey has defined a USGS Alert-Notification Systems for Volcanic Activity based on four levels (Gardner & Guffanti, 2006) (Tab. 12.1). Similar alert-notification systems have been also proposed with the aim of standardization (e.g. Fearnley *et al.* 2012; Winson *et al.* 2014). However, as each volcano has a different history and behavior, and at long-dormant volcanoes, there are no, or very few, instrumental monitoring data on past unrests statistically useful to permit a quantitative evaluation of the unrest, the formulation of alert levels, and mostly the passage from one level to the next one, is a very challenging task. The construction of a conceptual working model of the volcano, together with the consideration, possibly by quantitative analyses including uncertainty, of all available data on potential precursors (and their changes in time) and the variety of expected eruptive phenomena appear the most promising way forward.

When a dormant volcano becomes an unrest state, a series of complementary monitoring systems have to be identified and installed in order to set-up ***early-warning systems***. For instance, instrumentation that permit the monitoring of volcanic plume gas and aerosol content and flux, such as Differential Optical Absorption Spectroscopy (DOAS), Fourier Transform Infrared (FTIR) spectrometer, and Light Detection and Ranging (LIDAR) laser spectroscopy, coupled with meteorological real-time information and dispersal forecasts (e.g. wind speed and direction, humidity, temperature, etc.), represent very valuable tools to mitigate volcanic risk associated to volcanic plumes (ash and gas) (Kantzas *et al.* 2010). Similarly, infrasound acoustic monitoring networks are able to provide valuable information on the time of onset of explosive events and to monitor and characterize the occurrence of PCDs and lahars flow units (Johnson & Ripepe, 2011).

Table 12.1 Alert levels adopted by the United States Geological Survey for volcanic activity. The table illustrates the USGS response, volcanic activity level and recurrence interval associated to each alert level (credit USGS).

CONDITION	USGS RESPONSE	ACTIVITY LEVEL	RECURRENCE INTERVALS
GREEN - No immediate risk	Normal operation plus information calls to local and other authorities for weak through strong unrest as appropriate	Background or quiescence	Most of the time
		Weak unrest	Days to weeks
		Minor unrest	Weeks to months
		Moderate-to-Strong Unrest	Months to years
YELLOW (WATCH)	Full call-down and EVENT RESPONSE	Intense Unrest	Years to decades
ORANGE (WARNING)	Full call-down and EVENT RESPONSE (if not already in place under YELLOW)	Accelerating intense unrest: Eruption likely within hours to days	Decades to centuries
RED(ERUPTION IN PROGRESS)	Full call-down and EVENT RESPONSE (if not already in place under YELLOW or ORANGE) Daily or more frequent updates on eruption levels	LEVEL 1: Minor eruption	Centuries
		LEVEL 2: Moderate explosive eruption	Centuries
		LEVEL 3: Strong explosive eruption	Centuries
		LEVEL 4: Massive explosive eruption	Centuries to millennia

Image credit: U.S. Geological Survey

12.3 EXAMPLES OF QUANTITATIVE VOLCANIC HAZARD ASSESSMENT

In this section, the application of some of the methods and practices described in previous sections are illustrated by few examples of hazard assessment at three active volcanoes, such as ***Vesuvio, Campi Flegrei and Santorini***, respectively. These volcanoes are representative of high-risk volcanoes located in a highly urbanized area (Vesuvio and Campi Flegrei located in the Campanian plain nearby the city of Naples) and in an island (Santorini), respectively. Due to their different nature (stratovolcano vs caldera) and settings (continental vs island volcano) these volcanoes pose different challenges in terms of hazard and risk assessment. The quite accurate knowledge of their eruptive history and behavior allows the illustration of the application of some of the methods described in Section 12.2. Of course, many other examples of hazard assessment could be presented at these and other volcanoes worldwide. Therefore, the selected examples should be seen just as illustrative of some of the approaches and techniques adopted nowadays in this field.

12.3.1 Development of event trees

As described in subsection 12.2.1.1, the current knowledge of a volcanic system can be effectively represented and quantified through the development of its ET. In this subsection, the ET developed for Vesuvio (Neri *et al.* 2008) is briefly presented, as representative of a specific ET describing the expected evolution of this volcano if an eruption could occur in the next decades. This approach differs from that of Newhall and Hoblitt (2002), since it reflects the specifics of Vesuvio rather than referring to a "standard" generic ET. Moreover, such ET aims to provide an overall framework for comprehensive hazard evaluation that should be used as background information for the development of short-term assessments once monitoring data are also integrated. The definition of the ***Vesuvio ET*** has required extensive discussion among different types of specialists who, based on all available knowledge on this volcano, have agreed on an overall structure of the ET as represented in Table 12.1.

The ***ET structure*** can be better described as formed by a number of "stages". Starting from a Precursory stage, indicating the starting of an unrest state, the ET evolves into the Initiation stage, when an antecedent sector collapse can potentially occur, and then into a Progression stage, at which an eruption either occurs or fails to occur. If the eruption occurs, the following branches of the ET describe the main potential characteristics of such event. For instance, the Dominant Eruptive stage indicates the prevailing character of the eruption (i.e. explosive or effusive), the Eruptive Category indicates the type of activity expected (for instance, for the explosive branch, the possibility to have a Plinian, sub-Plinian, Violent Strombolian and so-on event) as occurred in the Vesuvio record, and finally the Generic Hazard stage, i.e. the description of the hazardous phenomena expected for each eruptive category, is defined. Similar stages have been proposed for the effusive activity branch, as well as for the sector collapse branch.

Once the general structure of the ET is defined, the ET is "quantified", i.e. a quantitative basis for evaluating the ***relative likelihoods*** of alternative branching possibilities is adopted. In this case, SEJ techniques are applied to assign conditional probabilities, with the associated confidence bounds, to each branch of the ET (see subsection 12.2.1.1). Such probabilities are estimated by the Vesuvio specialists through careful consideration of the available field and volcanological data, historical chronicles, numerical modeling results, as well as any other strand of evidence available to them. The results of such elicitation of conditional probabilities of the ET branches are also reported in Figure 12.1. Three values for each probability are obtained for each branch corresponding to the modal (50 percentile) value in the center, flanked by two values defining the corresponding 90% credible interval spread around this value (i.e. the 5% and 95% quantiles). Similar probability estimates are assigned to the specific hazard phenomena characterizing each eruptive category (see Neri *et al.* 2008, for more details).

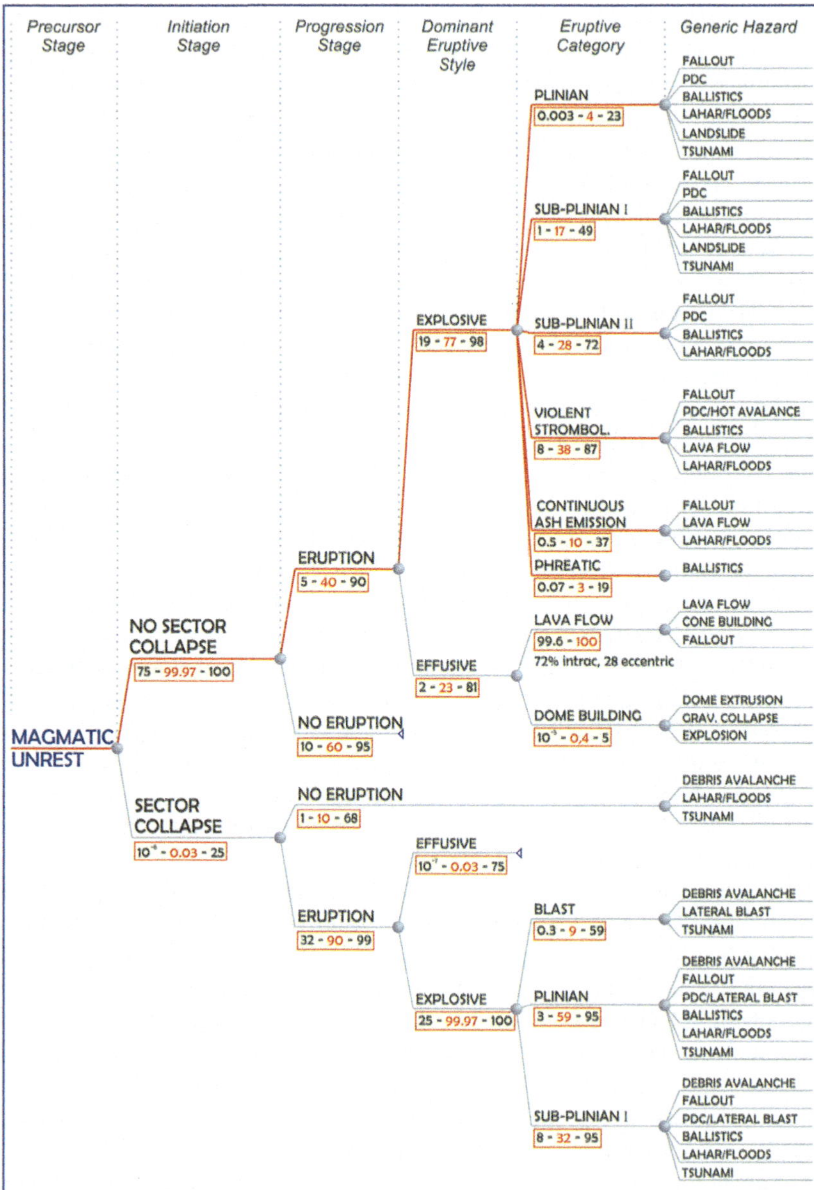

Figure 12.1 Event tree of Vesuvius together with itemized elicitation results for main branch probabilities giving the expert group's modal values and 90% credible interval values at each node (expressed as percentage probabilities in quantile form [5 percentile – 50 percentile – 95 percentile]) (from Neri *et al.* 2008).

The ET approach, in combination with expert judgment, can also be adopted to provide other key information to decision-makers. For instance, it is possible to provide guidance on the ***temporal evolution of individual scenarios*** as defined in the ET, as well as on the temporal properties of the different hazards. In other words, it is possible to define simplified scenario time-lines for eruptive activity and its phenomena. Figure 12.2, for instance, presents a schematic representation of the unrest, ash fall, PDC, and volcanic earthquake activity that has been proposed for characterizing the expected scenarios and their impacts. The six plots refer to the Plinian, Sub-Plinian I and II, Violent Strombolian, Phreatic and Continuous Ash Emission scenarios as described in Cioni *et al.* (2008). As it is evident from the plots, the six scenarios have remarkably different temporal scales and pose significantly different issues in terms of risk mitigation. Despite the fact that the exact temporal evolution of eruptive phenomena is certainly affected by a very large uncertainty, first-order information of the type shown in Figure 12.2, based on the available evidence of the past behavior of the volcano or on analogues activity, may be proved to be very useful for civil protection authorities.

12.3.2 Numerical simulation of eruptive scenarios

Numerical modelling of specific eruptive scenarios (e.g. most likely or maximum expected events) allows us to gain a deeper understanding of the possible dynamics and evolution of the eruption. This is, for instance, the case of ***column collapse and PDCs generation,*** which certainty represents some of the most complex and non-intuitive phenomena occurring during explosive eruptions. At Vesuvio, such phenomena have been investigated many years ago through the development and application of numerical multidimensional and multiphase flow models (e.g. Neri *et al.* 2007; Esposti Ongaro *et al.* 2008). It is worth to mention that such models have been extensively applied, for validation purposes, to well-known eruptions, such as the Vulcanian explosions of the Soufriere Hills volcano in Montserrat (UK, West Indies) (Clarke *et al.* 2002) and the volcanic blasts of Mount St. Helens (Washington, US) in 1980 (Esposti Ongaro *et al.* 2012).

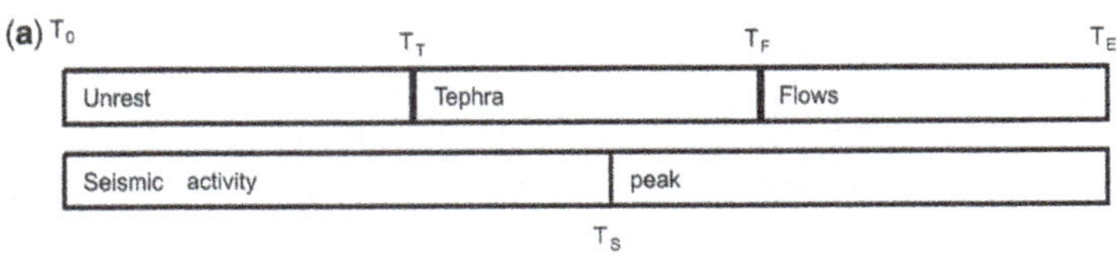

Figure 12.2 (**a**) Simplified time-line sequence for eruptive activity: T_0 is when unrest is first recognized as volcanic in origin; T_T is the time when tephra first falls on houses; T_F is the time when pyroclastic flows first affect built-up areas; T_E is the time when hazardous volcanic activity ends; T_S is the time when earthquake activity reaches its peaks.

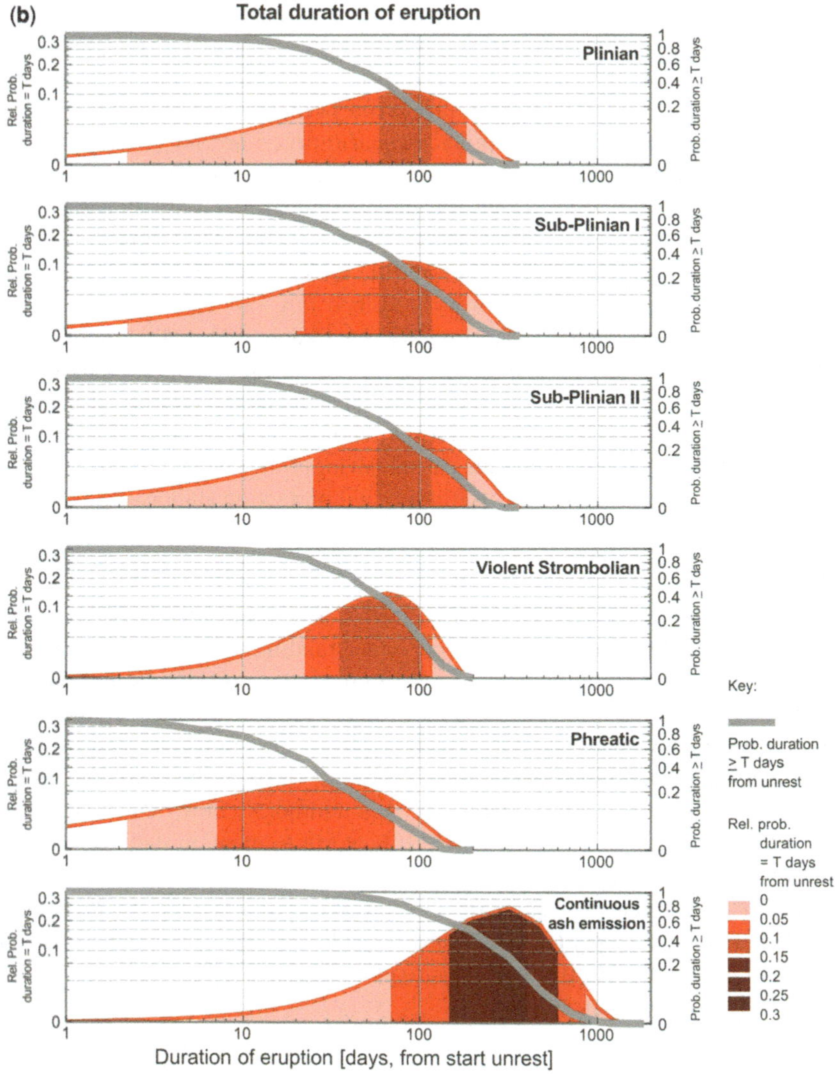

Figure 12.2 (**b**) Expert group elicitation outcomes for total duration of a future volcanic emergency at Vesuvio for Plinian, Sub-Plinian I and II, Violent Strombolian, Phreatic and Continuous Ash Emission events. In each panel the chart shows the cumulative probability function that the eruption duration will exceed T days (grey line — referenced to right-hand axis) for that eruptive category, while the red shaded curve below shows the relative probability of any specific duration (referenced to left-hand axis). The probability functions are Beta distributions based on pooled results using the elicited 5, 50 and 95 percentiles credible interval quantiles (from Neri *et al.* 2008).

Figure 12.3 illustrates the collapse of the volcanic column and the associated generation of PDCs for a sub-Plinian event similar to that occurred at Vesuvio in AD 1631 (Neri *et al.* 2007; Esposti Ongaro *et al.* 2008). This event has been assumed by the Italian Civil Protection authorities as the reference event for the design of the National Emergency Plan of Vesuvio. The dynamics is here represented through a temporal series of six snapshots of the distribution of particle volume fraction in the atmosphere and along the flanks of the volcano. The simulation has allowed the appreciation of the complexities of column collapse (characterized by a strongly unsteady dynamics of the fountain), as well as the effects of 3D topography on the flow propagation.

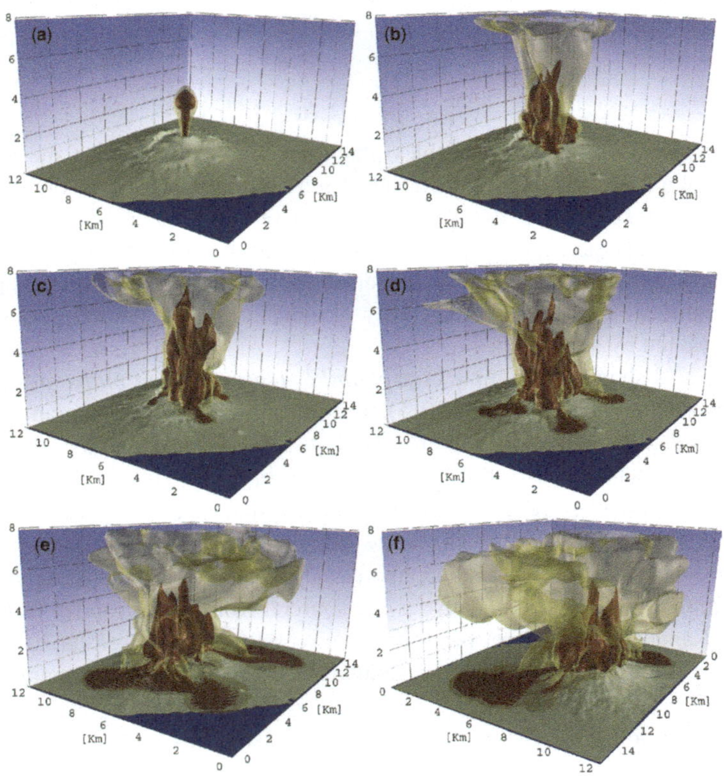

Figure 12.3 Temporal evolution of the large-scale dynamics of partial column collapse scenario at Vesuvio. The intensity of the eruption is representative of a sub-Plinian event. The two color isosurfaces represent the total volume particle concentration in the plume and PDCs and correspond to 10^{-4} (dark brown) and 10^{-6} (light brown). Plots refer to (**a**) 30, (**b**) 300, (**c**) 600, (**d**) 900, and (**e**) 1,300 s from the injection of the mixture into the atmosphere taken from a SW view. The (**f**) plot refers again to 1,300 s from the injection of the mixture into the atmosphere but is taken from N. The topographic relief of Mount Somma, on the N slope of the volcano, is clearly visible, as is its major influence on the propagation of the flows (from Neri *et al.* 2007).

Such kind of simulations also provide first-order estimates of some of the hazardous physical actions associated with PDCs (e.g. dynamic pressure, temperature, ash-in-air concentration, etc.), and of the extent of areas potentially invaded by the flows. These scenario simulations, together with the available geological evidence, have also been used as key input information for the definition of probability PDC invasion maps through expert elicitation sessions (see Neri *et al.* 2008, for more details). Similar numerical simulations have been produced for the definition of hazardous scenarios from other volcanic phenomena including the dispersal of volcanic ash (e.g. Barsotti *et al.* 2008; Beckett *et al.* 2015). In this latter case, numerical models can be coupled with weather forecasts in order to produce ash dispersal forecasts to inform air traffic operations.

12.3.3 Probabilistic mapping of hazards

Due to the large number of variables playing a main influence on the possible evolution of a future eruption, the consideration of the variability of some of them is often a useful first step for the achievement of a robust probabilistic hazard assessment. A classic example is the consideration of the variability of the wind field in the hazard assessment of ash dispersal and fallout from explosive eruptions. Indeed, the local and seasonal variability of winds largely control the areas most likely impacted by ash.

Figure 12.4 shows an example of ***probability tephra fallout maps*** produced at Santorini volcano (Jenkins *et al.* 2015). The maps refer to a sub-Plinian scenario with vent location on the Nea Kameni island, just in the center of the caldera. Such scenario is presently considered the largest possibly occurring in the short-term at Santorini, based on the past record of the volcano and its present conditions (Vougioukalakis *et al.* 2016). The maps are produced by using the advection-diffusion model TEPHRA2 (Bonadonna *et al.* 2005) and through a Monte Carlo procedure in which the wind statistics are sourced from the European Centre for Medium Range Weather Forecasts (www.ecmwf.int) and are based on 20 years of daily averaged regional ERA-Interim re-analysis data (1990 to 2011) sourced from a 0.7° grid.

The two maps show the probability to have ash loads on the ground of 10 and 100 kg/m^2 once the wind variability is considered, but keeping constant all other eruptive conditions at the source (in this case described by plume height, total mass erupted and distribution of mass along the column). From the maps clearly emerges that significant ash fallout hazard exists over the island, at least for the buildings with the poorer roofs. The two maps also show that the distribution of hazard is a function of the vulnerability thresholds considered. Dominant northwesterly and westerly winds above 1 to 2 km altitude are most likely to control the dispersal of the most fine ashes and, therefore, the hazard to the lower thickness thresholds (Figure 12.4 left). Vice versa, with increasing ash load threshold (Figure 12.4 right), higher hazard is orientated towards the south, since lower altitude winds in this direction become important for particles that fall out more quickly.

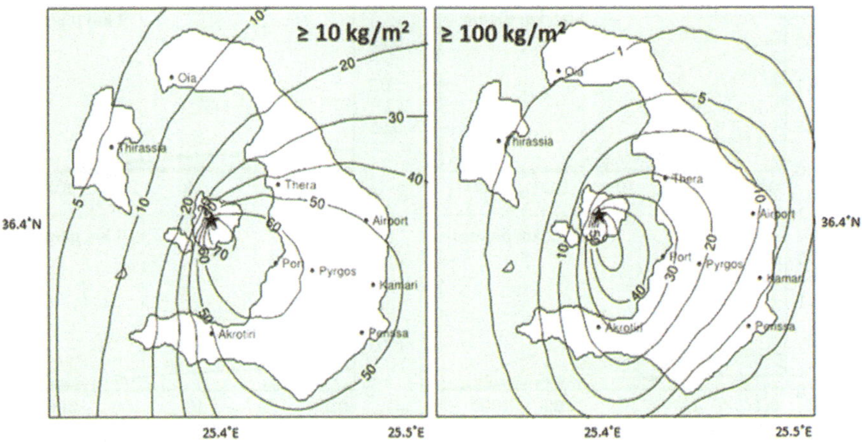

Figure 12.4 Probability tephra fallout maps at Santorini volcano for the largest considered sub-Plinian reference scenario. The two maps show the probability (as a %) of exceeding 10 (left) and 100 (right) kg/m^2 loading using the TEPHRA2 model. The black star marks the simulated vent location (from Jenkins *et al.* 2015).

Similar probability maps have also been proposed for assessing the ***areas potentially affected by airborne ash*** with concentrations potentially dangerous for air traffic or human beings. Figure 12.5 illustrates an example of hazard assessment from airborne ash (here represented by the concentration of particles of 15 micron, VP15) generated by different eruptive scenarios, in this case weak plumes with different heights, again for the Santorini island. Specifically, the plots illustrate the exceedance probability curves for fine ash concentration at ground level for different locations on the island. The ash concentration in the air are obtained by the numerical model VOL-CALPUFF (Barsotti *et al.* 2008), able to describe the transient and 3D dispersal and deposition of volcanic ash through an hybrid Eulerian-Lagrangian approach. It is worth to mention that, in contrast to the estimated ground deposit hazard, fixed plume height exceedance probability curves for fine ash concentrations at ground level show hazard, at some of the nine study sites, to be higher for smaller events, i.e. lower plume heights, than for larger events (see Figure 12.5). Indeed, releasing larger amounts of ash higher into the atmosphere supports the transport of the finer portion of ash farther from the vent, and vice versa.

The last example of probability hazard maps refers to the Campi Flegrei caldera (Figure 12.6). In this case, the hazard maps have been obtained by considering the physical variability of several variables of the volcanic system, as well as the presence of a few epistemic uncertainty sources. The maps refer to the ***long-term***

probability of invasion of dilute PDCs generated by explosive events in such caldera setting (Bevilacqua *et al.* 2015; Neri *et al.* 2015; Bevilacqua, 2016).

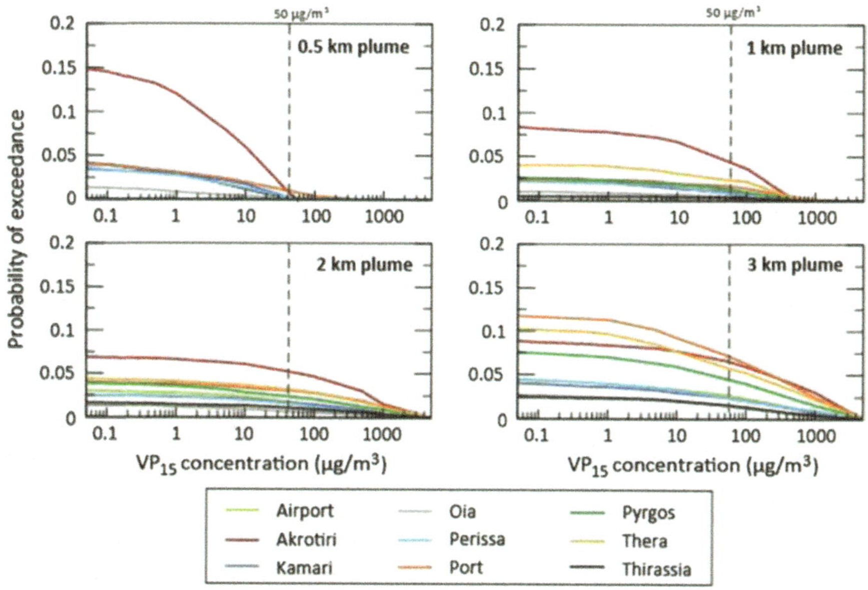

Figure 12.5 Exceedance probability curves for VP15 ash concentrations at nine sites on the island of Santorini as computed by the VOL-CALPUFF model for individual plume heights of weak plumes (corresponding to 0.5, 1, 2 and 3 km height of the mostly likely eruption scenario). The daily threshold for good air quality (50 µg/ m³) is shown by the vertical line. Note the maximum value of 0.2 on the y axis and that ash loads are shown on a base 10 logarithm scale (from Jenkins *et al.* 2015).

The maps consider the physical variability of vent location, which, in a caldera setting, is largely variable, and of event scale, which also typically spans over a few orders of magnitude. In parallel, the study has tried to quantify the epistemic uncertainty associated with the inaccurate reconstruction of the eruptive record, the uncertain localization of past vents, the uncertain role of faults and fractures in determining the exact vent location, as well as the approximate reconstruction of PDC deposits. Such sources of uncertainty have been quantified considering the available field evidences and by adopting expert judgment techniques for their estimation. Moreover, the effect of topography on the propagation of the flows have been described by a simple box model representing the current as a dilute sedimenting turbulent flow.

The consideration of these different sources of uncertainty lead to the development of a double stochastic procedure (Bevilacqua, 2016). As a consequence, the probability of PDC invasion of a specific site has not been described by a simple value, but through a probability density function, able to represent the uncertainty sources considered.

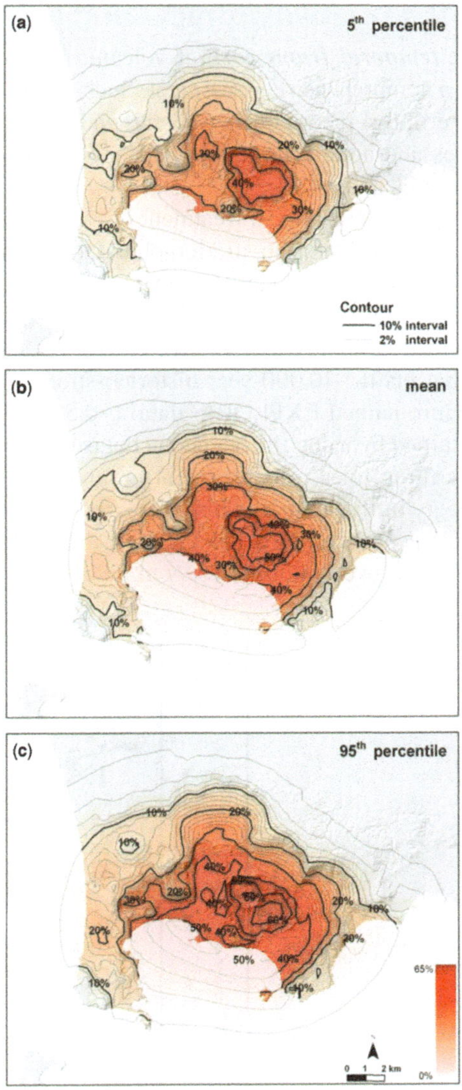

Figure 12.6 PDC invasion probability maps computed for dilute PDCs at Campi Flegrei as computed by Neri *et al.* (2015). The maps assume the vents located in the on-land part of the caldera and the possibility to have two simultaneous vents in an eruptive event. The maps also consider the variability of vent opening as described in Bevilacqua *et al.* (2015) and the variability of the event scale as reconstructed for the last 5 ka. Contours and colors indicate the percentage probability of PDC invasion conditional on the occurrence of an explosive eruption. The maps relate to **(a)** the 5th percentile, **(b)** the mean spatial probability, and **(c)** the 95th percentile (from Neri *et al.* 2015).

12.3.4 Temporal probability forecasting of hazards

Quantification of the ***temporal framework*** of volcanic hazards is also a crucial step for carrying out a comprehensive assessment. Such analysis can be developed by using the specific eruptive record of the volcano under study (typically derived from different sources and approaches) and/or the worldwide record of volcanic activity and analogous volcanoes.

For instance, at Vesuvio, a statistical treatment of the eruptive record has been carried out by comparing the historico-stratigraphic data (Cioni *et al.* 2008), with the post-1631 data (the time of its last sub-Plinian event, Scandone *et al.* 1993), and also with the relationship using a worldwide record of explosive eruptions (Simkin, 1993). Results are compared in Figure 12.7 (left). The results show that the slope of the power-law line that fits the 10,000-year historico-stratigraphic explosive data (herein and in the figure named EXPLORIS data) (−0.51) is flatter at lower VEI values than those obtained by using the Scandone dataset (−0.76) or Simkin's data (−0.78), although the three lines re-converge at around VEI 7, where the latter has a break in coverage. In this case, the historico-stratigraphic EXPLORIS data indicate a lower relative probability of small sized eruptions in relation to the numbers of moderate to large eruptions (Neri *et al.* 2008).

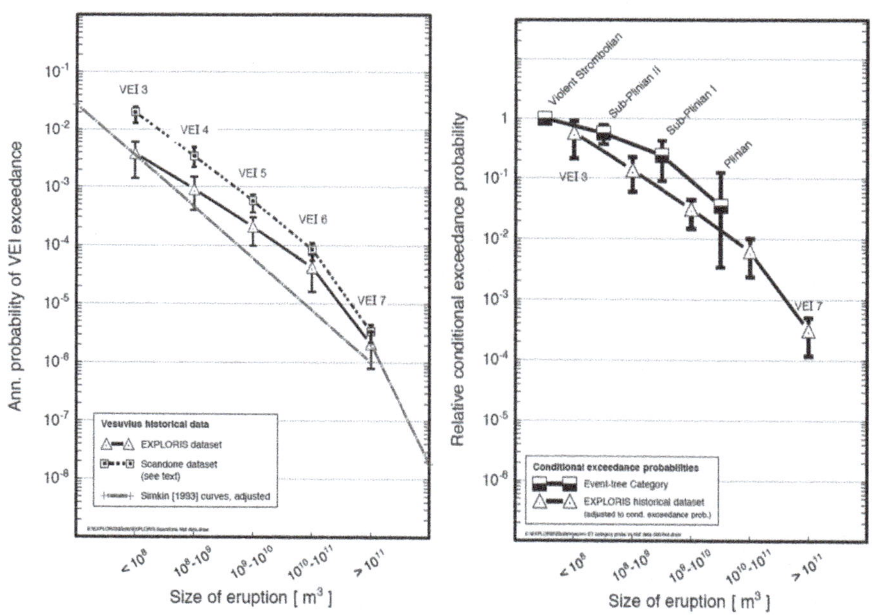

Figure 12.7 (left) Annual probability of eruption size (expressed by VEI) exceedance at Vesuvio and (right) relative conditional exceedance probability from historico-stratigraphic explosive data and expert judgment (modified from Neri *et al.* 2008).

Improvements to such estimates based on dataset analysis, can be obtained by considering the completeness issue for the record of low magnitude eruptions over long time periods and by adopting Bayesian updating given the observations available at the specific volcano under study (e.g. Marzocchi *et al.* 2004; Baxter *et al.* 2008; Neri *et al.* 2008).

The estimates of such analyses are also valuable in providing alternative perspectives for the experts to consider when exercising judgments in relation to occurrence likelihoods for the evolving Vesuvio ET (Figure 12.1). It is, therefore, feasible to attempt approximate comparisons between the elicitation outcomes as those presented in Figure 12.1, and those shown in Figure 12.7 (left). Indeed, once the latter annual probabilities have been adjusted to provide equivalent conditional occurrence probabilities (in this specific case by using the VEI 3 value as an anchor for the two curves), the two behaviors can be compared (see Figure 12.7 (right)). This plot also seems to signify that the experts' (conditional) likelihood of experiencing a larger magnitude event in the next eruption is slightly higher than the 'average' historical behavior would imply, although the relative uncertainty spreads are higher for future activity than for past behavior.

12.4 VOLCANIC RISK MANAGEMENT AND MITIGATION

Civil protection authorities (CPA), at national, regional and local levels, have the responsibility of volcanic risk management and mitigation. Based on the available hazard assessments and information about vulnerability and exposure of the populations, infrastructures and goods at risk, CPA need to implement activities and mitigation measures able to respond to the potential effects of a volcanic eruption (e.g. Altan *et al.* 2010; Baxter *et al.* 2008; Chester, 1993; Marzocchi *et al.* 2009; Mercer *et al.* 2012). These actions and measures need to be taken before, as well as during and after, the eruptive event.

12.4.1 Actions before the eruption

Actions taken before the eruption are by far the most important ones. Such actions can be focused on the emergency planning, as well as on the development of medium- and long-term prevention measures. All these actions need to be led and coordinated by CPA at the different levels and carried out in close collaboration with the scientific community, the institutions and the populations at risk.

During this preparedness phase, the recognition of the different roles of CPA and scientists is of primary importance, thus, avoiding conflicts and duplication and leading to the assignment of clear responsibilities to each part. In particular, it is very important that CPA are fully aware of the present main limitations in eruption forecasting. Similarly, people at risk should know that eruption forecasting always involves a major uncertainty. Keeping this uncertainty in mind, appropriate actions should be taken in order to protect people and, at the same time, minimize false alarms. Considering the uncertainty in forecasting, in case of an escalating unrest, the time window available to take decisions on counter-measures is often very narrow. It is, therefore, fundamental to be able to react fast, with a pre-established and flexible table of planned actions.

The ***emergency planning*** starts with a careful evaluation of the possible future eruptive scenarios expected at the volcano (e.g. Baxter *et al.* 2008; Vougioukalakis *et al.* 2016). In some cases, one or more scenarios are taken as references (together with their relative probability of occurrence), whereas, in other cases, probability hazard maps, integrating a full range of possible scenarios, are used as a guide. In all cases, emergency plans have to be prepared well in advance of the volcano unrest. The time interval between the onset of an eruption and its most violent peak may range from a few hours to several weeks or even months. On the other hand, the time required for reacting and setting up effective mitigation measures depends on the size of the eruption, the extension of the area at risk, the availability of transport and communication facilities, as well as the general level of technological development of the area. This can generally be measured in hours up to days. Two main types of action are usually planed: a *gradual phased response,* when a progressive development of a volcanic crisis is expected or an *immediate response,* when the volcanic crisis is strongly accelerating, thus requiring an immediate preventive evacuation of the population.

During this phase, ***early warning systems***, aimed at informing the authorities and the community about an impending danger should drive specific mitigation actions and assist the evacuation of the areas at risk. These systems can be built using modern or more traditional technologies, depending on culture, economical sustainability, presence of power supply and similar aspects. Of course, it is fundamental to design appropriate protocols for the release of warnings to the public, identify which organizations should be notified, and define the procedures that should be followed.

Exercises are a fundamental component of emergency planning with the aim of testing procedures and preparedness and to maintain the attention on the spot. Exercises should be scheduled frequently and depending on the state of the volcano, the social-economic context and the level of risk perception. Table-top exercises greatly help to validate, consolidate and test the command and control chain of emergency response, test the multiple functions of an organization's operational plan and clear attribution of roles and responsibilities of participants. Such exercises can be also used to identify weak points and solutions to remediate them. Moreover, they can help people to learn how to work together, and can also be very useful to test communication protocols and procedures at every level of administration. *Full- or partial-scale exercises* typically test the delivery times and implementation of capacity on the field. They are also fundamental to test the whole civil protection response, to validate the plan, and to consolidate procedures about road access, enforced evacuation, communication systems and similar aspects. *Reduced exercises* can be organized in public buildings, schools, universities, among others. They are relatively easy to organize and can be carried out periodically, depending on the different contexts and players.

Training and communication are also fundamental components of volcanic risk mitigation. Training implies continuous and sustainable programs, with a frequency of training depending on the cultural and socio-economic context at risk. In particular, CPA should take measures and actions to guarantee the development of people's awareness about risk, the education about the correct behavior to keep during the eruption and the actions to take in case of emergency. CPA should also establish relations with international organizations on civil protection in order to be able to request and use international support in case of a main crisis and deal with media regularly and not only during or after an event. Similarly, scientists should be trained to know how to react and communicate with CPA, institutions, social agencies and media. In particular, the *media* represent one of the most strategic and useful means to help authorities to cope with a crisis or disaster. A well-informed journalist represents a main chance to communicate correct information to the public. Training of *journalists,* i.e. share with them experience, knowledge and strategy in communication, can significantly help to avoid misunderstanding, panic and false alarms. If relationships with media are set in advance, i.e. before the crisis, communication during the emergency will likely be much more effective. Finally, training of pupils and students is fundamental: it teaches them the correct behavior in case of emergency, helps them to understand civil protection concepts and allows the transfer of this knowledge to their family.

Medium-term ***prevention measures*** can also be taken for buildings, infrastructures and networks, depending on their vulnerability, degree of exposure and expected phenomena. These can include reinforcement and protection of buildings exposed to ashes, structural reinforcement of strategic buildings against lateral dynamic and static pressures, recommendations for roof characteristics, use of refractory materials, burying of infrastructures and networks or planning of their relocation, among others. ***Reducing functional vulnerability*** of governance, transport, health care system and communications is another important target. All these systems have a crucial role in facing an emergency and their functionality should be guaranteed as much as possible.

Finally, long-term ***land-use and urban planning***, based on the available hazard and risk maps, represents the most appropriate measure to reduce, or possibly avoid, any future losses, this being the most effective mitigation strategy to favor the depopulation of the regions at risk.

12.4.2 Actions during and after the eruption

Once an eruption takes place, the two main unknowns are the duration of the eruption and its evolution in time, i.e. the type, magnitude and intensity of the future phenomena. In this condition, the eruptive scenario observed and reconstructed by scientists is continuously updated, based on the evolution of the ongoing activity. CPA are continuously informed in order to operate the necessary changes in the emergency plan. A main difference between volcanic eruptions and other natural disasters, is that volcanic phenomena can last for a long time, in some cases even for several years. Another main difficulty is that it is not possible to accurately predict the eruption duration. All these factors make the management of long volcanic crises and events particularly challenging.

Activities carried out by CPA after the end of the eruption are quite similar to those carried out for any other natural disaster, i.e. assistance of people until the normal situation is restored, re-establishment of the functionalities of the systems that have been damaged, estimation of damages and losses, and similar aspects. In the case that relocation of population is also carried out, social, economic and psychological issues need also to be faced.

12.5 SUMMARY

Volcanic eruptions are one of the most impressive and destructive natural events and represent nowadays a potential threat for hundreds of millions people worldwide. In this chapter, the basics of volcanic activity have been summarized, as well as the main features of primary and secondary volcanic hazards (e.g. lava flows, tephra fall, PDCs, gas emissions, lahars, sector collapses and landslides, volcanic earthquakes and tsunamis) and their associated impacts. In particular, the methods and approaches currently adopted to carry out a long-term volcanic hazard assessment (i.e. reconstruction of the eruptive record and construction of volcanological databases, modelling and simulation of volcanic processes, probabilistic assessment and uncertainty quantification) have been presented, as well as some of the tools and challenges of the short-term hazard assessment (i.e. multiparametric monitoring networks, early-warning systems, alert levels). A few applications of such methods have been illustrated with specific reference to the high-risk Vesuvio, Campi Flegrei and Santorini volcanoes. Finally, the challenges and key actions for effective volcanic risk management and mitigation have been briefly reviewed.

REFERENCES

Aki K. and Ferrazini V. (2000). Seismic monitoring and modelling of an active volcano for prediction. *Journal of Geophysical Research*, **105**(B7), 16617–16640.

Altan O., Backhaus R., Boccardo P. and Zlatanova S. (eds). (2010). Geo-information for Disaster and Risk Management. Examples and Best Practices. Joint Board of Geospatial Information Societies (JBGIS), UNOOSA, Copenhagen, Denmark. UN web site: http://www.un-spider.org/about/portfolio/publications/jbgis-unoosa-booklet.

Aspinall W. P. (2006). Structured elicitation of expert judgment for probabilistic hazard and risk assessment in volcanic eruptions. In: H. M. Mader, S. G. Coles, C. B. Connor and L. J. Connor (eds), Statistics in Volcanology, Geological Society, London. Special publications of IAVCEI, pp. 15–30.

Barsotti S., Neri A. and Scire J. S. (2008). The VOL-CALPUFF model for atmospheric ash dispersal: I. Approach and physical formulation. *Journal of Geophysical Research*, **113**(B03208), 85.

Baxter P., Boyd R., Cole P., Neri A., Spence R. and Zuccaro G. (2005). The impacts of pyroclastic surges on buildings at the eruption of the Soufriere Hills Volcano, Montserrat. *Bulletin of Volcanology*, **67**, 292–313.

Baxter P. J., Aspinall W., Neri A., Zuccaro G., Spence R. S. J., Cioni R. and Woo G. (2008). Emergency planning and mitigation at Vesuvius: a new evidence-based approach. *Journal of Volcanology and Geothermal Research*, **178**(3), 454–473. doi:10.1016/j.jvolgeores.2008.08.015 (1).

Beckett F. M., Witham C. S., Hort M. C., Stevenson J. A., Bonadonna C. and Millington S. C. (2015). Sensitivity of dispersion model forecasts of volcanic ash clouds to the physical characteristics of the particles. *Journal of Geophysical Research Atmospheres*, **120**(22), 11636–11652.

Bevilacqua A. (2016). Doubly Stochastic Models for Volcanic Hazard Assessment at Campi Flegrei. Publications of the Scuola Normale Superiore, Vol 21, Scuola Normale Superiore Pisa, Springer, ISBN 978-88-7642-556-1.

Bevilacqua A., Isaia R., Neri A., Vitale S., Aspinall W. P., Bisson M., Flandoli F., Baxter P. J., Bertagnini A., Esposti Ongaro T., Iannuzzi E., Pistolesi M. and Rosi M. (2015). Quantifying volcanic hazard at Campi Flegrei caldera (Italy) with uncertainty assessment: I. Vent opening maps. *Journal of Geophysical Research*, **120**, doi:10.1002/2014JB011775.

Bignami C., Bosi V., Costantini Ch. and Lavigne Fr. (eds). (2012). Handbook for Volcanic Risk Management. Prevention, Crisis Management, Resilience. MIAVITA team, Orleans, p. 198.

Blong R. J. (1984). Volcanic Hazards: A Sourcebook on the Effects of Eruptions. Academic Press, Australia, p. 424.

Bonadonna C., Connor C. B., Houghton B. F., Connor L., Byrne M., Laing A. and Hincks T. K. (2005). Probabilistic modeling of tephra dispersal: hazard assessment of a multiphase rhyolitic eruption at Tarawera, New Zealand. *Journal of Geophysical Research*, **110**(B03203), 1–21.

Cerminara M., Esposti Ongaro T. and Neri A. (2016). Large-eddy simulation of kinematic gas-particle decoupling and turbulent entrainment in volcanic plumes. *Journal of Volcanology and Geothermal Research*, in press. http://dx.doi.org/10.1016/j.jvolgeores.2016.06.018

Chester D. K. (1993). Volcanoes and Society. E. Arnold, London.

Chouet B. (2003). Volcano-seismology. *Pure and Applied Geophysics*, **160**, 739–788.

Cioni R., Bertagnini A., Santacroce R. and Andronico D. (2008). Explosive activity and eruption scenarios at Somma-Vesuvius (Italy): towards a new classification scheme. *Journal of Volcanology and Geothermal Research*, **178**, 331–346.

Clarke A., Voight B., Neri A. and Macedonio G. (2002). Transient dynamics of vulcanian explosions and column collapse. *Nature*, **415**, 897–901.

Costa A., Suzuki Y. J., Cerminara M., Devenish B. J., Esposti Ongaro T., Herzog M., Van Eaton A. R., Denby L. C., Bursik M., de' Michieli Vitturi M., Engwell S., Neri A., Barsotti S., Folch A., Macedonio G., Girault F., Carazzo G., Tait S., Kaminski E., Mastin L. G., Woodhouse M. J., Phillips J. C., Hogg A. J., Degruyter W. and Bonadonna C. (2016). Results of the eruption column model inter-comparison study. *Journal of Volcanology and Geothermal Research*, doi:10.1016/j.jvolgeores.2016.01.017.

Crosweller H. S., Arora B., Brown S. K., Cottrell E., Deligne N. I., Ortiz Guerrero N., Hobbs L., Kiyosugi K., Loughlin S. C., Lowndes J., Nayembil M., Siebert L., Sparks R. S.

J., Takarada S. and Venzke E. (2012). Global database on large magnitude explosive volcanic eruptions (LaMEVE). *Journal of Applied Volcanology*, **1**, 4.

Dean K. G. and Dean K. G. and Dehn J. (eds). (2016). Monitoring Volcanoes in the North Pacific: Observations From Space. Springer -Verlang, Berlin, Heidelberg, p. 363. ISBN: 978-3-54024125-6.

de'Michieli Vitturi M., Neri A. and Barsotti S. (2015). PLUME-MoM: a new integral model of volcanic plumes based on the method of moments. *Geoscientific Model Development*, **8**, 2447–2463. doi:10.5194/gmd-8-2447-2015.

Druitt T. H., Edwards L., Mellors R. M., Pyle D. M., Sparks R. S. J., Lanphere M., Davies M. and Barreiro B. (1999). Santorini Volcano, Geological Society Special Memoir, 19, London, p. 165. ISBN-13: 978-18622390485.

Dvorak J. J. and Dzurisin D. (1997). Volcano Geodesy: the search for magma reservoirs and the formation of eruptive vents. *Reviews of Geophysics*, **35**, 343–384.

Esposti Ongaro T., Neri A., Menconi G., de'Michieli Vitturi M., Marianelli P., Cavazzoni C., Erbacci G. and Baxter P. J. (2008). Transient 3D numerical simulations of column collapse and pyroclastic density current scenarios at Vesuvius. *Journal of Volcanology and Geothermal Research*, **178**(3), 378–396. doi: 10.1016/j.jvolgeores.2008.06.036.

Esposti Ongaro T., Clarke A. B., Voight B., Neri A. and Widiwijayanti C. (2012). Multiphase-flow dynamics of pyroclastic density currents during the May 18, 1980 lateral blast at Mount St. Helens (USA). *Journal of Geophysical Research*, **117**, doi:10.1029/2011JB009081.

Favalli M., Pareschi M. T., Neri A. and Isola I. (2005). Forecasting lava flow paths by a stochastic approach. *Geophysical Research Letters*, **32**, L03305. doi: 10.1029/2004GL021718.

Fearnley C. J., McGuire W. J., Davis G. and Twiggs J. (2012). Standardisation of the USGS Volcano Alert Level System (VALS): analysis and ramifications. *Bull Volcanol*, DOI 10.1007/s00445-012-0645-6.

Fisher R. V. and Schmincke H.-U. (1984). Pyroclastic Rocks. Springer, Berlin, Heidelberg, New York, p. 472.

Gardner C. A. and Guffanti M. C. (2006). U.S. Geological Survey's Alert-Notification System for Volcanic Alert. USGS Fact Sheet 2006-3139 (reprinted in 2014).

Global Volcanism Program, 2013. Volcanoes of the World, v. 4.5.1. Venzke, E (ed.). Smithsonian Institution. Downloaded 05 Oct 2016. http://dx.doi.org/10.5479/si.GVP.VOTW4-2013.

Harris A. J. L. (2013). Lava flows. In: Modeling Volcanic Processes: The Physics and Mathematics of Volcanism, S. A. Fagents, T. K. P. Gregg and R. M. C. Lopes (eds), Cambridge University Press, New York, pp. 85–106.

Heiken G. and Wohletz K. (1985). Volcanic Ash. University of California Press, Berkeley.

Iverson R. M. and Denlinger R. (2001). Flow of variably fluidized granular masses across three-dimensional terrain: 1. Coulomb mixture theory. *Journal of Geophysical Research*, **106**(B1), 537–552.

Jenkins S. F., Barsotti S., Hincks T. K., Neri A., Phillips J. C., Sparks R. S. J., Sheldrake T. and Vougioukalakis G. (2015). Rapid emergency assessment of ash and gas hazard for future eruptions at Santorini Volcano, Greece. *Journal of Applied Volcanology*, **4**, 16. doi:10.1186/s13617-015-0033-y.

Johnson J. B. and Ripepe M. (2011). Volcano infrasound: a review. *Journal of Volcanology and Geothermal Research*, **206**, 61–69.

Kantzas E. P., McGonigle A. J. S., Tamburello G., Aiuppa A. and Bryant R. G. (2010). Protocols for UV camera volcanic SO2 measurements. *Journal of Volcanology and Geothermal Research*, **194**, 55–60.

Marzocchi W., Sandri L., Gasparini P., Newhall C. and Boschi, E. (2004). Quantifying probabilities of volcanic events: the example of volcanic hazard at Mount Vesuvius. *Journal of Geophysical Research*, **109**, B11201. doi:10.1029/2004JB003155.

Marzocchi W., Mastellone M. L., Di Ruocco A., Novelli P., Romeo E. and Gasparini P. (2009). Principles of multi-risk assessment. Interaction amongst natural and man-induced risks. European Commission EUR23615, p. 72. http://ec.europa.eu/ research/ environment/pdf/multi-risk_assessment.pdf.

Mastin L. G., Guffanti M., Servranckx R., Webley P. W., Barsotti S., Dean K. G., Durant A. K., Ewert J. W., Neri A., Rose W. I., Schneider D. J., Siebert L., Stunder B. J., Swanson G., Tupper A., Volentik A. and Waythomas C. F. (2009). A multidisciplinary effort to assign realistic source parameters to models of volcanic ash-cloud transport and dispersion during eruptions. *Journal of Volcanology and Geothermal Research*, **186**(1–2), 10–21. doi: 10.1016/jvolgeores.2009.01.008.

Mercer J., Gaillard J.-C., Donovan K., Shannon R., Alexander B., Day S. and Becker, J. (2012). Cultural awareness in disaster risk reduction: lessons and opportunities. *Environmental Hazards*, **11**(2), 74–95.

Neri A., Esposti Ongaro T., Menconi G., de' Michieli Vitturi M., Cavazzoni C., Erbacci G. and Baxter P. J. (2007). 4D Simulation of esplosive eruption dynamics at Vesuvius. *Geophysical Research Letters*, **34**, doi: 10.1029/2006GL028597.

Neri A., Aspinall W. P., Cioni R., Bertagnini A., Baxter P. J., Zuccaro G., Andronico D., Barsotti S., Cole P. D., Esposti Ongaro T., Hincks T. K., Macedonio G., Papale P., Rosi M., Santacroce R. and Woo G. (2008). Developing an event tree for probabilistic hazard and risk assessment at Vesuvius. *Journal of Volcanology and Geothermal Research*, **178**(3), 397–415. doi: 10.1016/j.jvolgeores.2008.05.014.

Neri, A., Bevilacqua A., Esposti Ongaro T., Isaia R., Aspinall W. P., Bisson M., Flandoli F., Baxter P. J., Bertagnini A., Iannuzzi E., Orsucci S., Pistolesi M., Rosi M. and Vitale S. (2015). Quantifying volcanic hazard at Campi Flegrei caldera (Italy) with uncertainty assessment: II. Pyroclastic density current invasion maps. *Journal of Geophysical Research*, **120**. doi:10.1002/2014JB011776.

Newhall C. G. and Hoblitt R. P. (2002). Constructing event trees for volcanic crises. *Bulletin of Volcanology*, **64**, 3–20.

Newhall C. G. and Pallister J. (2014). Using multiple datasets to populate probabilistic volcanic Event Trees. In: Volcanic Hazards, Risks and Disasters, P. Papale, J. Eichelberger, S. Loughlin, S. Nakada and H. Yepes (eds), Elsevier Publisher, Amsterdam, The Netherlands, pp. 203–232.

Patra A. K., Bauer A. C., Nichita C. C., Pitman E. B., Sheridan M. F., Bursik M., Rupp B., Webber A., Stinton A. J., Namikawa L. M. and Renschler C. S. (2005). Parallel adaptive simulation of dry avalanches over natural terrain. *Journal of Volcanology and Geothermal Research*, **139**, 1–22.

Roman D. C. and Cashman K. V. (2006). The origin of volcano-tectonic earthquake swarms. *Geology*, **34**, 457–460.

Scandone R., Giacomelli L. and Gasparini P. (1993). Mt. Vesuvius: 2,000 years of volcanological observations. *Journal of Volcanology and Geothermal Research*, **58**, 5–26.

Scarpa R. and Tilling R. (eds). (1996). Monitoring and Mitigation of Volcano Hazards. Springer-Verlag, Berlin, Heidelberg, p. 840.

Sigurdsson H., Houghton B., McNutt S., Rymer H. and Stix J. (eds). (2015). Encyclopedia of Volcanoes, 2nd edition, Elsevier Publisher, Amsterdam, The Netherlands, p. 1421. ISBN: 978-0-12-385938-9.

Simkin T. (1993). Terrestrial volcanism in space and time. *Annual Review of Earth and Planetary Sciences*, **21**, 427–452.

Sparks R. S. J. (2003). Forecasting Volcanic Eruptions. Earth and Planetary Science Letters Frontiers in Earth Science Series 210, pp. 1–15.

Sparks R. S. J. and Aspinall W. A. (2004). Volcanic activity: frontiers and challenges in forecasting, prediction and risk assessment. In: R. S. J. Sparks and C. J. Hawkesworth, (eds) The State of the Planet: Frontiers and Challenges in Geophysics. Geophysical Monograph 150, IUGG Volume 19. American Geophysical Union, Washington DC, pp. 359-373.

Spence R. S., Kelman I., Baxter P. J., Zuccaro G. and Petrazzuoli S. (2005). Residential building and occupant vulnerability to tephra fall. *Natural Hazards and Earth System Science*, **5**, 477–494.

Suzuki Y. J., Koyaguchi T., Ogawa M. and Hachisu I. (2005). A numerical study of turbulent mixing in eruption clouds using a three-dimensional fluid dynamics model. *Journal of Geophysical Research*, **110**(B8), B08201. doi:10.1029/2004JB003460.

Thierry P., Stieltjes L., Kouokam E., Nguéya P. and Salley M. P. (2008). Multi hazard risk mapping and assessment on an active volcano: the GRINP project at Mount Cameroon. *Natural Hazards*, **45**, 429–456.

Thordarson T. and Self S. (2003). Atmospheric and environmental effects of the 1783–1784 Laki eruption; a review and reassessment. *Journal Geophysical Research*, **108**, 4011, doi: 10.1029/2001JD002042.

Tilling R. I. (2008). The critical role of volcano monitoring in risk reduction. *Advanced Geosciences*, **14**, 3–11.

Venezky D. Y. and Newhall C. G. (2007). WOVOdat Design Document: The Schema, Table Descriptions, and Create Table Statements for the Database of Worldwide Volcanic Unrest (WOVOdat Version 1.0). U.S. Geological Survey Open File Report 2007e1117.

Vougioukalakis G., Sparks R. S. J., Pyle D., Druitt T., Barberi, F., Papazachos, C. and Fytikas, M. (2016). Volcanic hazard assessment at Santorini volcano: a review and a synthesis in the light of the 2011–2012 Santorini unrest, Bulletin of the Geological Society of Greece, vol. L, 2016. In: Proceedings of the 14th Intern. Conference, May 2016, Thessaloniki.

Wilson T., Kaye G., Stewart C. and Cole J. (2007). Impacts of the 2006 Eruption of Merapi Volcano, Indonesia, on Agriculture and Infrastructure. GNS Science Report 2007/07, p 69.

Winson A. E., Costa F., Newhall G. N. and Woo G. (2014). An analysis of the issuance of volcanic alert levels during volcanic crisis. *Journal of Applied Volcanology*, **3**, 14.

Woo G. (1999). The Mathematics of Natural Catastrophes. Imperial College Press, London, p. 292.

Zuccaro G., Cacace F., Baxter P. J. and Spence R. (2008). Impact of explosive eruption scenarios at Vesuvius. *Journal of Volcanology and Geothermal Research*, **178**, 416–453.

Chapter 13

Technological hazards

John M. Logan, Nicos V. Spyropoulos and Nicolas R. Dalezios

13.1 TECHNOLOGICAL HAZARDS CONCEPTS

13.1.1 Nature of technological hazards

It is clear that environmental hazards include a wide range of hazard types from natural or geophysical events, through technological or man-made events to social or human behavior events. The extent to which hazards are voluntary or involuntary may be important. Moreover, the degree of human responsibility for a disaster increases significantly from the accidental geophysical hazards, such as earthquake or tsunami, to man-made technological hazards, such as transport accidents, industrial explosions, water and air pollution or radioactive fallout, to the man-induced social hazards, such as civil riot, food additives, smoking or mountaineering. Indeed, it is recognized that technological hazards constitute a new kind of threat, where the consequences are due to accidents rather than to different levels of exposure to toxic substances. It is true that most environmental hazards have both natural and human components. For example, flood events may be exacerbated by climate fluctuations, such as increased frequency and intensity of storms, as well as by human activities, such as land drainage and deforestation. Moreover, the loss of life caused by tropical cyclones depends to a certain extent on storm severity, although it can be reduced significantly by means of early warning systems.

On the other hand, the impacts of a man-made nuclear accident are affected by the prevailing weather conditions, which control the downwind path and the fallout rate from the radioactive plume. There is thus a growing recognition of hazards as hybrid events, which originate from a combination of environmental, technological and social processes (Jones, 1993). Initially, technological hazards were restricted to the developed countries, however, the risk currently extends to developing countries, since technology grows steadily and continuously. Moreover, it has been stated that technological hazards have been created by the spread of high-risk technologies, which, at the present time, incorporate complex systems and it is becoming very difficult to assess all the potential interactions and failures. Nevertheless, since technology is dealing with ever more toxic substances and operating to increasing levels of performance, accidents can be considered as inevitable in an analogous way that natural hazards are a 'normal' part of geophysical processes.

Technological hazards are mainly man-made accidents. If the triggering event arises from human initiative, there are links with social hazards, such as crime or terrorism. Nevertheless, the term technological hazards attracts a wide interpretation. Technological hazards may range from a single toxic chemical to an industrial nuclear power. Moreover, the degreed of human involvement may also vary, ranging from life-style to occupational risks. Specifically, occupational risks may include health issues involving long-term exposure to chemical pollutants or low-level hazardous waste, as well as the safety issues. There is also attention to 'na-tech' or 'hybrid' disasters, occurring when natural hazards, such floods or earthquakes, constitute a release mechanism for oil spills, chemicals or radiological materials.

13.1.2 Taxonomy of technological hazards

Technological hazards are characterized by a causal sequence of events and that an essential distinction could be made between energy and material releases. A general taxonomy classifies technological hazards into seven major classes based on the level of severity, namely multiple extreme hazards, extreme hazards and hazards (Hohenemser *et al.* 1983). Nevertheless, most rapid-onset technological disasters stem from failures in three areas of human activity, which focus on public safety, rather than public health. These areas constitute the subject of this chapter and are the following:

Large-scale structures. They include public buildings, bridges and dams, where risk is usually defined as the probability of failure during the life-time of the structure.

Transport. They include road, sea, air and rail, where risk is usually defined as the probability of injury or death per kilometer traveled.

Industry. They include manufacturing, power production, storage and transport of hazardous materials, where risk is usually defined as the probability of injury or death per person per hours exposed.

The above three areas can provide releases of energy and materials, which are hazardous to humans. In particular, the energy may be released in either mechanical impact form, such as dam burst, or chemical impact form, such as explosion or fire. Moreover, the most hazardous materials are high-level radioactive materials and explosives. In addition, many chemicals are considered hazards, since they are explosive, flammable, corrosive or toxic. It is recognized that hazardous materials are most likely to cause a disaster, if transferred through air or water pollution and affect a population. It is true that this transfer justifies the links between technological hazards and natural environment. Moreover, if 'na-tech' events are excluded, the 'trigger' for a technological disaster may originate from one of the following causes, namely defective design, inadequate management or sabotage or terrorism. Based on the above, technological hazards may be considered as 'accidental failures of design or management affecting large-scale structures, transport systems or industrial activities which present life-threatening risks on a community scale' (Smith, 2013).

Failure leading to technological hazards is a diachronic phenomenon. Gradual improvements of engineering design and increasing awareness of health and safety issues implemented through legislation have contributed to safer engineering structures. For example, at the present time, about one-third of all dam failures are caused by overtopping due to inadequate spillway design, about one-third are caused by seepage through the dam and one-third due to foundation problems (Lagatec, 1987). Moreover, in public transport, trains, ships, aircrafts and cars are safer than in the past. However, transport-related deaths steadily increase, due to the rise in the distance travelled (Yagar, 1984). Indeed, mobility has increased in developed countries due to increase in business travel and amounts for leisure time. Similarly, in the industry sector, there have been significant advances through safety laws in most developed countries. In particular, there have been developments in occupational safety due to increased industrial risks. In addition, the development of the nuclear industry has also created new risks leading to strict regulations. Furthermore, the demand for other energy forms has also led to industrial hazards. Specifically, oil and gas explosions have contributed to many technological hazards. Indeed, oil and gas have been gradually developed offshore in areas, such as Alaska and North Sea, due to technological innovations in drilling, although they are vulnerable to human safety. Moreover, the increased transportation of hazardous materials, including radioactive waste, constitute another factor, which contributes to the growth of technological hazards.

13.2 TECHNOLOGICAL HAZARDS IN LARGE-SCALE STRUCTURES

This section focuses on the potential hazards in buildings having stones on vertical surfaces, particularly exterior ones exposed to environmental conditions. Emphasis is on the history of structures particularly those covered with marble, granite and sandstone, as these are ones that have undergone documented hazards.

13.2.1 Technological risk identification in buildings

The potential technological hazards presented are focused on buildings whose primary exterior veneer is composed of stone panels. Such buildings may be tall in metropolitan areas (Figure 13.1) or shorter in less dense regions. The technological risk is increased when the buildings are close to the center of urban areas. Large buildings and structures such as major office buildings, museums, libraries, and entertainment centers located in heavily populated areas are often the focus of attention because of the potential loss of life if failure occurs. This arises as the price of land in such locations often leads to multi-story, high-rise buildings.

Figure 13.1. Photograph of Amoco building, Chicago, USA showing vertical windows and marble columns with panels on both sides. Inspection platform is shown part way up building.

The issues of potential hazards come not from the foundations but from the exposed areas of the structures. The buildings generally have steel frameworks whose exteriors veneers are often panels of natural stones mixed with windows, and metal framing. Recently, composite materials have replaced some stone panels. Although stone panels may be mixed with windows (Figure 13.1), it is the stone panels that have been found to form the major risk. The stone panels are generally marble, granite or sandstone. It is these stone panels that pose the greatest issue for public safety as they often deteriorate with the formation of fractures. The fractures may form as arrays of small ones within the panel or develop completely across a panel. Either, offer the potential for failure of a whole panel or pieces falling from high places on the building and endangering vehicles and pedestrians. Panels have fractured with pieces being carried in some cases by the wind for kilometers.

The basic question is the ability of the stone panels to withstand the wind loads generated by storms, hurricanes and tornados. Although there is often warning of approaching hurricanes and tornados, the high winds sometimes associated with seasonal storms are not often given as much public notice.

Stone veneers particularly those formed by marble panels have received the most attention. Marble panels on architectural icons such as the Amoco Tower in Chicago, USA, the Finlandia Concert Hall in Helsinki, Finland and the Grad Arche de la Defense in Paris, France all began bowing to the extent that they posed risks of fracturing and falling off the building. The potential for public injuries or deaths and the resulting wide publicity has lead to extensive research regarding the physical properties and the loss of strength of marble (Raleigh, 1934). Cycles of freeze-thawing which could result in fractures on the microscopic or even macroscopic scale, thus reducing the marble strength, and possibly leading to panel failure have also been proposed (Bortz *et al.* 1993). Instances of this as a major causative factor have not been widely corroborated presently.

Chemical interaction during weathering leading to mineral alteration, dissolution and possibly fracture corrosion with subsequent loss of strength has been postulated, especially in environments of acid rain (Winkler, 1967). It has generally been found that rainwater does not penetrate more than a few grain diameters into rock panel faces on vertically mounted pieces because of subsequent drying by wind and sun. However, localization of apparent alteration and accompanying effects has prompted investigation in some cases. A building with granite and glass panels in Toronto, Canada was investigated as some granite panels were observed to have some "veining" defined by some dark-colored minerals and very small fractures. Measurements were made of tensile strength, porosity, induced fracture orientation and microscopic observations of minerals within and on either side of the "vein". The results showed no evidence of a loss of strength or mineral alteration that would alter the future strength of the granite if left on the building. Such apparent insitu observations require investigation at a number of representative locations to determine their significance.

Thermal expansion due to temperature diurnal cycles has been extensively researched especially with respect to buildings where the vertical panels do not retain significant rain or snow fall (Logan, 2006; Koch & Siegesmund, 2004). Considerable study has been given to marble as it has been extensively used on buildings because of it's generally white color and aesthetic appeal and from it's use in past years in churches and sculpture in countries such as Italy and Greece.

The principal factor here is that the mineral calcite, which is the primary component of marble, is thermally anisotropic. It expands parallel to the c-crystallographic axis but contracts perpendicular to it when subjected to temperatures above the average ambient state, which makes it a unique mineral. This differential expansion during elevated temperatures followed by relaxation as the temperatures return to ambient, produces fatigue within the grain-to-grain contacts eventually weakening the marble. The extent of the thermal cycling is contingent upon exposure to the sun but is higher in the rock than in the air.

Such weakening if continued for extensive time may lead to macro-fracture development but generally another factor may be significant. Specifically this is the release of residual elastic strain stored in the rock as a result of incomplete annealing upon re-crystallization of the limestone to form marble. This results with weakening from the thermal effect (Logan, 2006; Siegesmund *et al.* 2000). This strain release produces increases in the dimensions of a marble panel. If the panel is mounted vertically on a building and anchored top and bottom, the weight of the panel may produce a bow from its original flat shape (Figure 13.2). To accommodate this change in shape microfractures will occur which further weaken the panel. The bow amplitude may become sufficient to induce bending stresses in the panel, exacerbating the condition potentially leading to macroscopic fracture and failure of the panel at about the middle of the panel height. Additionally, as the panels are anchored to retain them on the building, the bowing may load the anchors and produce failure there (Figure 13.3). To remove the possibility of moisture accumulating behind a panel and possibly producing corrosion of the metal framework and anchors, caulk is installed around the edges of the panel. Even if a horizontal fracture forms leading to complete loss of panel strength, the caulk may have sufficient strength to keep the pieces from falling for a limited time.

Figure 13.2. Photograph of marble panels with convex bows that developed from originally flat pieces. Panels are about 78 cm on a side and approximately 3 cm thick.

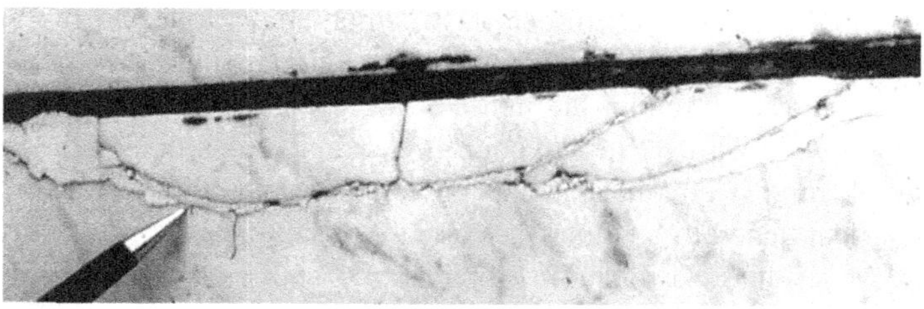

Figure 13.3. Photograph of fractures along upper anchor in marble panel prior to removal from building. The fractures were the result of stress concentrations at the anchors resulting from bowing of the panel.

Granite has not exhibited the extreme deterioration from environmental conditions that marble has although laboratory studies have looked at the bowing potential (Siegesmund *et al.* 2007). As it is multimineralic, the major components quartz, feldspar and micas all have different responses to the atmosphere, mitigating the effects seen in calcitic marble. It does often have, however, a pronounced strength anisotropy due to the crystallization process where the crystals often have a preferred orientation parallel to each other. Although this is not an evident layering it often results in a "rift" with the rock weaker parallel to this feature, and stronger perpendicular to it. Specifications for granite exterior panels stipulate that the rift be parallel to panel face to minimize failure due to wind loads. It is frequently presumed that the rift is consistent within a quarry so that granite may be removed with respect to the assumed rift orientation, but geological mapping of quarries has shown that the rift orientation may change considerably (Logan *et al.* 1993).

This variation of physical and resulting mechanical properties is common in rocks. Although buildings with marble exterior cladding showing extreme signs of weathering have received considerable attention, there are other buildings with little evidence of deterioration. Marble from the Apuane Alps region of Italy is widely used in buildings because of its white color and the fame that it achieved from sculptures such as those produced by Michelangelo. But there are over fifty varieties of marble all commonly known as Carrara Marble and all with different physical and mechanical properties (ERTAG, 1980). For example, the marble types conducive to sculpture have porosities of 6% or greater which makes them very susceptible to weathering. Marble from all over the world has differing properties of composition, grain size and orientation, porosity, residual strain, and others. Variations in properties are also found in other materials such as glass due to potential differences in their processing. To ensure public safety extensive testing of all material installed is highly desirable.

13.2.2 Technological risk assessment in buildings

Suggestions of strength loss in panels come from a variety of sources. In granite the development of fractures that occur either adjacent to anchors or extend across a panel leads to removal of selected panels and strength measurements to determine the magnitude of change (Figure 13.4). The situation with marble panels is different. It has been argued that the exposure to thermal gain can lead to strength loss and development of bows in panels on buildings. The issue is: can the strength loss become extreme enough to threaten panel stability?

To assess this hypothesis, tensile strength was measured by point-load tests (ISRM) on over 90 panels from one marble-clad building using specimens 2.5 cm in diameter with supplemental uniaxial, Brazilian and C880 flexural strength tests (Logan, 2006). The panels were divided into groups. Eight virgin panels had not been installed on the building and were stored under controlled temperature conditions so they presumably approximated the strength of material that was originally installed. Eighty-three access panels that were on the building for 25 years were removed to be tested. Most of the panels that were removed had measured bows of 18 mm or more. These showed strength losses compared to the virgin panels ranging from 40 to 90%. The strength distribution in panels containing significant bowing is not uniform. At the top and lower edges of panels the strength loss is only about 19%, however, where the amplitude of the bow is largest the loss averages about 53% (Figure 13.5). Additionally, test specimens from areas of bow curvature show the extended or outside portion of the bow to be 25 to 30% weaker than the inside part. Petrographic observations document that microfractures are concentrated in the extended portions contributing to the significant strength loss in those areas.

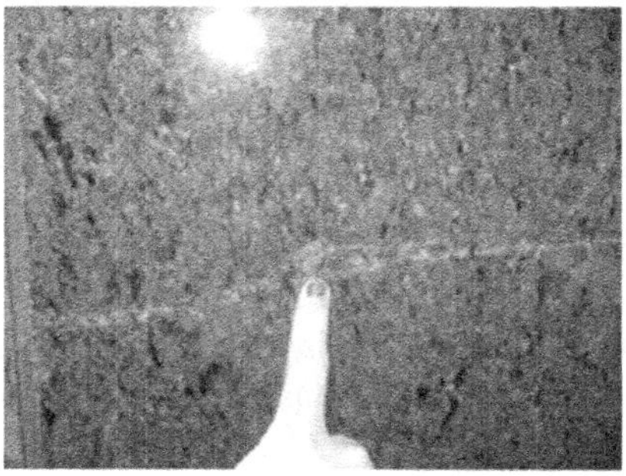

Figure 13.4. Photograph of horizontal fracture spaning a granite panel, which resulted in its removal.

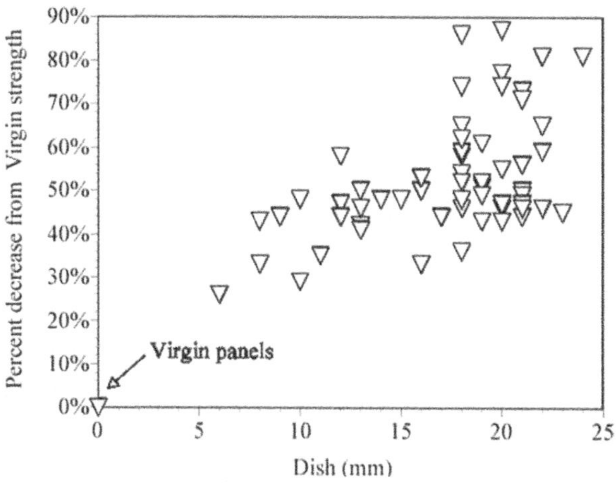

Figure 13.5. Tensile strength measurements where each data point is an average of ten specimens measured from the center of 83 panels from a building and 8 virgin panels. The plot shows the percent decrease in strength from the virgin panels versus the panel dish reflecting the amount of bow.

The development and extent of bowing in marble panels has been demonstrated to relate to their thermal exposure. In building risk assessment, are there portions of a building that require more attention? To determine if this is an issue on one building, the initial step on one building was to measure the distribution of all 43.000 panels. One chevron is shown to characterize some 14 on each side of the building. The strong correlation with thermal exposure is shown in Figure 13.6. Most of the panels were bowed convexly outward, with just 2% concave bows.

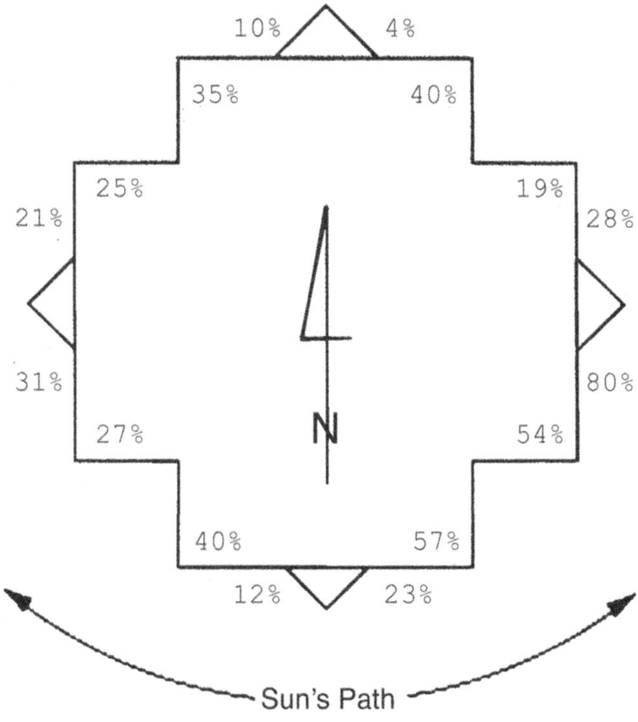

Figure 13.6. A diagram showing the sun's general travel path and the percentage of marble panels bowed 1.13 cm or larger. The values are for the total number of panels for each side of building or corner. Each chevron characterizes 14 on a given side.

Given that stone panels on buildings may degrade in strength and develop bows which lead to further loss of strength a basic issue is their ability to sustain wind loads. Panel dimensions, especially thickness are designed to meet a safety factor, which is panel strength divided by anticipated force generated by the wind. The wind direction, velocity, duration and seasonal variation are assessed from the local meterological records. Additionally for high-rise buildings and those in metropolitan areas, a question of surrounding buildings modifying the wind characteristics has to be considered. The result is that critical locations on a building can be potentially recognized.

Given recognition of potentially high wind loads, consideration of the safety factor then leads to minimum allowable strength for panels to remain on the building. As the panel strength is directly influenced by the amount of bow particularly in marble panels, and the fracture presence in granite and all panels criteria for identification of such panels and potential replacement. Commonly in marble clad buildings, measurements of bow are made and panel strength measured so a correlation can be established. Panel strength is sometimes made on the building by artificially loading the panel hydraulically to failure. In many cases the possibility of broken pieces falling from the building, precludes such methods. In most cases the sizes of bows are measured, the panel is then removed and tested on the ground. From this a data base can be established relating amplitude of bow to strength, and criteria developed for panel replacement.

Such criteria establish potential critical locations for potential panel replacement, thus the next step in risk assessment is generally to survey the building to determine the extent and location of panel depredation. This can be done by simply visual inspection of panels deemed to be in critical locations first, but eventually all panels. More quantitative is physical measurement the size, amplitude and location of bows in marble panels. These measurements coupled with the strength measurements would stipulate one group of panels to consider for replacement. Additionally, fractures especially in location of anchors would be mapped and such panels could be included in the replacement list (Figure 13.3).

For granite, sandstone and other stones similar criteria are established such as fracture size, location, density, and possible displacement and are documented, primarily with photographs, and removed and strength tested. Potential discoloration has also been note in some cases, with the inference that the induced fractures have increased porosity and facilitated chemical changes in minerals.

The presence of fracturing in all stones will result in an increase in porosity on a range of scales. This opens another non-destructive technique for assessing panel strength, namely the timing of acoustic wave paths across the stone (Figure 13.7). With increasing fractures, the porosity increases and the travel time of the acoustic wave also increases as grain to grain contact is decreased. Thus an increase in travel time indicates a loss in strength. Removal of panels and measuring acoustic travel time and strength provides quantitative criteria for panels evaluation and potential replacement. The benefit of the acoustic technique, as mentioned is that it is non-destructive and thus measurements on specific panels can be made spanning weeks, months and years to plot changes in panel strength.

13.2.3 Technological risk management in buildings

Once risk estimates are made the issues become risk evaluation. Clearly this is based on estimates of panel failure, all or pieces falling from a building and potentially damaging property or people. The issue of public safety becomes paramount when evidence of deterioration of the exterior materials becomes evident. In some cases surveys of the exterior may determine that the problems exist locally and can be remedied by replacement in those areas. Such a step was taken on the Amoco Building when it was decided that the problems were only extreme at the corners of the building where wind loads were highest. (Figures 13.8a) This building was the third-tallest building in the world in 1989 and the tallest marble-clad one. It was decided to install straps across each marble panel so that if failure did occur the pieces would be retained until they could be removed. In this case, the installation of the straps severely damaged the panels (Figure 13.8b). Even though no panels fell from the building the result was a decision to remove all of the 43,000 marble panels from the building to avoid any panels or partial ones falling from the building and endangering the public. The building was about 15 years old at that time with cost estimated at that time to range from $60 to $100 million, to ensure public safety, and made the front page of the Chicago newspaper (Figure 13.9). It was decided to replace the marble with granite panels, which due to extensive quality control during replacement have remained without problems.

Figure 13.7. Photo of acoustic delay meter on marble with travel time in microseconds of waves between transducers.

(a)

(b)

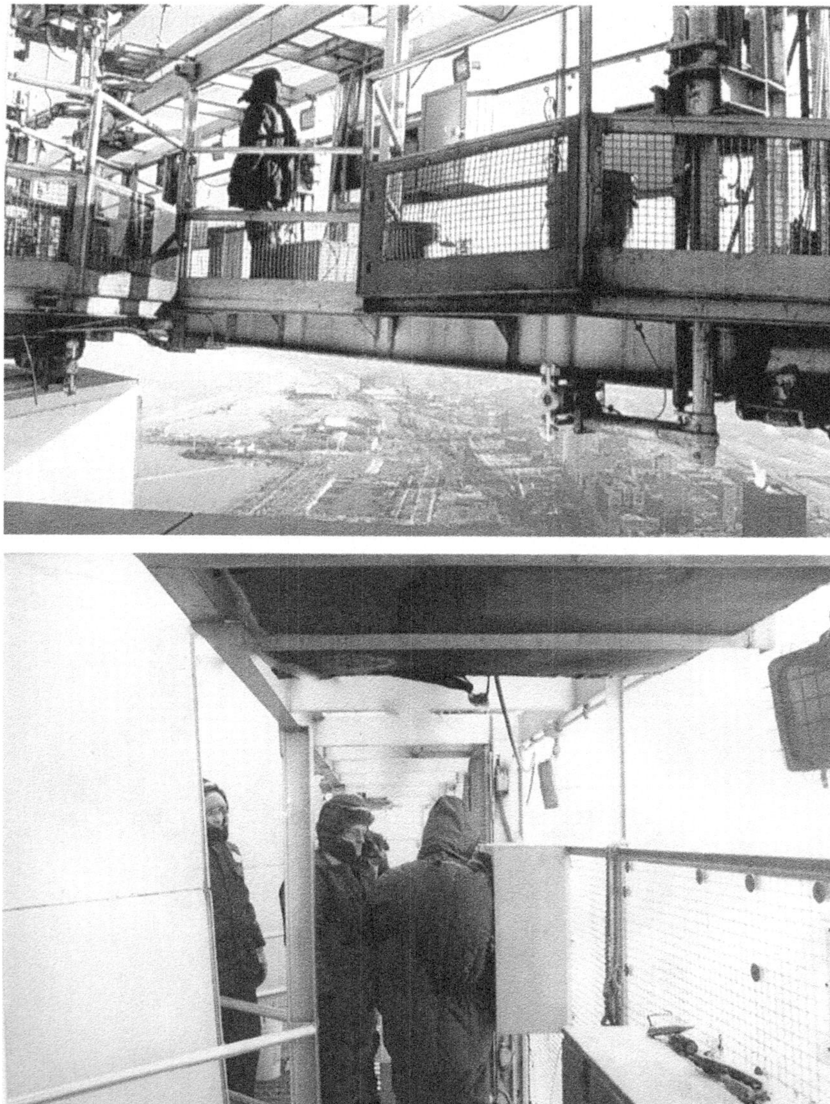

Figure 13.8 (**a**) Inspection platform about to drop down outside of Amoco building, Chicago to measure, test and inspect panels. Platform is at top of the 80 story, tallest marble clad building in 1990, 380 meters. (**b**) Same platform as in figure 8a, but now enclosed because of winter weather. Emergency inspection is underway due to imminent possibility of marble panel failures.

Chicago Sun-Times

35¢ HOME DELIVERY: 1-800-545-9000

COLD Pages 2, 32 TUESDAY, MARCH 7, 1989 5★ Sports Final

$60 million for Amoco Bldg.

Costly face-lift

SUN-TIMES / Jim Frost

Shelby C. Pierce, manager of engineering for Amoco Oil Co., with a sheet of the granite that will be used to repanel the Amoco Building.

By Jerry C. Davis and Andrew Herrmann

A face-lift costing $60 million to $80 million will begin next month on the towering lakefront home of Amoco Corp., with light-colored granite replacing 43,000 white marble panels that are deteriorating.

Amoco officials said the marble selected for the facade of the 80-story skyscraper at 200 E. Randolph in the early 1970s has lost too much strength and must be replaced. Stainless steel strapping was installed last year to hold the marble panels to the building.

'Outrageous' marble is out; Page 4.

The face-lift will cost at least half of the $120 million construction price of the building in 1975. At 1,136 feet, the Amoco Building is the tallest marble-faced building in the world.

As with any face-lift, the "surgeons" are promising the operation will leave the patient looking better—but not unrecognizable.

Up close, the replacement granite has a "salt and pepper" appearance, but it is whiter than the existing marble, said Shelby C. Pierce, manager of engineering for Amoco Oil Co.

"From any distance people won't even recognize that it is not entirely white," Pierce said.

Dating back to 1973, when a 350-pound slab tumbled from the 82nd floor and struck the

Turn to Page 4

Figure 13.9. Newspaper article reflecting the impending re-cladding of the Amoco building as a result of the potential hazards possible of marble falling from the building.

When it is decided to replace any stone pieces with those from storage, there is a question of the physical properties being similar to those originally designed. This is a function of the natural variability of the stone from any quarry. Within a given quarry, regardless of where it is or it history stone in naturally variable in composition, fabric, porosity and other features which are reflected in the strength and durability. In the case of the Amoco building, after visiting granite quarries in a number of countries, one was selected which was subsequently mapped, samples tested from various quarry locations and additional strength tests made on selected representative samples before shipping.

In contrast, the marble-clad building at First Canadian Place, Canada's tallest skyscraper has also decided to remove all of its 45,000 panels as a result of two incidents in which whole or portions of panels fell from the building. In the second event, one panel weighing 115 kilograms fell from the 51st story, landed on a third floor roof and fortunately no one was injured. The first event also did not result in damage. As a result of the second accident, and because the building is in the center of the down town financial district, surrounding streets were closed to vehicle traffic for a number of days leaving some 48,000 streetcar riders and thousands of car drivers dealing with major delays (Toronto Star, May 17, 2007, front page, p. A1). The marble suffered from the same deterioration processes that affected the Amoco Building. It was decided to remove all the marble cladding and replace it with other material besides stone.

In both of these examples, the cost was high, but without a management decision to replace the stone, possible consequences were high. Alternatively, many buildings which are smaller and in less populated locations have replaced panels as they become obviously deteriorated.

In summary, this section discusses the potential hazards in buildings with exterior stone panels on vertical faces, especially tall ones in urban centers. The focus is on marble, granite and sandstone panels and their susceptibility to lose strength when subjected to environmental conditions such as such heat gain and wind loads. This may lead to bowing and fracturing in marble panels, and a further loss of strength, and fracturing in granite and sandstones. Recognition of these features and their association with strength measurements on panels leads to criteria for potential panel replacement. This may involve the entire building or selected areas at periodic intervals. In either case emergency plans if whole panels or even pieces fall need to be in place. In all cases continuing monitoring is critical.

13.3 INDUSTRIAL AND TRANSPORTATION HAZARDS

The global natural and technological hazards have been steadily increased throughout the past fifty years and will continue to grow as a result of population clusters increase against vulnerable agriculture and food security (Dalezios *et al.* 2012), climate change and stress of existing infrastructure and constructions.

Section 13.3 deals with natural disasters causing industrial hazards and industrial and transportation hazards. In this chapter three major disasters occurred in the last 30 years are described in a high level approach. The first is based on the Nature's abnormal power which emerged with an earthquake and subsequent tsunami damaging part of the Fukushima' nuclear power station in the north east cost of Honshu Island threatening Japan and the rest of the world. The other two are based on industrial activity (the Chernobyl accident) and on the oil tanker Exxon Valdez which slammed into Bligh Reef and spilled millions of gallons of crude oil into the cold, clear waters of Alaska's Prince William Sound.

13.3.1 Fukushima

It was March 11th 2011 where Japan experienced three mega disasters in a matter of minutes. A 9.1 earthquake was immediately followed by a 15-metre tsunami. Both affected six nuclear reactors in Fukushima Daiichi (Figures 13.10 and 13.11).

Not only Japan, but the world has been attempting to avert an all-out nuclear melt-down there. Approximately 18,500 people died due to the earthquake and tsunami and hundreds of thousands of homes destroyed, inflicted billions of dollars in damage, and unleashed a nuclear crisis the likes of which has not been seen since the Chernobyl disaster in 1986. There have been no deaths or cases of radiation sickness from the nuclear accident, but over 100,000 people were evacuated from their homes to ensure this. Fukushima is one of Japan's biggest prefectures, and is split into three regions: Hamadori to the east, home to the crippled nuclear reactor, Nakadori in the central part and Aizu in the west. Japanese authorities implemented a 20 km exclusion zone around the power plant, and the continued displacement of approximately 156,000 people as of early 2013.

Figure 13.10. Fukushima nuclear facility (Image courtesy of www.wikipedia.com).

Figure 13.11. Fukushima nuclear facility after earthquake and tsunami hit (Image courtesy of www.wikipedia.com).

In Japan there are 54 nuclear plants located mainly in coastal areas. These nuclear plants supply more than 30 percent of the nation's electricity. The potential for a problem in any one of these plants is mind-boggling. Some of the risk factors that played a role at Fukushima were the location on tsunami-prone east/west coast or near earthquake faults, aging plants and backup electrical systems relying on diesel generators and batteries, which could fail in extreme circumstances.

If a natural disaster or large scale industrial accident strikes communities might not have access to food, water and electricity for days or even weeks (Hurricane Katrina). Obviously, if homes are lost, then emergency food will be also. But if homes are spared, there might be days or weeks to have to basic necessities. Ten days after Japan's triple disaster, there are millions of people, without food, water or power. Having an ample supply of clean water is a top priority in any emergency. A normally active person needs to drink at least two quarts (half gallon) of water each day. There is also need for water for food preparation and hygiene.

13.3.2 Chernobyl

It was April 25th 1986 where abnormal radiation levels, apparently originated in the Ukraine, had been detected by the Swedish Power Board. The radiation source was traced to the Chernobyl nuclear power generating facility, which is located approximately 80 miles north of Kiev, in Ukraine (Figure 13.12).

Figure 13.12. A chimney towers over the sarcophagus that covers destroyed Reactor No. 4 in the Chernobyl. (image courtesy of www.wikipedia.com).

The accident destroyed the 4th Chernobyl reactor, killing immediately more than 30 operators and firemen. More than 200 people involved with the clean-up diagnosed with acute radiation syndrome (ARS). A hundred of people died between 1987 and 2004, but their deaths cannot necessarily be attributed to radiation exposure. Nobody off-site suffered from acute radiation effects, although a large proportion of childhood thyroid cancers diagnosed, since the accident is likely to be due to intake of radioactive iodine fallout. The accident caused the largest uncontrolled radioactive release into the environment ever recorded for any civilian operation, and large quantities of radioactive substances were released into the air for about 10 days. This caused serious social and economic disruption for large populations in Belarus, Russia and Ukraine (Figure 13.13).

Figure 13.13. This image of the abandoned city of Pripyat, home to the Chernobyl Nuclear Power Plant was taken by Landsat-5 three days after the April 26, 1986. (image courtesy of www.wikipedia.com).

Two radionuclides, the short-lived iodine-131 and the long-lived caesium-137, were particularly significant for the radiation dose they delivered to members of the public. It is estimated that all of the xenon gas, about half of the iodine and caesium, and at least 5% of the remaining radioactive material in the Chernobyl 4 reactor core (which had 192 tons of fuel) was released in the accident. Most of the released material was deposited close by as dust and debris, but the lighter material was carried by wind over Ukraine, Belarus, Russia and to some extent over Scandinavia and Europe. The casualties included firefighters, who attended the initial fires on the roof of the turbine building. About 200,000 people ('liquidators') from all over the Soviet Union were involved in the recovery and clean-up during 1986 and 1987. They received high doses of radiation, averaging around 100 mill Sieverts. Some 20,000 of them received about 250 mSv and a few received 500 mSv. Later, the number of liquidators swelled to over 600,000, but most of these received only low radiation doses. The highest doses were received by about 1000 emergency workers and on-site personnel during the first day of the accident.

13.3.3 Exxon valdez

On March 24, 1989, the oil tanker Exxon Valdez had just entered Alaska's Prince William Sound, after departing the Valdez Marine Terminal full of crude oil (Figure 13.14). At 12:04 am, the ship struck a reef, tearing open the hull and releasing 11 million gallons of oil into the environment. Initial responses by Exxon and the Alyeska Pipeline Company were insufficient to confine much of the spill, and a storm that blew in soon after, spread the oil widely. Eventually, more than 1000 miles of coastline were fouled, and hundreds of thousands of animals perished. Exxon tried to transfuse the oil from the entrails of the ship and ended up paying billions in cleanup costs and fines, and remains tied up in court cases to this day (Figure 13.15).

Figure 13.14. The damaged oil tanker Exxon Valdez, towed out of Alaska's Prince William Sound by a tugboat and a U.S. Coast Guard Cutter, on June 23, 1989. On March 24, 1989, the tanker ran hard aground on Bligh Reef, spilling 11 million gallons of crude oil into the sound – at the time, the largest oil spill disaster in U.S. history.

The captain, Joseph Hazelwood, was acquitted of being intoxicated while he was at the helm, and he convicted on a misdemeanor charge of negligent discharge of oil, fined with $50,000, and sentenced to 1000 hours of community service. Though the oil has mostly disappeared from view, many Alaskan beaches remained polluted up to this day and crude oil is just buried a few inches below the surface. The sound's coastal ecosystem was permanently damaged. Thousands of gallons of oil spill polluted the beaches and this oil is still toxic and still hurting the ecosystem near the shore.

Figure 13.15. The Exxon Baton Rouge (smaller ship on left) attempts to offload crude oil from the Exxon Valdez after the Valdez ran aground in Prince William sound near Valdez, Alaska, on March 26, 1989, (image courtesy of www.wikipedia.com).

The US Government considers, as of 2010, only 13 of the 32 monitored wildlife populations, habitats and resource services that were injured in the spill as fully "recovered" or "very likely recovered." Some are still listed today as "not recovering." This includes a pod of orcas, which lost 15 of its 22 members after the spill, and has not produced a calf since. Given only one older female is left, scientists appear certain that this unique pod of orcas will go extinct — it's just a matter of time. The government conclusion is that "there appears to be no hope for recovery." The "not recovering" list also includes Pacific herring, one of the Prince William Sound keystone species. Although the Pacific herring was a vibrant commercial fishery activity, after the event it was rapidly declined and finally fishery closed, and has not reopened yet. The eight inches long, herring was everywhere, swam in millions or more, but a sudden flash of their silver undersides confused the predators. In April, their spawning turned and the bays and lagoons became milky white. More than 40 species, such as bald eagles, brown bears, seals, humpback whales, tufted puffins, murres, are dependent on these small fish. The example of Exxon Valdez event (Cutter, 1997) and also the September 11th created another dimension of the role of transportation as hazard itself and in large scales.

13.3.4 Transportation of hazardous materials

The means and systems of transportation have been developed to facilitate the move of goods, services and people in a cost-effective and secure way from point A to point B. Following the growth of population and technological developments, the means and systems of transportation have been dramatically expanded in the last 100 years and became vulnerable in natural phenomena, such as extreme weather conditions, earthquakes, tsunamis, landslides and similar disasters. These phenomena disrupted and damaged transportation in various spatio-temporal scales (Cova & Conger, 2004). Moreover, the transportation of hazardous materials is also a debatable issue, simply because it places substantial involuntary risks on proximal people and the environment. The transportation of hazardous materials has steadily increased throughout the past thirty years associated with an increased public and governmental concern regarding hazardous materials management, and an affiliate increase in activities associated with designing and using hazardous material management systems.

Satellite based and Radio Frequency Identifier (RFID) technologies have been developed to track extremely dangerous shipments and provide direct location-based information projected onto high accurate electronic charts, which are essential for both transportation and public safety agencies. Governmental and local transportation authorities, as well as public safety agencies, have to have continuously updated hazardous material transportation recordings in order to plan, train and equip responders to meet the identified threat and run operations and simulations. The tracking and identification technology must be consistent within national and transboundary territories. Satellite weather information that cover beyond the horizon territories, associated with Geographical Information Systems (GIS), Global Position and Automatic Identification Systems (AIS) signals together with multi-temporal satellite imagery, for example to map the thickness of ice in northern sea routes, provide a wealth of information available to transportation users, planners, and emergency responders.

It has to be mentioned that means and systems of transportation that are damaged by a hazardous event also play a critical role in emergency management and evacuation procedures. Reliable transportation lifeways are generally considered as critical in an emergency, because of their vital role in the rebuild of all other lifeways. Emergency decision makers must re-route personnel to an accident site, remedy lifeways, relocate or evacuate threatened populations, and provide aid and relief, all of which rely on transportation (Markatos *et al.* 2007). Transportation lifeways could help to reduce the impact and increase the mitigation of reaction in the above industrial accidents. Transportation lifeways plays an integral input in operational planning tools (the preparedness phase) and by using them a simulation of redeployment resources can be developed to maximize readiness and train emergency personnel, as well as the public (Spyropoulos *et al.* 2010).

13.4 OPTIMAL OPERATIONAL RESPONSE PLANNING IN NATURAL AND TECHNOLOGICAL HAZARDS

Section 13.4 covers the simulation and management of these phenomena through operational planning in a web based mapping service environment to provide an optimal operational response planning.

13.4.1 Simulation and management of phenomena through operational planning in a web based mapping service environment

The recent increase of industrial and natural disasters frequency and severity has stimulated a great awareness on behalf of both the general public and local governments, in environmental information management systems that are able to collect, process, visualize and interpret geospatial data and value-added application workflows in order to support decision making in case of emergency (Markatos *et al.* 2007). Since 1900 there have been 35,000 natural catastrophic events on the planet resulting into more than 8 million life loses.

In order to make efficiently handling, forecasting, mitigation and prevention of such disasters, it is necessary to create new, open, scalable and distributed service platforms. Instead of leaving this task to specialized and stand-alone and fragmented systems, integrating such services in global web mapping platforms has a huge potential to create a totally new area or market of applications for such platforms, which is expected to be addressed not only to the general public, but also to governments and local authorities. Geospatial data, real time and historical measurements, wireless sensors, traffic information, population distribution data, and value added services can all be integrated into current web mapping services by creating a new set of Application Protocol Interfaces (APIs) and enhancing already available and paid for, geospatial data that these services currently incorporate. Such services can address several emerging markets, such as those mentioned below:

- the general population in order to increase awareness and environmental sensitivity;
- the wireless sensor applications community, by creating a uniform, global service for wireless sensor network (WSN) applications, data management and dissemination, with countless possibilities;
- local and country authorities, which already spend significant amounts to purchase stand-alone, custom-made systems to handle natural disasters and hazards;
- independent entities, such as schools, private country communities and companies that maintain plants vulnerable to natural hazards;
- stakeholders in the sea transportation and energy sectors.

Global web mapping services is a new evolving trend on the web, that consumes a great amount of resources, and requires maintenance, updates, new services and new business concepts, in order to become economically profitable. Providers of such services, such as Microsoft, Google, Amazon and Yahoo, compete to achieve the best user experience, the most powerful API and the most flexible application platform in order to attract not only end users, but also, and perhaps more importantly, developers of value-added services. By attracting such services, it is becoming more appealing to end users, and there is an attempt to increase their share in the evolving advertising and value-added services market. In doing so, several "basic" services are offered for free, such as routing, distance calculators, and other essential services that although they do require up-to-date data, they alone do not seem able to bring in any income.

To produce income, two main threats are discrete in web mapping services. At first, integration of traditional search engines with geographical systems is required. Geographical information in this context may improve the quality of service that the end user experiences and may also generate some income. Typical examples are those services and meta-services relating to travel, such as tickets, hotel bookings, car rentals and similar aspects. Second, new services, such as location-based services that may or may not make use of real-time information, such as traffic, webcams, price information, and more. There is already significant technology development and many stakeholders come into this play, such as mobile operators, handheld operating systems, web advertising platforms, spatial data providers, and many more. Although there is not yet any hard evidence, it is felt that, up to now, this market is a very hard one, which consumes significant resources. Indeed, competitors push hard to get a small share of a geo-tagged version of what in the past has not been as profitable as promised, such as search, travel, financial information, integration, and personalization. New ideas, such as "green" energy, are in very early stages of implementation yet.

13.4.2 Reducing the impact and increasing the mitigation of reaction in the above industrial accidents using operational planning tools

By setting a cluster of integrated tools, a combined simulation of accidents and optimal response planning can be developed. Simulation tools could be the hazardous elements, nuclear contamination and similar aspects. Optimal planning or response operations concepts could be evacuation assessment, safe evacuation buffer zones, optimal routing integrated with simulation and optimal placement of emergency response vehicles. These tools are used either to create hypothetical scenarios or to support the emergency response to a real incident. The setup could be a Web-based GIS environment (GIS data, Digital Elevation Models (DEMs), thematic maps), the inputs could be pointers to a specific event, meteorological parameters, time intervals, whereas the output could be contours that show the dispersion of hazardous elements or nuclear radiation over space and time.

Simulation uses detailed surface structure and updated status by satellite providers together with case-specific inputs and event-specific parameters. It calculates the evolution of the dispersion of hazardous substances or radiation over space and time (Figure 13.16). Simulation results are used to feed the routing and evacuation algorithms. For example, a platform that gathers data from large scale, high density wireless sensor networks (WSN), depicts them on an open spatial data representation platform, such as Google maps, and produces "fire" or "toxic" alarm messages. This data can either be real world data gathered directly from the environment through a WSN, or simulation data acquired from the output of a WSN simulation tool (Kolega *et al.* 2009).

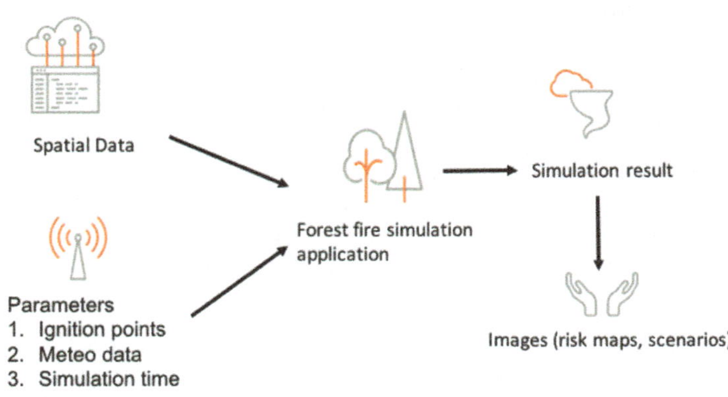

Figure 13.16. Simulation Workflow (forest fires example).

Optimal planning tools: evacuation assessment. These tools are needed to make decisions on where to deploy forces that will keep the roads open so as to avoid congestions that may create casualties. It is critical that the evacuation is not done independently of the event that forces the evacuation. The simulation runs will determine which roads will be unusable and use the rest of the network for the evacuation. Close-to-real-time calibration is done by satellite providers image updates. An evacuation buffer is a geographical area that when reached by the radiation, then evacuation of the area of interest is recommended. The buffer indicates how much time is available for the evacuation. As a result, there is one buffer for 24-hour evacuation, one for 18, 12, 6, and so on (user-defined). When, for example, the hazardous substance reached the 12-hour buffer for area-of-interest X, then there are 12 hours to evacuate area X, and so on (Figure 13.17).

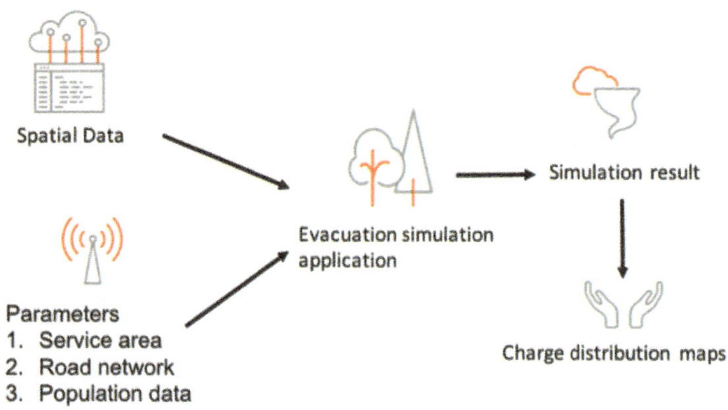

Figure 13.17. Evacuation simulation workflow.

Optimal planning tools: Routing. Normally, traditional routing would send emergency response units through the colorful GIS-based routes. Integrated routing runs simulation and routing together and will suggest a safe alternative approach.

Optimal planning tools: response unit placement. The question is: Where to place N emergency response vehicles in order to be able to reach 'any' point of the road network in the minimally achievable time? OR How many emergency response vehicles are needed and where to place them in order to be able to reach 'any' point of the road network in X minutes? This could be shown by different colors on the road network in a web-based GIS environment.

13.4.3 Natural and Industrial Hazards Assessment, Environmental Modelling and Operational Planning (NIHAEMOP)

Natural and Industrial Hazards Assessment, Environmental Modelling and Operational Planning can be amplified to a maximum extent using the Global Web Mapping Services (GWMS). The GWMS services are offering cloud based such as:

- maps, including 3D, aerial, satellite, street view, walk-through, and more;
- services such as routing, travel info, search integration, entertainment, traffic, location-based services, social networks, and more;
- generic platforms and APIs on which more applications can be built.

Moreover, GWMS require huge amounts of spatial data and updates, such as satellite images, vector data, population, businesses, as well as real time data, such as weather, traffic, personal location information and agreements with spatial data providers, authorities, mobile phone carriers, and service providers (Figure 13.18). Indeed, GWMS is an integral part of the NIHAEMOP, since the natural and industrial hazards are becoming an increasingly significant issue and it is expected to become even more significant in the coming years. In fact, the wireless sensor networks (WSNs) and the early warning systems are dramatically exploited, the contingency planning by National and local authorities is becoming more and more demanding, the energy and carbon credits market is becoming more popular, the need for training and awareness by the general population is increased and last, but not least, the human safety and environmental security need not to be restricted in the military domain.

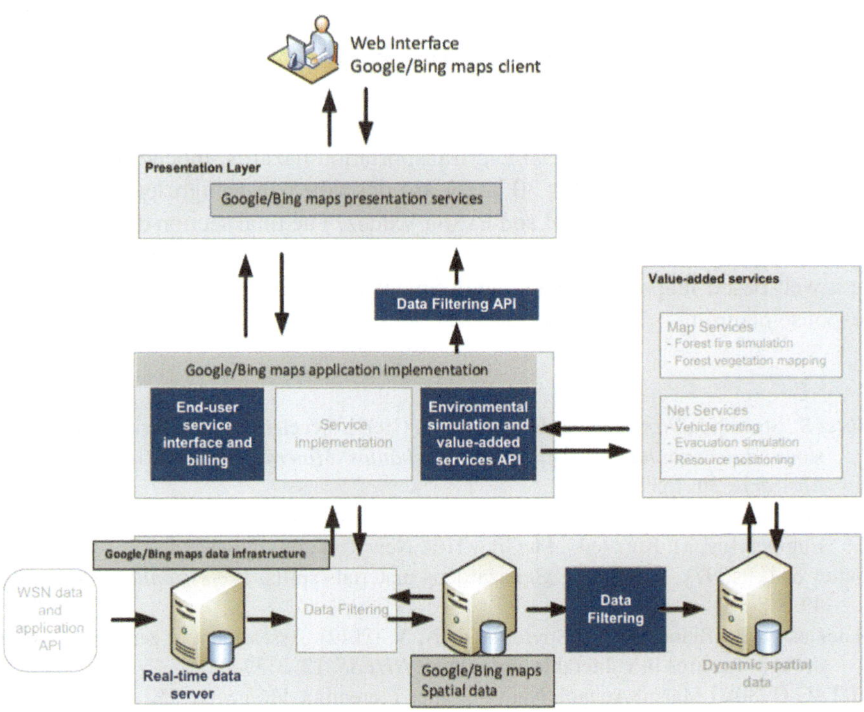

Figure 13.18. A generic architecture of a natural hazard management system for forest fires over Google and/or Bing maps (courtesy Optum Ltd) Workflow (forest fires example).

Natural and Industrial Hazards Assessment, Environmental Modelling and Operational Planning can be analyzed and investigated through the Global Web Mapping Services (GWMS) by:

(1) developing an integrated platform over Google or Bing Maps or on any other cloud based geospatial platform using APIs for wireless sensor networks of all kinds; APIs for environmental modeling and simulation applications and services; APIs for operational logistics for natural hazards management; and integration of all the above;
(2) exploring a Google or Bing Maps or on any other cloud based geospatial platform for WSN applications simply because the WSN market is expected to grow beyond imagination in the next few years; there are countless applications in all domains; only proprietary solutions currently exist; and there are huge amounts of collected data, which enable new applications.

13.5 SUMMARY

Technological hazards are mainly man-made accidents. Technological hazards are characterized by a causal sequence of events and that an essential distinction could be made between energy and material releases. At first, the nature and classification of technological hazards is described. Then, focus is on the potential hazards in buildings having stones on vertical surfaces, particularly exterior ones exposed to environmental conditions. The following section covers natural disasters causing industrial hazards and industrial and transportation hazards. Indeed, three major disasters occurred in the last 30 years are described in a high level approach, namely Fukushima, Chernobyl and Exxon Valdez. The final section describes how these phenomena can be simulated and managed through operational planning in a web-based mapping service environment to provide an optimal operational response planning.

REFERENCES

Bortz S., Stecich J., Wonneberger B. and Chin I. (1993). Accelerated weathering in building stone. *International Journal Rock Mechanics Mineral Science Geomechanical Abstracts*, **30**, 1559–1566.

Cova T. J. and Conger S. (2004). Transportation hazards. In: Handbook of Transportation Engineering, M. Kutz (ed.), McGraw Hill, New York, pp. 17.1–17.24.

Cutter S. L. (1997). Trends in U.S. hazardous materials spills. *Professional Geographer*, **49**, 318–331.

Dalezios N. R., Blanta A. and Spyropoulos N. V. (2012). Assessment of remotely sensed drought features in vulnerable agriculture. *NHESS*, **12**, 3139–3150.

ERTAG (1980). I Marmi Apuani. Nuova Grafica Fiorentina, Firenze, p. 55.

Hohenemser C., Kates R. W. and Slovic, P. (1983). The nature of technological hazards. *Science*, **220**, 378–384.

Jones D. K. C. (1993). Environmental hazards in the 1990s: problems, paradigms and prospects. *Geography*, **339**, 161–165.

Koch A. and Siegesmund S. (2004). The combined effect of moisture and temperature on the anomalous expansion behavior of marble. *Environmental Geology*, **46**, 350–363.

Kolega E., Vescoukis V., Spyropoulos N. and Douligeris C. (2009). A large scale Wsn-based forest fire sensing and management system. *Scientific Journal*. Annals of the "Constantin Brancusi" University of Targu Jiu. Engineering Series. **2**, 207–229.

Lagadec P. (1987). From Seveso to Mexico and Bhoral: learning to cope with crises. In: Insuring and Managing Hazardous Risks, P. R. Kleindorfer and H. C. Kunreuther (eds), Springer-Verlag, New York, pp. 13–27.

Logan J. M. (2006). On-site and laboratory studies of strength loss in marble on building exteriors. In: Fracture and Failure of Natural Building Stones: Applications in the Restoration of Ancient Monuments, S. K. Kourkoulis (ed.), Springer, Berlin, Heidelberg, New York. pp. 345–362.

Logan J. M., Hastedt M., Lehnert D. and Denton M. (1993). Variations of rock properties within a quarry. *International Journal of Rock Mechanics and Mining Sciences and Geomechanical Abstracts*, **30**, 1527–1530.

Markatos N., Vescoukis V. and Kiranoudis C. (2007). Firementor: Operational Planning and Decision Support System for Forest Fires. Hellenic Secretariat for Research & Technology. Project code FP4.

Raleigh L. (1934). The bending of marble. *Proceedings of the Royal Society A*, **144**, 266–279.

Siegsmund S., Ullemeyer K., Weiss T. and Tschegg E. K. (2000). Physical weathering of marbles caused by thermal anisotropic expansion. *International Journal of Earth Science*, **89**, 170–182.

Siegsmund S., Mosch S., Scheffzuk C. H. and Nikolayev D. I. (2007). The bowing potential of granitic rocks: rock fabrics, thermal properties and residual strain. *Environmental Geology*, **50**, 254–265.

Smith K. (2013). Environmental Hazards: Assessing Risk and Reducing Disaster. Springer Berlin, Heidelberg, New York, 6th edn, p. 478.

Spyropoulos N., Valmis S., Ganas A. and Metaxas G. (2010). Land reclamation of surface mining based on imagery intelligence and spatial decision support systems. *World Transactions on Engineering and Technology Education*, **8**(1), 115–119.

Winkler E. M. (1967). Weathering and weathering rates of natural stone. *Environmental Geology Water Science*, **9**, 85–92.

Yagar S. Ed. (1984). Transport Risk Assessment. University of Waterloo Press, Waterloo, Ontario.

Part 5

Epilogue

Chapter 14

Climate change and climate extremes

Spyros Rapsomanikis, Aikaterini Trepekli and Nicolas R. Dalezios

14.1 CLIMATE CHANGE AND MODELLING

14.1.1 Climate variability and change

There is a general agreement within the scientific community that the Earth's climate is undergoing changes in response to natural variability, including solar variability, and to increasing concentrations of greenhouse gases (GHG) and aerosols. Climate change is caused largely by human activities and poses significant risks for natural and human systems. According to the World Meteorological Organization (WMO), climate change refers to statistically significant variation either in the mean state of the climate or in its variability, persisting for an extended period (typically decades or longer). Climate variability is defined by WMO as variations in the mean state and other statistic metrics (variations, the occurrence of extremes, frequency) of the climate on all temporal and spatial scales beyond that of individual episodic events.

The consensus on global climate change is based on scientific observations summarized in Intergovernmental Panel on Climate Change's (IPCC, 2013):

(1) The Earth is warming. The globally averaged land and ocean surface temperature data, as calculated by a linear trend, show a warming of 0.85°C (0.65 to 1.06°C) from 1880 to 2012; under the high emission scenario (A2), it is projected to increase by 4.6°C by the 2090s in comparison with measures from the 2000s. The upper 75 m layer of ocean has warmed by 0.11°C (0.09 to 0.13°C) per decade from 1971 to 2010. Permafrost temperatures have increased in most regions since the early 1980s in response to increased surface temperature.

(2) Other changes related to global warming are of great concern, including the increase in the frequency of intense rainfall over the mid-latitude land

areas of the Northern Hemisphere; decreases in Northern Hemisphere spring snow cover and in the average thickness and extent of Arctic sea ice; the shrinkage of glaciers worldwide; the 26% increase in ocean acidity. The sea level will continue to rise by an additional 0.18 to 0.59 m by 2100. A recent study found that if GHG emissions continue unabated, Antarctica has the potential to contribute more than 1 m of sea-level rise by 2100 and more than 15 m by 2500 (DeConto & Pollard, 2016).

(3) Natural processes produce substantial seasonal, annual, and even decadal variations that are superimposed on the long-term warming trend, but it is very unlikely that natural variations could have given rise to the observed levels of global warming, especially over the last several decades.

(4) Anthropogenic forces emerging in the mid-20th century have contributed to surface temperature increases, to increased surface melting of the Greenland ice sheet, to the melting of glaciers, to the enhancement of global upper ocean heat content (up to 700 m) and the global mean sea level rise observed during the last 3–4 decades, and have affected the global water cycle during the last 2–5 decades. Most of the warming over the last several decades can be attributed to human activities that release carbon dioxide (CO_2) and other heat-trapping greenhouse gases (GHGs) into the atmosphere. The major portion of anthropogenic emissions of green-house gases is emitted from the burning of fossil fuels-coal, oil, and natural gas—for the production of energy. Humans emit about 30 billion tones of CO_2 each year through energy consumption. However, changes in agricultural and livestock farming practices, deforestation, land use changes, such as replacing forests with cropland, and certain industrial activities also had significant effects. Transport and cement production are also large emitters of GHGs. Human activities have also increased aerosols emissions. Aerosols have a wide range of environmental effects, but on average they increase the amount of sunlight that is reflected back to space, a cooling effect that offsets some of the warming induced by increasing GHGs.

(5) Human-induced climate change and its impact will continue during this century and beyond. Individually and collectively, and in combination with the effects of other human activities, these changes pose risks for a wide range of human and environmental systems, including freshwater resources, coastal environments, ecosystems, agriculture, fisheries, human health, and international security, among others.

14.1.2 Climate emissions scenarios

14.1.2.1 Special report on emissions scenarios

IPCC Special Report on Emissions Scenarios (SRES) (IPCC, 2000) was established to explore future developments in the global environment, with special reference to the production of GHG and aerosol precursor emissions. Major driving forces of climate change, including population growth and socio-economic development, encompass various future scenarios that might influence GHG sources and sinks, such as energy system and land use change. Four narrative storylines, A1, A2, B1 and B2, represent different demographic, social, economic, technological and environmental conditions.

A1: Globalization, with an emphasis on globalized human wealth and intensive market forces. The A1 storyline and scenario family anticipates a future world of very rapid economic growth, a global population that peaks in mid-century and declines thereafter, and a rapid introduction of new and more efficient technologies. The A1 scenario family is subdivided into three groups, which are distinguished by their technological emphasis: fossil intensive (A1FI), non-fossil energy sources (A1T), or a balance across all sources.

B1: Globalization, with an emphasis on sustainable development. The B1 storyline and scenario family anticipates a convergent world, with a global population pattern as in the A1 storyline, but with the expectation of rapid changes in economic structures toward a service and information economy, with reductions in material intensity, and the introduction of clean and resource-efficient technologies. The emphasis is on global solutions to the problems of economic, social, and environmental sustainability, without including additional climate initiatives.

A2: Regionalization, with an emphasis on human wealth and regional conflict. The A2 storyline and scenario family anticipates a substantially heterogeneous world. Fertility patterns across regions are expected to converge very slowly, resulting in a continuously increasing global population. Economic development is primarily regionally oriented and per capita economic growth, and technological changes are more fragmented and slower than in other storylines.

B2: Regionalization, with an emphasis on sustainability and equity. The B2 storyline and scenario family anticipates a world in which the emphasis is on local solutions to economic, social, and environmental sustainability. It describes a world with a continuously increasing global population at a rate lower than A2, with intermediate levels of economic development, and less rapid, more diverse technological change than in the B1 and A1.

14.1.2.2 Representative concentration pathways

The Fifth Assessment Report (IPCC, 2013) introduced a new set of scenarios, called Representative Concentration Pathways (RCPs), which incorporate climate change mitigation policies to limit carbon emissions. The RCPs include GHG concentrations resulting from economic, demographic, climate, land use and energy-use determents. These GHGs are converted to emissions trajectories using carbon cycle models. The RCP2.6 scenario assumes strong mitigation of GHGs in the 21st century reaching maxima of 3.0 W m^{-2}, which is then reduced to 2.6 W m^{-2} by 2100. The RCP4.5 and RCP6.0 scenarios stabilize at 4.2 W/m^2 and 6.0 W/m^2, respectively, by 2100. The SRES B1 is comparable with RCP4.5 and scenario, while the RCP6.0 lies between the SRES B1 and A1B scenarios. The RCP8.5 scenario is similar to the 'business as usual' scenario of fossil fuel use and to the SRES A2 by 2100. In all RCPs global population levels off or starts to decline by 2100, with a peak value of 12 billion for RCP8.5 scenario. Also, gross domestic product (GDP) is considered to be increased in all RCPs; the RCP2.6 has the highest GDP and the least dependence on fossil fuel sources. Carbon dioxide emissions for all RCPs except the RCP8.5 scenario are assumed to be maximized by 2100.

14.1.3 Modelling climate variability and change

14.1.3.1 Global climate models (GCMs)

Projections of future climate changes or extreme events rely on an increasingly sophisticated set of models and statistical methods. Applications of global climate models (GCMs) include simulating sensitivity, predicting near-term climate variability and change and downscaling projections of future climate change at the regional or local scale. Details of different models and methodologies reported in this paragraph are mainly adapted from the Chapter 9 of IPPC (2013). The GCMs could be subsets of Atmosphere–Ocean General Circulation Models (AOGCMs), Earth System Models (ESMs), with a typical horizontal grid spacing ranging roughly from 1 to 2 degrees for the atmospheric and around 1 degree for the oceanic component, and the Earth System Models of Intermediate Complexity, with a comparative lower resolution and complexity (e.g. LOVECLIM1.2; MIROC-lite-LCM, MIROC-lite, Bern3D-LPJ; IAP RAS CM, GENIE).

Uncertainties of Global Climate Models. The response of a climate model and a climate system over a set period depends on the length of the period examined (Marotzke & Forster, 2015). For longer periods, the change in external forces produces a change in temperature, which is in turn amplified through other responses of the climate system, primarily related to water phases. This amplification process is termed climate feedback. If climate feedback factors are too strong in a model, then the model may overestimate the imposed response over a given period (Risbey, 2015). Marotzke and Forster (2015) underlined that decadal scale trends do not provide grounds for revising the sensitivity of the climate models to forcing, such as increased concentrations of GHGs, solar cycles, and volcanic eruptions, because climate feedback plays little part on this time scale. Recently, a hydro-climate reconstruction for the 20th century based on proxy data was diverged from model outputs (Ljungqvist *et al.* 2016), and some authors now advocate that climate models should be validated through comparison with, and integration of, measured or proxy-based climate data (Ault *et al.* 2014). For projected changes by the end of the 21st century, there are different types of uncertainties attributable to: (1) the quasi-random internal climate variability (Huber & Knutti, 2014); (2) the applied climate model parameters (Schmidt *et al.* 2014); (3) the emissions scenarios used. For certain precipitation-related extremes, the uncertainty in projected changes is more the result of uncertainties in climate models. For temperature extremes, the emissions scenarios are the main source of uncertainty in projections.

Atmosphere–Ocean General Circulation Models. AOGCMs simulate the dynamics of the physical responses of the climate system, and they are applicable for climate projections in which biogeochemical feedback is not critical. In such case, the time-evolution of atmospheric composition, such as the amount of CO_2, is prescribed as an input to the model from a given scenario, and follows a predefined pathway regardless of changes in the climate, which are the outputs of the model (concentration-driven simulations).

Earth System Models. ESMs account biogeochemical processes interacting with the physical climate, modifying its response to external forcing (Flato, 2011). For example, there are ESMs, which simulate land and oceanic carbon cycle processes and feedback (e.g. MPI-ESM-LR, IPSL-CM5A-LR). Numerical simulations with ESM models can be configured either as concentration driven, where the output does not affect the predefined input, or 'emissions driven' wherein CO_2 emissions are prescribed as the input and the subsequent changes in CO_2 is an internally calculated element of the ESM (IPCC, 2013). Overall, some EMSs include atmospheric chemistry processes (e.g. CESM1 (FASTCHEM), CNRM-CM51,GFDL-CM3, GISS-E2-H), while others include ocean biochemical processes (e.g. CESM1(BGC), CMCC-CESM, GFDL-ESM2G, IPSL-CM5A-LR), whereas some models consider both processes (e.g. HadGEM2-ES; HadGEM2-CC, CanESM2, NorESM1-ME, MRI-ESM1). Several climate models are currently capable of simulating the microphysical characteristics and the mixing state of interacting aerosol particles (e.g. the MATRIX model in the GISS climate model). Many AOGCMs and ESMs include the sulfur cycle (e.g. in HadGEM2).

Dynamic Global Vegetation Models. Changes in biogeochemical processes and biodiversity (including changes in albedo due to land use changes) also have the potential to feedback on the climate system on a variety of time scales. Models and experiments that integrate knowledge about ecosystem processes, plant physiology, vegetation dynamics, and disturbances, such as fire are needed, and such models should be linked with climate models. ESMs incorporating a dynamic component representing vegetation and the associated carbon cycle (e.g. JULES in HadGEM2-ES) can show how scenarios that consider both land-use change and climate change lead to substantial regional discrepancies in where and how much vegetation cover expands or decreases. Some models include processes relevant to non-CO_2 trace gas, as also interactions between fires and other the climate system factors are now being introduced into ESMs (e.g. model LM3V).

14.1.3.2 Regional climate models

Downscaling of GCM simulations is required for climate scenario development and impact analysis at a finer spatial scale. Many climate-related decisions, especially those related to adaptation, demand information on regional to local scales. Implications of projected global mean temperature changes tend to be underestimated at regional levels (Seneviratne *et al.* 2016) and for certain extreme conditions, such as floods and droughts, regional projections might indicate larger changes than is the case for projections of global averages. Regional modeling efforts have been limited by constraints on computing resources, uncertainties and complexities associated with data assimilation and parameterization, and by limitations inherent to the large-scale simulations on which the downscaling is performed (IPCC, 2012). An additional challenge for regional projections is representing regional modes of variability, such as the El Niño Southern Oscillation (ENSO) and the North Atlantic Oscillation; these modes have a strong influence from local to global climate change. Climate force scenarios that project human influences on local and regional climate, such as regional aerosol loading and land use change, are needed because these factors may have a large influence on local and regional climate change (NRC, 2010). In Figure 14.1 is described an example of using a climate –chemistry transport model to estimate GHG, aerosols emissions and related effects after wildfire events. The key processes affecting the chemical concentrations represented in this model (CHIMERE) are: emissions, transport (advection and mixing), chemistry and deposition. External forcing is required to run a simulation as: meteorological fields, primary pollutant emissions and chemical boundary conditions. Using these input data the model calculates the atmospheric concentrations of tens of gas-phase and aerosol species over local to continental domains.

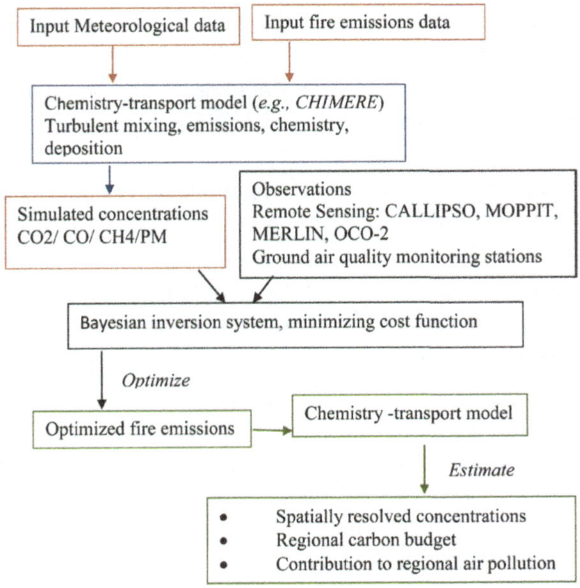

Figure 14.1. Regional climate –chemistry transport- model to estimating GHG, aerosols emissions and related effects after wildfire events.

Spatial downscaling approaches can be divided into dynamic, higher resolution modeling, referred as Regional Climate Models (RCMs) and empirical or dynamic statistical downscaling

Dynamic downscale global model simulations. A high resolution RCM, from ca. 50 km to ca.10 km, is nested within the driving GCM, in order to provide a better representation of local terrain, simulating climate features and physical processes over the region of interest whose boundary conditions are drawn either from global reanalyses or from global climate model output. The main advantage of RCMs is that they account for complex topographical features and land-cover inhomogeneity, in a physically based way. Table 14.1 summarizes the main features of dynamic downscaling, as well as the respective advantages and disadvantages

Table 14.1 Features of dynamic downscaling.

Dynamical Downscaling	Limited-area model supplied with large-scale (GCM or reanalysis) fields at its lateral boundaries. High-resolution "time slice" using a GCM. Global models with variable resolution (such as ARPEGE).
Advantages	**Disadvantages**
1. Transformation factor (largely) based on understood numerical methods.	1. Computationally expensive. Therefore difficult to apply to a large ensemble of hindcasts (as needed for bias-correction)
2. Full set of internally-consistent downscaled variables (multi-level & high time frequency).	2. Requires a large amount of driving GCM data.
3. Not (directly) dependent of availability of observations (e.g. applicable anywhere).	3. Systematic errors also exist in RCMs.
4. Encompasses non-stationary relationships between large and small scales, and potential changes in regional forcing.	4. Does not (directly) produce impact-relevant parameters.

Statistical downscaling (SD). Empirical relationships are derived by linking large-scale atmospheric variables (predictors) and local/regional climate variables (predictands). To establish these empirical relationships, regression models may be used, calculated by multiple linear regression, principal components analysis (PCA), and artificial neural networks (ANNs). Rather than using large scale predictor variables and statistical relationships between the weather types, or atmospheric circulation patterns, one can use the local weather types or patterns and calculate observed station data. These types or patterns may be based on existing derived weather classes, or they may be derived using techniques, such as ANN. Once the weather types have been identified, statistical models between the types and local station data are calibrated and then can be verified. Further, there are stochastic weather generators that may be conditioned on the large-scale state in order to derive site-specific weather. The first step in the use of a stochastic weather generator is to calibrate it using observed daily weather data. Parameter sets, which contain information about the statistical characteristics of the observed data, result from the calibration process and are subsequently used to simulate artificial weather data that have the same statistical characteristics as the observed data, but differ on a day-to-day basis. An increasing number of studies on extremes are on the SD approach, and on features, such as river flow and discharge, sediment, soil erosion and crop yields. SD capabilities are also examined through secondary variables like runoff, river discharge and stream flow. The main features of SD are presented in Table 14.2.

Table 14.2 Features of statistical downscaling.

Statistical Downscaling	Weather classification: Group the record into a defined set of states, such as synoptic types Regression methods: Produce relationships between local predictand and model output Weather generators: Relate statistical behavior of predictands to model variables
Advantages	**Disadvantages**
1. Computationally cheap	1. Assumes stationarity of large-small scale transformation factors
2. Can be applied to a large number of ensemble realizations	2. Transformation factors not always based on well understood physical mechanisms
3. Requires limited number of input GCM fields	3. Does not capture systematic changes in regional forcing
4. Can downscale GCM simulated variables directly into impacts-relevant parameters	4. Downscaled variables limited in number and not (always) internally consistent

14.2 CLIMATE EXTREMES

Long term observations of past decades suggest that characteristic recurrence frequencies, intensities and durations of certain extreme events have already increased (IPCC, 2013). The extreme event designates the initial and consequent physical phenomena, including some that may have causal factors issuing from human activity in addition to factors related to climate. Many extremes are the result of natural climate variability (e.g. El Niño episodes). However, for certain classes of regional, long-duration weather extremes, the climate model outputs suggest that the occurrence of such conditions has changed due to anthropogenic climate forces.

Temperature extremes. At a global scale, it is very likely that there have been trends toward warmer and more frequent warm days and nights, and a decrease in the number of cold days and nights (IPCC, 2007b). At the regional scale, changes in extreme temperatures display a rather linear scaling with the global mean temperature difference (Seneviratne *et al.* 2016). Since the middle of the 20th century, a tendency toward greater length or number of warm spells or increased heatwaves like in large parts of Europe, Asia and Australia has been identified. Human contribution to the frequency and intensity of daily temperature extremes (IPCC, 2012) and the occurrence of a few prominent heatwaves has been demonstrated (Fisher & Knutti, 2015).

Heavy Precipitation. Most regions have experienced an increase in the number of heavy precipitation events, whereas fewer regions have been subjected to a respective decrease of rainfall events. "In a warming world, wet regions become wetter and dry regions drier". The water holding capacity of air increases by about 7% for 1°C change in temperature (Trenberth, 2011), leading to increased water vapor in the atmosphere. Current research provides more evidence that human activities have a positive contribution to the intensity of precipitation extremes (Min *et al.* 2011).

Extreme winds. Changes in wind extremes may arise from changes in their associated phenomena or in the movement of large-scale circulation patterns. Some studies have identified a recently high correlation between hurricanes' intensity and sea surface temperature over the Atlantic and elsewhere (Elsner & Jagger, 2010). The intensity of pre-monsoon Arabian Sea tropical cyclones has been increased from 1979 to 2010, attributable to a simultaneous upward trend in anthropogenic black carbon and sulphate emissions (Evan *et al.* 2011). The Inter-Tropical Convergence Zone (ITCZ) moved southward, leading to weaker monsoons across Asia (Haug *et al.* 2001).

Drought is a complex environmental impact and its effects are not linear (Seneviratne *et al.* 2016). There is medium confidence that since the 1950s some regions of the world have experienced more intense and longer droughts. The primary cause of drought is the lack or deficit of precipitation. However, increased evapotranspiration induced by the interaction of increased radiation, wind speed, or vapor pressure deficit, can contribute to the deficit of soil moisture causing negative anomalies in stream-flow and/or groundwater levels. Land use changes have potential impacts on droughts (Arneth *et al.* 2014) and the anthropogenic forcing has contributed to the global trend towards increased drought in the second half of the 20th century.

Flood. During the 20th century the frequency of great floods has increased substantially. Changes in floods might be expected since they are influenced by changes in soil moisture status, precipitation, snowmelt, ice cover, which in turn have been detected among various regions. Climate change has the potential to change river flood characteristics, as so land cover and the river engineering (e.g. large dams).

Extreme sea level rise. Based on a quasi-global tide-gauge dataset, it has been detected an increasing trend in extreme sea levels, consistent with trends in mean sea level (after IPCC, 2012). This relationship has been also confirmed at regional scale (Haigh *et al.* 2014). Currently, there are insufficient evidence, that anthropogenic climate change was the main contributor of any observed changes in coastal erosion and inundation.

14.3 FUTURE TRENDS IN CLIMATE EXTREMES

Extreme climate variables are projected to experience significant changes over the 21st century, just as they have during the past century. The following projections are the main outcomes evaluated by an ensemble of climate models described also in IPCC (2012). Table 14.3 presents observations and future trends of climatic variables and climate extremes (IPCC, 2012).

Temperature extremes. Models project substantial increases in the frequency and magnitude of warm daily temperature extremes by the end of the 21st century at the global scale. However, regional changes in temperature extremes will often differ from the mean global temperature change. Based on the A1B and A2 scenarios, a 1-in-20 year hottest day is likely to become a 1-in-2 year event by 2100 in most regions. The 1-in-20 year extreme daily maximum temperature will likely increase by about 2°C to 5°C by the late 21st century, depending on the region and emissions scenario. Similarly, there is considerable confidence that the frequency of cold extremes will decrease and that the number of frost days will decline in the middle and high latitudes (NRC, 2010). The frequency and intensity of heatwaves is projected to continue to increase over most land areas.

Heavy Precipitation. Intensification of daily precipitation extremes is projected to continue for the rest of the century. This trend will not be apparent in all regions because of natural variability and possible influences of anthropogenic aerosols (IPCC, 2013). A robust projection, resulting also from the fact that warmer air can hold more moisture (IPCC, 2007a), is that the frequency and the fraction of rainfall falling in the form of heavy precipitation events will increase in many regions in the 21st century. With medium confidence it is expected that in some regions, increases in heavy precipitation will occur despite projected decreases in total precipitation in those regions. Donat *et al.*, (2016) observed that increases in total and extreme precipitation in dry regions are linearly related to the model-specific global temperature change, explaining the spread in precipitation intensification by the late 21st century.

Table 14.3 Observations and future trends of climate extremes (from IPCC, 2012).

(I)	Observed Changes (since 1950)	Projected Changes (up to 2100) with Respect to Late 20th Century
Temperature	1. Trends toward warmer and more frequent warm days and nights. 2. Increased heat waves like in large parts of the world.	Virtually certain decrease in frequency and magnitude of unusually cold days and nights and increase of unusually warm days and nights, globally. Very likely increase in duration, frequency, and/or intensity of warm spells or heat waves over most land areas.
Precipitation	Increase in extreme rainfall intensity in most parts of the world. Increase in extreme winter rainfall in Europe.	Likely increase in frequency of heavy precipitation events the high latitudes and tropical regions.
Winds	Low confidence in trends (insufficient evidence).	Low confidence in projections of extreme winds.
Monsoons	Low confidence in trends (insufficient evidence).	Low confidence in projected changes in monsoons.
Extra-tropical Cyclones	Likely poleward shift in extratropical cyclones. Low confidence in regional changes in intensity. Medium confidence in an anthropogenic influence on poleward shift.	Medium confidence in a reduction in the numbers of mid-latitude storms. Likely impacts on regional cyclone activity Regional increases over the Eastern North Atlantic, Western and central European area.
Tropical Cyclones	Increase in the number and proportion of hurricanes reaching categories 4 and 5.	High confidence in projected increased mean tropical cyclone maximum wind speed.
Modes of Variability and El Nino	Increase in ENSO variability and frequency (medium confidence). Likely trends in Southern Annular Mode (SAM).	Low confidence in projections of changes in NOA, SAM, Indian Ocean Dipole, and ENSO variability. Increase in the relative frequency of central equatorial Pacific events.

(Continued)

Table 14.3 Observations and future trends of climate extremes (from IPCC, 2012) (*Continued*).

(I)	Observed Changes (since 1950)	Projected Changes (up to 2100) with Respect to Late 20th Century
Droughts	More intense and longer droughts, in Europe and West Africa (medium confidence). Less frequent, intense, or shorter droughts in central America or western Australia.	Medium confidence in projected increase in duration and intensity of droughts in some regions of the world.
Floods	High confidence in trend toward earlier occurrence of spring peak river flows in snowmelt- and glacier-fed Rivers. Limited to medium evidence available to assess climate-driven observed changes in the magnitude and frequency of floods, regionally.	Low confidence in projections of changes in fluvial floods. Some models project large increase in flood frequency in Southeast Asia, Peninsular India, eastern Africa and the northern half of the Andes.

Extreme Winds. Extreme winds are often considered in the context of the extreme phenomena with which they are associated, such as tropical and extra-tropical cyclones. In this framework there is low confidence that any observed long-term (i.e. 40 years or more) increases in tropical cyclone activity are robust, after accounting for past changes in observing capabilities.

Tropical Cyclones. When projections of changes in extreme winds are generally characterized with low confidence, an increased mean tropical cyclone maximum wind speed can be projected with high confidence. Specifically, some studies predict that the globally averaged intensity of tropical cyclones will shift towards stronger storms, due to GHG warming, with intensity increases of 2–11% by 2100, while the globally averaged frequency of tropical cyclones, will be decreased by 6–34% (after IPCC, 2012). There is a potentially large risk for tropical cyclone storm surge in the Persian Gulf, and larger-than-expected threats in Cairns, Australia, and Florida (Lin & Emanuel, 2016).

Thunderstorms and Climate Change. Climate change or, more specifically, global warming might affects thunderstorm frequency, severity and distribution. A warmer planet would appear to be more conducive to the development of intense thunderstorms that comprise hail-producing convective systems. But thunderstorm updrafts require the presence of strong vertical temperature gradients; any warming that occurs at the surface likely also would occur aloft. Thus, the net change in instability and the net change in the potential for strong updraft development may be minimal. In addition, the increased moisture also would likely yield more widespread low-level clouds, and such cloudiness would negatively impact solar heating and resultant storm initiation and thunderstorm development. Increased moisture theoretically would be available for cloud and storm development in a warmer world. Increased cloud water content, in turn, generally enhances downdraft strength (through "water loading"). But it does not necessarily follow that storms with water-enhanced downdrafts would be more favorable building-blocks of thunderstorms. It seems that the maximum thunderstorm frequency likely would shift poleward with time. This is because the bands of fast upper-level winds (i.e., the jet stream), that arise from the equator-to-pole temperature gradient, would shift poleward in a warmer world. The areas most favored for thunderstorm development also would shift poleward. Because the most favorable wind shear environments (changes in wind speed and/or direction with height) are tied to the location of the jet stream, it seems reasonable to conclude that, overall, potential changes in shear would not appreciably impact thunderstorm incidence or severity, just their location. In general, a warmer atmosphere capable of carrying more water vapor will be more energetic. Therefore, global warming will more likely increase the risk of thunderstorms and hail (IPCC, 2012).

Extra-tropical Cyclones. An increased anthropogenic forcing will lead to a reduction in the number of mid-latitude cyclones averaged over each hemisphere, and to a poleward shift of the tropospheric storm tracks. Southern Hemisphere extra-tropical cyclone frequency and intensity is projected to be reduced in mid-latitudes and slightly increased in high latitudes. Recent studies show regional increases in the storm track activity over the Eastern North Atlantic/Western European area.

Dust Storms. There is low confidence in projected future changes in dust storms, since it is difficult to project future land use, although an increase could be expected where aridity increases. Probably, all dune fields, from northern South Africa to Angola and Zambia will be highly dynamic and the sand will be exposed and move as a result of 21st century climate warming.

Climate Modes. There is low confidence in projections of changes in monsoons, in NOA, Southern Annular Mode, Indian Ocean Dipole, ENSO variability and the frequency of El Nino episodes. However, most models project an increase in the relative frequency of central equatorial Pacific events. Also there is low confidence in projections of small-scale phenomena, such as tornadoes, since competing physical processes may affect future trends.

Drought. There is medium confidence that the duration and intensity of hydrological droughts will increase in the 21st century in some seasons and areas, due to reduced precipitation and/or increased evapotranspiration (ET), but other factors leading to a reduction in river flows or groundwater recharge are changes in agricultural land cover and upstream interventions. This projection applies to regions including southern Europe and the Mediterranean region, central Europe, central North America, Central America and Mexico, northeast Brazil, and southern Africa. In a future warmer climate the models simulate summer dryness in most parts of the northern subtropics and mid-latitudes with an associated increased likelihood of drought (IPCC, 2007a).

Floods. Projected precipitation and temperature changes imply possible changes in floods, although overall there is low confidence in projections of changes in fluvial floods. Hirabayashi *et al.*, (2013) examined the global flood risk for the end of this century based on the outputs of 11 climate models. The results showed a large increase in flood frequency in Southeast Asia, Peninsular India, eastern Africa and the northern half of the Andes. However, in certain areas flood frequency is projected to decrease. Runoff will increase in regions, where precipitation increases faster than evaporation, such as at high northern latitudes (Milly *et al.* 2005).

Extreme Sea Level. There is high confidence that locations currently experiencing adverse impacts, such as coastal erosion and inundation will continue to do so, due to increasing sea levels, all other contributing factors being equal.

14.4 IMPACTS OF CLIMATE EXTREMES

Changes in atmospheric conditions can be translated to impacts on physical, ecological, and human systems. Human dependence on ecosystems is clear since they provide food, fuel, and freshwater. The ecosystems can regulate climate through the global carbon cycle and the hydrologic cycle, thus, negative impacts on ecosystems from climate extremes affect also human well-being. Impacts on the natural physical environment (droughts, floods, extreme sea levels and coastal impacts), as well as other physical consequences, are often the result of compound events. It is critical to clarify that an extreme climatic event does not necessarily have extreme impacts. As was defined from IPCC (2012), an extreme-significant impact is the aggregation of major consequences to humans, ecosystems or physical environment, due to the occurrence of a single extreme or successive extremes or even non-extreme events simply due to the persistence of conditions, which can promote an extreme event.

14.4.1 Exposure and vulnerability

Currently, there is high confidence that climate change will affect disaster risk, not only through changes of some events, but also through indirect effects on vulnerability and exposure. Humans are differentially exposed and vulnerable on physical events based on inequalities resulted from diverse economic, political, cultural characteristics and environmental conditions. Population growth, skewed development processes, such as rapid and unplanned urbanization in hazardous areas, international financial pressures, failures of governance, and environmental degradation are all significant drivers of risk and are particularly related to vulnerability creation. Poverty and the scarcity of livelihood options for the poor will aggravate or affect vulnerability levels irrespective of the type of hazard (IPCC, 2012). Similarly, rapidly re-habilitating livelihoods or re-constructing infrastructure, often leads to recovering in ways that recreate or even increase existing vulnerabilities. The determents of vulnerability can be discriminated to social and environmental-physical determents.

Environmental-physical determents. Rapidly growing urban populations may affect the capacity of developing countries to cope with the effects of extreme events, mainly due to the inability of governments to provide the requisite urban infrastructure. When the 11% of people in low development countries and the 15% of people in high developed countries are expected to be exposed to hazards each year, the average numbers killed is 53% and 1%, respectively (after IPCC, 2012). Globally, the pressure for urban areas to expand onto coastal zones has resulted in increased exposure of populations to coastal flood risk, as will be discussed later. Similarly, potentially vulnerable natural systems are low-lying islands and drylands. This does not make the inhabitants of such locations vulnerable per se, because they may have capacities to resist the impacts of extreme events. In urban areas, the lack, or continuous loss, of green space may increase the exposure to extreme climate events through decreasing runoff amelioration and exacerbate the urban heat island phenomenon. Again, urbanization itself is not always a driver for increased vulnerability, but is the type of urbanization (e.g. expansion of informal settlements) that determines the social vulnerability.

The social dimension of vulnerability includes demography, education, health and well-being, culture, institutions, and governance aspects. Particular groups may have differential exposure or vulnerability to extreme events, including ethnic minorities, immigrants, residents of group living facilities; people with mobility and cognitive constraints; the homeless, poor households, renters and women-headed households (IPCC, 2012). There are also different effects within a given category. For instance, in Hurricane Katrina, citizens were differently exposed and vulnerable in accordance to their colour, gender (women are often at greater risk of dying in disasters), age, their class (upper or working class) (after IPCC, 2012). However, one of the most vulnerable social groups facing climate extreme events and climate change is the so-called 'environmental and climate refugees' (Myers, 2002), which consists of economically and environmentally displaced persons.

14.4.2 Major impacts of climate extremes on several sectors

The identification of spatial and temporal trends and differences in vulnerability in all sectors (environmental, social, production, infrastructure) and exposure is essential in order to design and implement effective adaptation and disaster risk management (DRM) strategies. It is possible to be exposed to an extreme event, but not be vulnerable, but to be vulnerable it is also necessary to be exposed (after IPCC, 2012). Below, are described the main impacts of climate extremes on various sectors.

Plant/Forest ecosystems. As it was summarized by Seddon *et al.*, (2016) currently areas with high sensitivity to climate variability and the related increases in extreme events, are the Arctic tundra, parts of the boreal forest belt, the tropical rainforests of South America, western Africa and southeast Asia/New Guinea, alpine regions worldwide, steppe and prairie regions of central Asia and North-South America, and eastern areas of Australia, Caatinga deciduous forest in northeast Brazil. Generally, tree species are expected to shift their ranges northward or upslope, with some current forest types, such as spruce-fir disappearing, oak-hickory expanding, and maple-beech contracting. Floods reduce the stability in diverse plant communities (Wright *et al.* 2015). Winter storms can impact forests in pre-alpine and alpine areas. Droughts associated with summer drying could result in regional vegetation die-offs, but also threatens tropical rainforests (e.g. Doughty *et al.* 2015). Under long period of reduced soil moisture availability, combined with high air temperature and low humidity, the tropical rainforests are expected to be subjected to a transition to a lower biomass forest state, due to elevated mortality of the largest trees (Rowland *et al.* 2015). On the basis of mid-range climate-warming scenarios for 2050, the 15–37% of species and taxa in the 20% of the terrestrial surface will be 'committed to extinction' (Thomas *et al.* 2004).

Wildfires. Earlier snowmelt, temperature changes, and drought leading to changes in phenology are important contributors to the increased frequency of forest fires (after IPCC, 2012), challenging the capacity of ecosystems to restore. For example, large and long-duration forest fires have increased fourfold over the past 30 years in the American West. Moreover, the length of the fire season has expanded by 2.5 months and the size of wildfires has increased several-fold.

Biodiversity. In the Northern Hemisphere, range shifts are almost wholly northward and up in elevation as species search for cooler temperatures (NRC, 2010). With medium confidence it is estimated that approximately 20 to 30% of plant and animal species assessed so far are likely to be at increasingly high risk of extinction as global average temperatures exceed a warming of 2°C to 3°C above preindustrial levels (IPCC, 2007b). Importantly, however, whole forest communities or ecosystems will not shift their ranges intact. Extreme events also contribute to population-level changes. As an example, an extreme flood event affected a desert rodent community by inducing a large mortality rate and resetting long-term population trends.

Marine ecosystems. The impacts of climate change on marine-based food systems could be severe given that a sizeable fraction of the world's fisheries are already overexploited (Worm *et al.* 2009). Over recent decades, there has been widespread poleward shifts in species distributions and under these geographical shifts fish could move beyond the range of national fisheries (NRC, 2010). Changes in magnitude and timing of costal upwelling have been noticed during the late 20th century, such as in northwest Africa, Peru, Portuguese coast or northern California current causing alterations (low nutrient levels and low primary productivity) in the coastal marine systems (Wang *et al.* 2015). Nevertheless, changes in ocean chemistry may be more important than changes in temperature for the performance and survival of many organisms (Wang *et al.* 2015). Increase of acidification may disrupt coral, whereas warm temperatures can result in coral bleaching in tropics. Coral reef ecosystems have been compromised by diverse activities including overfishing and eutrophication (NRC, 2010).

Agriculture. Climate change and extreme events can have negative impacts or cause marked damage to food system infrastructure, with the potential to destabilize food systems and threaten local to global food security, affecting the availability of different agricultural products, the access by changing prices of transportation costs, and utilization via potential spoilage and food-borne diseases generated by increased temperatures or flooding. According to FAO (2015) 21.8% of all damage and losses from natural hazards from 2003 to 2013 has been in the agricultural sector in developing countries. Globally over the period 1964–2007, droughts and extreme temperature reduced national cereal production by 9-10%, with more damage observed in developed countries (Lesk *et al.* 2016). Agricultural systems cannot adapt to high annual variations in rainfall. Instead, agricultural systems respond only to dry periods by reducing growth, supporting the view that they can be strongly affected by future extreme events (Rammig & Mahecha, 2015). Biodiversity in grass communities increases ecosystems' resistance for various climate events (wet or dry, moderate or extreme) and stabilizes ecosystem's productivity. Therefore, anthropogenic environmental changes via modifications in biodiversity affect ecosystem stability (Isbell *et al.* 2015).

Livestock. Livestock respond to climate change through heat and water stresses, and diseases. Production of milk, beef and pork, is projected to decline with warming temperatures and heat stress, especially those above 3°C. Heatwaves are expected to increase livestock losses, with the extreme heat exacerbated by rising minimum night-time temperatures, as well as increasing difficulties in providing sufficient water and food (NRC, 2010).

Water Availability. The global water cycle of the 20th century has intensified, and the rate of intensification for the 21st century is expected to accelerate severe impacts in human and natural systems (Pendergrass & Hartmann, 2014). Changes in precipitation and evaporation patterns, runoff, stream flow and incursions of sea water into coastal aquifers as sea levels rise, will all affect the rate of groundwater recharge, the absolute volume of groundwater available, and its quality (NRC, 2010). Direct drought impacts affect all forms of water supply. Indicators of water resources drought impact include lost production, the cost of alternative or replacement water sources, and altered human well-being. The resulting economic and environmental impacts highlight the vulnerability to droughts of both developing and developed countries. The most severe human consequences of droughts are often found in semiarid regions, where water availability is already low.

Damage from floods. As vulnerability in the water sector increases when demands are placed on the resource, exposure increases when more property, assets, and lives encounter flooding. The salinization of wells after the 2004 Indian Ocean tsunami had a highly negative consequence for those communities that had no alternative access to freshwater resources. At the global scale, currently 20 to 300 million people per year are affected by floods and they projected that beyond 2060, at least 300 million people could be affected by substantial flooding even in years with relatively low flooding (IPCC, 2012). Despite the direct impacts of extreme flood events, the indirect effects would include changes in the wider economy. A global economic exposure, resulting from both river and coastal flooding, was simulated to increase by circa 200% between 2010 and 2050 (Jongman *et al.* 2012). Using global historical spatial data and projections the largest absolute exposure changes are simulated in North America and Asia. However, without considering any changes in economic development among European countries, large increases are projected in both number of people flooded and economic damage in central Europe as also in U.K. (after IPCC, 2012).

Damage from storms. Storm surges are responsible for much of the damage and loss of life associated with land-falling hurricanes (Lin & Emanuel, 2016). Recent flood disasters in the USA (2005, 2008, 2012), the Philippines (2012, 2013), and Britain (2014) illustrate how vulnerable coastal cities are to storm surge flooding (after Aerts *et al.* 2014). $81 billion was the total damage cost from hurricane Katrina (NOAA) and the loss of insurance reached approximately – $41 billion (Insurance Information Institute). Inundation (surge plus tide and wave action) and shoreline change may be some of the severe impacts on small, often low elevated, tropical island states. Sea level rise and more intense storm surges are of concern for the 635 million people worldwide, who live less than 10 m above sea level, many of them in coastal cities.

Urban infrastructure and function. Heatwaves increase periods of peak energy demands and, in conjunction with other climate changes, are expected to worsen urban air quality. Water limitations in some regions, and increased demand for water for other uses, could result in less water for use in cooling at thermoelectric plants. During 2003, droughts in Europe created stress in energy systems, and affecting the power plant cooling efficiency and shutting down six power plants. The economic losses for the agriculture sector in the European Union were estimated at €13 billion (IPCC, 2012). Cities and adjacent to fire-prone habitats are projected to confront increasing threats of fire, and desert cities. Urban infrastructure and related transportation functions are threatened by extreme events, such as: heavy precipitation causing flooding of roadways, rail lines, runways, and subterranean, or road washout, stimulating scouring of pipeline roadbeds; storms affecting bridge stability and decks, disrupting utilities, such as natural gas, electricity, and telephone lines; heatwaves causing rail-track deformities, thermal expansion on bridge joints and paved surfaces, and migration of liquid asphalt; rising sea levels, combined with storm surges can cause erosion and inundation of roads, rail lines and airport runways and change port facilities to accommodate higher tides and storm surges (NRC, 2010).

Human health. Changes in water quality and quantity, climate impacts on agricultural production and drought, all may have result in additional human health and nutrition challenges (NRC, 2010). In 2005, hurricane Katrina made landfall along the Gulf Coast USA causing at least 1,836 deaths from sequential floods. The tropical storms in Yemen in 2008 caused damages and losses of about $1.6 billion. The exposure of Sub-Saharan African countries to drought and floods accounts for 80% of loss of life (African Union *et al.* 2008). Extreme events may also lead to increased incidence of respiratory symptoms, particularly in developing countries, to increased exposure to disease vectors, or even mental health impacts (NRC, 2010). Also, there is medium confidence that the observed warming has increased heat-related human mortality and decreased cold-related human mortality in some regions.

Migratory movements. In Syria the devastating drought 2007–2010 caused a mass migration of farming families, ca. 1.5 million people, to suburbs. In 2010, Indus valley floods in Pakistan resulted in over 4 million displacements, and a total of 20 million people affected directly or indirectly (Lau & Kim, 2011). In 2007, floods in Mozambique displaced 163,000 people, and in 2008, 102,000 more; in Niger, large internal movements of people are due to pervasive changes related to drought and desertification trends (Afifi, 2011). The United Nations Office for the Coordination of Humanitarian Affairs and the Internal Displacement Monitoring Centre estimated that 1.6 billion people have been affected by droughts and storms over the last 3 decades.

14.5 MANAGEMENT OF CHANGING RISKS OF CLIMATE EXTREMES

Basic methodologies to serve both disaster risk reduction and climate change adaptation are: early warning systems, risk transfer in developing countries, effective legislation, and public awareness.

14.5.1 Effective risk management

The main components of an effective risk management, which could integrate disaster risk management (DRM), climate change adaptation and lead to a sustainable development are: (1) Risk identification. (2) Assessment and evaluation of the responses to risk by a range of stakeholders. (3) Iterative decision making and deliberate learning. (4) A portfolio of actions customized to specific local circumstances, to reduce and transfer risk and to respond to events and disasters. (5) Ensure durability: Governance and institutions must design mechanisms that can ensure the long-term durability of policies, provide stability for investors and society. Financing mechanisms of DRM and adaptation could be coordinated through development agencies and national organizations. Without competent regulatory bodies that assure conditions for both insurers and clients, the market cannot provide sustainable insurance contracts (Linnerooth-Bayer and Hochrainer-Stigler, 2015). (6) Effective communication: Climate researchers and educators must recognize legitimate disagreements about the risks of climate change, and should support informed dialogue about value-laden choices (Stern, 2016). Different cultural groups put their trust into different organizations. This cultural diversity may prove a pressing concern. Recent studies, as Wong-Parodi *et al.* (2016), argue that agent-based modelling should be embedded in a decision-science approach. Environmental education programs among children and adults may have benefits for public understanding of risk from extreme events.

14.5.2 Risk sharing and transfer

The basic probabilistic risk analysis, used by risk governance regards risk as the product of the probability of an event(s) multiplied by its consequence. While risk reduction aims to reduce exposure and vulnerability, risk transfer options, such as insurance, micro-insurance, reinsurance, and national to global risk pools involving risk sharing, as well as innovative insurance mechanisms (after IPCC, 2012), aim to compensate losses by those, who are directly affected by an extreme event. However, there are limitations of the effectiveness of risk financing instruments, especially in high vulnerable countries. Donor institutions provide only a small percentage of post-disaster capital and currently there is reluctance on the part of insurers to cover catastrophic events from climate-weather extremes. As Linnerooth-Bayer and Hochrainer-Stigler (2015) reviewed, the traditional instruments (i.e. pre-disaster savings, post-disaster credit and informal reciprocal financial arrangements) can be ineffective for catastrophic risks that affect savings and micro-finance institutions, households and regions. Traditional insurance can be directly linked to reducing risks and losses if premiums reward investments in risk, but can be costly especially at high level risks. Index-based micro-insurance targets the reduction of transaction costs and it can service low income markets, since claims are independent of losses. Further, if forecasts are combined with index based insurance payouts, the clients can have the flexibility to take preventive measures to reduce damages. But these programs need the support of the governance, when poor societies are considered. Non-traditional risk financing instruments, such as micro-risk transfer programs, are too expensive and highly risk for low-income farmers or households. Instruments that transfer sovereign risk to the capital markets can enhance the flexibility of government and donor financing, but they can be very costly for poor countries.

14.5.3 'No or low regrets measures'

The sequences or steps needed to develop adaptation and DRM plans at the national level could be divided in vulnerability-thresholds-first or scenarios-impacts-first approach (after IPCC, 2012) The former approach is recommended for detecting areas that need immediate intervention when planning horizons are short, resources are very limited, or uncertainties about future climate impacts are very large (Hallegatte, 2009). 'No or low regrets measures' are suggested for implementation when such constraints prevail. Examples of low regrets measures and case studies where they have, or could be, applied are listed below.

Early warning systems can reduce loss of lives by providing adequate information on risks regional and local before extreme events occur (e.g. wildfires, heat waves, cyclones, floods). For example, improved early warning systems, including evacuation plans and infrastructures, would help to cope with increasing losses from hurricanes in the USA. In such cases local leaders are asked to notify community members of dangerous extreme weather event. Also, warning systems integrating seasonal forecasts with drought projections and improved communication involving extension services to compensate droughts, would enhance the food security in Syria or West Africa.

Risk communication. Effective risk communication between decision-makers and local citizens, particularly vulnerable groups is vital to attain exposure or vulnerability. Public campaigns for risk awareness providing information and behavioural advice, on what to do during extreme events could attain the inundation impacts on people living in tropical small island developing states; or the European heatwave-impacts, especially to elderly people. Use of social care networks can be used to reach vulnerable groups. Public health measures to reduce the exposure to heatwaves can include identification of vulnerable populations, clinical treatment guidelines for heat-related illness and surveillance activities. These approaches were employed, among others, to attain heatwave effects in

France. Communication and education strategies are most effective when they are community-based and promote adaptation measures as 'shared responsibility'. This can be achieved through educational material and training to enhance public awareness.

Water use efficiency. Upgrading water supply sanitation, drainage and irrigation systems, at local or regional scale is an effective measure to address floods or droughts in the context of food security. Negative effects of inundation events related to extreme sea levels could be eliminated by using hard technologies to limit saltwater contamination of groundwater or maintaining drainage systems. Also, city-wide drainage and sewerage improvements could be applied to cope with the risk of floods in informal settlements. Design to protect reservoirs providing water supply in a city could be useful against wildfire events.

Sustainable land use planning and management. Development of heat stress tolerance agricultural practices by altering the timing or location of cropping activities, or changing cropping patterns or diversifying production systems into higher value, would address some of the negative effects of droughts to food security. In low-lying coastal zones saltwater intrusion and coastal flooding are already leading to the adoption of saltwater tolerant crops or a shift from agriculture to aquaculture (after IPCC, 2012). In urban regions, heat-related illness could be reduced by land use modifications, which increase albedo and grass cover area.

Ecosystem conservation measures. Conservation agriculture, including actions of no-tillage, drilling, seeding, crop rotation, and livelihood diversification, can mitigate drought's adverse effects like mitigation of farmers from rural areas. Mangrove restoration can address inundation related to extreme sea levels in low-lying islands.

Building codes and standards. A relative cost-effective measure for risk disaster strategies and mitigation policies is the development and improvement of building codes and standards reflecting the learning from the outputs of current disaster risk. Adoption or enforcing the building codes could compensate the impacts of tropical cyclones in the USA. Enforcing building design and regulation could eliminate the effects of flash floods in informal settlements (e.g. in Nairobi).

Micro-insurance and weather indexed insurance. Regional risk pooling can be serviceable to compensate the effects of inundation related to extreme sea levels, hurricanes or droughts threatening food security. Providing loans and risk loss insurance to people affected by drought, like farmers in Syria, can increase resilience.

14.5.4 Win to win adaptation actions

Win to win options may provide opportunities for GHG reduction, adaptation, risk reduction, and development benefits. These actions include: sustainable land and water use planning, energy efficiency, adaptation actions towards ecosystem improvement, carbon sequestration, and food self-sufficiency. Multi-hazard risk management approaches are supposed to be implemented to both rural and urban environments, since they share interdependence vulnerability. Nevertheless, both systems are interlinked at local and global scale through the flow of investments, remittances and material transfers.

Megacities and climate adaptation. Currently, most efforts for adaptation management of the cities have been focused in structural or engineering measures, such as improving water treatment infrastructure, water supply, drainage, or protecting existing development and infrastructure against sea level rise. However, a reduction in industrial and transportation GHG emissions through interventions on infrastructure, energy consumption and travel behaviour can provide an option for multiple co-benefits, like reduction of heat related illnesses and improvement of local air quality. Actions that could address these challenges are: development and implementation of energy systems that emit few, or at least fewer, GHGs or aerosols; direct capture of GHGs during or after fossil fuel combustion; redesigning "smart" or "green" cities that foster less use of automobiles (although with varying impacts on water and energy consumption) (NRC, 2010). Moreover, reductions in energy demand can be achieved through changes in habits and perspectives. Numerous scientific and engineering sectors can contribute to this effort, such as pollution control, chemical engineering, meteorology and biological sciences, as well as behavioural science to enhance the capacity of social agent actors to evaluate these urban transformations through changes of current behaviours and perspectives of stakeholders in the climate context.

Megacities and risk management. Evaluation of trade-offs by risk management sector can help to avoid unintended consequences of such transformations to more vulnerable communities (e.g. disproportional carbon) and markets (e.g. private firms should consider climate change risk disclosure and revise risk insurance programs). Recent flood management strategies combine probabilistic risk assessment of storm surge with vulnerability determination of exposed assets at a census level, accounting for sources of uncertainty and the timing of investments in storm-surge flood-risk protection (Aerts *et al.* 2014). Given this risk, it is important to ensure the risk of catastrophic structural failures due to under-design or over-design, is eliminated (Haigh *et al.* 2014). If during re-construction after a disaster, building standards and land use planning are applied and the related legislation is improved, resilience can be achieved in the short terms and can create 'green' communities in the long term (IPCC, 2012).

Rural areas and climate adaptation. In rural areas the mitigation of GHGs can be achieved using other types of energy production, like hydroelectric power, solar and wind farms, biofuel crops, which can impact both rural live hood and global markets (de Fraiture *et al.* 2008). Further, development of new fertilizer application technologies can also reduce N_2O fluxes. Alternatively, minimizing soil tillage yields may create multiple benefits, such as: maintenance of soil structure and moisture; increase of soil carbon storage; reduction of energy costs and mechanization (NRC, 2010). Provision of better control of natural resources can lead to sustainability of rural livelihoods and ecosystem services.

Rural areas and risk management. Ecosystems management, agro-ecology, agro-forestry and restoration activities, such us watershed rehabilitation, and forest landscape restoration offer co-benefits for disaster risk reduction from drought, floods and coastal storms, while they contribute to adaptation to climate changes by creating economic opportunities, increased carbon sequestration, energy savings, as well as energy and food self-sufficiency (after IPCC, 2012). For example, restoration of wetlands in the Gulf of Mexico has proved to reduce damage from hurricanes by damping wave action and diminishing wind penetration (Day *et al.* 2007). Land-use change is both a cause and consequence of many biophysical and socioeconomic changes (Vogel *et al.* 2012). While a measure like converting lands to agriculture or modifying agricultural practices may provide co-benefits, it may also reduce the adaptive capacity of other systems. For instance, dike systems can reduce flood exposure by offering immediate protection, but also encourage settlement patterns that may increase risk in the long term. All these trade-offs should be evaluated. Institutions can support the ability of farmers to adapt either by protecting their property rights or by informing them for new technologies. Therefore, designing agricultural strategies will require analyses that, among others, incorporate the response of farmers and markets not only to production and prices, but also to policies and institutions. After all, transformation cannot be approached without understanding related ethical and governance dimensions.

14.6 SUMMARY

Undoubtedly, climate change, and extreme climate or weather events can have adverse effects in human well-being, currently and in the near future. But the severity of impacts is not shaped by the nature of a hazard per se. It is the level of exposure and vulnerability of a system to a climate event that determines whether an impact is extreme or not. Vulnerability is influenced by a wide range of factors, including anthropogenic, climate change and socioeconomic development. Scientific progress, coordinated with the extended literature of climate adaptation and DRM can provide valuable solutions to some of the challenges facing societies and natural systems. National development plans and adaptation strategies must be primarily focused on vulnerable areas and groups and institutionally less-diversified countries. However, the implementation and sustainability of the resulting transformations is a challenge by itself. The projects which aim to reduce disaster risk and enhance resilience (e.g. multi-hazardous management) should be multidisciplinary. Natural sciences should be integrated with behavioural science, sociology, economics, public policy management, and communication. The role people play in the determination of extreme impacts must be re-conceptualized in order to generate sustainable behaviour. Through collaboration, these sciences must identify and employ the mechanisms by which individuals and societies become agents of change rather than abettors of stasis.

REFERENCES

Aerts J. C. J. H., Botzen W. J. W., Emanuel K., Lin N., de Moel H. and Michel-Kerjan E. O. (2014). Evaluating flood resilience strategies for coastal megacities. *Science*, **344**(6183), 473–475.

Afifi T. (2011). Economic or environmental migration? The push factors in Niger. *International Migration*, **49**, e95–e124.

African Union (2008), UNISDR (United Nations International Strategy for Disaster Reduction), and World Bank. 2008. Status of Disaster Risk Reduction in the Sub-Saharan Africa Region, by Rakhi Bhavnani, Seth Vordzorgbe, Martin Owor, and Franck Bousquet. Washington, DC: World Bank, p. 85.

Arneth A., Brown C. and Rounsevell M. D. A. (2014). Global models of human decision-making for land-based mitigation and adaptation assessment. *Nature Climate Change*, **4**(7), 550–557.

Ault T. R., Cole J. E., Overpeck J. T., Pederson G. T. and Meko D. M. (2014). Assessing the risk of persistent drought using climate model simulations and paleoclimate data. *Journal of Climate*, **27**(20), 7529–7549.

Day J. W. *et al.* (2007). Restoration of the Mississippi Delta: lessons from Hurricanes Katrina and Rita. *Science*, **315**(5819), 1679–1684.

de Fraiture C., Giordano M., Liao Y. (2008). Biofuels and implications for agricultural water use: blue impacts of green energy. *Water Policy*, **10**, 67–81.

DeConto R. M. and Pollard D. (2016). Contribution of Antarctica to past and future sea-level rise. *Nature*, **531**(7596), 591–597.

Donat M. G., Lowry A. L., Alexander L. V., Ogorman P. A. and Maher N. (2016). More extreme precipitation in the world's dry and wet regions. *Nature Climate Change*, **6**(5), 508–513.

Doughty C. E. *et al.* (2015). Drought impact on forest carbon dynamics and fluxes in Amazonia. *Nature*, **519**(7541), 78–82.

Elsner J. B. and Jagger T. H. (2010). On the increasing intensity of the strongest Atlantic Hurricanes. In: Hurricanes and Climate Change: Volume 2, B. J. Elsner, E. R. Hodges, C. J. Malmstadt and N. K. Scheitlin (eds), Springer, Dordrecht, Netherlands, pp. 175–190.

Evan A. T., Kossin J. P., Eddy Chung C. and V. Ramanathan (2011). Arabian Sea tropical cyclones intensified by emissions of black carbon and other aerosols. *Nature*, **479**(7371), 94–97.

FAO (2015). The Impact of Natural Hazards and Disasters on Agriculture and Food and Nutrition Security – A Call for Action to Build Resilient Livelihoods. Food and Agriculture Organization of the United Nations, Rome, Italy, p. 16.

Fischer E. M. and Knutti R. (2015). Anthropogenic contribution to global occurrence of heavy-precipitation and high-temperature extremes. *Nature Climate Change*, **5**(6), 560–564.

Flato G. (2011). Earth system models: an overview. *Wiley Interdisciplinary Reviews, Climate Change*, **2**(6), 783–800.

Garcia R. A., Cabeza M., Rahbek C. and Araújo M. B. (2014). Multiple dimensions of climate change and their implications for biodiversity. *Science*, **344**(6183).

Haigh I. D., MacPherson L. R., Mason M. S., Wijeratne E. M. S., Pattiaratchi C. B., Crompton R. P. and George S. (2014). Estimating present day extreme water level exceedance probabilities around the coastline of Australia: tropical cyclone-induced storm surges. *Climate Dynamics*, **42**(1), 139–157.

Haug G. H., Hughen K. A., Sigman D. M., Peterson L. C. and Röhl U. (2001). Southward migration of the Intertropical Convergence Zone through the Holocene. *Science*, **293**(5533), 1304–1308.

Hirabayashi Y., Mahendran R., Koirala S., Konoshima L., Yamazaki D., Watanabe S., Kim H. and Kanae S. (2013). Global flood risk under climate change. *Nature Climate Change*, **3**(9), 816–821.

Huber M. and Knutti R. (2014). Natural variability, radiative forcing and climate response in the recent hiatus reconciled. *Nature Geoscience*, **7**(9), 651–656.

IPCC (2000). N. Nakicenovic and R. Swart (eds), Cambridge University Press, UK, p. 570, Available from Cambridge University Press, The Edinburgh Building Shaftesbury Road, Cambridge CB2 2RU ENGLAND, Summary for Policymakers, IPCC, Geneva, Switzerland. 20pp.

IPCC (2007a). Climate Change 2007: The Physical Science Basis. In: Contribution of Working Group I to the Fourth Assessment Report of the Intergovernmental Panel on Climate Change, S. Solomon, D. Qin, M. Manning, Z. Chen, M. Marquis, K. B. Averyt, M. Tignor and H. L. Miller (eds), Cambridge University Press, Cambridge, UK, and New York, NY, USA, p. 996.

IPCC (2007b). Climate change 2007. Impacts, adaptation and vulnerability. In: Contribution of Working Group II to the Fourth Assessment Report of the Intergovernmental Panel on Climate Change, M. L. Parry, O. F. Canziani, J. P. Palutikof, P. J. Van Der Linde and C. E. Hanson (eds), Cambridge University Press, Cambridge, UK, p. 979.

IPCC (2012). Managing the risks of extreme events and disasters to advance climate change adaptation. In: A Special Report of Working Groups I and II of the Intergovernmental Panel on Climate Change, C. B. Field, V. Barros, T. F. Stocker, D. Qin, D. J. Dokken, K. L. Ebi, M. D. Mastrandrea, K. J. Mach, G.-K. Plattner, S. K. Allen, M. Tignor and P. M. Midgley (eds), Cambridge University Press, Cambridge, UK, and New York, NY, USA, p. 582.

IPCC (2013). Climate Change 2013: The Physical Science Basis. In: Contribution of Working Group I to the Fifth Assessment Report of the Intergovernmental Panel on Climate Change, T. F. Stocker, D. Qin, G.-K. Plattner, M. Tignor, S. K. Allen, J. Boschung, A. Nauels, Y. Xia, V. Bex and P. M. Midgley (eds), Cambridge University Press, Cambridge, United Kingdom and New York, NY, USA, p. 1535.

Isbell F. *et al.* (2015). Biodiversity increases the resistance of ecosystem productivity to climate extremes. *Nature*, **526**(7574), 574–577.

Jongman B., Ward P. J. and Aerts J. C. J. H. (2012). Global exposure to river and coastal flooding: long term trends and changes. *Global Environmental Change*, **22**(4), 823–835.

Lau W. K. M. and Kim K.-M. (2011). The 2010 Pakistan flood and Russian heat wave: teleconnection of hydrometeorologic extremes. *Journal of Hydrometeorology*.

Lesk C., Rowhani P. and Ramankutty N. (2016). Influence of extreme weather disasters on global crop production. *Nature*, **529**(7584), 84–87.

Lin N. and Emanuel K. (2016). Grey swan tropical cyclones. *Nature Climate Change*, **6**(1), 106–111.

Linnerooth-Bayer J. and Hochrainer-Stigler S. (2015). Financial instruments for disaster risk management and climate change adaptation. *Climatic Change*, **133**(1), 85–100.

Ljungqvist F. C., Krusic P. J., Sundqvist H. S., Zorita E., Brattström G. and Frank D. (2016). Northern Hemisphere hydroclimate variability over the past twelve centuries. *Nature*, **532**(7597), 94–98.

Marotzke J. and Forster P. M. (2015). Forcing, feedback and internal variability in global temperature trends. *Nature*, **517**(7536), 565–570.

Milly P. C. D., Dunne K. A. and Vecchia A. V. (2005). Global pattern of trends in streamflow and water availability in a changing climate. *Nature*, **438**(7066), 347–350.

Min S.-K., Zhang X., Zwiers F. W. and Hegerl G. C. (2011). Human contribution to more-intense precipitation extremes. *Nature*, **470**(7334), 378–381.

Myers N. (2002). Environmental refugees: a growing phenomenon of the 21st century. *Philosophical Transactions of the Royal Society B: Biological Sciences*, **357**(1420), 609–613.

NRC (2010). Advancing the Science of Climate Change America's Climate Choices: Panel on Advancing the Science of Climate Change, Board on Atmospheric Sciences and Climate Division on Earth and Life Studies, The National Academies Press, Washington, DC, p. 528.

Pendergrass A. G. and Hartmann D. L. (2014). Changes in the distribution of rain frequency and intensity in response to global warming. *Journal of Climate*, **27**(22), 8372–8383.

Rammig A. and Mahecha M. D. (2015). Ecology: Ecosystem responses to climate extremes. *Nature*, **527**(7578), 315–316.

Rowland L. *et al.* (2015). Death from drought in tropical forests is triggered by hydraulics not carbon starvation. *Nature*, **528**(7580), 119–122.

Schmidt G. A., Shindell D. T. and Tsigaridis K. (2014). Reconciling warming trends. *Nature Geoscience*, **7**(3), 158–160.

Seddon A. W. R., Macias-Fauria M., Long P. R., Benz D. and Willis K. J. (2016). Sensitivity of global terrestrial ecosystems to climate variability. *Nature*, **531**(7593), 229–232.

Seneviratne S. I., Donat M. G., Pitman A. J., Knutti R. and Wilby R. L. (2016). Allowable CO2 emissions based on regional and impact-related climate targets. *Nature*, **529**(7587), 477–483.

Stern, P. C. (2016). Sociology: impacts on climate change views. *Nature Climate Change*, **6**(4), 341–342.

Thomas C. D. *et al.* (2004). Extinction risk from climate change. *Nature*, **427**(6970), 145–148.

Trenberth K. E. (2011). Changes in precipitation with climate change. *Climate Research*, **47**(1–2), 123–138.

Vogel A., Scherer-Lorenzen M. and Weigelt A. (2012). Grassland resistance and resilience after drought depends on management intensity and species richness. *PLoS ONE*, **7**(5), e36992.

Wang D., Gouhier T. C., Menge B. A. and Ganguly A. R. (2015). Intensification and spatial homogenization of coastal upwelling under climate change. *Nature*, **518**(7539), 390–394.

Wong-Parodi G., Krishnamurti T., Davis A., Schwartz D. and Fischhoff B. (2016). A decision science approach for integrating social science in climate and energy solutions. *Nature Climate Change*, **6**(6), 563–569.

Chapter 15

Hazards information management and services

Pavol Nejedlik and Nicolas R. Dalezios

15.1 HAZARDS INFORMATION

Each hazard involves a certain level of risk for the societal system, which can be hardly completely eliminated. There are two features influencing the character of disaster risk management during the previous century, namely the population growth, which steadily requires more space, and second the rapidly growing aggregations caused by migration from rural to urban areas. The negative point of these facts is that the major portion of the population increase takes place in the third world, where the migration of poor population to urban agglomeration often ends up in marginal lands in low laying plains or on steep slopes with poor infrastructure often subjected to the permanent threat of floods, landslides and other hazards. These features define the vulnerability of the system. Regarding the time scale, there is a distinction between slow onset and rapid onset disasters. Large scale fatal disasters are connected to the slow onset events (mostly drought related famines) rather than to rapid onset threats (www.cred.be/emdat). This induces different requirements for hazard information.

Each effective strategy to manage a disaster risk must begin with the identification of the hazard and with the identification of the vulnerability of the respective object. Hazard mapping by utilization of different techniques are among the ways in which risk information may be organized for the benefit of potential users. Hazard analysis has to: identify the hazard; provide warning of the developing threat; and work out the community, i.e. environmental and social, vulnerability. It is not possible to prevent the threat of coming natural hazards. Nevertheless, the impacts of the event can either be completely avoided or at least minimized by alerting the society to the threats and by raising awareness of preparation need including the possible evacuation.

Based on the factors causing environmental hazards, natural hazards are considered and handled in this chapter, e.g. geophysical phenomena that in their extreme forms threaten life, property or cause an environmental degradation. Natural hazards are grouped into geophysical hazards, hydrometeorological hazards and biophysical hazards, which tend to be more common and cannot be prevented (Rivera & Kapucu, 2015). Following the definition of a hazard as a situation that poses a level of threat to life, health, property, or environment, the hazard information has to be considered in both potential and already existing hits caused by the particular phenomenon. Different hazards are perceived differently. For example, people respond to the slow onset and less visible effects of drought or permafrost thaw much differently than to the rapid and destructive impact of an avalanche or tornado.

Most hazards are dormant or potential phenomena with only a theoretical risk of harm. However, once a hazard becomes "active", it can create an emergency. A hazardous situation that has occurred is called an incident. Hazard and probability interact to create a risk (Zalatynskyi, 2013). An environmental hazard is recognized if a certain phenomenon is able to bring the potential to threaten and adversely affect the surrounding both natural and artificial environment. Different hazards bring different types of danger. To identify the potential danger coming from a natural hazard the identification of the hazard itself has to be conducted. This involves the identification of the possible geographical occurrence, its extent, intensity and further features of the hazard. The identification of environmental hazards is usually tightly connected with the geographical position and general characteristics of the particular locality. The occurrence and the temporal development of the particular hazard seems to be rather unique. The review of climate science for six atmospheric hazardous phenomena shows that the ability to detect and attribute the trends varies, depending on the phenomenon (Kunkel *et al.* 2013). The center of the hazard affect can also change in time. For example, drought analysis in Europe in the recent six decades has shown that the center of drought hotspots, which was recognized in Scandinavia and Russia in 1950s and 1960s, has moved rather to Mediterranean area and Carpathian region in 1990s and 2000s. The north-eastern regions experience decreasing drought trend, while an opposite trend is recognized in south-western regions of Europe (Spinoni *et al.* 2016).

15.1.1 Characteristics of information

The overall risk of natural hazards flows from the relationship between the hazards, vulnerable systems, and existing natural capacity of the particular system to sustain the impact. A number of factors characterizing the amount, extent and event duration of the potential environmental hazard determines its general feature as magnitude or severity and define its possible impact. Each environmental hazard, which can potentially occur in its acute catastrophic or chronic form can be categorized in five categories, namely chemical, physical, mechanical, biological and psychosocial.

The natural environment affected by the hazards includes all living and non-living organisms, which occur naturally on the Earth. Complete ecological units, which include all living and non-living natural phenomena, as well as the universal natural physical phenomena that occur in many ecological units and their features, such as air, water and climate, are affected by the above-mentioned classes of hazards. The vulnerability of each system towards environmental hazards depends on its sensitivity, exposure and resilience towards a concrete hazard and on which part or which activity can be potentially affected. The potential risk of different systems and activities and their parts like natural resources, population, ecosystem goods, land use and development, economy, cultural assets and infrastructure and facilities depends on their ability and their resources to mitigate and prepare to potential and to respond, adapt and recover from already existing hit of the hazardous phenomenon.

Each environmental hazard has its internal characteristics, which are significant for the understanding of the respective hazard and for the possible reaction towards the hit of the hazard. Among the most important are: past recurrence intervals; seasonality; regular patterns; future probability; speed of onset; magnitude; duration; spatial extent; relation to other natural phenomena; and relation to the general and/or specific geographical feature.

15.1.2 Types and classification of information

The information needed for the possible reaction to the environmental hazard depends on the characteristics of the respective relation in between the hit of possible hazard and the object or activity being hit. Many of the environmental hazards are closely interrelated. That is why the information about a possible hazard could come from indirect sources. There is a relatively high interaction and conditionality in between seemingly different hazards. For example, tsunamis are caused by submarine earthquakes, and coastal flooding caused by a hurricane can bring soil erosion. Final goal of getting the detail information about the hazards is to reach the complex integrated description of the hazards, which could lead to both the vulnerability assessment of the particular region or sector towards the natural environmental hazards and to setting up some hazard control and protection system against the concrete hazard. Gill and Malamud (2014) made a broad overview, characterization, and visualization of the interaction relationships between 21 natural hazards, drawn from six hazard groups (geophysical, hydrological, shallow Earth, atmospheric, biophysical, and space hazards). They identified 90 interactions given information about the primary hazards, their relations and further consequences. This implies the need for rather complex information about the single hazard with possible multiplied consequences. Some studies have also brought interesting information that some natural hazards might be intertemporarily correlated (Graves & Bresnock, 1985). They surprisingly found by studying the statistical occurrence of hails and tornadoes, that these hazards appear to be more likely in the year t if an event was experienced in the year t-1. Though the degree of positive serial correlation appears to vary a great deal regionally, this also brings an information about a single hazard behavior.

Different criteria are applied to the information, which characterizes environmental hazards and provides the opportunity to handle the actions. Any environmental hazard is characterized by its physical performance and the time of occurrence. Regarding the temporal information about hazards, two temporal characteristics are necessary, namely time frequency and time scale. The information needed ranges in the scale from minutes to years depending on the character of the hazard parameter. Despite of the very concrete nature of a particular hazard the information required can be rather complicated, such as in hail, as none of the hazardous threat acts in the nature or in any environment as a single isolated phenomenon. Although the hazards originate from different systems with different sources many of them are interconnected and some of them practically originate from another hazard system as its product. For example, cyclonic storm can produce hail though this is not an inevitable part of the process of the development and decay of the storm cloud. On the other hand, there are general hazards like climate change, which do not directly act as hazards, but they can include a great number of hazardous phenomena in changing their extent, magnitude and frequency. Some of the most frequent natural hazards with a brief description of the information needed are listed in Table 15.1.

Further to a single hazard information, integrated hazard information of potential hazard impacts are often used. Such information is attached either to a defined territory or to a sector or human activity and combines a threat of more hazards. Widespread methodology uses the environmental vulnerability index (EVI), (Barnett *et al.* 2008), which enables a single-figure expression of their relative environmental vulnerabilities.

Hazard maps or risk maps use different techniques and different interpretation modes. Vulnerability assessments are increasingly used to identify hazard zones, thereby forming the basis for pre-impact and hazard mitigation planning (Brooks *et al.* 2005). Hazard zones mapping introduces the move from single stressors (hazards) to multiple stressors (like global change) and incorporates the influence of cross-scalar dynamics, both spatial and temporal, on vulnerability of a particular area towards an existing natural phenomenon. Warning services use symbols for the hazards. Frequently used symbols are shown in Table 15.2.

Table 15.1 Natural hazards and information needed.

Hazard	Definition	Information Needed	Frequency of the Information Needed
Geological			
Avalanche	The sudden fall or slide of a large mass of material down the side of a mountain. Avalanches may contain snow, ice, rock, soil, or a mixture of these materials. Avalanches can be triggered by changes in temperature, by sound vibrations, or by vibrations in the earth itself. (*The American Heritage® Science Dictionary.* http://dictionary.reference.com)	General "materialpack" conditions, slope steepness, slope orientation, terrain, vegetation.	Years to days.
		Weather, seismic activity.	Operational, close to real time information in days to minutes.
Earthquake	The sudden release of stored elastic energy in the Earth's lithosphere, caused by its abrupt movement or fracturing along zones of preexisting geological weakness, and resulting in the generation of seismic waves (Smith & Petley, 2009).	Seismic source zones	Ten years
		Ground motion prediction equations	Years
		Seismic activity	Real time
Costal erosion	A process of wearing away material from a coastal profile due to imbalance in the supply and export of material from a certain section. It takes place in the form of scouring in the foot of the cliffs or dunes or at the subtidal foreshore. (Mangor, 2004)	Geomorphology and geological make up of the coast.	Hundreds of years to years.
		Tide range, sea-level fluctuations, wave and swell conditions. Rainfall patterns.	Years

(Continued)

Table 15.1 Natural hazards and information needed (*Continued*).

Hazard	Definition	Information Needed	Frequency of the Information Needed
Lahar	A wet mass of volcanic fragments flowing rapidly downhill. Lahars usually contain ash, breccia, and boulders mixed with rainwater or with river or lake water displaced by the lava flow associated with the volcano. (*The American Heritage® Science Dictionary.* http://dictionary.reference.com)	Geomorphology and geological make up of the volcano relief. Vegetation. Velocity, viscosity, impact force, the dimensions of the gorges	Hundreds of years to months. Years. Real time information.
Landslide	The downslope displacement of surface materials (predominantly rock and soil) under gravitational forces (Smith & Petley, 2009).	Landslide dimensions, rate, activity, age, genesis, repeatability, morphology, number and types of objects under threat, hazard and risk degree.	Online or periodic monitoring.
Sinkholes	A natural depression in a land surface formed by the dissolution and collapse of a cavern roof. Sinkholes are roughly funnel-shaped and on the order of tens of meters in size. (*The American Heritage® Science Dictionary.* http://dictionary.reference.com)	Dimensions, age, evolution stage, degree of karstification, number and types of objects under threat.	Online or periodic monitoring.
Tsunami	The displacement of a significant volume of water, generating a series of waves with large wavelengths and low amplitudes (Alexander, 1993).	Coastline features, local bottom topography. Seismic activity, increase of the wave height.	Years. Hours to minutes.

Volcanic eruption	The subterranean movement of magma and its eruption and ejection from volcanic systems under the influence of its confining pressure and superheated steam and gases (Alexander, 1993), together with associated tephra, ash, and gas.	Seismic activity, ground deformation, volcanic gases, thermal emissions, changes in water levels and chemistry.	Days to minutes.
Meteorological			
Blizzard	A violent snowstorm with winds blowing at a minimum speed of 56 km (per hour and visibility of less 400 m for three hours. (*The American Heritage® Science Dictionary.* http://dictionary.reference.com)	Wind speed, visibility, snowstorm, snowfall.	Hours to minutes.
Drought	A long period of abnormally low rainfall, especially one that adversely affects growing or living conditions. (One of many definitions)	Precipitation, drought duration, drought intensity, drought stress, water stress.	Weeks to days.
Hail	Precipitation in the form of rounded pellets of ice and hard snow that usually falls during thunderstorms. (*The American Heritage® Science Dictionary.* http://dictionary.reference.com)	Cumulonimbus, radar reflectivity, convection, precipitation phase, hail size, hailfall intensity, hailstorm motion, wind.	Minutes.
Extreme temperature (Heat wave)	A period of at least 48 h during which neither the overnight low nor the daytime high H_i falls below the NWS heat stress thresholds. (Robinson, 2001)	Maximal temperature, average temperature, tropical days, tropical nights days, heat period.	Days.
Extreme temperature (Cold wave)	A prolonged period of temperatures below the normal average for that period of time (either short or long term, local, regional, or global).	Minimal temperature, average temperature, frost day, ice day, cold period.	Days.

(*Continued*)

Table 15.1 Natural hazards and information needed (*Continued*).

Hazard	Definition	Information Needed	Frequency of the Information Needed
Cyclonic storm	A low-pressure atmospheric disturbance resulting in strong winds accompanied by rain, snow, or other precipitation and often by thunder and lightning. (*The American Heritage® Science Dictionary.* http://dictionary.reference.com)	Cumulonimbus, radar reflectivity, convection, flash, precipitation, storm motion.	Minutes.
Ice storm	A storm of freezing rain that deposits a glaze of ice on the ground. (*The American Heritage® Science Dictionary.* http://dictionary.reference.com)	Freezing rain, supercool drops, temperature, glaze ice.	Hours.
Tornado	A violently rotating column of air extending from a cumulonimbus cloud to the Earth, ranging in width from a few meters to more than a kilometer and whirling at speeds between 64 km and 509 km per hour or higher with comparable updrafts in the center of the vortex. (*The American Heritage® Science Dictionary.* http://dictionary.reference.com)	Helicity (wind parameters), potential energy, convection.	Minutes.

Climate change	A long-term significant change in the weather patterns of an area. (*The American Heritage® Science Dictionary.* http://dictionary.reference.com)	Global temperature changes, greenhous gasses, climate extremes, precipitation patterns, sea level rise, glacier melting.	Decades, years.
Geomagnetic storm	A worldwide disturbance of the earth's magnetic field induced by solar flares, solar wind, or sunspot activity. (*The American Heritage® Science Dictionary.* http://dictionary.reference.com)	Magnetic intensity.	Minutes.
Hydrological			
Flood	A temporary rise of the water level, as in a river or lake or along a seacoast, resulting in its spilling over and out of its natural or artificial confines onto land that is normally dry. (*The American Heritage® Science Dictionary.* http://dictionary.reference.com)	Water level, water flow. Precipitation, ice phenomenon.	Daily, hourly.
Biophysical			
Wildfire	An uncontrolled fire fuelled by natural vegetation. (Smith & Petley, 2009)	Fuel type, weather, wildfire front, forward rate of spread.	Days to hours.

Table 15.2 Symbols of most frequent natural hazards.

Environmental hazard

Meteorological hazards

Thunder storm	Rain	Fog	Icings	Ground frost	Snow	Low temperature	High temperature	Wind	Snow drift

No Severe Weather	Be Aware	Be Prepared	Take Action

Awareness level

Hydrological hazards

Flood warning	Severe flood warning	Flood	Flash flood	Snow melt flood	Regional rainfall flood	rainfall flood Snow melt and	Ice flood

Tsunami

Tsunami	Tsunami hazard zone	Tsunami evacuation route

Other geophysical hazards

Avalanche	Earthquake	Volcano activity	Landslide

15.2 EARLY WARNING SYSTEMS AND TYPES OF COMMUNICATION

15.2.1 Early warning systems and monitoring of hazards

Early warning system means the provision, where the appropriate information about existing or emerging threat enables to take a reasonable action to reduce the risk involved. UNISDR (United Nations International Strategy for Disaster Reduction) defines early warning as 'the provision of timely and effective information, through identified institutions, that allows individuals exposed to a hazard to take action to avoid or reduce their risk and prepare for effective response' (ISDR, 2004). Early warning systems develop rapidly based on the technical development. International Strategy for Disaster Reduction (ISDR) set as one of its priorities for the period 2005–2015 in Hygo Framework (2005) the enhancement of early warning, while in Sendai Framework (2015) for 2015–2030 one of the priorities was "people-centered multi-hazard, multisectoral forecasting and early warning systems at national and local as well as global and regional levels". Sendai Framework for Disaster Risk Reduction 2015–2030 approved by the General Assembly of the UN in June 2015 is the major document of the UN regarding disaster and climate risk management in the context of sustainable development.

An extensive survey of the early warning systems done by the UNISDR in 2006 outlined the general features of an effective people-centered Early Warning System, which includes knowledge of the risks faced, technical monitoring and warning service, dissemination of meaningful warnings to those at risk and public awareness and preparedness to act. Failure in any one of these elements can mean failure of the whole early warning system. It is to be stressed that many of the Early Warning Systems actually do represent the category "system of systems". J. Moßgraber (Moßgraber *et al.* 2013) described well the general architecture of an early warning system. Such systems should involve independent operational and managerial elements with scalable and resilient communication layers.

The protection against any hazard is expressed by the ability of the system to respond to the disasters either by completely avoiding them or by absorbing the hazard impacts alleviating some of them and by possessing the ability of the adaptive process to facilitate the recovery from the impacts. These are the features of resilience concept, which assumes the knowledge of the vulnerability level of the system and respects it. Early warning systems became a part of the resilience concept of social systems, which brings further global knowledge about a single or a group of hazards and the possibility of immediate respond to the threat. They always involve the monitoring module for early detection and possible prediction of the hazard impact. A demonstrative concept of disaster resilience of a place was brought by S.L. Cutter (Cutter *et al.* 2008), in Figure 15.1.

Early warning systems are expressed (positioned) in DROP model in the part characterizing the event and preparing the response. Based on the factors determining the final goals of prevention or mitigation of hazards impacts early warning systems are designated to identify the situation and to trigger in time response to the coming threat. Timely information can help either to avoid or at least to reduce a possible impact of a coming hazard. The basic features of hazard monitoring should include: (1) the knowledge of current hazard monitoring capabilities and procedures; (2) automatic processing unit which will prepare the alerts at different levels; (3) a systems that will provide early warning of developing threats and regularly updated information on their characteristics and progress; (4) a system that ensures that the warnings reach the right people in a timely manner; and (5) arrangements for communicating threat information to the general public accompanied by appropriate background information and action recommendations.

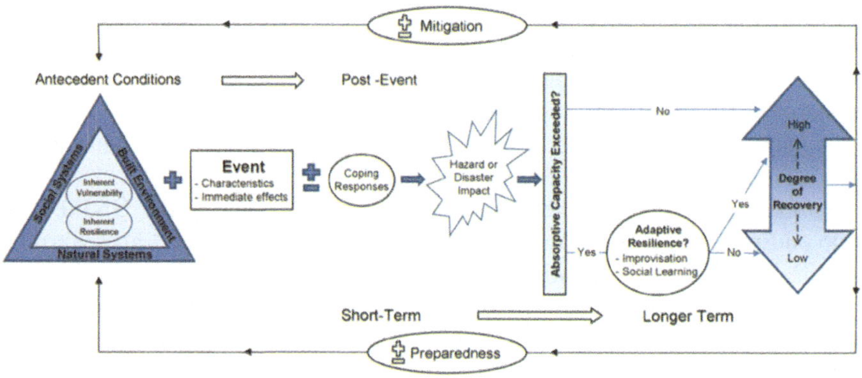

Figure 15.1. Schematic representation of the disaster resilience of place (DROP) model. Source: Cutter *et al.* 2008. A place-based model for understanding community resilience to natural disasters. Global Environmental Change 18 (4): 598–606.

Therefore, monitoring environmental hazards implies various applications of both routine continual measurements conducted with different frequency in standard networks like meteorological or seismographic operational networks and the measurements being done during the seasons sensitive to specific hazard like avalanches or monitoring, which requires random sampling. The monitoring part of the early warning systems forms the core for further data integration and processing. All data about the particular hazard do involve a certain level of inaccuracy, either in spatial or in temporal dimension, as well as the inaccuracy in the description of concrete phenomenon or its part. Some research conducted over the environmental hazards databases, which form the basis of different early warning systems, showed that "the nature of the regulatory structure used to collect environmental hazards data introduces questions about positional and attribute accuracy" (Scott *et al.* 1997). Monitoring networks detecting the environmental conditions of different phenomena are mostly, either fully or to a certain level, automated and comprise a plurality of remotely located environmental sensors and equipment of different nature. Nevertheless, in many cases they are supplemented also by manual observations (for example ice phenomenon at hydrological stations). Introduction of new monitoring techniques brings great opportunities in detecting the occurrence and the development of the particular phenomenon. However, the development of the observing network influences the monitoring of the concrete hazardous event and can introduce difficulties when analyzing its time series in order to get its statistical characteristics and trends.

Monitoring of any hazards usually concentrates on supporting the short-term forecasting needs of the early warning systems bringing the latest technologies. Innovation and improvements in instrumentation are often implemented without sufficient overlap or calibration against existing methods to document the impact of the changes on the longer term climate record. Kunkel *et al.* (2013) monitored trends in extreme storms in the USA. A strong evidence was shown for an averaged upward trend in the frequency and intensity of extreme precipitation events and the COOP network was considered adequate to detect such trends. They found that the U.S. observing network is better suited for the assessment of changes in very heavy precipitation than for any other class of extreme storm. National Weather Service Cooperative Observer (COOP) network has largely employed the same standard 8-in. non-recording precipitation gauge throughout its history. On the other hand, detection of long-term changes in tropical cyclones activity was found to be hindered by a number of issues with the historical records. Heterogeneity introduced by changing technology and methodology was the major issue. The same standard of the instruments in the network throughout its history minimizes time-dependent biases resulting from changes in instrumentation. Homogeneity of the time series plays the decisive role in the climatology of the particular hazard.

The development of the environmental databases has risen considerably the spatial and temporal accuracy in recent decades. Nevertheless, as data from ground-based stations together with the areal information from orbiting and stationary satellites flowing through the downlink interface and the information from all sources are integrated in one database system the data consistency has to be watched. Historical and geographical data are used to prepare the first element of the early warning system represented by our knowledge of the particular hazard and possible risk in the general description of early warning system adopted by the EWCIII (2009). It forms a base for decision processes showing the risk level and further the possibility to issue an alarm message and to response. Such process has to be secured by the knowledge base and with the proper technical equipment and local capacities (see Figure 15.2). Monitoring part of the system is based on available technologies and modelling.

The final decision making process and issuing the warning situation is usually left with the person responsible. Nevertheless, when the trigger levels of defined measured parameters or indices are exceeded automated messages can be issued. The most common structure of current early warning systems run is based on a closed line from technical monitoring and detection through data collection and processing to forecasting module to transmitting warnings to the users. Basic principle of the operational process of the warning system can be designed in a similar way as the scheme shown in Figure 15.3.

The monitoring part of any warning system makes an essential starting point and has to be based on the basic knowledge of the particular hazard and all its features. The performance and analysis of routine measurements aim to detect changes in the environment or the health status of the population in order to recognize a possible threat. Such recognition should imply intervention indicated by monitoring data. Monitored data are presented at various levels being grouped by the geographical region, different time scale and/or by the type of selected parameters during the processing period. The processing part of the warning system is in many cases naturally connected with the prediction and the alerts communicated to the end users and have the characteristics of nowcasting.

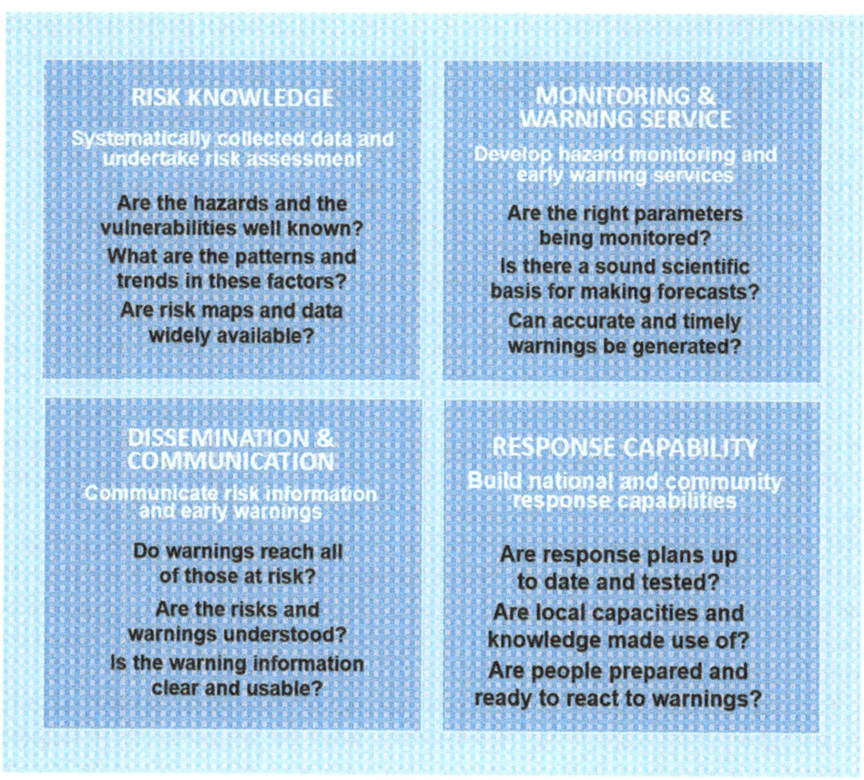

Figure 15.2. Four Elements of people-centred Early Warning Systems.
Source: UN/ISDR Platform for the Promotion of Early Warning.

Figure 15.3. Early warning system operational scheme.

15.2.2 Types of communication

Technical development has brought an expansion of global early warning systems. Nevertheless, national or regionally based early warning systems are needed not only for the protection of the citizens, but they also provide the building blocks of the global early warning system. The modern systems of emergency alert are built on a design structured after the radio signal became a possible medium to transfer the information. An over-the-air broadcasting became the best available technology for widely disseminating emergency alerts. Such systems developed mostly in the 1950`s. At the present time, web-based technologies are used, which are built on such structures that enable either to send the information to specified customers or just to retrieve quasi on line information at any time.

Nevertheless, the way that a warning message can differ depends on the users, their ability to receive and to consume the warning message in an appropriate way, as well as on the available technical facilities on both producer and user site. Some new processes in communicating the awareness information using the so-called microblogging service via Twiter were described (Vieweg *et al.* 2010). Table 15.3 lists standardly used media as part of Early Warning Systems.

One of the examples is the Emergency Alert System (EAS) used in the USA (ref. 81 FR 27342). It is part of an Integrated Public Alert and Warning System, which is an emergency population warning system in the United States. It replaced the former Emergency Broadcast System (EBS) and CONELRAD system originally designed for defense purposes (Figure 15.4).

EAS is designed as an integral system, where the major stakeholders include state, local, territorial and tribal governments. EAS is based on its regulations and standards governed by the Public Safety and Homeland Security Bureau of the Federal Communications Commission (FCC) by using the EAS Protocol. This

Protocol is a method of constructing an alert message that can be effectively, efficiently and uniformly processed by EAS equipment. Protocols are used to distribute alerts through a hierarchical, broadcast-based alert message distribution system in which an alert originator at the national, state or local level formats a message. Alert message cascades via multiple lines to the state and local Emergency Operational Centers, National Weather Service and via Federal Emergency Management Agency (FEMA) and Primary Entry Points are also disseminated to the public. FCC coordinates the EAS together with FEMA and the National Weather Service. The alert information is spread in different ways ensuring that all the recipients in the territory have the chance to receive the alert message. A number of operators are involved in alert dissemination, such as broadcasters (over 13 thousands); cable operators and wireline service providers (almost 3 thousands); and satellite systems.

Table 15.3 Media used for warning message delivery.

Media	Benefits	Limitations
Radio and television	Widespread signal, flexible modified information	Conveyed information, limited use in night.
Satellite radio/ television	Highly reachable signal	Limited usage
Local/Community ratio	Available for remote communities	Limited program ratings
Telephone	Quick direct message	Limited number of users, limited signal area
SMS	Quick direct message for groups	Limited number of users, limited signal area, language problems
Cell broadcasting	Quick direct message for groups	Limited to subscribed users, language problems
Internet/e-mail	Flexible, continuous delivery from multiply sources	Limited usage by rural remote communities
Social media (Facebook, Twiter, etc.)	Quick delivery to many customers	Limited usage by rural remote communities, multiplied information for internet users
Sirens	No time limitation in signal delivery	Signal covers limited area, limited content of message

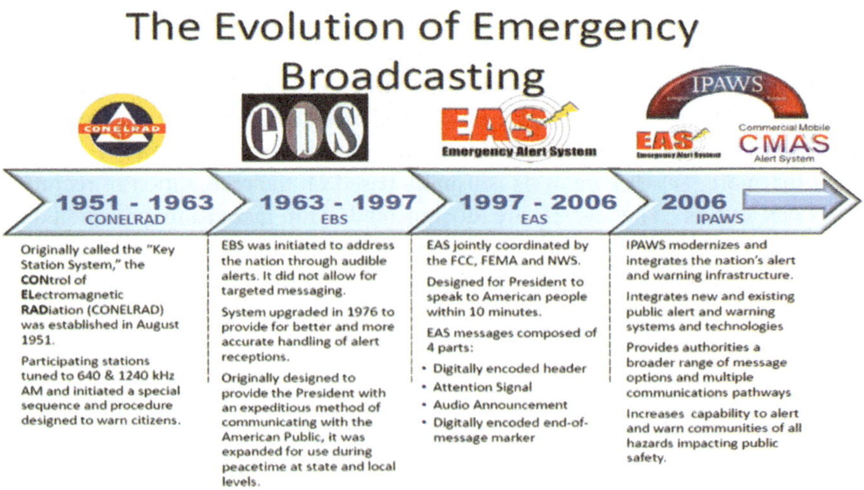

Figure 15.4. Evolution of emergency broadcasting systems in the USA. *Source:* FEMA (By FEMA - http://www.fema.gov/emergency/ipaws/ipaws_history.shtm).

Further to that, a wireless emergency alert system based on smartphones (Wireless Emergency Alerts, WEA) has been introduced in recent years. All major cell phone providers support WEA. This system works independently from EAS, but it can broadcast identical alert information. It distributes free messages directly to the mobile phones. They warn about severe weather and other threats of the safety. WEAs messages are sent by state and/or local safety officials, the National Weather Service, the President and by some special safety agencies. The reliability of the system has increased considerably. The analysis of a false alarm of EBS in 1971 showed that many stations did not receive the message and the majority of those that did receive it, they ignored it. EAS test in 2011 showed 82% success rate. Moreover, analogue systems are mostly replaced by the digital systems, which considerably improve the communication capacity, such as, for example, the RAKEL system used by Swedish Civil Contingency Agency.

In communication of the early warning message, some resilience between regional and national warning systems is important to ensure that critical messages reach the users in the right time and not being lost or delayed. During the message exchange and delivery process, system failures like link failure or other technical failure, as well as client failure, may occur due to either running out of underlying system resources or due to poor quality of service. Message loss or delay can eliminate any possible appropriate prevention action towards the hazard impact. Communication media used in Early Warning Systems are in permanent development, for example the use of femtocells for message delivery, but only an effective communication between the producers and users of the early warning message at the institutional level can ensure that the message is recognized and well understood and can be communicated further to the units responsible for response activities. Multiple receivers base, time location of the information, and in time control algorithms increase the chance of in time delivery and reduce the burst messages in the network and identify the message status in time.

Furthermore, any risk analysis is usually based on the recognition of recorded hazardous events. The events are documented with increasing precision and increasing number of parameters. Nevertheless, only a complete description of the event and the documentation of the full process from the occurrence and impact of the event from the first signal recognizing the event to monitoring the applied recovery activities helps to improve efficient response planning.

15.3 SERVICES FOR HAZARDS INFORMATION

Services for hazard information respond usually to the so-called critical infrastructure of the particular country or region. Critical infrastructure is the system, which keeps the health, safety, security and well-being of the society. Each component of the infrastructure, which comprises land, facilities and businesses for goods and services, environment, networks and social well-being has certain risk points with critical limits. It refers to both human and natural hazardous processes and natural hazards are one of the sources of possible disruptions of critical infrastructure components. National critical infrastructure can differ in some components or their relevancy from country to country. Relevant information on critical infrastructure is needed before the services dealing with hazards are planned. The task for hazard information services is to identify and to evaluate potential natural hazards, to develop a system to recognize and respond to the hazard and to recommend protective measures and recovery plans. The set of hazard information creates the basis for reduction or elimination of possible impacts. Information services form strategic instrumental tool anticipating the occurrence and magnitude of future disaster in order to reduce the level of disaster risk and to improve resilience of the system. The failures of early warning systems have been failures rather of communication of the information at institutional level than science. Governments at all levels are required to create institutions, which assess the disaster risks based on the scientific knowledge and to guide the disaster risk reduction by providing the hazard information services.

General responsibilities to manage the risk information and to create and develop the appropriate institutional and technical services for this purpose lies with the central governments. Official direct state regulatory policy should lead the local governments to incorporate risk reduction activities into their security planning. Nevertheless, new features have emerged in the general governance in recent decades. Multilevel governance could be recognized in different regions. In Europe, for example, it refers to the loss of functions of the central governments upwards to the European Union, downwards to special-purpose bodies and outwards to agencies (Rhodes, 1997). One of the examples is the weather warnings system for Europe, known as meteoalarm (meteoalarm.eu). Border changes also play an important role at the regional scale as new central state institutions are created (Baltic countries, Balkan countries, former Czecho-Slovakia). These processes directly influence risk management of natural hazards in the region. There are different bases for integration both the basic information dealing with the disaster risk reduction, as well as the distribution of operational information at various levels, such as regulations of EU parliament, Hyogo and recently Sendai Framework for Disaster Risk Reduction, or various portals integrating operational information.

In order to make the hazard information generally understandable, the responsible services use a predefined way of communication in the form of certain standards and protocols. They help to manage the impacts caused by natural hazards and provide best practice for an incident response system. Standardization in the field of security tends to enhance the safety and resilience of the society. The hazard communication standards establish uniform requirements to make sure that the hazards are communicated in a recognizable way after they are observed firstly to the responsible management services and consequently distributed to the bodies responsible for response activities. Well enforced standards and regulations are an important part of hazard management strategy (Daniels *et al.* 2006). The natural hazard safety issue has already appeared in ISO standards in ISO37120 in 2014. It defines and establishes methodologies for a set of 100 indicators to steer and measure the performance of city services and quality of life. Safety issue is one of the main systematic themes of ISO37120. After implementation of ISO37120, UNISDR and WCCD (World Council on City Data) started to work on incorporation of essentials for making cities resilient into the ISO standard for resilient cities. The essentials focus on issues, such as good urban planning, land-use, a risk-informed citizenry, safe schools and health facilities, protection of eco-systems, early warnings and emergency management. Concrete hazard information services are designed according to the particular conditions, which are created by the geographic features, population, human activities and general situation.

The services dealing with the hazard information differ in their scope based on the general aim of the service, the operator, the target groups and similar aspects. Different criteria can be used to group the services. Generally, the following types of services providing the hazard information can be distinguished: (1) strategic services building the general risk assessment, tailored products risk plans and vulnerability maps including early warning systems and response and rescue plans concentrated on either geographical units or defined sector or group of hazards; (2) operational services providing continuous monitoring and producing close to real time alert messages; (3) hazard response services including aftermath activities; (4) educational services as a tool for public to discover hazards in their area and to learn steps to reduce the personal risk level. Strategic services form the base for preparedness programs. Hazard related services are built on similar structure, which includes the database and registers, and process oriented applications usually arranged as a set of projects. Table 15.4 describes the common design of such services. The operator of the service must ensure that the service has the capabilities needed to accomplish its mission at the level of planning, as well as at the operational use. Risk information handling is based on the IT systems, which have to provide a certain level of secured mission support in the face of the risks to the IT system. These security capabilities must be determined and controlled by the operator. In terms of technology, many different parts form information management, ranging from web content management to documents and record management.

Table 15.4 Components and activities of natural hazard information services.

Components of the Service	Activities Provided by the Services
Critical infrastructure register	–Updating
Hazard register	–Hazard reports design –Hazard information assessment –Vulnerability assessment –Updating
Risk register	–Risk assessment –Primary and secondary impacts identification –Disaster risk reduction planning –Risk management planning –Resilience assessment –Sector resilience planning –Protection planning
Operational database	–Early warning systems design –Emergency information management –Updating, maintenance
Communication infrastructure	–Framework for emergency communication –Protocols, compliance with local and national standards –Identification of Primary entry points –Maintenance
Response and recovery	–Disaster risk reduction planning –Assessing dependencies –Environmental impact assessment –Disaster response measures –Recovery planning –Community engagement plans

Information flow represents the decisive part of any disaster management. Although no managerial action averts or eliminates the threat, proper management of risk information creates the possibility to take an adequate response action. Thus, communication of the incident occurrence and its assessment makes a critical point for further activity. Risk information management involves two major issues: first the creation of the information and second its proper and secure distribution, use and storage. The birth of the hazard information starts on the line: forecasting, monitoring, early warning. The frequency of the information needed in such line depends on the type of the natural hazard and some examples are presented in Table 15.1. Further communication of the information depends on the receiver. Hierarchical prioritization is usually applied in the risk information management organized by official structures, such as president and the respective bodies responsible for response action are the first informed in EAS system. Once an emergency has been identified and assessed, further information (alert) should be designed in the way in which it is able to activate the respective steps in emergency plan at any level. The information communication, with some exceptions like sirens, entirely depends on communication technology of the respective IT systems. IT security is defined with security of the information within the boundaries of the network infrastructure technology domain. Cyber security technologies of nowadays are under rapid development and offer the solutions that minimize the opportunity for the loss of information availability and integrity. Nevertheless, each of the delivery systems should include also its risk assessment and provide the backup`s lines.

15.4 EXISTING HAZARDS INFORMATION SYSTEMS

Natural hazard information systems represent an interaction between man and nature. They try to bring a buffer to harmful impacts of natural processes in the society and environment. The institutionalization of such systems brought complex information service, public awareness and deep interaction with civil protection bodies able to take protection measures. Nevertheless, new feature in the society like rapid increase in population and wealth and complicated infrastructure, as well as some changes in the nature like climate change, keep absolute damages high and generally rising. Diachronically, the systems have been developed from the basic radio broadcasting like EBS to fully web-based specialized technologies concentrated on a particular region (Gogu *et al.* 2005).

A number of information portals provide different levels of natural hazard information from general, often interactive, maps of the potential hazards to semi real-time alerting systems. General claims being put on the hazard information systems involve: accessibility and reliability; accuracy; automatic emergency alert; flexible technological upgrade; interactive; prediction capability; time and spatial coverage and expandability; and track record.

The most effective way of natural hazard management is in the context of integrated development planning. Depending on the objective and application, there can be distinguished into the basic services dealing with the natural hazard assessment and development of planning process, as well as the operational information services including early warning systems. They use different data sources and implement different management schemes. The data for basic services usually come from national and international natural phenomena monitoring and disaster management bodies, multi-sectoral and sectoral agencies, ministries, and public utilities. Many of the data are in the form of maps, reports, proceedings, different historical records and other similar forms. Data retreatment and preparation make the first step for their further processing.

Data management in early warning systems is different as the data usually represent well-prepared records including the meta-data often in the form of predefined protocols. Geographical data play a decisive role in this context. Among the techniques of geographical data processing and interpretation Geographical Information System (GIS) is mostly used. It consists of systematic means of geographically referencing number of "layers" of information, which facilitate the overlaying, quantification, and synthesis of data. GIS software is able to combine the incidental data with maps in different layers and to display hazard impact areas in space. This techniques further enables to interpret multiple hazards in one composite map or chart together with more hazard related data for analysis. One area can suffer from more than one hazard, with different primary and secondary impacts so that the cascade events producing different interactions have to be accommodated into the risk assessment. Indeed, multiple hazard maps help to analyze the impacts of different interconnected hazards, which can be originally triggered by one natural phenomenon, such as, for example, landslides caused by heavy rains connected with severe thunderstorms. The methods in natural hazard risk assessment develop approaches towards complex multi-hazard and multi-risk evaluation. Existing single hazard and vulnerability assessment models and methods are being harmonized within a multi-risk scenario (MATRIX project reporting, 2016).

There are portals specialized on hazard information services, which run as "self-standing" bodies, but a number of such portals is devoted to the natural hazard information services and are hosted by more general portals of wider scope. Many of the general portals provide simple, often aggregated, statistics of the natural hazards, either at global or national scale, like www.archive.org, www.statista.com, www.emdat.be, www.weatherdb.com, www.australia.gov.au. Different information services dealing with natural hazards further to the tailored products bring also interactive tools with different statistics providing the basic information about natural hazards within a certain area, either at global or regional scale (Swedish natural hazard information system in Table 15.5). Many of natural hazard information services are usually hosted by more general services (portals), which also deal with further hazards like industrial or other human-born hazards. They either provide directly natural hazard information or provide the link to the respective services making a frame for national global security service respecting the national security policy. One example is FEMA in the USA. FEMA portal (www.fema.gov) does not serve as early warning centre. It cumulates the services preparing the strategies for complex disaster management at any level providing the guidelines, as well as the administrative assistance.

Nevertheless, a number of specialized portals dealing with natural hazard services are in run. They are bringing different types of information services dealing with natural hazard statistics, natural hazard on line monitoring and early warning systems and services providing assistance to disaster management. Natural hazard

Table 15.5 Examples of different portals providing natural hazard information services.

Characteristic	Web Address
Natural hazard statistics (global and/or national)	
Aggregated natural hazards statistics	https://ourworldindata.org/natural-catastrophes
Global seismic hazard information	https://www.ngdc.noaa.gov/hazard/
General natural hazards statistics	http://www.statista.com/statistics/236504/number-of-natural-disasters-in-the-us-by-type/
US national disaster statistics	http://www.disaster-survival-resources.com/us-disaster-statistics.html
Australia national disaster statistics	http://www.australia.gov.au/about-australia/australian-story/natural-disasters
Weather natural hazards statistics	https://natural-hazards.weatherdb.com/
International disaster database	http://www.emdat.be/
Swedish natural hazard information system	www.ndb.msb.se/Default.aspx?l=EN
General early warning systems, close to real time alerts	
Global alert system GDCAS	http://www.gdacs.org/
General alerts	http://earthalerts.manyjourneys.com/web/
Complex hazard information service	http://www.disastercenter.com/
Global EWS information service	http://hisz.rsoe.hu/alertmap/index2.php?/
Global natural hazard information	www.usgs.gov
Global natural hazard information service	http://www.pdc.org/
Specific early warning systems, close to real time alerts	
Atmospheric hazards alert service for Europe	www.meteoalarm.eu
Country drought monitoring system (CZ)	http://www.intersucho.cz/en/
Multihazard warning service (AU)	www.bnhcrc.com.au
Geological hazard information (NZ)	https://www.geonet.org.nz/volcano
International Tsunami information centre	http://itic.ioc-unesco.org/index.php
Drought monitoring system for Europe	http://edo.jrc.ec.europa.eu
Natural hazard portal (CH)	http://www.natural-hazards.ch

statistics and early warning systems are at close look by public authorities and public itself. Table 15.5 presents some examples of the services provided at different levels and with different scopes. Multi-hazard information systems, either nationally or globally designed, are complemented with the single hazard systems monitoring one specific hazard, such as drought, with the multiplied impact assessment. Concentration on the specific hazard increases the accuracy of the alert message and the estimation, i.e. forecast, of hazard appearance issued by early warning systems. A distinct position in this way occupy the weather services. All of them provide further to the weather monitoring also weather forecast including the weather related hazards. Unification of the methods of presentation of the weather hazards lead in Europe to deep cooperation within the meteoalarm service (see Table 15.5.).

Different scope of hazards requires different planning, monitoring and alert service. With the exception of a few natural hazards, for example landslides, they are not space limited and can threaten the areas on many countries or even continents, such as hurricanes, tsunamis, or volcanic activity, and their impacts are multi-sectoral. Indeed, volcanic activity represents a big threat for the neighborhood, but also for air traffic thousands kilometers away, hurricanes threaten any traffic within their reach, as well as coast and inland areas infrastructure. Any hazard information service is based on multilateral cooperation of different bodies and agencies at national and international levels. This rises from the collaboration of national agencies at specific early warning systems, such as drought monitoring in the Check Republic, to the cooperation of the international agencies at global scale represented by the GDACS portal (Table 15.5). GDACS is a cooperation framework between the United Nations, the European Commission and disaster managers worldwide to improve alerts, information exchange and coordination in the first phase after major sudden-onset disasters.

Natural hazard information services became a firm part of national and world security. Scientific knowledge opens new possibilities in monitoring and reporting on natural hazards, which enables better preparation of disaster management. Despite of the large number of existing services, hazard knowledge is not being put to full use in warning preparedness. Existing programs and regulations anchored in laws and many governmental and other official documents dealing with risk management. However, weak law enforcement in many, especially, but not exclusively, in developing countries is resulting in slow introduction of the required technologies, many times just after a catastrophe has occurred.

15.5 SUMMARY

Natural hazards management is one of the major challenges for the society. This fact is further stressed with ongoing climate change, which brings in many places of the world increased extremity of weather both in frequency and magnitude. The impacts of natural hazards on the society and environment are fully recognized by the scientific community and it is well employed in respective research programs. There are ways of using research to help maximize the potential rewards and many sophisticated technologies dealing with disaster management caused by the natural hazards were implemented. Nevertheless, many difficulties were recognized in legislating national policies related to natural hazards, especially to climate related hazards, and low level of implementation of practical applications to eliminate the impacts of such events. This rises from often inadequate institutional arrangement and political failure to take the action. There is no adequate support given to developing countries in technical capabilities, resources, expertise, early warning services, which limits the possibility to take the necessary action in time. Legislative problems in international data-exchange policies and procedures limits the development of modelling procedures and also the warning and forecasting systems. All the above mentioned problems lead to low measures of response and recovery. The policy steps have to be taken to move the public attention from the financial compensation of the damages to the prevention and preparedness of the natural hazard impacts.

REFERENCES

81 FR 27342. Amendment of the Emergency Alert System. A Rule by the Federal Communications Commission on 05/06/2016. 47 CFR 11. Available at: https://federalregister.gov/a/2016-09059.

Barnett J., Lambert S. and Fry I. (2008). Hazards of indicators: insights from the environmental vulnerability index. *Annals of the Association of American Geographers*, **98**(1), 102–119.

Brooks N., Adger N. W. and Kelly M. P. (2005). The determinants of vulnerability and adaptive capacity at the national level and the implications for adaptation. *Global Environmental Change Part A*, **15**(2), 151–163.

Cutter S. L., Barnes L., Berry M., Burton C., Evans E., Tate E. and Webb J. (2008). A place-based model for understanding community resilience to natural disasters. *Global Environmental Change*, **18**(4), 598–606.

Daanish M., Gioli G., Qazi S., Waraich R., Rehman A. and Zahoor R. (2015). Gendering flood early warning systems: the case of Pakistan. *Environmental Hazards*, **14**(4), 312–313.

Daniels R. J., Kettl D. F. and Kunreuther H. (2006). *On Risk and Disaster. Lessons from Hurricane Katrina*. University of Pensylvania Press, Philadelphia, Pensylvania 19104-4112. ISBN-13: 978-0-8122-1959-3 SBN-10: 0-81221959-7.

Early warning detection and notification network for environmental conditions. Available at: https://www.google.com/patents/US6023223.

EWCIII. Developing Early Warning Systems: A checklist. In: Third International Conference on Early Warning. 27–29 March 2009. Available at http://www.unisdr.org/2006/ppew/info-resources/ewc3/checklist/English.pdf.

Gill J. C. and Malamud B. D. (2014). Reviewing and visualizing the interactions of natural hazards. *Reviews of Geophysics*, **52**, 680–722, doi:10.1002/2013RG000445. Global Survey of Early Warning Systems. UNISDR, 2006. Available at: http://www.unisdr.org/files/3612_GlobalSurveyofEarlyWarningSystems.pdf.

Gogu C. R., Freimark H., Stern B. and Hurni L. (2005). Web based information system for natural hazard analysis in an Alpine Valley. In: Geo-information for Disaster

Management. van Osterom P., Zlatanova S and E. M. Fendel, SBN 3-540-24988-5, Springer, Berlin, Heidelberg, New York.

Graves P. E. and Bresnock A. E. (1985). Are natural hazards temporally random? *Applied Geography*, **5**(1), 5–12.

Hyogo Framework for Action 2005–2015: ISDR International Strategy for Disaster Reduction. (A/CONF.206/6). Available at: http://www.unisdr.org/2005/wcdr/intergover/official-doc/L-docs/Hyogo-framework-for-action-english.pdf.

ICT for Disaster Management. Available at: https://en.wikibooks.org/wiki/ICT_for_Disaster_Management.

ISDR (2004). Terminology: Basic Terms of Disaster Risk Reduction. International Strategy for Disaster Reduction secretariat, Geneva. Available at: http://www.unisdr.org/eng/library/lib-terminology-eng%20home.htm.

Kunkel K. E., Karl R. T., Brooks H., Kossin J., Lawrimore H. J., Arndt D., Bosart L. *et al.* (2013). Monitoring and understanding trends in extreme storms: state of knowledge. *Bulletin of the American Meteorological Society*, **94**(4), 499–514. © 2013 American Meteorological Society.

Mangor K. (2004). Shoreline Management Guidelines. DHI Water and Environment, p. 294.

MATRIX (New Multi-HAzard and MulTi-RIsK Assessment MethodS for Europe). Project reference: 265138. Permalink: http://cordis.europa.eu/result/rcn/159827_en.html © European Union, 2016.

Moßgraber J., Chaves F., Middleton E.S., Zlatev Z. and Tao R. (2013). The seven main challenges of an early warning system architecture. In: Proceedings of the 10th International ISCRAM Conference, T. Comes, F. Fiedrich, S. Fortier, J. Geldermann and T. Müller (eds), May 2013, Baden-Baden, Germany.

Rivera F. I. and Kapucu N. (2015). Disaster Vulnerability Hazards and Resilience Perspectives from Florida. Environmental Hazards, Springer, Cham, Heidelberg, New York, Dordrecht, London, ISBN 978-3-319-1645, DOI: 10.1007/978-3-319-16453-3.

Robinson P. J. (2001). On the definition of a heat wave. *Journal of Applied Meteorology*, **40**, 762–775.

Scott M., Cutter S. L., Menzel C., Ji M. and Wagner D. (1997). Spatial accuracy of the EPA's environmental hazards databases and their use in environmental equity analyses. *Applied Geographic Studies*, **1**, 45–61. doi: 10.1002/(SICI)1520-6319(199721)1:1<45::AID-AGS5>3.0.CO;2-V.

Sendai Framework for Disaster Risk Reduction 2015 – 2030. UNISDR/GE/2015 – ICLUX EN5000 1st edition. Available at: http://www.unisdr.org/files/43291_sendaiframeworkfordrren.pdf.

Spinoni J., Naumann G., Vogt J. and Barbosa P. (2016). Meteorological Droughts in Europe: Events and Impacts – Past Trends and Future Projections. Publications Office of the European Union, Luxembourg, EUR 27748 EN, doi:10.2788/450449.

Vieweg S., Hughes A. L., Starbird K. and Palen L. (2010). Microblogging during two natural hazards events: what twitter may contribute to situational awareness. In: Proceedings of the SIGCHI Conference on Human Factors in Computing Systems pp. 1079–1088. New York, NY, USA ©2010. ISBN: 978-1-60558-929-9 doi: 10.1145/1753326.1753486. www.cred.be/emdat.

Zalatynskyi V. (2013). Danger - a subjective evaluation of objective reality. *Science & Military*. – L. Mikulas, Slovak Republic. Armed Forces Academy of General Milan Rastislav Stefanik, **8**(1), 53–62. EV 2061/08, ISSN 1336-8885.

Index

W

IWA Publishing's authorised EU representative for General Product Safety Regulations is Diane D'Arras, 15 rue Duret, 75116 Paris, France, e-mail: safety@iwap.co.uk.

Printed and bound by CPI Group (UK) Ltd, Croydon, CR0 4YY
16/06/2026
02140358-0002